Emergency Incident Management Systems

Emergency Incident Management Systems

Fundamentals and Applications

Mark S. Warnick, Ph.D.
Louis N. Molino Sr.[†]

Second Edition

This edition first published 2020
© 2020 John Wiley & Sons, Inc.

Edition History
John Wiley & Sons Inc. (1e, 2006)

All rights reserved. No part of this publication may be reproduced, stored in a retrieval system, or transmitted, in any form or by any means, electronic, mechanical, photocopying, recording or otherwise, except as permitted by law. Advice on how to obtain permission to reuse material from this title is available at http://www.wiley.com/go/permissions.

The right of Mark S. Warnick and Louis N. Molino Sr. to be identified as the authors of this work has been asserted in accordance with law.

Registered Office
John Wiley & Sons, Inc., 111 River Street, Hoboken, NJ 07030, USA

Editorial Office
111 River Street, Hoboken, NJ 07030, USA

For details of our global editorial offices, customer services, and more information about Wiley products visit us at www.wiley.com.

Wiley also publishes its books in a variety of electronic formats and by print-on-demand. Some content that appears in standard print versions of this book may not be available in other formats.

Limit of Liability/Disclaimer of Warranty
In view of ongoing research, equipment modifications, changes in governmental regulations, and the constant flow of information relating to the use of experimental reagents, equipment, and devices, the reader is urged to review and evaluate the information provided in the package insert or instructions for each chemical, piece of equipment, reagent, or device for, among other things, any changes in the instructions or indication of usage and for added warnings and precautions. While the publisher and authors have used their best efforts in preparing this work, they make no representations or warranties with respect to the accuracy or completeness of the contents of this work and specifically disclaim all warranties, including without limitation any implied warranties of merchantability or fitness for a particular purpose. No warranty may be created or extended by sales representatives, written sales materials or promotional statements for this work. The fact that an organization, website, or product is referred to in this work as a citation and/or potential source of further information does not mean that the publisher and authors endorse the information or services the organization, website, or product may provide or recommendations it may make. This work is sold with the understanding that the publisher is not engaged in rendering professional services. The advice and strategies contained herein may not be suitable for your situation. You should consult with a specialist where appropriate. Further, readers should be aware that websites listed in this work may have changed or disappeared between when this work was written and when it is read. Neither the publisher nor authors shall be liable for any loss of profit or any other commercial damages, including but not limited to special, incidental, consequential, or other damages.

Library of Congress Cataloging-in-Publication Data

Names: Warnick, Mark S., author. | Molino, Louis N., author.
Title: Emergency incident management systems : fundamentals and
   applications / Mark S. Warnick, Ph.D., Louis N. Molino.
Description: Second edition. | Hoboken, NJ : Wiley, 2020. | Revision of:
   Emergency incident management systems / Louis N. Molino. c2006. |
   Includes bibliographical references and index.
Identifiers: LCCN 2019036716 (print) | LCCN 2019036717 (ebook) | ISBN
   9781119267119 (paperback) | ISBN 9781119267126 (adobe pdf) | ISBN
   9781119267133 (epub)
Subjects: LCSH: Emergency management–United States. | Crisis
   management–United States. | Disaster relief–United States. | Risk
   management–United States.
Classification: LCC HV551.3 .M65 2020 (print) | LCC HV551.3 (ebook) | DDC
   363.34/80973–dc23
LC record available at https://lccn.loc.gov/2019036716
LC ebook record available at https://lccn.loc.gov/2019036717

Cover Design: Wiley
Cover Images: Courtesy of Mark S. Warnick; Background © Brisbane/Shutterstock

Set in 10/12pt WarnockPro by SPi Global, Chennai, India

10 9 8 7 6 5 4 3 2 1

This book is dedicated to those that individuals who have or are responding to emergencies every day. This includes past and present first responders. This book is also dedicated to my dear friend who wrote the original version of this book, but is no longer with us, Lou Molino Sr. You are sorely missed my brother!

Greater love hath no man than this, that a man lay down his life for his friends.
John 15:13

Contents

Foreword *xvii*
Preface *xix*
Acknowledgments *xxi*
About the Book *xxiii*
About the Companion Website *xxv*

1	**Introduction** *1*	
1.1	The Revolutionary War *3*	
1.2	The Big Burn of 1910 *5*	
1.3	The Military Connection *9*	
1.4	The Birth of IMS Method *12*	
1.4.1	No Single Person in Charge *13*	
1.4.2	No Formal Protocols or Policies *14*	
1.4.3	Conflicts and Egos *15*	
1.4.4	Integrating Multijurisdictional Response *15*	
1.4.5	No Collaborative Organizational Structure *16*	
1.4.6	Strictly Enforced Intra-agency Command Structure *16*	
1.4.7	Command Based on Home Rule *16*	
1.4.8	Too Many Subordinates Reporting to a Single Supervisor *17*	
1.4.9	Lack of Accountability *17*	
1.4.10	No Interagency Planning *18*	
1.4.11	Lack of Common Terminology *18*	
1.4.12	A Lack of Interoperable Communications *18*	
1.4.13	A Lack of Logistics *19*	
1.5	California's Solution *19*	
1.6	Creating the Incident Command System *20*	
1.7	Evolution of IMS Methods *21*	
1.8	The "Big Three" of IMS *23*	
1.9	The Melding of the IMS Concepts of Today *24*	
1.10	The National Incident Management System (NIMS) *26*	
1.11	Presidential Directives *27*	
1.12	The NIMS Mandate *29*	
1.13	NIMS Updates/Changes (2008) and Training *31*	
1.14	NIMS Updates (2017) *33*	
1.15	Conclusion *34*	

2	**A Case Study of Incident Management** *37*	
2.1	The Lifecycle of an Incident *37*	
2.2	Common Attributes of an Incident *38*	
2.3	The Importance of Knowledge and Experience *39*	
2.4	Case Study: Tokyo Versus Oklahoma City *40*	
2.4.1	Tokyo Subway Attack *41*	
2.4.2	Oklahoma City Bombing *45*	
2.4.2.1	At the Scene of the Explosion *46*	
2.4.2.2	At the State Emergency Operations Center *53*	
2.5	Comparing and Contrasting These Incidents *54*	
2.5.1	Command *54*	
2.5.2	Control *54*	
2.5.3	Cooperation *57*	
2.5.4	Collaboration *58*	
2.5.5	Communications *60*	
2.6	Conclusion *61*	
3	**Incident Management in Other Countries** *65*	
3.1	The United Nations *65*	
3.2	Australia *67*	
3.3	Bermuda *68*	
3.4	Burma/Myanmar *68*	
3.5	Bangladesh *69*	
3.6	Brunei *70*	
3.7	Cambodia *71*	
3.8	Canada *72*	
3.9	China *72*	
3.10	Germany *73*	
3.11	Haiti *74*	
3.12	India *76*	
3.13	Indonesia *77*	
3.14	Iran *78*	
3.15	Iraq *80*	
3.16	Japan *80*	
3.17	Maldives *82*	
3.18	Malaysia *82*	
3.19	Mexico *82*	
3.20	New Zealand *83*	
3.21	Palestine *84*	
3.22	Philippine Islands *85*	
3.23	Russia *86*	
3.24	Singapore *87*	
3.25	United Kingdom *87*	
3.26	Vietnam *91*	
3.27	Other International Uses *92*	
4	**The Five Cs of Crisis (or Incident) Management** *95*	
4.1	Command *95*	
4.1.1	Situational Awareness *97*	

4.1.1.1	Who Is in Command? *97*	
4.2	Control *99*	
4.3	Communications *101*	
4.4	Responder Communication Problems *102*	
4.4.1	Terminology *102*	
4.4.2	Interoperability *102*	
4.4.3	Current Communications Facilitation *102*	
4.5	Integrated Responder Communications *104*	
4.6	Creating a Communications Unit for Responders *105*	
4.7	Radio Networks *105*	
4.8	Stakeholder Communications *106*	
4.8.1	Government Stakeholders *107*	
4.8.2	Media Stakeholders *108*	
4.8.3	Social Media *109*	
4.8.4	Local Utility Companies *110*	
4.8.5	Local Businesses *110*	
4.8.6	Civic Organization and Advocacy Groups *111*	
4.8.7	Houses of Worship *112*	
4.8.8	Volunteer Organizations *113*	
4.9	Communications Wrap-up *114*	
4.10	Cooperation and Coordination *114*	
4.11	Cooperation and Coordination in the State of Illinois *116*	
4.12	Private Sector Cooperation and Coordination *117*	
4.13	Strengthening Intelligence/Information Sharing with Coordination and Cooperation *118*	
4.14	Cooperation and Coordination During an Active Incident *119*	
4.14.1	Joint Information Center – Cooperation and Coordination *119*	
4.14.2	Liaison Officer – Cooperation and Coordination *121*	
4.14.3	Agency Representative(s) – Cooperation and Coordination *122*	
4.14.4	Non-command Cooperation and Coordination *123*	
4.15	Conclusion *124*	
5	**The National Incident Management System (NIMS)** *129*	
5.1	NIMS Method Guiding Principles *130*	
5.1.1	Flexibility *130*	
5.1.2	Standardization *130*	
5.1.3	Unity of Effort *131*	
5.2	Key Terms and Definitions *132*	
5.3	Understanding Comprehensive, Flexible, and Adaptable *133*	
5.3.1	Comprehensive *133*	
5.3.2	Flexible *134*	
5.3.3	Adaptable *135*	
5.4	NIMS Components *136*	
5.5	The Importance of Preparedness with NIMS *136*	
5.5.1	Cycle of Preparedness as a Part of NIMS Incident Management *137*	
5.5.2	NIMS Drills and Exercises to Support Preparedness *138*	
5.5.3	Seminar *139*	
5.5.4	Workshops *139*	
5.5.5	Tabletop Exercise (TTX) *139*	

5.5.6	Games	*140*
5.5.7	Drills	*141*
5.5.8	Functional Exercises (FEs)	*141*
5.5.9	Full-Scale Exercises (FSEs)	*142*
5.5.9.1	Use of Nonresponse Personnel	*144*
5.6	NIMS Method of Resource Management: Preparedness	*144*
5.6.1	Which Agreement Should Be Used?	*147*
5.6.2	What is a Mutual Aid Agreement (MAA)?	*148*
5.6.3	What is a Memorandum of Understanding (MOU)?	*151*
5.6.4	Identifying and Typing Resources	*151*
5.7	NIMS Qualifying, Certifying, and Credentialing Personnel	*152*
5.8	NIMS Method of Resource Management Response and Recovery	*154*
5.8.1	Identify the Resource	*154*
5.8.2	Order and Acquire the Resource	*154*
5.8.3	Mobilize the Resource	*155*
5.8.4	Track and Report Resources	*155*
5.8.5	Demobilize and Reimburse the Resource	*155*
5.8.6	Restock Resource(s) in an Incident	*156*
5.9	NIMS Multiagency Coordination Systems	*156*
5.9.1	Emergency Operations Centers (EOCs)	*156*
5.9.1.1	Which Stakeholders Should Be Represented in the EOC?	*158*
5.9.1.2	EOC Organizational Structure and Management	*158*
5.9.1.3	EOC Activation and Deactivation	*159*
5.9.1.4	What Triggers Levels of Activation?	*160*
5.9.1.5	Multiagency Coordination (MAC) Group	*161*
5.9.1.6	Joint Information Center	*163*
5.10	Conclusion	*164*
6	**An Overview of the Incident Command System**	*169*
6.1	Taking Control with ICS	*170*
6.2	Common Components of Incident Management Systems	*174*
6.2.1	The ICS Component of NIMS	*175*
6.3	ICS and NIMS-Differences and Commonalities	*176*
6.4	Incident Management System and NIMS Integration	*177*
6.4.1	Common Terminology	*178*
6.4.2	Modular Organization	*180*
6.4.3	Integrated Communications	*180*
6.4.4	Consolidated Incident Action Plans	*181*
6.4.5	Manageable Span of Control	*181*
6.4.6	Predesignated Incident Facilities	*182*
6.4.7	Comprehensive Resource Management	*182*
6.5	Conclusion	*182*
7	**Command Staff, General Staff, and Their Functions**	*187*
7.1	Incident Commander (IC)	*187*
7.2	Unified Command	*188*
7.3	Command Staff	*190*
7.3.1	Safety Officer (SOFR) Function	*191*
7.3.2	Public Information Officer (PIO)	*192*

7.3.3	Liaison Officer (LOFR) *193*	
7.3.4	Investigations and Intelligence Gathering Officer (IO) Alternative Placement *194*	
7.4	General Staff *195*	
7.4.1	Operations Section Chief (OSC) *196*	
7.4.2	Logistics Section Chief (LSC) *200*	
7.4.3	Planning Section Chief (PSC) *201*	
7.4.4	Finance (and Administration) Section Chief (FSC) *205*	
7.4.5	Investigations and Intelligence Section Chief (ISC) Alternative Placement *206*	
7.5	Expanding the Hierarchal Structure *207*	
7.5.1	Modular Organization Supports ICS Expansion *208*	
7.5.2	Organizational Flexibility *208*	
7.6	Conclusion *209*	
8	**Expanding the Operations Section** *213*	
8.1	Operations Section *213*	
8.1.1	Operations Branches, Divisions/Groups, Strike Teams/Task Forces *215*	
8.1.2	Branches *216*	
8.1.3	Additional Branch Considerations *218*	
8.1.3.1	Health and Welfare Branch *218*	
8.1.3.2	Construction/Engineering Branch *219*	
8.1.3.3	Air Operations Branch (AOB) *220*	
8.1.4	Divisions/Groups *221*	
8.1.4.1	Divisions (Geographic) *222*	
8.1.4.2	Groups (Functional: Jobs They Perform) *222*	
8.1.5	Single Resources *223*	
8.1.6	Strike Team *224*	
8.1.7	Task Force *225*	
8.2	Conclusion *225*	
9	**Expanding Logistics** *229*	
9.1	Logistics Section Expansion *229*	
9.1.1	Logistics Branch Structure *230*	
9.1.2	Support Branch *231*	
9.1.2.1	Facilities Unit (Support Branch) *231*	
9.1.2.2	Ground Support Unit (Support Branch) *235*	
9.1.2.3	Supply Unit (Support Branch) *236*	
9.1.3	Service Branch *236*	
9.1.3.1	Food Unit (Service Branch) *236*	
9.1.3.2	Medical Unit (Service Branch) *238*	
9.1.3.3	Communications Unit (Service Branch) *244*	
10	**Expanding Planning and Intelligence** *265*	
10.1	Planning and Intelligence Modular Expansion *265*	
10.1.1	Situation Unit *267*	
10.1.1.1	Field Observer *269*	
10.1.1.2	Geographic Information System Specialist (GISS) *270*	
10.1.1.3	Display Processor *272*	
10.1.1.4	Infrared Interpreter *272*	

10.1.1.5	Fire Effect Monitor	*273*
10.1.1.6	Fire Behavior Analyst	*273*
10.1.1.7	Long-Term Fire Analyst	*274*
10.1.1.8	Technical Specialists	*274*
10.1.2	Resources Unit	*275*
10.1.2.1	Status Recorder	*276*
10.1.3	The Documentation Unit	*281*
10.1.4	The Demobilization Unit	*286*
10.1.5	Two Optional Units	*290*
10.1.5.1	Strategic Operational Planner (SOPL)	*290*
10.1.5.2	Training Specialist (TNSP)	*291*

11 Expanding Finance and Administration *295*
11.1 Compensation Claim Unit *298*
11.1.1 Compensation for Injury Specialist (INJR) *301*
11.1.2 Claims Specialist (CLMS) *302*
11.2 Cost Unit *303*
11.3 Procurement Unit *304*
11.4 Time Unit *308*
11.4.1 Personnel Time Recorder (PTRC) *309*
11.4.2 Equipment Time Recorder (EQTR) *309*

12 ICS Investigations and Intelligence (I/I) *313*
12.1 Historical Overview *314*
12.2 More than Law Enforcement *316*
12.3 Investigations and Intelligence Gathering (I/I) Information Sharing *317*
12.4 Placement Consideration of Investigations and Intelligence Gathering (I/I) *319*
12.4.1 Investigations and Intelligence Gathering (I/I) as Command Staff *320*
12.4.2 Investigations and Intelligence Gathering (I/I) as General Staff *321*
12.4.3 Investigations and Intelligence Gathering (I/I) Section Chief *322*
12.4.3.1 Investigative Operations Group *324*
12.4.3.2 Intelligence Group *328*
12.4.3.3 Forensic Group *331*
12.4.3.4 Missing Persons Group *333*
12.4.3.5 Mass Fatality Management Group *335*
12.4.3.6 Investigative Support Group *338*
12.4.4 Investigations and Intelligence Gathering (I/I) in the Operations Section *342*
12.4.5 Investigations and Intelligence Gathering (I/I) in the Planning Section *343*
12.5 Conclusion *344*

13 The Agency Administrator, Common Agency Representatives, and a Basic Overview of the Planning Process *349*
13.1 The Agency Administrator *349*
13.2 Agency Administrator Representatives *351*
13.2.1 Resource Advisor *351*
13.2.2 Historic Preservation Advisor *352*
13.2.3 Incident Business Advisor *353*
13.2.4 Buying Teams *355*
13.2.5 Administrative Payment Team *356*

13.2.6	Other Teams	*356*
13.3	An Overview of the ICS Planning Process	*357*
13.3.1	The Incident Action Plan (IAP)	*357*
13.3.2	General and Command Staff Planning Responsibilities-the Basics	*358*
13.3.3	The Benefits of the Planning Process	*361*
13.3.4	The Planning Process-an Overview	*362*
13.3.4.1	Initial Understanding of the Situation	*362*
13.3.4.2	Establishing Incident Objectives and Strategies	*363*
13.3.4.3	Develop a Plan	*365*
13.3.4.4	Prepare and Disseminate the Plan	*365*

14 Management by Objectives – SMART Goals *369*
14.1 Underlying Factors for Determining Incident Objectives and Strategies *372*
14.2 Establishing Immediate Incident Objective Priorities *373*
14.2.1 Life Safety *374*
14.2.2 Incident Stabilization *376*
14.2.3 Property Conservation *377*
14.2.4 Environmental and Economic Protection *378*
14.3 Management by Objectives *379*
14.4 Writing Goals and Objectives for the Incident Action Plan *381*
14.4.1 SMART Objective Worksheets *384*
14.5 Management by Objective for Never-Ending Incidents *384*
14.6 The Importance of SMART Objectives in the Planning Process *385*
14.6.1 Completion of SMART Objectives *387*

15 The Planning P-In Depth *391*
15.1 The Beginning of the Incident and Notifications *391*
15.2 Initial Response and Assessment *392*
15.2.1 Complexity Analysis-Incident Typing *392*
15.2.2 Incident Complexity *393*
15.2.3 Typing the Incident – Incident Management Team (IMT) Typing *395*
15.3 Incident Briefing – Preparing for a Transfer of Command *396*
15.3.1 Filling Out the ICS 201 *397*
15.3.2 Potential Other Forms for Transfer of Command *400*
15.4 Delegation of Authority (DOA) *400*
15.4.1 Delegation of Authority Briefing *401*
15.4.2 Transfer of Command *405*
15.4.3 Initial Incident Command/Unified Command Meeting *407*
15.4.3.1 Establish Core Planning Meeting Principles for the Incident *408*
15.4.4 Facilitating (Ongoing) Meetings *408*
15.4.4.1 Ground Rules *410*
15.4.4.2 Agenda *410*
15.4.4.3 Vetting Visitors *411*
15.4.4.4 Documentation *412*
15.4.5 Initial or Ongoing? *412*
15.5 Incident Command Objective Meeting *413*
15.6 The Command and General Staff Meeting-The Basics *413*
15.6.1 Business Meeting Prior to Command and General Staff Meeting *415*
15.6.2 The Initial Command and General Staff Meeting *417*

15.6.2.1	The Command Staff Briefing Within the Command and General Staff Meeting *418*	
15.6.2.2	The General Staff Briefing Within the Command and General Staff Meeting *419*	
15.6.2.3	Optional Command or General Staff if Activated *420*	
15.6.3	The Closing of the initial Command and General Staff Meeting *420*	
15.6.4	Preparations for the Ongoing Command and General Staff Meeting *420*	
15.6.4.1	Planning Section Chief (PSC) Preparations *420*	
15.6.4.2	Incident Commander (IC)/Unified Command (UC) *423*	
15.6.4.3	Situation Unit Leader (SITL) *423*	
15.6.4.4	Operations Section Chief (OSC) *423*	
15.6.4.5	Safety Officer (SOFR) *424*	
15.6.4.6	Logistics Section Chief (LSC) *424*	
15.6.4.7	Finance/Administration Section Chief (FSC) *424*	
15.6.4.8	Intelligence and Investigations ([I/I] if activated at the Command or General Staff Level) *425*	
15.6.4.9	Public Information Officer (PIO) *425*	
15.6.4.10	Liaison Officer (LOFR) *426*	
15.6.5	The (Ongoing) Command Staff and General Staff Meeting *426*	
15.6.5.1	The Command Staff Briefings *428*	
15.6.5.2	General Staff Briefings *429*	
15.6.5.3	Optional Command or Staff If Activated *430*	
15.6.6	The Closing of the Command and General Staff Meeting *430*	
15.7	The Tactics Meeting *431*	
15.8	Preparing for the Planning Meeting *435*	
15.8.1	Planning Section Chief (PSC) *435*	
15.8.2	The Situation Unit *435*	
15.8.3	The Resource Unit *436*	
15.8.4	Technical Specialists *436*	
15.8.5	Incident Action Plan Preparation and Approval *437*	
15.8.5.1	The Planning Meeting *437*	
15.9	Printing the Incident Action Plan *440*	
15.9.1	The Incident Action Plan (IAP) Cover Sheet *442*	
15.9.2	ICS Forms Integration with the Incident Action Plan (IAP) *444*	
15.9.2.1	The ICS Form 202-Incident Objectives *444*	
15.9.2.2	The ICS Form 203-Organization Assignment List *444*	
15.9.2.3	The ICS Form 204-Assignment List *445*	
15.9.2.4	ICS Form 205-Communications Plan *445*	
15.9.2.5	ICS Form 206-Incident Medical Plan *445*	
15.9.2.6	ICS Form 208-Safety Message *446*	
15.9.2.7	ICS Form 220-Air Operations Summary *446*	
15.9.2.8	Incident Action Plan Map *446*	
15.9.2.9	The Traffic Plan *447*	
15.9.2.10	Weather Forecast *448*	
15.9.2.11	Additional Information *448*	
15.9.3	Early Distribution of Incident Action Plan *448*	
15.9.4	Regular Distribution of the Incident Action Plan *448*	
15.9.5	The Operational Period Briefing *449*	
15.9.6	Beginning of the Operational Period *450*	
15.9.7	Special Planning Meetings *450*	

15.9.7.1	Transition Meetings *451*
15.9.7.2	Debriefing/Close-Out Meeting *451*
15.9.7.3	Public Meetings/Press Conferences *451*
15.9.7.4	Demobilization Planning *451*

16   Integrating Incident Management into Hospitals *455*

16.1	Hospital Emergency Incident Command System (HEICS) *455*
16.2	HICS *459*
16.2.1	Triage Briefly Described *461*
16.3	HICS Does Work for Incident Management *463*
16.3.1	Joplin MO Tornado *463*
16.4	The Fundamental Elements of HICS *467*
16.5	Chain of Command *469*
16.6	Command and General Staff *469*
16.6.1	HICS Operations Section *470*
16.7	Staging Manager *471*
16.7.1	Medical Care Branch Director *471*
16.7.1.1	In-Patient Unit Leader *472*
16.7.1.2	Outpatient Unit Leader *472*
16.7.1.3	Casualty Care Unit Leader *472*
16.7.1.4	Mental Health Unit Leader *472*
16.7.1.5	Clinical Support Services Unit Leader *472*
16.7.1.6	Patient Registration Unit Leader *473*
16.7.2	Infrastructure Branch Director *473*
16.7.3	Security Branch Director *473*
16.7.3.1	Access Control Unit Leader *474*
16.7.3.2	Crowd Control Unit Leader *474*
16.7.3.3	Traffic Control Unit Leader *474*
16.7.3.4	Search Unit Leader *474*
16.7.3.5	Law Enforcement Interface Unit Leader *475*
16.7.4	HazMat Branch Director *475*
16.7.5	Business Continuity Branch Director *476*
16.7.5.1	IT Systems and Applications Unit Leader *476*
16.7.5.2	Services Continuity Unit Leader *477*
16.7.5.3	Records Management Unit Leader *477*
16.7.6	Patient Family Assistance Branch Director *477*
16.7.6.1	The Family Reunification Unit Leader *478*
16.7.6.2	Social Services Unit Leader *478*
16.8	HICS Planning Section *479*
16.9	HICS Logistics Section *479*
16.9.1	Services Branch *479*
16.9.2	Support Branch *479*
16.9.2.1	Employee Health and Well-Being Unit Leader *480*
16.9.2.2	Labor Pool and Credentialing Unit Leader *481*
16.9.2.3	Employee Family Care Unit Leader *481*
16.10	Finance and Administration Section *481*
16.11	The Planning P/The HICS Planning Process *482*
16.12	Emergency Operations Plan *484*
16.12.1	An All-Hazards Plan *486*

16.12.2 Who Should Create the Emergency Operations Plan (EOP)? *487*
16.12.2.1 Acknowledgement of the Incident *487*
16.12.2.2 How an Incident Should be Managed *487*
16.12.2.3 Preplanning Actions and Resources Needed *488*
16.12.2.4 Situational Awareness *489*
16.12.2.5 Roles and Responsibilities *489*
16.12.2.6 Communications *490*
16.12.2.7 Staffing *490*
16.12.2.8 Credentialing *491*
16.13 Volunteer Management *493*
16.14 Health and Medical Operations *494*
16.14.1 Fatality Management *495*
16.14.2 Decontamination *495*
16.14.3 Health and Medical Advisories *495*
16.14.4 Interjurisdictional Relationships *496*
16.14.4.1 Patient Management *496*
16.14.4.2 Logistics *497*
16.14.4.3 Finance and Emergency Spending Authorizations *498*
16.14.4.4 Resource Management *499*
16.14.4.5 Donations Management (Solicited and Unsolicited) *499*
16.14.4.6 Infrastructure Management (Building, Grounds, Utilities, Damage Assessment) *500*
16.14.4.7 Evacuation *500*
16.14.4.8 Safety and Security *501*
16.14.4.9 Coordination with External Agencies *503*
16.15 Conclusion *507*

References *513*

Index *527*

Foreword

Since the first days, when the wonderful men and women firefighters in Southern California (Firefighting Resources of Southern California [FIRECSCOPE]) decided that it was in the best interest of the community to jettison their egos and any stovepiping outside of firefighting and embraced the "Incident Management System" (IMS) and later "Incident Command System" (ICS), there has never been such a strong piece of literature on ICS. This change introduced the most effective formalization of interoperable response for collaboration, cooperation, and communication in the US national preparedness system. I am Dr. Michael A. Brown, and it is my pleasure to introduce you to the most comprehensive and thorough piece of emergency management literature ever written on the topic of the National Incident Management System (NIMS) and its component ICS. Through my career, I have read and experienced the debates from both supporters and naysayers regarding the need for and use of the ICS model. This debate is often partisan with academics sometimes noting disparaging literature about the ICS and practitioners heralding the wonderful attributes of the hierarchical operational system. I think this most comprehensive work by Dr. Mark S. Warnick will dispel any misunderstandings or doubts about what the ICS is, and is not. Further, this textbook will increase the knowledge of even the most seasoned practitioner with facts and historical relevance that left me speechless. For example, the reference to "Buffalo Soldiers" and the thorough examination of ICS around the world.

The fact that I was asked by Dr. Warnick to write the forward is noteworthy as well and somewhat ironic. When we first met, Mark was a doctoral candidate with a reputation for thorough and "out of the box" inclinations that sometimes rattled his peers, but his foresight always proved invaluable with hints of Edgar Cayce futuristic predictability. Little did we know that he and I had served together in the unit in the military, and the same barracks some 40 years prior, and of course "fast FORWARD," (play on words) we continue to serve. I ask you to read this masterful emergency management literature and marvel at the precision, depth of research and evaluative beauty within each chapter. Learn about Australia, Canada, and even the Maldives' ICS. The information provided on social media is cathartic, and his insight on ICS for houses of worship a must read. Understand that the proliferation, intensity, and size of focusing events will continue, which means the need for ICS and its closely tethered interoperability is needed more than ever. Is ICS a panacea for disasters? No! But it is a most effectual operational tool for coping with and helping to respond to all manner of hazards. In closing, I wish to thank Dr. Mark S. Warnick for providing the disaster preparedness and emergency management community with a jewel of literature that will change the courseroom and emergency operations centers for the better in the future.

Michael A. Brown
Core Faculty Emergency Management
Capella University School of Public Service

Preface

A dear friend of mine, Louis N. Molino Sr. wrote the first edition of this book. As his health began to fail, he asked if I would help co-author the second edition with him. I am sad to say that at the time, I was finishing up my dissertation for my doctoral degree, and I put him off for six months so that I could focus on my degree first. During those six months Lou, who took care of others instead of himself, passed away.

After the passing of Lou Sr., and the completion of my degree, I called Louis N. Molino Jr., and told him about our plans. Lou Jr. was gracious and put me in contact with Bob Esposito from Wiley Press, and after negotiating the contract, the second edition of this book was borne. This book was not only created to help public safety personnel but to honor Louis N. Molino Sr., and to give back to his children in some way.

July 4, 2019   *Mark S. Warnick*

Acknowledgments

There are so many people to acknowledge. First and foremost, I need to acknowledge my wife, Cleo, who has been a beacon of light when I felt as if I would never complete this book. When my head became bogged down with technical aspects, she was there to help me clear my head and get me back on track. She is the love of my life, and she helped me to actually finish this book through her gentle and kind encouragement and her gentle nudges when I no longer felt like writing.

I also need to acknowledge my content editors Angie Bowen, Jim O'Neil, and Joe Campbell. These individuals were chosen because they have worked in the field for many years, and they were there for me to ensure that I did not miss anything. While I am sure there is something we missed (or that I got wrong), these individuals were gentle, honest, and encouraging!

Angela (Angie) Bowen has devoted 31 years to public safety and 28 years to emergency nursing. Angie is a nurse/paramedic with a background in rescue, EMS, and pediatric emergency/trauma/transport nursing. Angie currently utilizes her skills with the Radiation Emergency Assistance Center/Training Site (REAC/TS) and Jefferson County (TN) EMS. Her dedication to helping others is only surpassed by her hunger for more knowledge so that she can help more people. She was once a student of mine, but now I am proud to call her my friend!

Jim O'Neill and I have been thick as thieves for somewhere around 30 years. Jim has over 40 years, starting out as an E.M.T in Boston, MA, and then he eventually went into real work, firefighting (love ya Jim!). For 11 years, Jim served as a Fire Chief in Suwannee County, FL, and he was a darn good one! When Hurricane Katrina struck, he was one of the individuals who responded. He found that he loved it so much; he has spent 14 years with the US Department of Homeland Security/Federal Emergency Management Agency (FEMA). Presently, he is a Disaster Recovery Center manager with the Federal Emergency Management Agency. His dedication to helping others is amazing!

Joe Campbell currently works as a Boating Investigator for the Tennessee Wildlife Resources Agency. Joe has 28 years in public safety and has worked in law enforcement, firefighting, emergency management and is a Paramedic, most of them simultaneously. Essentially, he has worked in almost every major field that uses the Incident Command System (ICS) method. Joe has become a good friend, a confidant, and someone I can vent to, even if he does answer my phone calls with "What's up peckerhead?". He too is a brother in arms, and I am thankful for him and for his help with this book.

I want to acknowledge my pastor, Thomas Black, and my church family at Fame Evangelical Church, whose prayers and encouragement helped me along the way. Their prayers of intercession were regularly felt, and they were answered.

Finally, I want to acknowledge my Lord and Savior Jesus Christ. If not for his love and guidance, I have no doubt that I would not be the person I am. Anyone who knows me will know that I am not perfect, but I am a work in progress, and I am forgiven. I readily acknowledge that Jesus died on the cross for my sins, and he cleansed me of my sins, even though I am still a sinner.

About the Book

This book will likely be different from any book that you have read before. It intentionally goes outside of the norms, so that you can learn, and hopefully retain the information that is provided. Probably, the first thing you will notice is the acronym or shortened version of a title is not used just one time, and then the acronym is used throughout the rest of the book. These were left together throughout the book to help you remember both the title and the acronym. The only exceptions are the Incident Commands System (ICS), Incident Management Systems (IMS), Hospital Emergency Incident Systems (HEICS), and Hospital Incident Command Systems (HICS).

As you read through this book, the first two chapters are there to give a historical perspective and to show that the Incident Command System does work. The third chapter shows what incident management is used in various places around the world. Chapters 4–16 are in place to describe the methods and systems used throughout the United States, and in many instances, to some degree, around the world. Special consideration was given to the Incident Command System (ICS) and the Hospital Incident Command System (HICS).

As you go through this book, you will read about real-life scenarios, and you will see analogies. You will see quizzes at the end of each chapter, and you will see resources that will guide you to more information on the topic of that chapter. More than anything, I hope that this book is presented in such a way that you can learn and retain this information. I am sure that I have made a mistake or two in this book, but the vast majority of it is correct. If you find mistakes, PLEASE let me know so they can be corrected in the next edition.

About the Companion Website

This book is accompanied by a companion website:
www.wiley.com/go/Warnick/EIMS_2e

Scan this QR code to visit the companion website.

The website includes:
- Solution manual

1

Introduction

Those who cannot remember the past are condemned to repeat it.

George Santayana

It has been said that necessity is the mother of all invention, and this holds especially true with incident management. Out of necessity, the creation of what we now call **Incident Management System** or **IMS** was born so that first responders could be more effective in responding to every kind of incident that may occur. IMS methods have been modified to help manage nearly every type of emergency situation that can occur. Whether it is a disaster in a hospital, a traffic collision, an armed hostage situation, or a disaster that utilizes the military and various other federal government entities, IMS is a method that helps ensure everyone is on the same page and working together.

Unlike most creations, IMS is not a physical thing that can be concretely viewed, picked up, or even handled. These IMS systems are concepts. More accurately described, IMS methods are a set of philosophies, policies, and procedures that help the users (of the system) to control and manage chaotic emergencies. These IMS methods are also used to provide structure and safety measures during a planned event. When it is employed properly, it allows those managing an incident or an event to divide larger tasks into much smaller tasks and to assign individuals who will manage specific functions within the response or event. In essence, IMS methods are divide and conquer structures that help to ensure that no details are overlooked.

In the United States, there is an overarching IMS method that is known as the National Incident Management System (NIMS). The NIMS method will be explained more comprehensively in later chapters, but this system was essentially designed to prepare for, prevent, and manage the response and recovery to disaster (and emergency) situations. Utilizing NIMS increases the coordination among individuals who respond to disasters on the local, state, and federal levels. While an oversimplification, NIMS was designed to integrate every stakeholder into an organized system that is ready to respond. When resources are ordered by an Incident Commander (IC) utilizing NIMS, it ensures that they will more than likely get the exact resource they need, and that resource (including personnel) will be on the same page as the local government.

Beyond the use of the overarching NIMS method, there is also what can essentially be called the command and control system of managing an incident in the United States. This is known as the Incident Command System (ICS). ICS allows local governments and local first responders to control an incident while having the insurance policy of being able to integrate outside resources if the local agency is overwhelmed. This IMS method allows the Incident Commander to divide the responsibility of managing the incident into as many as five different tasks, each

Emergency Incident Management Systems: Fundamentals and Applications, Second Edition.
Mark S. Warnick and Louis N. Molino Sr.
© 2020 John Wiley & Sons, Inc. Published 2020 by John Wiley & Sons, Inc.
Companion Website: www.wiley.com/go/Warnick/EIMS_2e

being managed by someone else, while the Incident Commander manages those supervisors and the overall incident. It is important to note that ICS is a component of NIMS, meaning that NIMS is the all-encompassing IMS method and ICS supports NIMS. While it may sound confusing, especially this early in the book, it is not difficult to learn.

It should also be noted that there are other IMS methods used in the United States including Hospital Incident Command System (HICS). HICS was designed much like ICS so that it integrates with NIMS. HICS also integrates with ICS as well, and there is usually a seamless, or near seamless integration of the three IMS methods. Much like ICS, HICS facilitates command and control of an emergency or a disaster while enabling the seamless integration of outside resources, should they be needed.

While one could get the impression that IMS methods are useful in overwhelming and large-scale disasters, they would be wrong. IMS methods can, and should, be utilized in daily emergencies. Local first responders and local nongovernmental agencies (e.g. American Red Cross and Salvation Army) should become proficient in the day-to-day use of IMS methods. By doing so, they will more effectively manage small incidents and become more proficient in the use of NIMS and ICS for when the "Big One" hits.

IMS methods can also be useful in nonphysical or catastrophic emergencies. Public safety agencies and other entities that are familiar with IMS can also utilize these same concepts to manage other types of nontraditional emergencies. These might include issues such as managing their day-to-day operations and nonphysical or nonemergent situations. IMS systems are extremely useful in organizing what may seem overwhelming.

IMS methods are also practical when an agency is facing a negative media blitz over a negative interaction. Most IMS methods can assist in making sure that all angles of public relations are covered. Consider the amount of crisis management that would be needed in an officer-involved shooting, a negative story about an ambulance (or fire truck)-involved collision, or even personnel caught in a scandal. Using an IMS method can and will help manage the crisis in the media, and if structured and utilized correctly, the IMS method can help initiate and provide ongoing damage control.

IMS can also be used to manage planned events such as sporting events, festivals, concerts, and similar events. Properly employed, IMS can be used to manage and control a multitude of incidents, including most nonemergency situations. We only need to look at past historical (major) sporting events to see how an IMS method can be used. It will also help to identify how resources can be organized so that they are highly effective and available.

If we look at Super Bowl LII that took place on 4 February 2018, we can see how managing a large planned event is possible when utilizing an IMS method. As you might imagine, the use of the Incident Command System (ICS) component, integrated with the National Incident Management System (NIMS) was used to manage this major sporting event. Those that oversaw the public safety aspect of this overwhelming planned event made sure that security was of the utmost importance. In their effort to provide top-notch public safety, they utilized and managed over 40 federal agencies, approximately 840 different officers from 60 local and state agencies, and they incorporated the Minnesota National Guard (Moore, 2015). Beyond the previously mentioned assets, there were also a large contingency of processes and equipment integrated into the overall management of this planned event.

Not only was ICS component and NIMS used to manage a large security presence (to prevent some type of chaos from happening), but it was also used behind the scenes. There were emergency managers who planned, and who were on standby for a mass casualty event, Emergency Medical Services (EMS) personnel were strategically placed and prepared to jump into action should someone become ill or injured, or in the event a mass casualty incident occurred. Hazardous Material (HazMat) teams were also staged and ready to respond if a Chemical, Biological, Radiological, Nuclear, or Explosives (CBRNE) incident occurred. Additionally, a multitude of other very specific teams were managed through the utilization of ICS component and NIMS. Some security teams were scanning for problems, while separately, other teams were ready to jump into action and take the steps needed to mitigate the effects of an emergency situation. Trying to manage a planned event, or even an emergent incident that uses a lot of personnel and equipment, would be a nightmare if not for IMS methods.

In the United States, the ICS component and its supportive companion, NIMS, has become the standard in recent years. Many people have attempted to historically date the initial development of ICS, and there has been much discussion about where the main underpinning began. No concrete date or circumstance has ever been agreed upon, even though it is usually taught that Firefighting Resources of California Organized for Potential Emergencies (FIRESCOPE) was the first official iteration of the formal system known as ICS.

In retrospect, we could say that it was a long progression of events that eventually led to the IMS methods that we now see in today's world. Many different concepts, and many different incidents have used similar methods throughout the last few hundred or more years, and some of them have been concretely identified as playing a significant role in the systems of today. Various incidents have very distinct bits and pieces of current IMS methods.

As we move onward through this chapter, there will be varying incidents discussed. These incidents may have contributed to the IMS we use today, then again, it may be coincidence that these methods were used. It should be noted that some of the history and incidents that may have contributed to modern-day IMS methods is nothing more than speculation. On the other hand, some of these incidents are concretely documented as contributing to the current IMS methods being used today.

1.1 The Revolutionary War

It does not take much of a stretch of the imagination to say that IMS methods may have come from early military campaigns that were undertaken in the United States. Some might say that the initial fledglings may have come from other countries. Since this book primarily focuses on the United States, we will discuss only instances in the United States which may, or potentially may not have led to these systems.

Part of the reason for focusing primarily on the United States is also based on expediency. An individual could spend a lifetime trying to make connections between the gazillions (or is it quintillions) of incidents that have occurred worldwide in the last few hundred years. Most of us do not have the time, resources, or even the inclination to take review it so in depth.

We only need to look at the history of the Revolutionary War to understand that some of the principles we see in modern-day IMS method were used during this war. Initially, the Revolutionary War had no strategy. It was nothing more than a few haphazard militia's fighting against the British when they came into or occupied their geographical location. Initially, there was no centralized commander, no strategy, and no single person, or group of people, in charge. There was no one single person that was charged with coming up with a singular or overarching strategy. In fact, most of the resistance to the British were organized locally and was not part of a larger tactic or main plan. While many of the battles were bloody, they did not follow a strategic battle plan. The objective was quite often to drive the British from the area rather than looking at the bigger picture of defeating the British in all of the colonies.

The Continental Congress began to realize that an organized response was not in place, and some would say they began to wonder if they could win these battles and claim their independence. After much debate and discussion, the Continental Congress appointed George Washington as the General, and commander, of the Continental Army. This occurred on 10 May 1779 (Stockwell, n.d.). By historically reviewing information, we see that Washington began to organize a comprehensive plan, and when implemented, Washington had named it *The Grand Strategy* (Stockwell, n.d.).

Washington, in his infinite wisdom, unified his battles using the militia from the colony's as the main force, and he integrated the French into specific battles where he felt the militias needed bolstering or reinforcements. He ensured that he provided specific orders for each battle, but he had chosen capable individuals manage the battles. Those capable individuals had the ability and the authority of General Washington to make on-the-spot decisions, based on the circumstances at hand.

Some historians have said that Washington's tactics were primarily defensive tactics. They believe that General Washington used a multitude of tactics to exhaust the British, which was sprinkled with hit-and-run attacks and a propaganda campaign. The purpose of the propaganda campaign was to undermine the will of the British citizenry and their soldiers (Brooks, 2017). There were some offensive actions taken against the British, but they were often rare.

Those that are learning about IMS methods for the first time should realize the similarities of the propaganda campaign of General Washington, and the IMS methods of today. In Washington's propaganda campaign, the successes of those fighting for independence (such as the militias and specific individuals) were touted. Additionally, the failures of the British were exploited, and their successes were rarely mentioned. This strategic release of information led to more support for the overall effort of war and independence.

When IMS methods are used in modern-day times, there is nearly always a component that covers public information. In most instances, this is a Public Information Officer (PIO) or the utilization of a Joint Information Center (JIC). Modern-day Public Information Officers (PIO's) fashion the release of information, but they do not use propaganda. Similar to the propaganda that was released by the Continental Army, the PIO and/or JIC strategically release information to garner support for their cause: the support for the management of an incident. While this is not propaganda per say, it is a similar method to garner support and provide information about an incident. It is easy to see that at least in some respects, it mimics the work that General Washington employed to help win the war.

In looking at the potential of a historical military connection to modern-day IMS methods, French General Jean-Baptiste Donatien de Vimeur, comte de Rochambeau arrived to assist with the struggle to gain independence in July 1780. He and General Washington would work together to fight the British, but prior to Rochambeau's departure from France, King Louis XVI advised him that he should be subordinate to General Washington. Essentially, General Rochambeau played a supportive role for the Continental Army, and he would work under the command of the Incident Commander (IC), General George Washington. While Washington was in charge, he also held planning meetings with Rochambeau. Historically, we see that these meetings were held between General Rochambeau and General Washington prior to an offensive to ensure that everyone was on the same page and to ensure everyone knew their role in these attacks (Covart, 2014).

If we compare the actions of Washington in the Revolutionary War to that of the modern-day IMS Method, we can see that there are multiple similarities. We can start with the appearance that a centralized command was initiated. That centralized command gave strategic instructions to subordinates on what should be done, but it also gave authority to a specific person (or persons) to command how that goal was met. This is a customary practice in all forms of IMS methods in use today. There is typically a centralized command on every incident, even if that centralized command is a group of individuals (Unified Command which will be explained in Chapter 7). This centralized command will always provide guidance and direction, but in most instances, the commander will provide the strategy and then give authority to a person who will make the field decisions.

From a historical perspective, we can see that General George Washington was assisted by an outside organization, and that organization was subordinate to the centralized command. We also see that General Washington held planning meetings, which utilized multiple individuals who helped in guiding the direction of the response, creating comprehensive strategies for the incident, or in this case, the war. A similar method is also undertaken in the IMS methods of today. In the planning phase, the centralized command utilizes the expertise of many (who are experts in specific areas) to come up with an overall comprehensive plan that is strategic. This type of planning helps to provide fewer surprises and creates contingency plans for if an incident does not go as expected.

Since the Revolutionary War, the United States Military has used similar methods to organize their strategies. It should be noted that this theory about the Revolutionary War contributing to modern-day IMS is not widely acknowledged. Nonetheless, it has striking similarities, which might support that the Revolutionary War at least played some type of role in the IMS methods that we use today.

1.2 The Big Burn of 1910

The Big Burn of 1910 is considered, by some, as another incident that helped contribute to modern-day IMS methods. As can be deducted by the name, this incident was an extremely large wildland fire. Many have called the Big Burn a major firestorm. This firestorm has been known by many different names, including the Big Blowup, the Great Fire of 1910, and the Devil's Broom fire.

To put the immensity of this incident into context, we first must understand the immense geographical area and the massive loss of life that is attributed to this fire. This enormous fire was so vast that it simultaneously affected the states of Montana, Idaho, and Washington, and in total covered a land mass of approximately 4700 mi^2, or approximately 3 million acres. This massive fire is credited with causing the death of 87 people (78 of them firefighters), and injuring countless more (Galvin, 2007).

According to historical accounts, the main firestorm burned over 3 million acres in just 36 hours. This occurred between 20 and 21 August 1910. While the main fire, which was driven by 70–80 miles per hour winds, the ongoing clean-up operations and firefighting efforts to address the smaller fires related to the Big Burn lasted well beyond October 1910.

Prior to the firestorm, various laws and agencies were put in place to protect the environment from man as well as protecting the environment from various types of disasters. In 1876, the US Congress created the office of Special Agent, as part of the Department of Interior. The Bureau of Forestry was created in 1905 by President Theodore (Teddy) Roosevelt, and the responsibility for taking care of those forests were transferred from the Department of Interior, where they were originally placed, to the Department of Agriculture. The new forest service was named the Division of Forestry and the President Roosevelt appointed Gifford Pinchot as the new director (Williams, 2005).

One of the first goals of Pinchot was to decentralize the Division of Forestry from the Washington, DC bureaucrats. Prior to his appointment, almost all decisions about the forests and their ongoing health were made and based in Washington DC. This allowed decisions to be made by mostly uniformed politicians who were well removed from the forests they were charged with protecting. Pinchot believed that decision-making should be done in the field. He accomplished this by creating forest districts and assigning District Foresters to monitor specific geographical areas (Williams 2005).

As part of his overall strategy, he hired and placed District Foresters in Denver, Colorado; Ogden, Utah; Missoula, Montana; Albuquerque, New Mexico; San Francisco, California; and Portland, Oregon. These District Foresters were charged with caring for these forests, based on the needs of the geographical locational, rather than using standing orders from a distance place. The previous method of standing orders inevitably protected some forests in the United States, while damaging others (Williams 2005). Pinchot believed that the newly formed districts would be more familiar with the unique needs of their geographical location, and those District Foresters would more familiar with local resources that could be utilized to meet those needs.

As part of the decentralization process, Pinchot built the forestry services first warehouse in a centralized location, Ogden, Utah. He filled it with a cache of equipment that might be needed to protect the forests, including firefighting equipment. Not long after creating the warehouse in Ogden, he began building other strategically located warehouses across the nation. He continued to build and fill warehouses in each of the forestry districts. In order, these additional warehouses were built in Ogden, Utah; Missoula, Montana; Albuquerque, New Mexico; Denver, Colorado; San Francisco, California; and Portland, Oregon. This ongoing strategy allowed the forestry service to have the firefighting equipment they needed within the district, and because they were located in locations with rail lines, supplies could be shipped to where it was needed in a relatively short amount of time (Williams 2005).

One of the District Foresters that Pinchot hired was William Greeley. Greeley, who was given the District Forester job in the Rocky Mountains, hit the ground running. Upon his arrival in the Rocky Mountains, he immediately tried to make agreements with lumber companies to help preserve and protect forests, much to the chagrin of Greeley's supervisor, Gifford Pinchot. Pinchot wanted to continue the practice of making verbal agreements with landowners in order to provide for the protection of forest lands, while Greeley wanted the forest industry to take a vested interest in protecting these lands (Williams, 2005).

Greeley continued down the path of trying to secure lumber company partners, and it did not take long before Greeley found a lumber company that bought into this strategy. George Long, who was running the old Weyerhaeuser Timber Company, was convinced that it was in the best interest of lumber companies to help preserve the forests. This decision by Long was partly based on the unbelievable loss of timber that Weyerhaeuser suffered a few years earlier. Because of losses, the company suffered in 1902. Long, and the company, were dedicated supporters of forging large private firefighting cooperatives. In June 1909, Long and Greeley made a formal agreement for cooperative forest management and firefighting efforts (Williams, 2005; "William B. Greeley," n.d.).

To the best of our knowledge, this was the first formal mutual aid agreement for firefighting (other than landowner agreements) ever made in the United States. This campaign of mutual aid agreements soon spread across the American West. Oregon, Washington, and California soon began creating more forest protection associations, and they looked more at collaborative and cooperative efforts to protect America's landscape.

Greeley also worked to forge agreements with the train companies. Through his efforts, and partially because he tied some fires to trains, Greeley worked to create expedited delivery of firefighting supplies whenever forest fires occurred. Agreements were signed that essentially encouraged those operating trains to provide a quick response of equipment, and if need be, the personnel needed to fight fires. This allowed Greeley to place the equipment near rail lines, as Pinchot envisioned, but it provided priority service, so the trains could rapidly transport equipment to where it was need, much like (Pinchot) envisioned (Williams, 2005).

If another forestry district had an emergency, both manpower and tools could be quickly shipped by rail in a relatively short amount of time. In the grand scheme of things, Greeley felt that animosity should not be at the center of protecting forests. His goal was to protect forests through cooperation and communication among those that had a vested interest in protecting the forests, the stakeholders of the Great American West ("William Greeley," n.d.).

Looking at IMS methods of today, it does not take much of a stretch of the imagination to see how these historical incidents fit into our modern-day IMS methods. The first similarity is that Greeley incorporated stakeholders, those with a vested interest in the protection of forest. This is a common practice that is currently used to ensure that planning, response, and recovery is entirely comprehensive. Second, the five basic components include command, control, cooperation, collaboration, and communication. In the years prior to the Big Burn of 1910, Greeley took command and control of the resources he had at his disposal, and he used cooperation, collaboration, and communications to build a method of dealing with almost any major fire. In a nutshell, Greeley worked to provide for the basic principles that make up IMS methods, and he was preplanning for a major event. In times prior to this, the standard procedure was to wait until the resources were needed, in place of preplanning.

These preplans that Greeley worked on would become the key in saving many lives, and in preserving massive amounts of property. In the spring and summer of 1910, the Great American West was suffering a severe drought. According to Petersen (1994–1995), there were either 1736 or 3000 fires in Montana and Northern Idaho prior to 20 August 1910. Many of those fires were blamed on the embers of passing locomotives and various lightning strikes that had occurred over the previous months. Due to the unusual amount of fires, Greeley being desperate for help proceeded to hire everyone he could to work fighting forest fires. While some would question the ethics of his hiring process, it was reported that he would hire anyone capable of fighting a fire. According to reports, this included hobo's, bums, and criminals. In some instances, local felons were released from jail early, or used in a form of work release to assist in fighting the fires (Pollak & Ives, 2015). The fires that plagued the region prior to the Big Burn were so taxing, that all fire crews, including the lumber company crews, were becoming exhausted.

Because of the exhaustion from fighting hundreds of smaller fires, Greeley made the unusual request for soldiers from the federal government to assist with firefighting operations. A group of Buffalo Soldiers from the United States Army's 25th Infantry Regiment, also known as the "Crack Black Regiment," were sent to the area to assist Greeley with firefighting operations (Pollak & Ives, 2015).

These soldiers were African-American men who were given the name Buffalo Soldiers by Native Americans. The Native Americans coined this name because of the fierce, brave, nature in which they fought. The Buffalo Soldiers arrived on or about 14 August 1910, with *Company G* arriving at Avery Idaho, and *Company I* arriving in Wallace Idaho (Cohen, 2010; Pollak & Ives, 2015).

In comparison to modern-day IMS Method, this request for soldiers mimics the request for National Guard troops and the military that is sometimes used in modern days disasters. We only need to look at incidents like Hurricane Katrina to see how the military can integrate into almost any disaster response. If we look at the Deepwater Horizon oil spill, we also see that the US Coast Guard was the lead agency, and they too integrated other agencies into their response. The use of Buffalo Soldiers resembles federal resources (such as Federal Emergency Management Agency [FEMA], the military, the Department of Energy and more) that we use in many of the Presidentially Declared Disasters that we see in modern times. Much like modern times, Greeley was the centralized command that these soldiers reported to, yet they had their own level of autonomy to make the decisions of how orders should be ordered.

Turning back to the Big Burn of 1910, Greeley and his crews were working on many small fires that were still burning when hurricane force winds blew into the region, causing instantaneous mass destruction. These hurricane force winds caused the smaller fires to combine into a singular massive fire (Wilma, 2003).

When this fire became a singular massive fire, Greeley and others who were in charge of fighting the Big Burn, used some of the same types of incident management that we see in modern-day IMS methods. Crews used the divide and conquer method, where some crews were sent to specific areas, while yet other crews were left behind to keep the peace. Additionally, certain select crews were charged with evacuating the towns to prevent the loss of civilian life. The agreements that Greeley forged only a year before were used to reduce the effects of the fires. Multiple trains were sent to various towns to assist in the evacuation of the citizenry. These organizing and actual operational actions that were undertaken were assumed under the direction of crews left behind to manage the evacuations, and those charged with setting backfires to save these cities and towns ("The Great Fire of 1910", n.d.).

Much like modern-day IMS methods, each crew was given a job that had specific duties affiliated with that job. This was much different from other responses during this time frame that usually required a crew to be responsible for multiple duties. The previously mentioned Buffalo Soldiers were one of those groups that had specific duties. They are accredited with saving many lives by managing evacuations and by specifically saving the town of Avery, Idaho. While some soldiers worked on evacuations, others set backfires to burn up any potential fire fuels that would in essence protect the cities and towns (Cohen, 2010).

According to historical data, it was recorded that over 10 000 men were utilized to fight this one massive fire. The area that was burned stretched from eastern Washington, into Idaho, and western Montana (Petersen, 1994–1995). It took extensive planning, probably much like what is done in modern IMS methods. In the planning phases, they had to dispatch and manage these firefighters (which included every able-bodied man that could be found) in the entire region. Over a period of month, these individuals worked cohesively until the incident was over. According to Petersen (1994–1995):

> Every able-bodied man fought the fire. Most were Idaho loggers, miners from Butte, Montana, and skid row bums brought in on trains from Spokane. The pay was 25 cents an hour, plus a bedroll, sourdough pancakes, coffee and canned tomatoes
> (para 25).

This too has similarities to the current trends in IMS. When major incidents occur, crews are brought in from multiple agencies. In most circumstances, they receive a pay for their services, and their nutritional needs are cared for. While most modern personnel are vetted, there are wildland firefighting crews on the West Coast who are serving prison sentences. This also is similar to the Big Burn because of the use of those who were in jail during the 1910 fires were used to help fight these fires. We also must recognize that crews on long incidents are provided a place to sleep, food, and pay, until the incident is over. In modern-day IMS methods, Base Camps are utilized to provide food, certain amenities (showering, shaving, etc.), and a place to sleep.

While the impact of the Big Burn on modern-day IMS is mostly supposition, one could say that these similarities, no matter how big or how small, could have played a role in helping develop current IMS methods. At the very least, the response to the Big Burn of 1910 played some part in the progression to the systems now in place.

1.3 The Military Connection

While the Revolutionary War may have played a role in helping to shape and form modern-day IMS methods, later military campaigns, plus research, and trial and error implementations have also been a contributor to modern-day IMS methods. Since the days of caveman, different countries and different regions have had disagreements. Those disagreements have (in some instances) grown into fights; some of those fights have gone even further and developed into wars (Molino, 2006).

As more wars developed, those who were leading and managing their assigned troops had a seemingly endless mission. That mission was to better their ability to actively engage in, and effectively win the battle, or the entire war. In the furtherance of that mission, the science of war has evolved. In some modern-day circles, it has become known as the *Art of War*, and it is often steeped in the writings of Sun Tzu. In the book *The Art of* War, written during fifth century, Tzu (republished 1772) talks about various facets that are needed to win a war. Many of those facets have traversed their way into modern combat and into the modern-day IMS methods.

Much of the art of war used today is loosely based on, and revolves around, four distinctive elements that assist in a war's management. Those elements are command, control, coordination, and communications, often known in the US Military as C4. As is common with the military, many acronyms are related to C-classifications that are often used. Included in these classifications are

- C2: Relates only to Command and Control
- C2I: Command, Control, and Intelligence
- C2ISR: C2I plus Surveillance and Reconnaissance
- C3: Command, Control, and Communication
- C3I: Command, Control, Communications, and Intelligence
- C4: Command, Control, Communications, and Collaboration
- C4I: Command, Control, Communications, and Collaboration plus intelligence
- C5I: Command, Control, Communications, Computers, Collaboration, and Intelligence

As can be seen from the *C* classifications above, some military officials also use the acronym *C4I*. The *I* in this acronym represents the element of *intelligence*. Intelligence is an important aspect that is often overlooked in public safety. This holds especially true when a crime was committed, or the incident involved a terrorist attack. Many would argue that in the post 9/11 era, this element is also an emergency services/public safety function and concern, which it is. However, the intelligence aspect of IMS methods, more specifically the ICS method, will be discussed in Chapter 12.

In looking at the military connection to IMS, the same C-classifications are equally important to the military as they are to first responders. In most instances, the C4 and the C4I are applicable in most modern-day IMS method situations. Suffice it to say that the same elements are needed to tackle and conquer nearly any type of emergency response situation. Consider the fact that public safety, whether fire agencies, law enforcement agencies, or EMS agencies are paramilitary organizations.

As we look at the military connection, we need to realize that the US military has evolved from men with French Muskets and swords (often marching by foot or on horseback) to the most sophisticated military force on the planet. These substantially more agile military units and branches now manage major wars that cover a multitude of geographical locations, sometimes even in different countries. The same concepts loosely apply to public safety.

Public safety agencies of the past would often tackle one incident, or one disaster, at a time. In more recent years, public safety agencies have evolved, and most are capable of monitoring, responding to, and making a difference through operational response for multiple ongoing incidents at the same time.

If you look at the organizational chart of a military organization, you can plainly see that there is a clear and apparent chain of command. When we talk about chain of command in this sense, we are talking about an official ladder of authority that declares who is in charge, and who a subordinate must approach to obtain permission for nearly anything they want to do in an official capacity. This chain of command creates leadership accountability.

While still writing about leadership accountability, it is important to realize that having leadership accountability is important in both the military and in public safety. Knowing who gave what order can be critical in correcting mistakes, and the military has known this since military conflicts have existed. It is critical to remember that no matter who we are, there is always a chain of command. Everyone has someone to report to, and that person usually holds them accountable for their actions.

Even if we look at the top dog in the military, the generals report to higher-ranking generals, the higher-ranking generals must report to the Secretary of Defense, who in turn reports to the President. You may think that the President reports to nobody, but you would be wrong. The President reports to the people of the United States, and he/she also reports to other world leaders when it comes to actions that may destabilize a region and various other issues. This is a basic example of the chain of command and that everyone has someone they report to.

If we put this chain of command in context to public safety, we can see that it holds true in all paramilitary organizations. In a fire department organization, firefighters report to lieutenants, who report to captains, who report to the fire chief (or another officer between themselves and the fire chief), the fire chief reports to the mayor, and the mayor reports to the city council and the citizens. Similarly, officer report to the shift commander in a law enforcement agency, who reports to the police chief. The police chief reports to the mayor who reports to the city councils and the citizens they serve.

In a military organizational chart, you can also see that there is a centralized command that directs "area" commanders to undertake tasks based on the conditions they are facing. They plan, and they call for mutual aid from another branch (or even the same branch) of military, and they work in teams, accomplishing different tasks that all lead to the greater good of the mission for which they are on. In some instances, they may even coordinate with foreign entities, in an effort to command, collaborate, coordinate, and communicate so that they can meet the goal of an agreed-upon outcome.

We only need to think about the wars in Afghanistan and Iraq to prove the point that the 4CI is needed. Whenever a ground team falls under heavy enemy fire, they call for assistance. The type of assistance they may receive can vary, but there is definitely a multitude of resources at their disposal. If a group of infantrymen are pinned down by small arms fire and/or mortar shells, they make a radio call to their command and advise them of their situation, possibly even identifying the resources that they think are needed. When these infantrymen request help, they give the location (coordinates) of where they are, and the location of where the enemy is. With that information, the commander can look to all branches of service and decide who is best suited to support the mission and provide support to these infantrymen. The command officer may determine that airstrikes from fixed wing aircraft are needed, so they may look at resources from the Navy, the Marine Corp, or the Air Force. They may decide that helicopters with airstrike capabilities are needed, so they may look to the Army, the Marine Corp, or the Navy may be needed. Perhaps the fixed wing and helicopter resources of the United States are stretched thin, and they need to ask for assistance from an ally such as the British, the Australians, or the French. Then again, the command officer may believe that ground troops are needed, and they need to be brought in by helicopter. In some instances, this may require multiple agencies (or even countries) because the individuals being brought in may be Army Rangers being brought in by Marine Corp or Navy Helicopters, or perhaps another country has the needed resources closer. In all of these instances, organizing the response from multiple entities, or perhaps even multiple countries can and does save lives. What if the response is a combination of field artillery, ground troops being brought in by helicopter, and air strikes, all delivered to the same incident? In thinking about the mutual aid, including the coordination, cooperation, and communications, we can begin to imagine how the military connection has played a role in the IMS methods that emergency services use.

Much like the military, emergency services in the United States has also evolved. The days of a single constable patrolling, or a bucket brigade to extinguish fires in a large city have developed into modern-day police forces with state-of-the-art equipment. We also see that many fire departments can do more with the state-of-the-art equipment that floods the market every year. This equipment allows firefighters to go beyond what was ever imagined in the early years of our country. These entities along with other contemporary delivery systems have provided a multitude of emergency response services to the citizens they serve.

Both the military and public safety utilizes new technological advances in their work, which in turn provides a safer and more complete response. In both the military and in emergency services, the technological revolution has begun to positively affect what these entities can do, at an amazing rate. The fact remains that technology by itself does not win a war, nor does technology physically respond to everyday emergencies. If we do not have a basic foundation to manage an incident, then all of the technological advances in the world may as well be rendered useless. Without humans to work alongside that technology, it is nothing more than a boat anchor.

The same holds true to major events like the 11 September 2001, Hurricane Katrina, or the tornado that devastated Joplin, MO. It was not technology that made a substantial difference, but rather men and women who use that technology and who perform the tasks that mitigate the incident. It does not matter if individuals are fighting a war or responding to a public safety emergency, without the human aspect of response, the technology is useless. On the other hand, in order for those men and women to do their jobs safely, efficiently, and effectively, they need these advanced tools and technology. The pair goes hand in hand, and the tools are part of the resources that both the military and public safety must manage during an incident.

When we talk about tools in public safety, we think of items like handcuffs, Kevlar vests, axes, hose, and protective gear, but not all tools are forged out of steel. Some tools are concepts and operations systems that allow the military and first responders to provide control in the midst of chaos. This is necessary whether they are engaged in war or a responding to a disaster.

Their primary tool to respond to any incident is not a weapon, nor is it a fire engine, but rather the concepts that has become known as an IMS. If we deliver equipment and the people to a war (or disaster), we need to realize that if we do not manage the response to the incident, that response may be reminiscent of the Keystone Cops. No matter how many individuals and/or how many millions or billions of dollars of equipment that arrives, it would be a futile attempt if we do not provide an organized response that manages the human responders.

To prove the utility of the IMS, a modern-day comparison is in order. Thinking about the military, if the management of ground troops and airstrikes are not properly managed, the ground troops could move into enemy territory only to suffer losses due to airstrikes from their own country. These ground troops could be accidently killed by those trying to reach the same objective. The same holds true in modern-day policing. If a building is being entered and there is not resource management to cover all exits, the perpetrator could escape through an exit that has no police presence. Only through organizing all aspects of a response will the commanding officer know that every exit is properly guarded.

Looking at the United States, we have proven repeatedly that our military might is one of the best on the planet. We have both the tactics and the technology that leave other nations envious of our power, yet the true reason for the United States' military success is not technology. The success that we enjoy is based on a mission-oriented, goal-driven mindset that is instilled in our leaders. This mindset is also reflected in the workers of the emergency services community in the United States (and abroad).

Technology is only a tool in a collective toolbox for our military. Much of that same technology is used on a daily basis in emergency response activities. The military of the twenty-first century has had to deal with technological leaps that are unprecedented in history. These leaps have undoubtedly caused many headaches and unseen problems that were unknown.

These same technological advances have been the cause of many problems to military leaders of the past. Modern warfare requires technology to become bigger and better, faster and stronger, but the warrior on the ground remains the most important part of war. Technology will allow him to win war with greater speed and ease, provided that the technology is managed and integrated into the plan of attack. Even with technology, it is the soldier on the ground fighting, and sometimes dying, in war that makes the true difference.

The same holds true in respect to emergency responders. It is not the soldier, but the firefighter, the police officers, the emergency managers, and the paramedic or Emergency Medical Technician's (EMT's) that serve their country on the home front. Technology will not, and cannot, do their jobs for them; it does however allow them to do their jobs in a more proficient manner and to undertake their tasks more efficiently. The men and women of emergency services make up the front line of Homeland Defense.

As we look back over time, it is plain to see that many of the fathers of incident command and incident management system had military backgrounds. They could see the obvious and sometime the unobvious needs of their emergency response agencies. They began to mold and modify military command and control structures learned while serving their country and developed those tools into systems that would allow their agencies to better respond to individuals who needed assistance during a time of crisis.

These individuals first adopted, then adapted, military command philosophy to be used in their day-to-day response activities as first responders. This adaptation process was not seamless nor was it an overnight success. Even after many years of honing these systems, incident management systems are still evolving, which creates a never-ending process. The emergency services community looks to the military for guidance in ever-increasing demands for response to incidents, and they rely on lessons learned in their own responses. The concepts that drive both emergency services incident management theory and military theories of tactical responses are still evolving. Often, those changes and improvements are on similar tracks, and for the most part, they always have been.

The beginnings of the modern-day Incident Management Systems for public safety were born almost simultaneously in vastly different areas. They were developed for a wide and somewhat diverse array of reasons with one common goal. That goal was to better serve the needs of the community and to save lives and protect property in superior ways. Where each of these individual principals originated is still in question.

1.4 The Birth of IMS Method

When a public safety student begins to study emergency management theory, they will realize that there is a "documented birth" of IMS. While other incidents may have helped form IMS methods, the focused creation as a public safety entity-based management system came about in the late 1960s. It was developed after the devastating wildfires that ravaged parts of Southern California. The fire service in California (as well as other state and local agencies) knew that they had to find ways to overcome a series of repeated deficiencies, mistakes, and negative events that seemed to occur during large-scale, statewide (or multiagency) emergency operations. The agencies who responded to these incidents began to identify common failures in their response efforts to these events. These ongoing problems were especially important when multiple agencies responded to the same incident, and the same issues were seen time, and time, again. The most apparent problems, at least in the initial stages of creating IMS methods, appeared to be the same (or at least similar) issues.

1.4.1 No Single Person in Charge

As larger incidents occurred, often multiple jurisdictions would gather at the emergency incident, and there would be no clear command or person in charge. Additionally, in most instances, each entity did not communicate, coordinate, or collaborate with other first responders. With no clearly identifiable incident leader, it became apparent that multiple different agencies responding to the same incident had their own goals and objectives. This often led to more chaos and confusion, and it put first responders at risk as each agency "did their own thing." Due to the fact that multiple entities were operating independently, competing tactics were sometimes contradictory to each other and dangerous to other crews operating at the scene of a major incident.

Old timers from the 1950s and the 1960s used to tell stories about how incredibly dangerous this was. While there is no way to confirm these stories, we often heard about firefighters who were directly attacking the flames with fire trucks and hoses, who would become caught in a wall of fire on all sides. From those stories, we learned that sometimes, later responding mutual aid crews did not know the first crew was actively fighting the flames, so they did what they felt was needed. These mutual aid crews would (sometimes) light backburn fires that would surround the first arriving crew, and the first arriving crew would be forced to run (or drive) for their life.

The old timers also told stories of crews helping to fight a structure fire, unaware that there were firefighters in the structure. Unbeknownst to the crews in the house, the mutual aid entity would arrive on scene, and not seeing anyone in charge, they would do what they thought was best. Not finding anyone in charge at a Command Post was typically because the command officers were usually with their crews fighting inside the fire rather than being at a Command Post and looking at the response objectively. This usually led to the mutual aid entity doing what they thought was best, and it would often cause issues. Sometimes those issues would turn life-threatening. In firefighting, this would sometimes case a situation where interior firefighters would be hit with a large bore stream of water from the outside, or ventilation that drew the fire toward firefighters rather than away from them. This often resulted in injuries to firefighters inside.

During police actions, the old-timers told stories of law enforcement officers nearly getting shot, primarily because the other agency had no idea the mutual aid agency was there. With tensions already high, the mutual aid agency unaware that a perimeter was partially in place, would come around the corner only to see a weapon pointed at them. There were times when officers did not realize that another officer was in the area when they were chasing a suspect, and in the pitch black, they would mistake another officer for the perpetrator. These are only a few of the dangerous issues that were caused by a lack of central command. If you want to hear some truly horrifying stories of near-death experiences, find an old retired firefighter or law enforcement officer, and ask them to share some of their stories. The time you spend with them will be very enlightening.

1.4.2 No Formal Protocols or Policies

In the old days, prior to IMS, there were often no formal protocols, policies, or even legal statutes, that clarified the responsibility of who was in charge at major incidents. There was no clear-cut identification of who oversaw the incident, and who reported to whom, and in what capacity. While there were some laws that could be cited (which somewhat addressed the issue) on a state-by-state basis, none of the laws in the 1950s through the 1970s was definitive enough to provide the clarity needed. They often failed to identify a single person or a singular agency that was in charge. Those laws that were often written were so vague, or incomplete, that they were usually left up to the interpretation of those responding, or by the entities they represented. They were also often based on the perspective of the mutual aid responding agency.

A lack of policies, especially in where a fire departments primary geographical response boundary was, caused many issues. In some instances, even when these response areas were defined, there was the occasional mutual aid department who would come in and refuse to follow orders from another agency. With the lack of clear and definitive laws and policies, many times those that were responding would have peeing contests to establish dominance over the incident. Some would use the mentality that this was their response area, while others would claim they knew how to better mitigate the situation than the original agency. While these were sometimes supported by other reasons given, the end result was usually the same. It usually caused anger, dismay, and verbal or physical altercations that essentially hampered the overall ongoing operations. If we dig into the history books, we can even see fights and even riots over agency response.

We only need to look back at the mid to late 1800s to see just how detrimental lack of protocols (and structure) can be. During that time, the local insurance companies only paid the first fire department to put water on the fire. Many urban volunteer fire departments would fight to defend the fire plug (or hydrant), and from historical stories, we see that sometimes two or more fire departments would break out into an all-out brawl while the fire raged on behind them. It became such an issue, that local volunteer fire departments hired prize fighters and/or tough guys who would respond immediately to the fire plugs or hydrants nearest to the fire, in order to defend it for their agency.

In one specific instance, the brawls that broke out over who could first spray water on the fire was a good thing for public safety. In 1851, a riot was caused by volunteer fire companies fighting over the hydrant in Cincinnati. Firefighters who preferred to fight like thugs (rather than fight the major fire that raged on) caused the City of Cincinnati to begin looking at fixing the problem. After considerable thought and consideration around this matter, the city determined that they needed to purchase a steam fire engine. This steam fire engine would be owned and operated by the city. The results, as it pertains to IMS, was a set of protocols that hired firefighters and gave them a jurisdiction for which they were responsible. This trend soon caught on, and over a period 20 or so years, most urban cities implemented paid fire departments that had set protocols and policies that identified who was responsible (National Volunteer Fire Council [NVFC], 2012). In some cases, outside volunteer fire companies were not allowed to fight fire or even respond within city limits, to reduce the brawls that had been seen previously.

This trend was slow to catch on in most rural areas, but this inclination seemed to have a natural progression from the urban areas, to the suburban areas, and eventually the rural areas. Between the 1870s and the 1970s, more states, counties, and municipalities began creating protocols and policies in place which defined borders of fire districts. Along with those defined areas came the set responsibilities of each of the agencies involved.

In law enforcement, this issue was barely seen. It is believed that part of the reason that it was not an issue with law enforcement was that most police officers were employed (even if volunteer employment) by a government entity. In most instances, from the time they were hired, basic policies and procedures as well as a geographical response area was already in place. Of course, this is just one person's speculation, but it may have merit.

1.4.3 Conflicts and Egos

Conflicts and egos have been a problem since the first governmental entity had to work with the second governmental agency, and probably even before that time. Let us face it, most people who work in public safety are *Type A* personalities. Type A personalities (in theory) are those with personalities that can be more competitive and more ambitious. Type A personality individuals can easily become impatient, and they are usually extremely aware of their time management. They usually have a schedule or task they want to achieve in a set amount of time, and they do their best to meet that schedule. They also tend to be more aggressive than the other types of personalities.

When we look at Type A personality in a public safety context, it is important to remember that this type of personality is usually needed to outthink the criminals, to overcome a disaster, or to be able to put the fire out while keeping everyone safe. While a Type A personality is important to ensure that public safety personnel are the victor, that personality can also cause conflict, and feed into an ego.

In the past, and even still today, clashes occurred between fire chiefs, police chiefs, and others who assumed command. Sometimes these clashes occurred because someone felt like they should be in charge (usually at the local government level). While rare, there would occasionally be verbal arguments, the occasional fist fight, and in a few incidents, arrests were made because an agency did not want to give up what control they had or assumed they had. These conflicts and egos would lead to animosity. This would often build and eventually create a hostile environment between the agencies. Sometimes, this animosity would manifest after the event or incident was over, creating even more tension among local agencies. This too led to problems when managing an incident.

Keeping egos in check was perhaps one of the more difficult problems that needed to be overcome. Much like a horse-drawn wagon, nobody got anything done without everyone pulling the same direction. Unfortunately, prior to an IMS method, this was a common problem that was seen, and sometimes still is seen in public safety today.

1.4.4 Integrating Multijurisdictional Response

Integration of multiple agencies from a myriad of local, state, and federal government entities often led to turf wars, or at the very least serious difficulties. This held especially true while in the response mode. Often, it could be attributed to the previously mentioned Type A personality. As more agencies would arrive on scene, the incoming agencies would often try to assert their authority, but at the same time wanted no part of command responsibility or liability. Not only would these individuals not take responsibility, but if something they suggested went wrong, they would often deny any responsibility because they were not "officially" in charge. This often led to distrust of, and among, other agencies.

In the aftermath of some incidents that went terribly wrong, there would sometimes be finger-pointing and the blaming of another agency. The person or agency responsible for making a bad decision would rarely be identified and/or held accountable. This led to more distrust of incoming mutual aid help, and sometimes lead to the attitude by the initial agency that it was "My way or the highway." This was not conducive when a collaborative effort was needed in an emergency.

1.4.5 No Collaborative Organizational Structure

Prior to the IMS methods, there would often be problems with incident wide organizational structures. This was because they were nonexistent in most instances. While each agency had their own chain of command, an overarching chain of command between agencies (especially on a major incident) typically was not part of the response protocols. Complicating the matter even more, there was usually no attempt made to form a collaborative organizational structure. More often than not, individuals responding to an incident would not collaborate to create a formal, or even an informal structure. They failed to recognize that this action would have increased collaboration, communication, and cooperation with each other.

The lack of collaboration, cooperation, and communication often led to freelancing of various agencies. Freelancing refers to agencies undertaking the actions that they felt were operationally necessary, while being oblivious to the needs (and tactics) of other agencies. These actions usually increased the risk to life safety for other first responders and led to taking longer to bring the incident under control. Moreover, the sense of accomplishment was not there. Sometimes agencies would do "their part" of the response and then feel as if other agencies did not do their part in bringing the incident under control. In some instances, the agency that accomplished their task first would chide, or even spitefully ask if they needed the other agency to do their job. As with most of the rest of the reasons for creating and IMS method, this created more conflict.

1.4.6 Strictly Enforced Intra-agency Command Structure

As agencies came together to respond to a single incident, it became obvious that each had their own command structure. That established command structure did not always coincide with the command structure of other responding agencies. On an agency-by-agency basis, the chain of command structure was often strictly enforced by their own agency, and it left no room for deviation. Rather than coming up with a universal command structure or assimilate to the command structure of the agency, they were there to support; some mutual aid agencies would unequivocally refuse to adapt to change their own structure. This refusal to change command structure occurred for a multitude of reasons, but in some instances, the underlying problem fell along the lines of holding on to traditions. In other instances, this defiant attitude was based on past (or current) turf wars between agencies.

The traditionalist agencies were most often the ones to create the biggest fuss over their command structure. They were often unwilling to even slightly change to match up with other agencies. Some believe that this led to a quote about some fire departments that states "150 years of tradition, unimpeded by progress."

1.4.7 Command Based on Home Rule

Whenever a major incident required multiple agencies, and a command structure was actually implemented, the jurisdiction in which the incident occurred would regularly attempt to fill all command positions with their personnel. Rarely was there a way to establish any type of leadership using individuals from the other agencies that were represented. On the surface, this seemed like a promising idea; however, there were often people in other agencies that had advanced qualifications and/or experiences dealing with the issue at hand, and their expertise was not utilized. Rather than using these experienced and qualified individuals to help mitigate the situation (by placing them in some type of leadership role), they were time and again utilized in undertaking manual labor. This essentially wasted that specialized talent they had. Ironically, that specialized talent probably would have made working conditions substantially safer, and the incident could have potentially come to an end quicker if those talents had been effectively used.

Beyond the command issue, the mentality of home rule also failed to take into consideration of creating liaisons between the home agency and the mutual aid organizations. In most instances, a liaison would be familiar with the command structure, the equipment, the qualifications, and the expertise of the mutual aid organization. A lack of ensuring a working relationship with a single person tasked for liaising between agencies often created more contention between agencies, and it usually added to the overall confusion on a major incident.

1.4.8 Too Many Subordinates Reporting to a Single Supervisor

The consideration of how many people a single person could effectively manage did not usually play into the decision-making process on many incidents. Numerous operations were carried out with only one leader or supervisor managing the entire incident. In some instances, a supervisor would manage an exorbitant number of individuals (on occasions, more than 100), which hampered the response and increased the potential of the death or injury to a first responder. This left many individuals (or groups of individuals) to improvise, or get off task, while in the operating theater. With no immediate supervisor to keep them on task, it was not uncommon for response crews to do as they wanted rather than following orders (freelancing). The idea of the military's system of platoons or other similar supervisory structures was rarely, if ever, considered. Even when an agency put a limit on how many people one person could supervise, it usually ended up being more personnel than they could effectively manage.

1.4.9 Lack of Accountability

Due to the complexity of most major incidents, and the inherent danger of multifaceted operations, being able to account for all personnel was, and still is, important. When a command structure was in place, agencies or one team of an agency, might freelance. Often, no immediate supervisor was assigned to documenting where each individual was assigned, what they were doing, and how they were doing it.

If a disastrous incident within the operational theater were to occur, the ability to account for each individual was compromised. This occasionally led to individuals unnecessarily being put in harm's way. It was not uncommon for someone to be injured or killed, and nobody was aware of it until much later. It also led to individuals being left behind when an evacuation was ordered, or when the operations ceased.

In 1972, a commission was formed by then President Richard Nixon. *The National Commission on Fire Prevention and Control* was formed to investigate how to reduce fire deaths, including firefighter fatalities. That commission released a report called *America Burning*. Among the many findings by the commission, two specific recommendations stand out, in relationship to accountability. The first was to create an administration to oversee and support firefighters, the second was to reliably record Line of Duty Deaths (LODDs) of firefighters. In 1976, the US Fire Administration was founded, and firefighter fatalities began to be credibly reported on 1 January 1977. Prior to 1977, there were not any credible numbers of how many firefighter fatalities were suffered each year. For the calendar year of 1977 there were 157 firefighter fatalities, and in 1978 there were 174 firefighter fatalities. Comparing this to firefighter fatalities statistics in 2016, the death toll fell to 69 (FEMA, 2018a).

1.4.10 No Interagency Planning

In the 1970s, interagency operational planning was overlooked time and again. If there was operational planning between two or more agencies, it was often done haphazardly and/or randomly. Clear and concise operational planning that utilized interagency integration was not usually a priority for the initial agency, or the mutual aid agencies. Tactics were rarely discussed before sending a first responder into characteristically dangerous and often hazardous, operations. Sometimes, this had negative and life-threatening results.

Prior to the formal creation of IMS methods, it was not unusual for mutual aid agencies to show up and to be told where they were most needed, with no further discussion. Even if they decided to follow those orders, because varying agencies had differing methods to mitigate a given circumstance, these entities could be working on the same incidents utilizing different methods. They were doing what they thought was best, rather than having a specific plan where everyone knew what tactics were being employed.

This would sometimes lead to a disastrous result. Because there are many methods to fighting a fire, each with their own risks, contradictory tactics could, and sometimes would, put firefighters at risk. The previous example about wildland firefighting also fits into this scenario. If one group of firefighters was undertaking a direct attack (using hoses and water) on a wildland fire, and another group of firefighters were fighting it indirectly (creating a fire line and backburning), then the firefighters involved with the direct attack could be trapped between the two fires. This is but one example of how lack of planning sometimes caused more life-threatening issues.

1.4.11 Lack of Common Terminology

Often, response agencies had developed their own terminology, jargon, or vernaculars over a given time. This resulted with words being used that often related to acronyms or the use of words not commonly found in the dictionary. The use of this vernacular, specific term, or acronym not only led to confusion, but in some cases, the term used by one agency would mean the exact opposite to another agency. This would sometimes lead to misinterpretations and miscalculations of how to proceed. In some instances, those undertaking tactics in the operational theater were put in harm's way because of the lack of clear and concise communications.

Another example in regard to common terminology is that some police agencies would use 10 codes, while others would use 11 codes. In one state, they may use alphanumeric codes to describe the reported crime, while in other states, they may use acronyms. If out of state resources were needed, it could be a major fiasco trying to talk to one another unless they all used plain English.

1.4.12 A Lack of Interoperable Communications

Communications incompatibilities were extremely evident. Even when refined and technologically advanced communications systems were in use for more than one agency, they were often mismatched with mutual aid organizations. Interoperability was often nonexistent. This lack of interoperability could be for a multitude of reasons including being on different frequencies, or in some instances, they were on different bandwidths (e.g. low band, UHF, VHF, ultrahigh band). When this occurred, there were no interoperable or interagency communications.

Agencies often purchased communications equipment and secured frequencies that fit their needs. These agencies rarely had any consideration for agencies that might come to their rescue. While there were a multitude of reasons for this, it is important to note that in some instances, agencies had used the same frequencies for years. In some cases, it was tradition over progress. Another factor that may have fed into the lack of consideration for other agencies is that it typically took a long time for the Federal Communications Commission to approve new frequencies, and many agencies did not want to wait. Additionally, radio equipment was expensive and to revamp an entire radio system would come with a large price tag. In all, the problem occurred because the systems that differing agencies used were technologically incompatible.

While operating at an incident, the agency might be able to literally see the other agency, but they did not have the capacity to (verbally) speak to each other over a communications device. The distance might only be a few hundred yards, but it might as well have been a couple of hundred miles. Lack of interoperability between communication devices was a major problem then, and it still is now.

1.4.13 A Lack of Logistics

Logistics is a critical issue that was often overlooked. It was common to see that as an incident expanded, so would the demand for additional resources. As large-scale or multifaceted incidents became more involved, more resources were needed to support operations. Unfortunately, the logistics to obtain the resources required to mitigate the incident were sometimes overlooked. Prior to the implementation of IMS methods, it was commonplace to commit all resources to operations. When all resources were committed to the operations, it often led to shortages in the availability of resources in the operating theater.

Postincident critiques, now commonly referred to as After-Action Reports (AARs), would regularly identify that a specific resource was needed but was not requested because it was thought it was inaccessible. In the AAR, they would find that the resource needed was not only available but that it was involved in the theater of operations, but operating in a different capacity (or not being utilized at all). The agency that had a need for that resource unnecessarily operated without it due to a lack of communication and collaboration, and a lack of logistical support.

Due to the lack of tracking and logistical planning, the needed resources were either not identified, or they were not ordered. This often led to more potential for injury and/or death as personnel attempted to adapt by using the resources they had on hand, rather what they really needed.

1.5 California's Solution

During the 1960s and 1970s, California was a state that regularly required multiagency responses, and these responses were (usually) substantially larger in scope than anywhere else in the United States. With large areas of underbrush, towering trees, vegetation covered hills, and dry winds, the fire load in the wildland areas of California was exorbitant. Following the wildfires, there were often torrential rains that often-caused mudslides on the bare hillsides that was stripped of vegetation by a fire. Beyond the major disaster of wildland fires, California would suffer the occasional earthquake and other disasters that affected that geographical region.

Whenever a major incident occurred, a multiagency response was needed. In these multiagency responses, many were seeing the same complications that seemed to plague other multiagency responses. The State of California began to investigate how to mitigate the negative effects of multiagency responses by promoting the positives and mitigating the negatives. California's answer to these, and other problems, was an experimental project named "Firefighting Resources of California Organized for Potential Emergencies." The project used the acronym FIRESCOPE.

The FIRESCOPE project began designing a basic standard for all personnel and response agencies. The idea was to help lessen the problems that were ongoing and apparent whenever it was required to manage these major emergency incidents. Those working on the FIRESCOPE project felt they had to come up with a way to better organize response efforts. Recommendations were made that addressed each issue individually. They then combined the mitigation measures for each individual problem and created an all-encompassing system that addressed all these issues.

After working out all the issues, and coming up with a basic design, they committed to undertaking a comprehensive field test of this new management system. When most people think of a field test, they often imagine a one-time test on a single day, but this was not the case with FIRESCOPE. The FIRESCOPE field test lasted multiple years. Each time a problem was identified, adjustments were made to the overall system so that it could become finely honed. This extensive field testing of the FIRESCOPE management system led to an extremely robust method of managing emergencies.

FIRESCOPE was the initial concept that, over a period of time, developed into the common procedure that we now know as Incident Command System (ICS). The information provided in this book is essentially just the tip of the iceberg, and those studying incident management would be wise to further their knowledge about FIRESCOPE. For a more complete historical review of FIRESCOPE and the transformation into ICS, you can visit www.firescope.org.

1.6 Creating the Incident Command System

Because the FIRESCOPE program began to show an amazing amount of promise in managing emergency incidents, it began to gain ground in various parts of the nation. Soon, an interagency task force of local, state, and federal agencies used the basic principles of FIRESCOPE and developed the Incident Command System (ICS). This creation of ICS was based on the FIRESCOPE structure and while not exact, it closely mimicked it. FIRESCOPE's success provided the platform for the ICS structure, and the ICS program was developed and is still widely in use in the United States today.

While the initial development of FIRESCOPE started in 1973, one year after the America Burning Report, it is important to realize that it took seven years of research and development to remedy complications found during the extensive field testing. Although a broad range of field testing was employed, it was realized that ICS would need to change as the incidents changed. After the adaption of this FIRESCOPE incident management system led to the ICS method, it was realized early on (in both programs) that the systems would need to be continually evolving.

While creating of this IMS method occurred nearly 50 years ago, the same holds true for the ICS method of today: as users of the ICS (now a component of the National Incident Management System [NIMS]), we need to realize that ICS is never complete due to the ever-changing fluctuations in the environment that first responders work in. This means that there must be a constant and ongoing research and development that ensures ICS is current and meeting the needs of first responders.

Early in the Incident Command System development, five essential requirements became clear

- The method needed to meet the needs of first responder's and be a tool for them to better organize all types of response.
- It would have to be flexible in order to meet the needs of the organization, regardless of size, as well as supporting the mitigation of incidents of varying types and sizes.
- Agencies would need to be able to use the method on a day-to-day basis. This meant the method would have to meet the needs of daily use, as well as catastrophic emergencies and disasters.
- The method would need to be simple enough to use that personnel from an assortment of agencies and assorted geographic locations could swiftly merge into a common organizational structure.
- The method would need to be cost-effective.

Once the structure of the system was completed, the ICS framework continued to change to meet the previously mentioned needs. The system began to evolve and gain acceptance, albeit slowly. It began to spread across the United States similar to the wildfires that it was designed to manage. Many larger agencies, and eventually states themselves, began to accept, perfect, and acclimate themselves with the work that originated in California for use in their agency.

In the mid-to-late 1980s, the National Fire Academy (NFA) began to fund regional classes that taught the Incident Command System to firefighters. These direct delivery classes typically utilized instructors that were already teaching fire classes. Usually, these instructors worked for a state training entity, to help create and integrate a standard for managing incidents. These instructors would usually teach the ICS classes on a county-by-county basis. By teaching ICS on a county-by-county strategy, the local firefighter would only have to travel to a class that was in their local county. Rather than going hundreds (if not thousands) of miles to receive training, they usually only had to travel under 30 miles. In most instances, one ICS class was offered per county, per year, throughout the United States.

1.7 Evolution of IMS Methods

The creation of the ICS method was a result of a chain of fires that challenged the California response community. Like the development of a physical item, history tends to remember the item itself rather than the development and the methods that were created to make it. As an example, if you were to attempt to remember the basic history of the light bulb, most individuals would say that Thomas Edison invented it in 1879. If you were to dig deeper, historical evidence would reveal that many different inventors played a role in the development process of the incandescent light bulb.

Much like the invention of the light bulb, the modern-day ICS component of NIMS being currently used by the first responder community was a well-designed result, or byproduct, of the Incident Command System. ICS was initially developed for the specific needs of large-scale wildland fires, although it should be mentioned that is was not the only IMS method developed. It became apparent to users of the early versions of ICS that the method (itself) was appropriate for use in other emergencies, including manmade and natural disasters.

More likely than not, the early development of ICS, beyond the use of wildland firefighting situations, was also carried out by Californians. The State of California seems to have a greater propensity for assorted disasters, which seems to occur somewhat regularly both on a large scale and a smaller scale. Sometimes the State of California has major disasters back to back.

The reason for further development of ICS made perfect sense. The first responder community quickly realized that the same underlying factors and problems that led the wildland firefighting community to develop ICS were quite often shared by other disciplines and in similar and dissimilar incidents. When it came to multidisciplinary response, the management of those other disciplines did not seem to integrate with the proven ICS method. If fire departments needed the assistance of law enforcement, EMS, emergency management, and public works, they were usually not on the same page as the fire department. This affected the management of these incidents both in terms of complexity and scope, not to mention frustration. Much like the problems that plagued multiagency response in California, similar commonalities were seen nationwide with multidisciplinary response, and usually revolved around

- Spans of control were too large (too many people reporting to one supervisor).
- Varying and dissimilar types of organizational structures (among different disciplines and agencies) responding to the same incident.
- A variety of types and levels of government who did not work well together.
- Little or no formal method to consistently share incident information.
- Incompatible and inadequate communication systems (and procedures) between disciplines.
- No formal method of coordinated planning among agencies and disciplines.
- Severe misinterpretations of lines of authority.
- Substantial terminology differences between agencies and differing response disciplines.
- Lack of formal, or in some cases unspecified, incident objectives.
- Lack of any incident action planning.
- Lack of backup plans in case the initial plans failed.

One typical factor in most major disasters is that they usually occur with no warning. These disasters and incidents tend to develop rapidly and can grow from the initial incident. A prime example of this might be as a small grass fire that evolves and expands into a major wildland forest fire. These incidents can transpire rapidly and become a multijurisdictional and multidisciplinary response that crosses multiple jurisdictional boundaries. Some common examples of these types of incidents might include an earthquake, a tsunami, severe thunderstorms, tornadoes, a large hazardous material spill (such as a pipeline break or train derailment), and other similar incidents.

It is important to realize that even the smallest of incidents, if not properly addressed and managed, can grow in size, proportion, and/or complexity. The conditions surrounding an event can rapidly change and increase in size when they are not properly managed. Without all of the response personnel being trained and proficient in ICS, it creates an environment that can allow the incident to expand in size and complexity.

It is also important to realize that even if everyone is on the same page, these incidents can increase substantially even when proper operational actions are being taken. This rapid increase in size creates a greater risk for first responders as well as individuals who may be present in the geographically affected area. Almost any incident that is not properly managed will usually increase this risk substantially. Unmanaged or mismanaged incident can also lead to extremely high property losses and loss of economic stability.

A great example of this was the wildland fires in Yellowstone National Park. In a single week during the summer of 1988, fires within the park encompassed more than nearly 99 000 acres. By the end of the month, dry fuels and high winds combined to make the large fires nearly uncontrollable. On the worst single day, 20 August 1988, tremendous winds pushed fire across more than 150 000 acres requiring a massive national-level response.

Similarly, in 2003, California succumbed to a major wildfire. The Cedar Fire was an especially destructive fire that was fueled by Santa Ana and Diablo winds. The fire was started by a lost hunter who tried to light a small signal fire, and before it was brought under control, the fire consumed over 280 000 acres. Surprisingly, 30 000 acres burned inside the city limits of San Diego. While there were numerous other fires burning in California in 2003, the Cedar Fire was the most destructive of the year, and according to some, the most destructive modern-day wildfire until the 2017 Thomas Fire (Tierney, 2018). At the peak of firefighting efforts during the Cedar Fire, there were a total of 4275 firefighters, and this does not include law enforcement, EMS, animal rescue personnel, and a whole host of other support agencies ("California Fire Siege 2003," n.d.). The Thomas Fire burned approximately 281 893 acres making it the largest wildfire in California history. As this book was in the authors final edit, the Thomas Fire was outdone by the Mendocino Complex Fire's Ranch Fire in August 2018. This complex fire burned a combined total of 459 123 acres (Incident Information System, 2018).

It is easy for even a novice to see that events such as the Yellowstone, the Cedar Fire, the Thomas Fire, and the Mendocino Complex Fire that an incident can quickly expand into a multiagency and a multijurisdictional response. Beyond the operational firefighting efforts, coordination is usually needed to divide and conquer the multifaceted response. Simultaneously, the Incident Commander and his/her staff must manage firefighting operations, evacuations, water supplies, finance, liaison, and a whole host of other issues that may be needed. These types of major incidents also have the capability of becoming a major news story. This type of media attention an incident may garner is based on the size and complexity of the incident. As will be mentioned in later chapters, the managing of the media is an important aspect, perhaps even a critical aspect, in almost every major incident, and even in some small incidents.

As first responders, we need to remember that the cost of response, recovery, and mitigation is a major consideration. Identifying the funding for a response may become problematic. Funding for response, recovery, and mitigation can come from various governmental and private stakeholders. Most municipal budgets usually cannot afford to fund these larger incidents. These federal and other stakeholder funding entities can (at times) substantially increase frustrations among those managing an operation. Increased frustration from not having what is needed to mitigate an incident can become further complicated when certain stakeholders require specific milestones be completed in a specific way.

Another issue can be when a funding agency requires response personnel to jump through certain hoops in order to receive support funding. One frustration that has come from past incidents has come from the federal government. It would seem that some agencies have been told that the incident was not substantial enough to receive federal assistance while they stand in front of what is left of your community. Utilizing an IMS method helps to ensure that everything is done according to the book so that you have a decreased chance of being denied the funding that you and your community desperately need.

1.8 The "Big Three" of IMS

Lack of nationwide collaboration to meet the needs of incident command and incident management needs initially led to three initial manifestations of IMS. Each manifestation played a pivotal role in creating the method widely used today. These three methods were a part of disorganized groups that became known as part of the "Big Three" command methods. Each of these three methods contributed in some way to the evolution of the ICS component of NIMS, and to NIMS itself. Later Incident Command System models were extensions of these three methods. The Big Three include

- *FIRESCOPE (also known as the Wildfire Incident Command System)*: FIRESCOPE evolved into a method that was used mostly by the wildland firefighting system in the United States. During 1970s and early 1980s, it was the most prevalent ICS Method in place. The FIRESCOPE program is still (somewhat) in use today in California, but it has seen a substantial decline in use since ICS and NIMS became the nationwide standard.
- *The National Fire Academy (NFA) Incident Command System (ICS)*: The Federal government recognized ICS as the model for the management of emergency incident scene in 1980. This version of ICS was slightly geared toward the "east coast" because it did not initially embrace wildland firefighting applications, and for the most part, it addressed emergencies faced by urban responders. In 1980s through early 2000s, these NFA-ICS classes were a mainstay of NFA delivery classes to local jurisdictions.
- *The Fire Ground Commander System (FGCS)*: The FGCS was established in the Phoenix Fire Department and gained ground as a method partially because of their well-known Chief, Alan Brunacini. This method used the same basic, or underlying, principles as the FIRESCOPE model but was made to be more practical for structural fire-based incidents.

These three methods would later meld in together into the NIMS method that is in place today, with ICS being a primary component of NIMS. The melding of the three methods was due, at least in part, to the 9/11 attacks on American soil. Some would argue that the differences in the big three are only semantic in nature, but each played a respective role in the creation of what we now call ICS and NIMS.

1.9 The Melding of the IMS Concepts of Today

The Big Three concepts of incident management were directed toward specific uses, but initially none of these concepts were perfected for all types of incidents. Some believe that the underpinnings of each of those "Big Three" evolved and melded together over the past 40 years. It is important to note that these concepts did not evolve in a linear format, where it was one dimensional. These three initial types of incident management seemed to evolve in a more three-dimensional fashion, and in a concurrent way. This three-dimensional growth began to incorporate more than just fire departments. As it developed, it began to show the utilitarian use of the system for myriad of agencies who might be utilized on an emergency incident.

The Incident Command System (ICS) was originally created for the fire service. Over the years, it became widely used, and accepted, by fire agencies throughout the United States. As other public safety disciplines began to understand the utility and the benefits of the system, its use became more prevalent in these disciplines. While some minor changes were made to make ICS more useful by other disciplines, the main concepts remained the same. The system soon began to be incorporated by law enforcement, hospitals, Emergency Management Agencies (EMA), public works, faith-based organizations, and other stakeholder agencies. As the effectiveness of the system became more realized, ICS gradually found its way into nonpublic safety function such as concerts, festivals, marathons, and other event management activities.

In 1980, the original FIRECSOPE method and other existing IMS methods transitioned into a national program called the National Interagency Incident Management System (NIIMS). NIIMS temporarily became the backbone of a wider-based method for all federal agencies having a role in wildland fire management. This was initially administered by the US Department of the Interior through its National Wildfire Coordinating Group (NWCG) which is based in Boise, Idaho. NIIMS was endorsed and used by multiple agencies in the United States until shortly after September 11 attack.

The most unthinkable incident occurred on 11 September 2001, changing the world as we knew it, forever. This catastrophic incident changed the way that we manage emergency incidents. This was partially due to the problems that developed in the aftermath of the 9/11 attacks, and many of these changes were made based on the shortcomings that were identified while trying to recover from this immense incident.

While the way of managing any incident was to utilize the ICS method, after the September 11 attacks, it became apparent that another way of more effectively integrating resources was needed. Even after the attacks, it was realized that ICS should remain the way to manage an incident at the local level because it worked, and worked well. Even so, the shortcomings of integrating resources from around this great nation of ours was quickly realized after the attacks. It was also realized that a national method, or national set of protocols, was needed to integrate outside resources in a more organized manner.

These set of concepts, known collectively as the National Incident Management System (NIMS) has become even more necessary since the September 11 attacks. This necessity was not only relegated to the United States but also around the world, especially considering the most recent terrorist attacks and major emergencies that continue to occur. The concepts that drive NIMS is still, and (most likely) will forever be, developing. The primary reason for this is because of the ever-changing threats that first responders face.

After the events of 9/11, the Federal Emergency Management Agency's (FEMA's), National Curriculum Advisory Committee on Incident Command Systems/Emergency Operations Management Systems recommended adoption of ICS as a multihazard all-agency system. The initial reaction to recognize ICS as the nationwide way of managing incident was swift, and it was supported by several key agencies and/or organizations. These actions included

- FEMA's National Fire Academy (NFA) adopted the ICS/NIMS as a model system for fire services. This is important because the NFA served as a focal point for federally based incident command and management training. The NFA accomplished this through a variety of models of training delivery both on its Emmitsburg, Maryland, campus and through its off-campus training delivery system.
- FEMA's Urban Search and Rescue (USAR) Response System, a component of the Federal Response Plan. At the time of the September 11 attacks, USAR used NIIMS as the basis of its onsite management structure. They utilized the NIIMS system to allow for seamless interface with the local Incident Management System. When the discussion turned to a nationwide single system, most USAR Teams welcomed it with open arms.
- The National Fire Protection Association (NFPA). The NFPA created life safety standards for many years. Prior to the September 11 attacks, many NFPA Standards referenced or in some cases directly called for the development and use of an incident management system. When the adoption of ICS and NIMS as the nationwide way of managing emergency incident was suggested, the NFPA wholeheartedly supported it.

The use of NIIMS was then relied upon by only certain federal government agencies, that is, until a Presidential Directive by President George W. Bush. President Bush mandated a standardized incident management system through Executive Order. Prior to this mandate, many federal agencies incorporated the structure of emergency incident management for which they felt best suited their objectives and needs.

These emergency incident management systems of that time created a hodgepodge of methods that, in many instances, did not integrate well with other types of emergency incident systems. Whether this lack of integration was caused by human error or was a product of poor incident management design (or a combination of the two) is still a matter of debate. After the

September 11 attacks, it became increasingly important to integrate, and to make sure that everyone was working from the same playbook.

1.10 The National Incident Management System (NIMS)

In February 2003, President George W. Bush issued Homeland Security Presidential Directive (HSPD) No. 5. This Presidential Directive was created to establish a single, comprehensive national incident management system to deal with domestic emergency incidents within the United States. HSPD-5 (and its companion document HSPD-8) used language that required NIMS to be used in more than just the response phase of emergencies; it required that NIMS would be used to prevent, prepare for, and respond to incidents, regardless of the size, cause, or the level of complexity. Furthermore, this Presidential Directive concluded that the system should be used in terrorist attacks, major disasters, and any other domestic emergencies, including day-to-day operations. The primary purpose was to ensure that all levels of government had the capacity to participate together in a collaborative, effective, and efficient manner by using a single command and management method.

NIMS is not, and was not, intended to be an operational incident management method. The operational aspect of incident management was relegated solely to ICS component. The components of NIMS were designed to integrate with ICS, thereby creating a more unified and collaborative response. The components of NIMS connected with ICS and work in harmony to form a comprehensive incident management system. While ICS is a component of NIMS, it is not the only component. The totality of NIMS components includes

- Preparedness
- Communications and Information Management
- Resource Management
- Command and Management
- Ongoing Management and Maintenance

It is important to note that the NIMS Methods required the use of the standardized Incident Command System (ICS) for operational management of an incident. NIMS provides a core set of principles, concepts, doctrines, terminology, and organizational practices to facilitate effective, efficient, and collaborative incident management on incident of all sizes and nature. It creates and promotes improved collaboration among all stakeholders.

NIMS is appropriate across the full range of potential incidents, hazards, and impacts, and is beneficial in managing resources regardless of size, location, or complexity of the incident. It is created and managed in such a way that it improves coordination and cooperation between both public and private entities, and it is effective in a large variety of incident management activities. Perhaps the most important aspect of NIMS is that it provides a communal standard for overall incident management.

The consistent practice of NIMS initiates and maintains the groundwork for efficient and effective responses. Whether it is a single-agency response, a multiagency, a multidisciplined, or a multijurisdictional response to a natural disaster or terrorism, NIMS is a way for everyone to be on the same page. If a single-agency response escalates into a multijurisdictional response, those entities who have integrated NIMS into their planning and incident management structure have a unique advantage. They can come to an incident after being dispatched with little notice, and still understand the overarching procedures and protocols governing the response.

Any personnel dispatched know, from the minute they are called, what the expectation is for equipment and personnel. The real beauty and actual utility of the system is the harmony in preparedness and response efforts which permits diverse entities to easily and comfortably integrate resources, and if needed to be able to quickly establish Unified Command (explained in Chapter 7) during a chaotic time of an incident where everyone needs a seat at the table.

1.11 Presidential Directives

Earlier in this chapter, it was discussed how HSPD-5 mandated NIMS to provide interoperability and compatibility among Federal, State, and local response and recovery agencies. Under this directive, NIMS also was required to provide a core set of concepts, principles, terminology, and technologies that integrated with the Incident Command System, but it did not stop there. NIMS was required to provide a multiagency coordination system, pave a system for the use of Unified Command (explained in Chapter 7), and was supposed to provide training. The NIMS method was also mandated to identify and manage resources, including how to classify resources, as well as provide qualifications and certifications to ensure that potential resources met specific NIMS proficiencies.

To confirm the NIMS method was meeting the goals of the Presidential Directive, there was a mandate that there should be a method of gathering, tracking, and reporting of incident information and incident resources. This was essential in providing quantitative reports and data to the President, as well as the public, to determine the preparedness and readiness level of the United States. As the saying goes, you cannot fix it if you do not know that it is broken. The data that was to be collected would be used for finding out where the system was broken.

HSPD-5 placed most of the responsibility for NIMS directly in the hands of the Secretary of Homeland Security. The Department of Homeland Security (DHS) was a new government department as of 25 November 2002. It was officially formed the day after HSPD-5 was signed by President Bush, and a little over a year after the September 11 attacks. The formation of DHS combined multiple federal resources under one federal entity. The creation of DHS either partially, or fully, incorporated 22 independent agencies or departments under this one umbrella agency. In the NIMS method, the federal government acknowledged the roles of local and state entities and they identified that in most instances, domestic incident management was not the federal government's responsibility.

This Presidential Directive clearly authorized the Secretary of Homeland Security as the responsible party for coordinating with federal agencies. Those agencies under the umbrella of DHS were required to integrate and coordinate with state, local, and tribal entities when directed by the DHS. Beyond that, the Secretary of Homeland Security was ordered to integrate and coordinate with private and nongovernmental organizations (NGOs). Essentially, the new DHS Director was mandated to work toward coordinating everyone together so that when it was needed, there was a more unified response. By doing so, they would help to ensure that there was the ability to provide seamless support to a multitude of incidents, including a massive coordinated response to a terrorist attack.

In undertaking these coordination efforts with these various organizations and entities, the Secretary was also charged with guaranteeing suitable planning, adequate equipment, acceptable training, and continued and ongoing drills and exercise activities. The Secretary was under mandate and required to provide assistance in developing an *all-hazards plan* as specified in HSPD-8.

This all-hazard plan was originally called the National Response Plan (NRP), however, it did not take long for the NRP to morph and transform into the National Response Framework (NRF). The underlying factor that helped to drive the NRF was the mandate to ensure that federal, state, local, private, and non-governmental organizations (NGOs) were organized, integrated, and that they met core capabilities. This was especially important (and difficult) because it included integrating those with varying specialties in disaster and emergency management. Those specialties usually related to mitigation, preparedness, response, and recovery. While this was a tall order, most realized that if the end result was accomplished, the United States would have better and more in-depth safeguards for the security and safety of the nation.

From the creation of the National Response Framework came a new term and a new concept that would change the way agencies planned. The use of *all-hazards planning* was a unique and new way of addressing potential threats to a community. Some agencies were not impressed with the term, and in some cases, the concept itself. As was often the case in public safety, some agency administrators began complaining about all-hazards planning before they even knew what it was, or how it would affect them or their agency. In some instances, these agencies operated on assumptions. Some agency administrators unwittingly assumed that they were expected to plan and prepare for every conceivable type of emergency, but in reality, all-hazards planning was based on two core beliefs.

The first core belief is that a risk analysis should be undertaken to identify potential hazards that a community may face. This means that the planning would be based on the unique risks inherent to the geographical location they served. A formal risk analysis would be performed to identify the unique locational priorities and resources that were needed to be effective. Without a risk analysis, and a risk-driven plan, the agency was essentially guessing what would be their biggest risks, or they are planning for everything and hoping for nothing. Either way, without using risk-driven planning, the agency put their citizens unnecessarily at risk.

The second core belief was that by undertaking a risk analysis, the agency could know the potential risks, and could develop the capacity to deal with these unique hazards through practical planning. What most did not immediately understand was that many of the same tools needed for the most prevalent risks, could also be integrated into the least prevalent risks. This was based on the assumption that a multitude of core jobs associated with the highest risks would also be useful in planning for lower risk disasters.

The same methods used for warning, evacuating, and sheltering for one type of incident would usually be effective in other incidents, no matter what type of incident it was. This would then assist them in creating a baseline capability that not only could be effective with known and expected risks but also could be modified easily modified to deal with the unexpected incident or disaster. This type of planning would lead to greater resiliency for the local government. As these core concepts were put to the test, it would be found that they did increase preparedness, and they helped to better manage catastrophic incidents.

Some may wonder how the National Response Framework (NRF) factors into incident management. When we preplan what resources are available, what services they can provide, and how they can be quickly and efficiently activated, we reduce the time frame of receiving the needed resources. Ask yourself this, which is quicker, identifying what is needed and requesting it, or identifying what is needed, assigning someone to find the said resource, negotiating the logistics involved (including cost), and arranging for transportation to the incident. It does not take a rocket scientist to see that the act of just requesting the resource is substantially quicker, and agreements are already arranged. Later in this book, we will discuss the Emergency Support Functions (ESFs) role in requesting resources.

1.12 The NIMS Mandate

In order to make NIMS work, every agency that might be involved in almost any incident would need to work together, and they all needed to be on the same page. It did not matter what phase of the incident these responders were in (mitigation, preparedness, response, and recovery), or if the incident was transitioning from one phase to another. What did matter was the seamless integration of all resources. NIMS accomplished this task. NIMS incorporates Non-Governmental Organizations (NGO's), local governments, state government, and the federal government into a unified response.

NIMS was mandated for every federal agency except one. The only federal agency that was exempt was the Department of Defense (DOD). While the DOD was exempt from following NIMS, the Secretary of Defense was specifically tasked with creating a companion method that would seamlessly integrate a military response into a domestic incident. This would ensure that in instances where military support was needed, they could seamlessly integrate with the other agencies who utilized NIMS and the ICS Component of NIMS.

After being given their marching orders, the DHS and the DOD each created the plan that President Bush requested. A lot of time and thought was put into the program, and eventually a finished product was ready. There was however a problem, how would they get every agency who responded to incidents to begin using NIMS? This was complicated by the fact that it involved stakeholders across a myriad of funding types and who had differing organizational structures.

After the federal governments' mandate to implement NIMS (in March 2004), it became apparent that some local agencies, especially fire departments and police agencies, were not so keen to integrate the NIMS method into their local jurisdictional response. A multitude of reasons were given as to why they were resistant to this change. The most prevalent issue appeared to be a misunderstanding of what NIMS was, and how the method could work in their favor. While this was the concern of some stakeholders, other stakeholders were upset because they had been using a different type or method of IMS. As is often the case, most that were not happy with the new method were resistant to change, no matter what that change was. Many smaller agencies also had their naysayers, and a common phrase that was heard from the old-timers who said "We ain't never done it that way before!" They were essentially saying that they did not see a reason to change, and they did not want, nor like, this change.

Had the federal government not insisted on mandating everyone to use NIMS, then the lack of total acceptance would have likely led to a hodgepodge of methods and more confusion during a response. This would most likely result in a lack of coordinated integration of resources, thereby making the response less effective. Because of the problems that could be caused by not adopting NIMS, the federal government decided they needed to find a way to entice organizations to willingly accept and operate under the NIMS Method.

Multiple methods were used to entice agencies to embrace the NIMS concepts and to become NIMS compliant. Perhaps the most effective was requiring any agency that provided emergency response (or supported emergency response), and who received government funding were required to become NIMS compliant in order to continue to receive federal funds (FEMA, 2004a; 2015a; 2008). Essentially, it was the concept of the "Golden Rule": he who has the gold makes the rules.

The ability to receive, or to continue to receive, federal funds (or contracts) was contingent upon the agency becoming "NIMS Compliant." This was first initiated in Fiscal Year 2006, shortly after Hurricane Katrina. It began with the Assistance to Firefighters Grant (AFG) program. While the concept of being NIMS compliant to receive federal funds appeared to be sound, the implementation was not well received by all.

In 2005, the Assistance to Firefighters Grant had 20 972 applicants with approximately 5990 grants awarded, but in 2006, only 18 160 applied for the same grant with approximately 5000 receiving grants. The trend of fewer applications held steady through 2007 where only 18 170 applied for grants. It was not until 2008 that the number of applicants returned to over 20 000 (FEMA, 2014). No empirical evidence was ever submitted which might identify the reason for the decline in the AFG applications, but there are multiple theories.

There were several trends of thought from multiple agencies in the fire service. Some of these agencies believed that the NIMS requirement played a role in the lower number of grant applications, however, it should be realized this was only conjecture. If the requirements did deter fire departments from applying for AFG Grants during this time, it could also be speculated that a certain amount of fire departments were in the process of becoming NIMS compliant and therefore did not apply. It could also be surmised that some departments were defiant about the requirement, as was evident through numerous discussions on the Internet. Of course, there were also a few conspiracy theorists who were worried about "Big Brother" sticking their nose into the local volunteer fire departments business. No matter what the reason, it was obvious that something affected the number of applicants for the AFG Grants over a three-year period, and it was likely some type of an aftereffect from the NIMS requirement.

Another way that the federal government inconspicuously pushed NIMS compliance was through a "No NIMS-no play" policy. Essentially, if you wanted to participate in an exercise that involved federal resources, or you wanted to be involved in a disaster response, you could not do so unless you were NIMS compliant. The reasoning behind this policy was simple. If your agency was not trained and certified in the National Incident Management System, it would add more confusion and chaos to the incident, drill or exercise. While not articulated, part of the reasoning behind this policy may be due to a concern that non-NIMS-compliant agencies may not understand resource typing, they could have potentially responded with equipment that was not needed, or the wrong equipment altogether.

Similarly, law enforcement agencies were required to be NIMS compliant. The dangling carrot that led NIMS compliance in fire department was also used in law enforcement. In order to receive federal funds, or even federal reimbursement for services, it was required that the law enforcement agency be compliant with NIMS. Much like the fire departments, law enforcement agency had a "No NIMS-No play" policy for law enforcement agencies.

This mandate also applied to the Superfund Amendment and Re-authorization Act title III. In accordance with 29 CFR 1910.120(q), it was mandated that all first responders who responded to a hazardous materials emergency must be properly trained in ICS component and NIMS, and they must be equipped to the basic minimum standard of NIMS (United States Department of Labor, 2008). This standard was Occupational Safety and Health Administration's (OSHA's) way of recognizing and implementing the 2004 ICS/NIMS mandate.

Additionally, hospitals, nursing homes, community clinics, and all healthcare facilities were informed that they too must meet NIMS compliance. Much like other disciplines, it was mandated that a healthcare facility be compliant with NIMS to receive federal funding (FEMA, 2007; "Research Brief", 2008). While the requirement of NIMS was not mandated to receive accreditation, or for the Center for Medicare and Medicaid Services, it was still quite effective because numerous healthcare facilities and research centers received federal funds.

With the federal government taking a hard stance on NIMS, it quickly evolved into a mainstay of emergency mitigation, preparedness, response and recovery. As a part of the Presidential Directive, it was required that NIMS change as the needs of public safety changed. To change with those needs, the Presidential Directive mandated that there should be ongoing research and development of the entire method, including the ICS component.

Even as this book was being written, changes were being considered to the NIMS method, and some changes were made. It is important to realize that NIMS is considered a living document, and a living system. This means that the system should change as the needs of those providing emergency services change. The NIMS method also changes as new research and systems advancements are modified.

1.13 NIMS Updates/Changes (2008) and Training

In 2008, ongoing research into the utility of NIMS led to changes that made the system more integrated and more useful. Many of these changes were made based on the identified difficulties that users faced in major incidents. These incidents included Hurricane Katrina, the September 11 attacks, and various other major (and minor) disasters.

The new proposed changes to NIMS did not modify the basic purpose, scope, or principles of NIMS. The reason that these were not changed was because they were deemed sound and effective principles. The changes that were made did however adjust the organization and the readability of the NIMS document, plus it put more of an emphasis on the importance of preparedness. While these were the mainstay of the 2008 changes, research and development continued well beyond the new implementations and led to additional changes in 2017.

It is important to realize that the research and development of NIMS was, and is, ongoing, and extremely important to keeping up with the changing times. Every year, new tools, and new innovative technologies are created that changes the way personnel respond. It would only seem to make sense that the NIMS method would become antiquated if it did not keep up with the changes in public safety. For years, crusty old firefighters have stated that "If you aren't keeping up with changes, you are falling behind." This statement could never be truer because public safety changes in some way from day to day, month to month, and year to year. Research and development of NIMS helps to keep the system current to the ever-changing times in public safety.

When the first changes to NIMS occurred in 2008, those changes appeared to facilitate better integration with nongovernmental organizations (including the private sector) as well as a more complete relationship between the newly implemented National Response Framework (NRF) and NIMS. While the majority of changes were straightforward and appeared to be a natural progression, one area revealed that NIMS was not only extremely important for natural disasters but also for manmade disasters as well.

With the recollections of 11 September 2001, terrorist attacks, the daunting challenges faced during the response and recovery to Hurricane Katrina in 2005, and other criminal and terrorist events in mind, NIMS added the ability of intelligence and investigations to hold its own position as a major (General Staff) section in the ICS component. In the previous version of the ICS component of NIMS, this function was not previously talked about or acceptable. The change was specifically added so that during a terrorist attack or a criminal event that the Incident Commander, at his or her discretion, could have more flexibility to create an Intelligence and Investigations as a section of the command method.

This gave the Incident Commander the flexibility they needed to either incorporate Intelligence and Investigations as part of planning section (as it was originally placed), or they could create a separate General Staff position for this function (FEMA, 2013c). Additionally, efforts were taken to ensure that those with different specialties or different fields in mitigation, preparedness, response, and recovery, could seamlessly work together without encountering major problems. These changes advanced the method to allow more elasticity in how it was set up based on the needs of the response. If the Intelligence and Investigations taking place were a minimal part of the response, it could stay where it was initially envisioned within the ICS component. On the other hand, if the incident was a major incident that revealed that a large contingency of personnel were needed in Intelligence and Investigations, then that could be accommodated as well.

NIMS could be used in major unexpected emergencies such as an earthquake, tornado, or hurricane, or NIMS could be used to manage planned events such as concerts, fairs, or visits from heads of major countries or businesses. It could be used in criminal or terrorist attacks, or it could be used to manage a major (or minor) investigations. The ability to use NIMS to manage almost any type of emergency soon drove the system to be more and more used, even beyond withholding funds. No matter what was being faced, it could, and did, assist in managing the task at hand.

Training was an important aspect of NIMS updates in the 2008 revisions. Both the Emergency Management Institute and the National Fire Academy offered NIMS training and certifications. In conjunction with other agencies that developed NIMS advancements, the 2008 update suggested more advanced training in using NIMS and the ICS component. In the 2008 update, NIMS was required to be inclusive of all public safety agencies. It was also strongly suggested that all agencies and individuals should receive NIMS training, with certain individuals within an agency receiving more advanced training. Beyond federal agencies that should be trained in NIMS, recommendations were released to include other agencies and government entities that should take NIMS training, including the following:

- Law enforcement
- Fire departments
- Emergency Medical Services (EMS)
- Hospitals
- Public health agencies
- Public works agencies
- Utility companies
- Skilled support personnel for the previously mentioned agencies
- All emergency management response, support and volunteer personnel.
- Local and state governments (including mayors, administrators, and city managers)
- NGOs

NIMS guidance did not stop there. The research and development team also made recommendations for nonemergency response individuals that should receive NIMS training. This guidance recommended that local, state, territorial, and tribal nations should provide training to:

- All managing supervisors
- Those with mission critical positions
- Individuals in charge of professional development
- Human resource managers (including those that oversee personnel policies)
- Individuals responsible for training and credentialing

It was also suggested that all federal policymakers, including elected and appointed officials, receive the training. While this may seem like an extensive list, there were multiple reasons that led to this decision. The primary reason revolved around the fact that when a major incident occurs, outside resources are usually needed. The NIMS training programs, and the NIMS document, was designed to support and create integrative collaboration among people who do not usually work together and/or may not even know each other. By these outside resources utilizing NIMS, these agencies and individuals improved their capacity to seamlessly respond to, and recover from, a disaster. It also allowed elected and appointed officials to understand what was involved in the processes, so they did not have unrealistic expectations or in the event of a disaster, they did not go outside of established standards which thereby created problems for response agencies.

Thousands of changes have been made to NIMS since HSPD-5 and HSPD-8 were first implemented. Some of those have occurred since the 2008 updates. In some instances, those changes were only a modification in the names used for resource typing (standardized names for specific equipment), while in other instances, moderate revisions were made to the NIMS Document. This includes items like the previously mentioned Intelligence and Investigation used to manage a crisis. The 2008 changes superseded the NIMS 2004 document, and most changes were integrated rather seamlessly and without issue.

1.14 NIMS Updates (2017)

Seven years after the 2008 changes, the research and development arm of NIMS made additional suggestions to help improve the system. These proposed changes maintained the key concepts and principles found in 2004 and 2008 versions, while incorporating lessons that were learned from the previous seven years. These lessons learned came from a multitude of sources including, but not limited to, drills, exercises, and real-life incidents. These suggested changes were derived from best practices that were learned, which eventually led to changes in national policy.

While writing this book, FEMA held a 30-day National Engagement Period in April and May 2016. During this period, stakeholders submitted nearly 3000 comments and provided feedback on the new draft of NIMS updates. This was done to ensure that those proposed changes reflected the collective expertise and experience of the whole community. Unlike the process of some federal entities, those managing NIMS wanted to ensure that those in the field, the individuals who actually use NIMS, had valuable input. This was done in part because the input from the users of NIMS to the proposed changes as was seen as "value added." After receiving feedback from those who utilize NIMS, final adjustments were made to ensure that the changes were improving the system instead of changing it for the sake of change.

In October 2017, FEMA released the refreshed NIMS doctrine. As was common in the previous iterations of NIMS, the changes that were made applied to all incidents, regardless of cause, size, location, or complexity. While changes were made to the living NIMS document, the key concepts and principles from 2004 and 2008 versions of NIMS remained the same.

Because this book was mostly written by the time that the 2017 iteration of NIMS was written, a few of those changes may have been missed in finishing the book. While every effort was made to incorporate these changes, there may have been some missed. There is no doubt that future changes will be made to the NIMS Doctrine. Both the student and the practitioner alike must remember that NIMS is a living document that changes as our professions and our methods of planning, mitigation, response, and recovery change. While there is little chance that the core functions presented in the NIMS document will change, it is very possible that multiple minor portions will change or be better clarified over the coming years.

1.15 Conclusion

Comparatively speaking, emergency incident management is a relatively new concept that was formally conceived within the last 60 years (or thereabouts), even though portions of incident management systems have been used hundreds of years. Current Incident Management Systems are a powerful tool for first responders, providing that they are used and used properly. In order to be a successful in your organization and to be a useful tool, IMS methods must be and remain, comprehensive, flexible, and adaptable.

Changes have been made to all types of Incident Management Systems since the inception of FIRESCOPE. While many versions have been created and used, it is important to note that the standard for incident management in the United States is NIMS. The use and mandate of NIMS is applicable at all jurisdictional levels and across all disciplines.

As was mentioned previously, NIMS is applicable across a full spectrum of potential incidents, hazards, and impacts, and is flexible enough to provide command and control of incidents regardless of their size, location, or complexity. This bears repeating because as users of NIMS, we need to be cognizant of why we utilize the NIMS method in the United States, instead of some other IMS method.

NIMS goes the extra step and goes beyond integrating first responders; it improves coordination and cooperation between public and private entities in an assortment of incident management activities. The NIMS method is designed to cover cascading events, but it is flexible enough to transcend jurisdictional boundaries. It allows for an incident management system that can be used to cover large areas, or even across state boundaries, or just the small local incident. The overall method provides a common standard to anyone who may be involved in the overall incident management. It encompasses all facets of mitigation, preparedness, response, and recovery both manmade and natural disasters as well as planned events. A more thorough explanation will be provided in Chapter 5, but emergency personnel in the United States need to realize that NIMS is the standard we must use and meet.

If different public safety agencies are expected to work together, a command system for managing people, resources, and equipment is necessary to create a successful outcome. This holds true regardless of how large, small, complex, or simple that incident may be. Use of the National Incident Management Systems allows integration of resources with both speed and efficiency.

It is also important to realize that in order for any IMS method to be truly useful and successful, the system must be implemented from the "top down" and the "bottom up" simultaneously. As we move forward, the use of ICS, NIMS, or any other method of incident management will not be effective unless we utilize and practice it on a regular basis. If you take this tool and hide it in the toolbox while waiting for the "big one" (a large-scale incident) to occur, you are hurting yourself, your agency, and most importantly the community you serve. If this amazing tool is not taken out of the toolbox and used regularly, it is very likely that when you do need this tool, it will probably be rusty, and the user will probably have problems trying to remember how to use it.

Chapter 1 Quiz

1 **True or False**: The Revolutionary war utilized several principles that are used in incident management systems today. Some of those principles include a centralized command, planning, mutual aid, and a chain of command.

2 **True or False**: Prior to the Big Burn, William Greeley thought the tactic of making agreements (similar to mutual aid agreement) with lumber companies would not work, so he did not forge any agreements with them for firefighting.

3 The military devised a system to manage wars known as the C4I. What are the five elements of C4I?

4 **True or False**: FIRESCOPE *did not* play a major role in the history of emergency incident management?

5 Which state took on the task of developing FIRESCOPE?
   - Utah
   - California
   - Illinois
   - New Hampshire
   - Texas

6 What reoccurring incidents(s) were the motivation for creating FIRESCOPE?
   - Riots
   - Earthquakes
   - Tornadoes
   - Fires
   - Hazardous Material Spills

7 What disaster was the driving force for HSPD-5 and HSPD-8?

8 What does the acronym NIMS stand for?

9 **True or False**: The federal government mandates the use of NIMS by stakeholders in disaster response.

10 **True or False**: A response agency does not need to use NIMS in order to receive federal funds.

11 Which types of agencies or individuals should take NIMS training?

12 **True or False**: State governors, mayors, and tribal chiefs are required to take NIMS training.

13 **True or False**: NIMS is useful for responding to most disasters but is not useful for events such as terrorist attacks.

14 **True or False**: NIMS is useful in disaster response and recovery, but is not useful in the mitigation and preparedness phase of emergency incident management.

15 **True or False**: It is a federal requirement that there is ongoing research and development of NIMS to ensure that this IMS system is up to date and that it incorporates lessons learned from utilizing it.

16 In what years were the NIMS method updated (Two answers)?
   (a) 2006
   (b) 2007
   (c) 2008
   (d) 2011
   (e) 2012
   (f) 2017

Self-Study

Read and understand Jessica Jensen's research regarding the county level behavior towards the mandated implementation of the National Incident Management Systems.

Jensen, J. (2011). The current NIMS implementation behavior of United States counties. *Journal of Homeland Security and Emergency Management*: 8:1, Article 20. https://doi.org/10.2202/1547-7355.1815.

2

A Case Study of Incident Management

The battlefield is a scene of constant chaos. The winner will be the one who controls that chaos, both his own and the enemies

Napoleon Bonaparte

As we consider the efficacy and utility of incident management, we need to understand why we use Incident Management System (IMS) methods. It is not until we understand the true importance of IMS that we as public safety agencies prioritize that it is needed. Unfortunately, many agencies consider IMS as an afterthought, or as something that is only used in large disasters. This type of thinking causes more problems than solutions.

Because the regular use of an IMS method is critical to reducing pain, suffering, loss of life, and economic loss, this chapter is devoted to helping understand just how important proper IMS methods truly are. Through looking at and understanding the lifecycle of an incident, we can see how our intervention can reduce pain, suffering, loss of life, and economic hardship. Furthermore, by comparing and contrasting two major incidents that occurred approximately a month apart, we can see how the use of an IMS method reduced the aforementioned pain and suffering, while not using an IMS method increased the pain and suffering.

2.1 The Lifecycle of an Incident

Incident management is the skill of making decisions that will mitigate the negative effects of an emergency incident. This form of management often occurs while the incident is still unfolding, and in most instances, the incident is still expanding or growing. If an incident is not managed, or if it is improperly managed, the conditions of this event will deteriorate even more. Through utilizing the principles and philosophies, and creating and enacting protocols, there can be a serious reduction in the negative effects that the incident can have. By reducing the lifecycle of an incident, we reduce the negative impact it has. An example of this can be seen in the graphic below which demonstrates the lifecycle of a crisis.

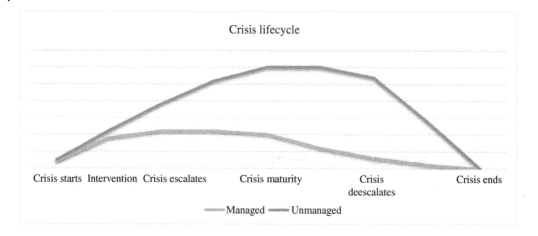

In this graphic, you can see that the lifecycle of a crisis is substantially reduced by introducing intervention. While this graphic is just an illustration, it provides a glimpse of how proper intervention can reduce the negative effects of an incident and reduce the life of the crisis.

The actions that are taken by an Incident Commander (IC) in conjunction with staff and ground personnel reduce the impact significantly, providing that the IMS method is used. If no IMS method is utilized, the incident will eventually peak and cause more death and destruction. This is not to say that all types of incident management will reduce the negative effects. A haphazard or unproven IMS method might allow this lifecycle to peak much in the same manner as no intervention. For that reason, a proven IMS method is needed to reduce the lifecycle of an incident.

2.2 Common Attributes of an Incident

As we consider managing an emergency incident, it should be realized that all emergency incidents have three common attributes (or themes) that remain synonymous with each and every crisis. They are

- Chaos
- Uncertainty
- Complexity

The objective of incident management is to create order where there was once chaos, to produce predictability and stability where there was uncertainty, and to build simplicity out of what was once complexity. While incident management can be used in planned and nonemergent situations, usually a crisis in public safety will involve an emergency. These crises will be from the daily emergency response calls to incidents that go well beyond the basic call. They may be as simple as a noise complaint, or an Emergency Medical Services (EMS) call that involves a person with an upset stomach, all the way to major events. Major events that require a response may include shootings, domestic disputes, fires, auto accidents, acts of terrorism, civil unrest, disasters, and other similar incidents. These principles that we use to manage everyday emergencies, as well as major incidents, can also be used in many different facets of your personal and professional life.

Incident management can be used, especially in public safety, to address an incident that is based on a public relations problem. This type of crisis could include accusations of racial profiling, improper investigations, maltreatment of a mentally disabled individual, sex in agency headquarters (or vehicles), or some other problem.

Another way incident management can be utilized revolves around being the new chief officer in charge of an agency, or a new division, and you want to change the status quo. You can use one of several IMS methods to help create an organizational hierarchy using a chart that identifies duties and responsibilities. Additionally, if you are the new chief officer, and you are changing the way business is done, you can write protocols for each level of the hierarchical chart, or for each position. In doing so, if someone is moved from one position to another, the job they are taking will have protocols, job descriptions, and duties. By utilizing IMS principles, it will allow the new chief officer to have a smooth transition when transferring people between positions, even in day-to-day management of an agency.

Thinking beyond the daily management, public safety agencies occasionally must deal with a disaster or crisis. During these times, you may have to make decisions under stress while experiencing inadequate or incomplete information. As the incident continues to go on, more intelligence (or information) will usually become available. Using an IMS method, this information should help guide you manage that specific crisis. It is critical to realize that no two incidents are exactly alike, and what may work for one incident may not necessarily work for a similar incident.

The principles you will learn in this book will assist you dealing with major incidents, small incidents, and even mundane calls. It does not matter if it is an earthquake, a hurricane, accusations of excessive force, a fire apparatus that hits a pedestrian, or even an accusation of violation of constitutional rights, incident management concepts are effective and can help us to manage a broad spectrum of public safety issues.

2.3 The Importance of Knowledge and Experience

In managing an incident, knowledge and experience plays an important role in enjoying a successful outcome. The knowledge and experience needed is not only related to your field, but in the use of incident management. As you gain further knowledge and experience in incident management, it will become increasingly more natural to utilize crisis response management as a tool. If you only use it on occasion, then your knowledge and experience may be insufficient. Perhaps the best way to describe the effects of using incident management as a method to get from a starting point of an incident to the finishing point is to compare IMS to driving an automobile.

When you first start to drive an automobile, it is an uneasy and awkward experience. In most instances, you have to think through what needs to be done to ensure that you do not have an accident. If you take driving lessons only once a month, your skills at driving only slightly improve; however, if you drive daily, your skills tend to improve more, and it becomes less awkward, less frightening, and more like second nature. By driving every day for an extended amount of time, the process becomes second nature, and you do not even think about what you need to do; you just do it while watching for hazards.

This same analogy can be applied to incident management. When you first begin to use an IMS method, it can feel awkward and uneasy. If you use incident management only on occasion, you have to think through what needs to be done to avert a disaster, or to reduce instances of mistakes. If you use those concepts regularly, over time it becomes less awkward, less frightening, and more like second nature. If you use it daily for an extended amount of time, it becomes second nature. Every first responder should strive to make incident management-second nature so that when a major incident does occur, they better prepared to manage the chaos that ensues.

Similarly, few people would take a single driving lesson, then decide to drive from Boston to Los Angeles. It also would not matter if the driver was a novice or an experienced driver, if they decided to drive from Boston to Los Angeles, it would normally take mitigation measures such as changing the oil, checking tires, and making sure the fluids in the engine compartment were topped off. By the same token, they would plan their route and the entire trip. This might include how much fuel it would take, how far they could drive in a day, what motels they will stay at, and of course, the route they should take.

Much like taking a long trip in a car, using an IMS method should somewhat mimic the process of mitigating problems in advance. We should not expect to be able to competently manage a major incident after only one lesson. We should practice, and practice even more, until we become highly proficient at IMS. Much like checking the oil and tires, we should include mitigation measures in our IMS method, and we should plan before an incident happens, not after we already in the thick of it.

Unfortunately, there are some agencies in public safety who have taken one or two classes, and they think they are competent and proficient enough to drive from Boston to Los Angeles. To put it bluntly, they think they can manage a catastrophic or Mass Casualty Incident (MCI) because they read about it and took a test. They fail to recognize that it takes planning, mitigation, and regular (and ongoing use) of an IMS method in order to be successful for the long trip they may have to face. IMS methods take practice, no matter what the system is named or the country where it originated.

2.4 Case Study: Tokyo Versus Oklahoma City

It is common knowledge that emergency incidents have a certain amount of chaos, uncertainty, and complexity. Usually, the larger the incident, the more these underlying factors present themselves. When we look at how to manage that chaos, uncertainty and complexity, we only need to look at an IMS method. Some might want to question the efficacy, or perhaps the legitimacy of IMS methods. For that reason, a case study was utilized to compare and contrast two incidents; one incident applied an IMS method (Incident Commands System [ICS]), while the other did not utilize any IMS method.

To help you to better understand the importance of using an IMS method for managing daily incidents, as well as rare and infrequent disasters, this review is of two separate incidents that occurred in 1995. The incidents are the Tokyo subway sarin gas attack and the Oklahoma City bombing. Both are similar in timeframe, type of incident, size of the response, and they have multiple other similarities. In the Oklahoma City bombing, the component that the first responders utilized was the ICS method, while in Tokyo, they did not utilize any IMS method. In looking at the efficacy of an IMS method, there are very few incidents that are as comparable in size, scope, and intent, which is the reason these two incidents were chosen.

The first incident to occur was the sarin attack in Japan during the month of March in 1995; the second incident, the Oklahoma City Bombing occurred in the United States in the month of April in 1995. Both incidents were considered mass casualty events. Both of these disasters were considered domestic terrorist attacks (manmade). While they occurred in different countries, both countries are considered developed countries. In both incidents, domestic terrorists used Weapons of Mass Destruction (WMD) to attack the citizens of their own country. These two events have multiple similarities, with the main differences being that in Japan it was a chemical attack (sarin), while the other was an Improvised Explosive Device (IED) bombing. Of course, while we want to compare the use of an IMS method versus none used, we must also consider that cultural differences might have played a role in these responses.

2.4.1 Tokyo Subway Attack

On 20 March 1995, members of the Aum Shinrikyo doomsday cult discharged a deadly nerve agent (sarin gas) in the world's busiest subway system; the Tokyo subway. The gas was developed by the cult at their compound, and the delivery system utilized was human delivery by members of the cult. The attack was timed to happen simultaneously on three different trains, using five devices. The group planned the attack so that they materialized as the trains converged on the main hub of the Tokyo subway system. This incident caused 12 confirmed deaths and 50 injuries specifically from the attack itself. Another 6252 injuries resulted from the subsequent panic, fear, confusion, and secondary contamination, with the secondary contamination making up the majority of the injuries (Murakami, 2000).

Preparedness is a key factor that is often overlooked in incident management, and in 1995, Tokyo officials and the federal government somewhat overlooked preparedness. While there were some disaster plans in place, those plans primarily covered what should be done when resources are overwhelmed in only the response phase. It did not take into consideration what to do in a Weapons of Mass Destruction (WMD) scenario. Nobody knew what to do on a Weapons of Mass Destruction (WMD) incident except for the federal government (Pangi, 2002). Additionally, *The Cycle of Preparedness* was not part of the Japanese disaster management methodology at the time of the attack.

As you will learn later in this chapter, preparedness is comprised of a continuous cycle of planning, organizing, training, equipping, exercising, evaluating, and taking corrective action. This is referred to as the Cycle of Preparedness (explained later in this Chapter 5). This cycle helps to work out any issues (or gaps) prior to an incident so that there will be active and effective coordination during a real incident response.

After reviewing extensive research, it appears as if the country of Japan and the local government entities in Tokyo undertook very few disaster preparedness efforts prior to the attack. This caused complications in the overall response because these entities did not integrate, cooperate, coordinate, or collaborate with each other before, or during, the incident. This in effect caused a substantial amount of problems and caused stove-piping (or the silo effect) of information sharing.

Stove-piping is when information is closely controlled (and sometimes choked down) in a horizontal fashion. Information can be shared up the chain of command within an organization, but cross-organizational or multijurisdictional sharing is not free-flowing. In most instances, this stove-piping is caused by failing to share information. While varying agencies might have critical information or intelligence, this information is kept for only that agency and is not shared with similar agencies who might be working on the same issue or issues. Because these agencies typically work on some level to keep some information secret, a stovepipe occurs. These stovepipes prevent all agencies from being on the same page with the same intelligence and situational awareness. Because many of the Japanese public safety agencies worked in their own circle, stove-piping existed. Prior to the attack, there were no preparedness efforts, and there was a serious deficit in interorganizational cooperation. This included being able to identify that the incident was an organized attack.

The initial signs of the attack were unclear. Initial first responders identified that attack as extremely chaotic and confusing, and what was causing the physical symptoms to people riding the trains was not known for several hours. Had there been information sharing in place, then the identification of a Weapons of Mass Destruction (WMD) incident most likely would have been discovered much sooner. This is based on the premise that a 360° view from multiple response agencies would have given an Incident Commander ([IC] if there were one) or a primary agency enough intelligence to determine that it was an organized attack. Considering that many passengers who had been on the subway were displaying the physical characteristics of choking, vomiting, and some even were even suffering blindness, should have been a quick indicator of a larger attack rather than assuming it was an isolated incident. Additionally, the report of individuals making their way from the subway station to the street only to collapse on the sidewalk should have been another quick indicator of something other than normal (Pangi, 2002).

As the local government and the various agencies began to respond to the incident, they were overwhelmed by chaos, confusion, and uncertainty. In the initial reports, the National Police Agency received information of several incidents at numerous stations and did not understand that they were all from a single organized attack. For quite some time, they had no idea that it was actually a Mass Casualty Incident (MCI). They were initially unaware of the magnitude, so they assumed that these reports were multiple smaller incidents rather than being interconnected. Had all agencies worked together, and reported suspicions to a centralized agency, the extent of this incident likely would have been realized substantially quicker; possibly within minutes (Pangi, 2002).

The use of incident management was significantly delayed. Pangi (2002) suggests that there were two problems that led to the postponement of using an IMS method; the first was the delay in identifying that it was an organized attack, and second, because of a lack of interagency interconnections. The agencies involved with this incident operated at the same time, but they had no centralized management to put forth a unified effort. Such an effort likely would have led to identifying additional resources that could have been utilized and integrated into the response, which would have likely reduced the pain, suffering, and the death toll.

Approximately 45 minutes after the first report, and 15 minutes after realizing that this was a singular attack, the National Police Agency requested assistance from the Self-Defense Force (SDF). The National Police Agency failed to shut down the subway system for at least one-half hour after asking for, and waiting on, assistance from the Self Defense Force. The total shutdown of subway system was not complete until one and a half hours after the incident began. For that hour and a half, bystanders could freely come and go through these contaminated areas, unrestricted. Most of the trains continued to run their normal schedule, while being contaminated from individuals that came from other trains. One train that suffered direct contamination from the attack was even allowed to run its entire route before it was stopped and isolated (Pangi, 2002).

This failure to act caused secondary contamination throughout the subway system and the city as well. Prior to this attack, there was no record found of government agencies nor subway personnel having discussed a Weapons of Mass Destruction (WMD) attack, nor had any of the entities prepared for one. Only the Japanese military had made plans of how to respond to a Weapons of Mass Destruction (WMD) attack, so even if those responding to the incident were able to miraculously identify what caused the incident, there were no plans in place to help direct them in the best way to handle such an incident (Pangi, 2002).

Whenever a Weapons of Mass Destruction (WMD) incident occurs, it is important that the response includes a multilevel, multidisciplined, and multiagency response. It will take a coordinated response to bring the incident to a quick and successful completion (Creamer, 2005). This was not the case during the sarin attack in Tokyo. Prior to knowing that the incident was a coordinated attack, response agencies worked in isolation from each other, and each agency made their own independent decisions. When it was realized that this was a coordinated attack, these same agencies still worked independently from other agencies. There was little to no coordinated response until the Self Defense Force arrived on scene; this was nearly two hours after the initial report. Even after they arrived, the cooperation, collaboration, and the use of an integrated response were only promoted between the Self Defense Force and the National Police Agency; other first responders were still managing the incident they responded to independently of any other public safety agency.

The Self Defense Force identified the agent used as sarin gas shortly after arriving on scene. First responders and other agencies were not notified of what the substance was for at least another hour. In looking at this detail, this equates to emergency personnel being on scene and unknowingly contaminated for a total of three hours before they were notified of what the chemical agent was. The primary reason for the delay was because there was no coordination between agencies and no overarching IMS method in place. Perhaps even more disturbing than the agencies and first responders suffering a long delay is that the hospitals never received official notification from a government agency. St. Luke's Hospital initially learned that the substance was sarin from television news reports around 11:00 a.m. (Pangi, 2002).

From the onset of the first call, the response to the incident was confusing, chaotic, and uncertain, not to mention extremely disorganized (Murakami, 2000; Pangi, 2002). Local and state governments were totally unprepared on multiple levels. The agencies involved had never practiced a multiagency response, and they had no IMS method in place to help integrate resources. This made the response haphazard at best. There was no overall guidance for the incident, and no singular or unified command. Those arriving on scene created their own strategies and decisions, which were usually based on what they saw, or based on what their agency thought was most viable way to proceed. Rather than having an overall strategy developed by a group of individuals who saw the whole picture, decisions were made viewing one small piece of the overall incident with the goals of only their agency in mind.

This lack of cooperation, collaboration, and communication also led to hospitals being overwhelmed. While some hospitals were overwhelmed from being inundated with a multitude of patients, other nearby hospitals had no patient, or only few patients. There was a haphazard approach in accounting for patient's destination, and the ability of the hospital to treat patients. As an example, St. Luke's Hospital was flooded with 150 patients that were transported by ground ambulance to this hospital in just the first hour and a half. This does not even take into consideration the walk-ins and the private citizen transports to the hospital. In all, St. Luke's Hospital had 641 patients show up at the emergency room on the day of the incident. It clearly overwhelmed this one hospital.

Communication with all hospitals, and between hospitals, was clearly lacking as well. In one eyewitness account, a television news van rushed several patients to a secondary hospital (not the closest hospital). They arrived an hour after the initial attack, only to find out that the hospital had no idea that an attack had taken place. The treatment of the patients that the news crew transported was delayed. They were initially denied help by a nurse because there was no doctor on duty. The news crew essentially begged for help, and eventually a doctor was brought in to treat those patients (Murakami, 2000, pp. 27–29).

In another eyewitness account, the secretary for the head of the School of Medicine at Shinshu University received a call approximately 30 minutes after the attack. It was a reporter asking if Dr. Nobuo Yanagisawa wanted to make a comment on what might have been used in the attack. Because he was unaware of the attack at this time, he turned on the television to gather more information. Having investigated a similar event of sarin gas that happened almost a year earlier in Matsumoto, Japan (on 27 June 1994), Dr. Yanagisawa was very familiar with the symptoms, and he thought that he knew what the substance, and what antidote should be used.

He immediately called in two doctors to assist him in getting the word out to the hospitals and the EMS providers. They also attempted to communicate with the fire department, so they could spread the word on the suspected type of attack, and the antidote. Unfortunately, contact was never made with the fire department. Initially, the three doctors began faxing the Matsumoto Report to the hospitals nearest to the incident and other nearby hospitals. The report was quite long, so it took a long time to send it by fax. Before the information was sent to all emergency rooms, they began getting requests for the information from hospitals not yet reached. In all, over 100 different hospital accepted patients (Murakami, 2000, pp. 220–223).

Even when this team of doctors led by Dr. Yanagisawa did send the report to the various emergency rooms, the lack of communication within the hospital itself led to delays. A prime example of this was St. Luke's Hospital. Dr. Yanagisawa called the hospital and requested to speak to the doctor in charge of the emergency room. While technically he should have gone through the person in charge of the hospital, he felt that time was of the essence, so he called direct. He had a brief discussion with the person he thought was in charge of the emergency department and told the person he would send the information via fax as soon as possible. He would later find out that several doctors were combing through the library, looking for what the substance might be until 11:00 a.m., and they found out the answer from news coverage of the incident (Murakami, 2000, p. 221).

While most of the other 100 hospitals were available and willing to assist, most received relatively few patients. A lack of communication, collaboration, and coordination, as well as a breakdown of the communication (hardware) system, led to the closest hospital to the incident being overwhelmed. With over 100 hospitals in close proximity, most saw less than 10 patients, while St. Luke's hospital saw over 600 patients.

The lack of preplanning, coordination, cooperation, and a lack of integration of resources led to more human suffering, and it caused the incident to last longer. It also placed emergency personnel at greater risk, and it allowed contamination to be spread citywide. Subsequently, more people (including first responders) needed to be seen as patients. Nurses, doctors, EMS, and many that had contact with any of the initial patients needed treatment, which overwhelmed the medical system for weeks.

Allocation of resources in this incident was disorganized as well. Because agencies had rarely worked together, there was an issue of trust. The culture among these response agencies was been described as isolationist (Pangi, 2002). This led to no information sharing and even more disorganization. Information in this incident only went from pier to pier, rather than going to a higher command, or in all directions so that all personnel were on the same page.

Another key factor that negatively affected the response was governmental bureaucracy. This bureaucracy not only added layers of approvals and direction but also compartmentalized agencies from each other. According to a study by Pangi (2002), this compartmentalization not only caused responding agencies to respond as separate units, it caused them to be in competition with each other. This competition caused information and expertise to not be shared. Rather than helping each other, these agencies made every effort to ensure that their agency knew more than the competing agencies; the same agencies they should have been working with and cooperating with.

As if these issues did not cause enough disorganization, then they added the failure of not using an IMS method to manage the response and recovery. It is easy to see that the Tokyo sarin attack lasted substantially longer than it needed to. Additionally, the chaos, confusion, and uncertainty increased because the responding agencies worked against each other rather than with each other. When one agency would employ a mitigation strategy, another agency might unintentionally do something that made that strategy less effective. Had there been coordination and collaboration, everyone likely would have been on the same page rather than being at odds with each other.

2.4.2 Oklahoma City Bombing

On 19 April 1995, in Oklahoma City, OK, the Alfred P. Murrah Building was bombed by a domestic terrorist. This bombing remained the deadliest terrorist attack on United States soil until the 11 September 2001 attack that was committed by foreign terrorists. The Oklahoma City bombing killed 168 people, including 19 children under the age of six, and physically injured more than 675 other individuals (Shariat, Mallonee, & Stidham 1998).

Beyond the human toll, the explosion destroyed or damaged 325 buildings, including the Murrah Federal Building. The blast was so strong that numerous buildings suffered damage in a 48-block area and the window glass of 258 nearby buildings was completely shattered. The blast destroyed or burned 86 cars and caused over $650 million in damage (Hewitt, 2003). This was the work of domestic terrorists, later identified as Timothy McVeigh and Terry Nichols (Oklahoma Department of Civil Emergency Management [ODCEM], n.d.). The homemade explosive was contained in seventeen 55-gallon drums and delivered to the Murrah Federal Building using a rented 24-ft Ryder delivery truck. The device was designed and positioned to inflict the most damage.

The response to the bombing of the Alfred P. Murrah Federal Building in Oklahoma City involved hundreds of public, nonprofit, and private organizations, as well as untrained spontaneous volunteers. According to the Oklahoma City Memorial Museum, (n.d.), this incident was the first time that the ICS was utilized in a major incident that was not fire-related. By utilizing the ICS method, there was a fully integrated response of federal, state, local, and tribal governments as well as nongovernmental resources. While there were some issues, overall, the ICS method worked extremely well in organizing and managing all responders.

In looking over the previous actions of Oklahoma City, the city government appeared to be moving in the right direction in preparedness as well. Not only had they created the plans, but they also had been implementing the Cycle of Preparedness. While their preparedness efforts had not considered such a large event, or an event that involved blowing up a federal building, those plans provided a basis and the key elements for any type of disaster response. Because they began undertaking this preparedness process, collaborative networks were already in place, there was already some cooperation that was ongoing, coordination was in place, and some of the communications issues had already been addressed. Additionally, many of the roles and responsibilities for each level of city government were already known and practiced, which reduced uncertainty. All of these factors combined created the foundation for a more organized and unified response and recovery (Manzi, Powers, & Zetterlund, 2002).

City agencies were also sporadically training together, undertaking exercises, evaluating plans, and taking corrective action. While they were moving in the right direction, there were still identified and unidentified gaps that needed to be fixed. Among those gaps were an effective way to evaluate exercises, and the occasional agency that did not think they needed to be involved in these preparedness efforts (Manzi et al. 2002).

As is often the case, some agencies in the area did not play well with others, so there were some additional gaps already in the plans prior to the incident. It should also be noted that planning for a catastrophic event such as this had never been broached prior to the bombing. Even so, the preparedness efforts that had been addressed would play an important role in how, when, and where resources were deployed (ODCEM, n.d.).

As you read further, you will see how specific discipline responded and how seamlessly they integrated with other agencies. You will see how there was coordination and cooperation, and some of the outcomes from using the ICS method. While the use of an ICS method was not perfect, it did provide for a more integrated response.

2.4.2.1 At the Scene of the Explosion

The Oklahoma City Fire Department was one of the key players in preparedness prior to the bombing. They knew about, trained on, and used the ICS system, and they encouraged others within their city to do the same. A year prior to the bombing, senior staff and senior management from Oklahoma City government attended a weeklong (Federal Emergency Management Agency [FEMA] sponsored) training course on disaster management. This course helped to forge strong interpersonal relationships between city officials and the tabletop exercise that was the culmination of the course. This training helped to open the eyes of many city officials. Long before the bomb was ever detonated, the city determined that everyone in city government would need to be on the same page, and they would need to integrate resources in a disaster (Manzi et al. 2002). This led to training and exercises that helped pave the way for a more unified response.

At the very second that the explosion occurred, many of the senior staff for the Oklahoma City Fire Department were having a meeting. After hearing the explosion and seeing the debris cloud, Oklahoma City Fire Station No. 1 self-deployed to the incident. Their response was initiated before the Oklahoma City Communications Center dispatched any units. While self-deploying is typically frowned upon, this was not the only fire station that self-deployed after hearing the explosion. Considering that Station No. 1 was only five blocks away from the Murrah Building, it was surmised that they probably arrived on scene in the first two minutes (ODCEM, n.d.).

The first arriving fire officer, a District Fire Chief, implemented the ICS system as soon as he arrived on scene. He immediately ordered that two triage areas be set up in strategic yet safe areas so that the influx of patients could be evaluated and prioritized for transport based on the severity of their injuries. Shortly after setting up ICS and triage areas, the Oklahoma City Fire Chief (the District Fire Chief's superior) arrived at the incident only moments later.

Upon arriving on scene, the Oklahoma City Fire Chief left his subordinate the District Chief in Command as the Incident Commander (IC). The Fire Chief supported the District Chief by performing intelligence gathering and reporting back to the District Chief. After observing the Murrah Building from all sides, the Oklahoma City Fire Chief made the suggestion to the Incident Commander (IC) that a Unified Command ([UC], discussed in detail in Chapter 7) be initiated. The initiation of a Unified Command (UC) would allow all responding agencies to provide direct input on the resources they possessed, the capabilities of their agency and personnel, other resources that were available, and any agency that was part of a Unified Command (UC) would have a seat at the proverbial table. It would also help to somewhat overcome communications barrier (Final Report, 1996).

After completing a quick evaluation of the damage to the area, the needs for initial response were reported to the Incident Commander (IC), and the Oklahoma City Fire Chief determined that all off-duty personnel should be called in and that reserve apparatus should be pressed into service. He made that recommendation to the Incident Commander (IC), the Deputy Fire Chief. Additionally, he suggested organizing mutual aid resources from nearby cities to help backfill stations. In doing so, daily emergency calls could still be answered in a timely manner while the Oklahoma City Fire Department was still dealing with the major disaster at-hand (Final Report, 1996).

Four minutes into the incident, the Incident Commander (IC) requested that the State Emergency Operations Center (EOC) be activated. This was important to the overall mission because it gave the Incident Commander (IC) one point of contact that could find and request additional resources. Once established, and at the Incident Commanders (IC's) request, they requested that the National Guard be requested and activated, that resources should be requested from Tinker Air Force Base, and that the American Red Cross and other voluntary organizations should be activated. There was also a request made to contact FEMA for support and to activate several Urban Search and Rescue (US&R) teams to assist with the response. Other requests were made, but these were the first round of requests, and they were made within minutes of arriving on scene (ODCEM, n.d.).

Like the self-deployed response by the Oklahoma City Fire Department, many ambulance crews from the Emergency Medical Services Authority (EMSA) self-deployed to the bombing incident. While they did not know it at the time, the explosion knocked out all incoming lines to the EMSA switchboard, and all emergency lines except for 911 were rendered useless by the explosion. Within minutes, cell phones were unable to make connections. The high call volume jammed the networks just from the sheer number of calls being placed (ODCEM, n.d.).

Prior to the incident, EMSA had integrated and worked with other agencies in the preparedness phase, so they were already aware of what their roles and responsibilities were. The first EMSA personnel that arrived began treating the walking wounded after arriving on scene. They set up at the designated triage areas and began sorting patients according to the seriousness of their injuries. Less than 15 minutes into the incident, they declared a Level Three mass casualty incident and called for EMS resources from mutual aid agencies from across the state (Nordberg, 2010).

EMSA also called for buses to transport those that were considered walking wounded, meaning they had nonlife-threatening wounds (Nordberg, 2010). Those that suffered more critical injuries would be transported by ambulance. Within the first hour after the incident, over 100 people were transported to one of 12 local hospitals in the area. Prior to sending patients to any hospital (or clinic), the vast majority of EMS crews checked with hospitals to identify that they were available and had the resources to treat the patient being transported. This allowed a more equal distribution of patients and it helped to prevent overwhelming only one or two hospitals. In total, over 670 patients were seen on the first day of the incident (Nordberg, 2010).

A few of the local hospitals were having communications issues in the first 20–30 minutes. The radio frequency used for disasters was turned off, or down, in most of these cases. This is a common practice when there is no disaster because interference, radio waves from distant areas bouncing off the earth's atmosphere (skip), and other similar issues can cause radios to crackle and make a lot of noise. Often, if that frequency or radio is not being used, the radio will be turned down to avoid this annoying noise, and then it is forgotten about. To help mitigate this issue of no disaster communications with hospitals, law enforcement officers were initially sent to the hospitals to advise them to turn the frequency on, or up, whatever the case might be. Additionally, these officers were ready to provide the needed communications by relaying information if needed (Nordberg, 2010).

Less than 10 minutes after the explosion took place, the American Red Cross arrived on scene. Not long thereafter, the Salvation Army arrived on scene to assist as well. The quick response of these two organizations suggest that they were in the process of self-deploying when the official request was sent. Both organizations were utilized immediately (ODCEM, n.d.).

While still within the first hour of the incident, numerous response agencies and volunteers arrived. In order to better organize them while keeping resources nearby, at least three staging areas were created. Individuals, equipment, and goods were kept separate in an effort to organize these resources while reducing confusion.

Operational response agencies and personnel were placed in the Personnel Staging Area. This was located just outside of the disaster zone. These were the people that would go into the active disaster zone and crawl through the rubble or treat patients as needed. They were staged close enough that they could quickly deploy (within minutes), but far enough away that they could not interfere with ongoing operations (ODCEM, n.d.).

The resources that arrived at the Personnel Staging area included local, state, and federal law enforcement officers, fire, and rescue crews, EMS personnel, the first of many federal government resources from Tinker Air Force Base, the first contingency of the Oklahoma National Guard, and other resources too numerous to list. They continued to arrive throughout the day, and staged resources were put into service as they were needed (ODCEM, n.d.).

An additional staging area was set up to receive donations that began coming in within the first hour. These donations included just about everything you can imagine (ODCEM, n.d.). From tools, to toilet paper, to work clothes, the community began to bring what they thought might help. At this staging area, the donated items were sorted and categorized. This allowed agencies to ask if there was the availability of a specific item, and this donation staging area could identify and provide the item if it was available. This would later be moved and turn into a donation's management warehouse.

Not long after the explosion, at least three different heavy equipment companies arrived at the Incident Command Post (ICP) offering all of their resources (ODCEM, n.d.). The Incident Commander (IC) tasked one company with setting up a staging area for heavy equipment and shoring materials, and then organizing the construction/demolition staging area. A liaison was placed in this staging area to ensure effective communication. These companies cooperated and collaborated together and with public safety. In one instance, a police escort was provided to one company that was bringing an extremely large crane from 15 miles north of the bombing site. This escort was given to provide a quicker response (Final Report, 1996). The After-Action Report (AAR) identified that there were some issues with staging, but Incident Command knew where their resources were and could check availability of a resources within seconds (ODCEM, n.d.).

The Oklahoma City Fire Department had already set up a Command Post on a street corner, and they had initiated ICS from the beginning of the incident. The fire department had put the key people in place to fill General Staff, and Command Staff positions. Because of the size of the incident, there were not enough law enforcement officers on scene to control the multitude of untrained people that came to help. An area of 48-blocks was affected, and it was determined that it would be a monumental task to try to keep people out. When it was thought that a secondary explosive device was found (90 minutes into the response), the entire 48-block area was evacuated (ODCEM, n.d.).

This provided an opportunity to regroup and move the initial Incident Command Post (ICP) to a safer location two blocks away. The new location was a small parking lot. Because it was still early in the incident, the only Mobile Command Centers that were present at that time were the Oklahoma City Fire and separately Oklahoma City Police command centers (Final Report, 1996).

As part of this regrouping and because of an earlier request for more law enforcement and the Oklahoma State National Guard, it was determined that enough resources were on hand to secure entry to the main site. This was based on the number of military personnel, National Guard personnel, and law enforcement that were staged, or who were already working top secure part of the perimeter. When it was determined that there was not a secondary device, rescuers were let back into the area, but access was restricted, and each individual trying to enter the area had to provide security personnel a valid reason of why they should be let in the area. This action substantially increased accountability throughout the disaster area (ODCEM, n.d.).

By 10:55 a.m., less than two hours after the explosion, the Urban Search and Rescue (US&R) teams from Phoenix Arizona and Sacramento California were activated by FEMA (ODCEM, n.d.). Nine additional Urban Search and Rescue (US&R) teams were alerted that they may be soon deployed to the incident. A total of teams would eventually assist in the response and recovery. The Urban Search and Rescue (US&R) teams that worked at the Oklahoma City bombing site included teams from New York City, NY, Montgomery County, MD, Virginia Beach, VA, Los Angeles, CA, Fairfax County, VA, Dade County, FL, Puget Sound, WA, Menlo Park, CA, and Orange County, CA (Oklahoma City National Memorial and Museum, n.d.).

Additionally, FEMA's Regional Director and his staff were deployed from Denton, TX, to Oklahoma City at approximately 11:00 a.m., so they could support what would be an ongoing operation. FEMA arrived in Oklahoma City within four hours of the bombing, and they offered the full cooperation of the federal government. This is important to note because they did not try to take over the response, they were there to support it (ODCEM, n.d.).

Around 11:00, the National Weather Service notified the Incident Command Post (ICP) that harsh weather was moving into the area. It was initially thought that a tent could be erected by the National Guard so that workers could get under cover in inclement weather; however, the lot in which the Incident Command Post (ICP) was located did not have sufficient room for it. This was due to the lot had begun to fill with other mobile command centers, and it became a problem that needed to be solved. The additional command centers that arrived were manned by the Federal Bureau of Investigation (FBI), the Bureau of Alcohol, Tobacco and Firearms (ATF), and the Drug Enforcement Administration (DEA), with each taking a fair amount of space (ODCEM, n.d.).

Southwestern Bell Telephone Company was also represented at this location. They set up a truck, so they could loan cell phones to all legitimate first responder's. This integration of a nongovernmental agency was voluntary. The company wanted to ensure that all personnel that responded to this disaster had working and valid communications. Any time these phones were used, they would take priority over all calls, including that of private citizens. While it took a few hours to erect, the phone company also erected a mobile cell tower to facilitate better communications (ODCEM, n.d.).

Due to the room constraints of this parking lot, it was decided to move the Incident Command Post (ICP) to the Southwestern Bell Telephone Company (headquarters) parking lot. The use of this parking lot was personally offered to the Incident Commander (IC) by the CEO of Southwestern Bell. The Southwestern Bell parking lot presented substantially more accommodations than the previous area. It provided a much larger parking area and a sheltered parking garage where volunteer organizations and private businesses could distribute food, and it provided a place that would protect and store the multitude of donations that were already arriving. The adjacent office building also met the sanitary needs of the emergency and relief personnel (ODCEM, n.d.).

Shortly after changing Incident Command Post positions, additional mobile command units began arriving. Additional mobile command units that arrived included two vehicles from the Oklahoma Department of Public Safety and one command center from the US Marshals Service (ODCEM, n.d.). This Command Post provided an area where all agencies meet and\or liaison with other agencies in an effort to have a more integrated response. If in the event one agency needed to coordinate with another, they only had to walk a short distance to their command center.

After the move to the new location, the Oklahoma National Guard and the Oklahoma City Public Works worked together to set up the National Guard's tent. The tent was used as an Incident Command Post for the National Guard operations. This tent was used for the forward operations for the National Guard until the end of the incident (ODCEM, n.d.).

Shortly after moving to the Southwestern Bell parking lot, the Fire Chief of the Oklahoma City Fire Department assumed command of the incident. Soon thereafter, the FBI and the new Incident Commander ([IC] the Fire Chief) met. Because this incident was a federal building, as well as other reasons, it was determined that the FBI would be the lead investigative agency. The FBI discussed the need for securing the scene to preserve evidence with the Incident Commander

(IC). It was determined that a chain link fence was needed to keep unauthorized personnel out of the area. This job was tasked to the Oklahoma City Street Department, who would erect the barrier fence around the scene and the Command Post. While the street department was its own agency, it would work under the direction of the FBI. It was also determined that traffic would be stopped at certain areas, so the Oklahoma City Traffic Management Department was given this task, also under the direction of the FBI, to blockade certain streets (Final Report, 1996).

Everything that has been described about this incident to this point occurred in the first two hours of the incident. While the remainder of the Oklahoma City Bombing incident could be reviewed in similar detail, we will only discuss key points about the remainder of the incident from the details found in the Oklahoma Department of Civil Emergency Management (ODCEM) After Action Report, ([AAR] n.d.) and interviews provided by the Fire Chief, Gary Marrs (Fire Engineering, 1995). Some key cooperation and collaborations that occurred after the first two to three hours included the following:

Day 1

- With the help of the City of Oklahoma elected officials, a vacant building was used to house the investigative arm of the incident, including the FBI, ATF, and the DEA.
- Ongoing liaison relationships were sustained with the Oklahoma Department of Public Safety, the National Guard, and the American Red Cross, in an effort to facilitate a more unified effort.
- The Oklahoma Medical Examiner's office and temporary morgue were set up in a nearby Methodist Church.
- Public events personnel set up facilities for housing and feeding rescue workers.
- A roped-off media area was set up to accommodate the influx of local, national, and international media. The media was updated regularly on the first day. After the first day, there were assigned times for press conferences throughout the duration of the incident.
- Public events personnel set up accommodations for press briefings.
- The Oklahoma Restaurant Association established a 24-hour food service operation to feed first responders at the Myriad Convention Center.
- Public works provided on-site sanitary facilities, lighting for nighttime work, and trash cans, plus developed a trash schedule for the duration of the incident.
- The FBI developed and issued a photo ID so that individuals not involved in the incident could be restricted from entering.
- Donation centers were set up to receive donated food, clothing, and other items. Eventually, three donation centers would be opened.
- Street cleaning crews began cleaning streets outside of the perimeter, so traffic could resume in the nearby area.
- A Forward Command Post for the Operations Section of search and rescue was set up in the loading dock of the Murrah Building.
- FEMA Incident Support Team arrived early in the afternoon of the first day and supported the Forward Command Post with electrical power, telephones, copiers, tables, chairs, and other necessary items.
- A Family Assistance Center was set up for immediate family members. Family received two briefings per day from the State Medical Examiner's Office. Along with being updated about those deceased or feared dead, there was counseling, and comfort provided. Organizations included in providing comfort were the American Red Cross, the Salvation Army, the Oklahoma Funeral Directors Association, clergy members, and mental health professionals.
- Security for the Family Assistance Center was provided by the Oklahoma County Sheriff's Office and the Oklahoma National Guard.

- The American Red Cross set up temporary shelters to provide for the needs of those who had been displaced by the explosion.
- A nonprofit organization based in Oklahoma, Feed the Children, provided food and disaster supplies for first responders. They also set up in the disaster area to distribute food and equipment to first responders. Feed the Children managed volunteers, and if they found a need not being filled, they would contact corporate donors to facilitate what was needed.
- The Myriad Convention Center became a central part of ensuring first responder needs were met. The convention center was set up on the first day, and services expanded over the next few days. Services provided to first responders included donated food, donated clothing, medicines, personal care items, optometry services, chiropractic services, podiatry care, massage therapy, and more.
- Perimeter security was initially provided by Oklahoma City Police and the Oklahoma County Sheriff's Office. A planning meeting for perimeter security was held with Oklahoma City Police, Oklahoma County Sheriff's Office, Oklahoma Department of Public Safety, Oklahoma Military Department, and Oklahoma Department of Civil Emergency Management to divide up perimeter security obligations. A permanent perimeter was decided, and a specific number of personnel to provide this security (from each agency) were assigned.
- An inner perimeter (for areas that may contain evidence) was secured by FBI personnel until the buildings inside the (overall) perimeter were cleared of evidence.

Subsequent Days

- At the Myriad Convention Center, special programs were set up for Urban Search and Rescue (US&R) Teams. This included free calls home from AT&T, free mail and parcel delivery from UPS, and other similar amenities. These amenities were managed, and it was decided who should be allowed to use these services under the ICS system.
- A military explosive ordinance disposal unit was requested. The unit does a sweep of the Murrah Building and finds no additional explosives.
- A Disaster Field Office (DFO) was established. In the Disaster Field Office (DFO), the Federal Coordinating Officer (FCO), the State Coordinating Officer, and emergency response teams worked together to meet the needs of the response and recovery.
- A Multi-Agency Coordination Center (MACC) comprised of federal and state representatives (who had the authority to commit resources) was created at the Myriad Convention Center. It is staffed by the Oklahoma County Emergency Management Agency, Oklahoma Department of Civil Emergency Management, FEMA advisors and specialists, the National Weather Service, the American Red Cross, and technical advisors from the US Forestry Service.
- A Joint Information Center (JIC) was set up at the Myriad Convention Center to provide for the organized release of any public information. All information would be vetted and cleared for release by all major agencies involved.
- Mental health services and Critical Incident Stress Debriefings (CISDs) were offered to first responders. These services were provided by the Oklahoma Department of Mental Health and Substance Abuse Services, the FBI, the American Red Cross, as well as volunteer private specialists. Additionally, crisis hotlines.
- A Congressional Liaison Office is established at the Disaster Field Office (DFO) to coordinate inquiries from US Senators and Representatives. By assigning this task to an individual, or individuals, it reduces interruptions that might slow operations.
- Crisis lines were created, and operated 24-7, to meet the needs of the general public.

- The Federal Coordinating Officer (FCO) activates numerous Emergency Support Function (ESF) from the federal government to assist with the incident. They include
  - ESF 2 Communications
  - ESF 3 Public Works and Engineering
  - ESF 4 Firefighting
  - ESF 5 Information and Planning
  - ESF 6 Mass Care
  - ESF 7 Resource Support
  - ESF 8 Health and Medical Services
  - ESF 9 Urban Search and Rescue
- Requests were made to the Defense Coordinating Officer (DCO) to provide six C-141 military aircraft to be used to transport the needed supplies and equipment. That request was approved almost instantly.
- A Disaster Mortuary Team (DMORT) from FEMA arrives to provide support for the coroner's office. Medical support for the Disaster Mortuary Team (DMORT) comes from US Public Health Service, with logistical support being provided by Oklahoma City Veterans Administration (hospital). They are assisted by 30 members of the Oklahoma Funeral Director Association and a military Graves Registration Unit.
- The Oklahoma Health Department is assisted by the Center for Disease Control in starting the Emergency Survivor Injury Registry. The registry allows the responsible agencies to identify victims, as well as document any injuries or continuing medical needs.
- An In-Kind Donations Coordination Team is created at the Disaster Field Office. One of the tasks they undertake is to create an "unmet needs" registry.
- A specialized Risk Assessment Team is requested and received. They evaluate the site for biological and chemical hazards. Instructions are given to rescuers on how to better protect themselves from environmental factors and decontamination procedures are developed. Four firefighters from a mutual aid fire department and six firefighters from Tinker Air Force Base will be responsible for decontamination.
- A joint Preliminary Damage Assessment (PDA) of the affected area is undertaken and completed by FEMA and Oklahoma Department of Civil Emergency Management.
- U.S Army Corps of Engineers sends requested specialists to provide input concerning structural integrity and blast damage distribution. They provide substantial input in the safest way to proceed with rescue and recovery missions.
- In an effort to restore federal services housed in the Murrah Building, the General Services Administration (GSA) conducts surveys to find temporary and permanent office space nearby.
- AmeriCorps is activated to assist with donations management. Their primary task is to keep inventory of all donations.
- A Recovery Service Center (RSC) is opened where applicants can personally meet with representatives of various aid programs. The Oklahoma Baptists provide daycare for those who come to the Recovery Service Center (RSC) to discuss aid assistance. Agencies involved include the following:
  - FEMA Disaster Housing
  - FEMA/State Individual and Family Grants
  - Oklahoma Division of Civil Emergency Management
  - Oklahoma Employment Security Commission
  - American Red Cross
  - Oklahoma Department of Human Services/Aging
  - Salvation Army

- Department of Area-wide Aging
- Small Business Administration
- Federal Employees Education and Assistance Fund
- Social Security Administration
- Veterans Administration
- Internal Revenue Service
- Oklahoma County
- FEMA/State Public Affairs
- Oklahoma State Insurance Commission

It is important to note that this is a brief synopsis of the overall effort. Through utilizing an IMS method, more specifically the ICS method, those managing the incident helped to assure that everyone was on the same page and were effectively working toward the same end goal.

2.4.2.2 At the State Emergency Operations Center

As was described earlier, the blast occurred at 9:02 a.m., and agencies on scene immediately began requesting additional resources. Documentation by the State Emergency Operation Center (EOC) revealed that the full activation of the EOC was requested at 9:04 a.m., only two minutes after the explosion. Within 25 minutes, the EOC was fully staffed and a representative was onsite of the explosion to be a liaison between the Incident Commander (IC) and the EOC.

The purpose of an EOC was to establish a centralized site where government (at any level) can provide interagency coordination and execute decision-making to support an incident response. In the EOC, there may be a multitude of experts in specific areas that can provide expertise and/or resources to the Incident Commander (IC) and the General Staff that are commanding the incident. As an example, when the state EOC was activated, the agencies that staffed it included the Oklahoma Department of Public Safety, the Oklahoma Department of Human Services, the Oklahoma Military Department, the Oklahoma Department of Health and the Oklahoma Department of Education. Not long thereafter, the National Weather Service, the Civil Air Patrol, and the American Red Cross were brought in to staff the EOC for their specialization.

On the ground, roles and responsibilities were quickly assigned by the Incident Commander (IC) with the majority of support being ordered by the EOC. While the search and rescue operations were initially managed by the Oklahoma City Fire Department, the outer security perimeter was initially managed by Oklahoma City Police Department and the Oklahoma County Sheriff's Department. Later, other Oklahoma State law enforcement officers and National Guard were integrated into security.

Eventually, the rescue operation included 11 federalized Urban Search and Rescue (US&R) Teams from local agencies across the nation. Firefighters from more than 75 Oklahoma communities and more than 35 departments from Texas, Kansas, Arkansas, and other states participated. In all, FEMA deployed more than 1000 of its own employees and hundreds from other federal agencies. The American Red Cross provided food and shelter for emergency personnel and support for victims and their families. Private firms ranging from building supply companies to funeral homes to restaurants supported the responders. The scale of the operation required resources from all levels of government and a wide variety of nongovernmental organizations.

The bombing was a federal crime involving a federal facility that resulted in the deaths of federal officers and employees. The legal jurisdiction for the investigation resided with the Federal Bureau of Investigation and other federal law enforcement agencies, who integrated local agencies. While the response had a few issues, they were relatively minor when compared to other incidents of similar size.

2.5 Comparing and Contrasting These Incidents

While neither the Tokyo sarin attack response (and recovery) or the Oklahoma City Bombing was incidents that went without problems, it is easy to see that the Oklahoma City Bombing was substantially more organized. While much speculation could be made about the reason behind the significant differences, there is no doubt that the IMS method being used in the Oklahoma City Bombing played a significant role in creating a better, more organized response. Understanding the positive role that ICS played is imperative to comprehending how much IMS is needed for managing any incident.

2.5.1 Command

The act of commanding an incident means that an individual or a group of individuals working together have the authority to direct operations, and they do so. In the Tokyo attack, there was not anyone commanding the incident. Each agency was working independently of other responding agencies. Due to the lack of command, the incident had no specified direction, security was lax or almost nonexistent, and each agency operated under their own authority. This equated to each agency making decisions and addressing issues as they saw fit, based on their view of the incident. Even after the realization that the Tokyo incident was a terrorist attack, there was no overall command presence to guide and direct these response and recovery efforts.

An overarching presence that provided direction for all personnel was not present during the Tokyo response. The lack of organization for the resources that responded to this attack caused each to work independently of other organization. Each entity managed the incident based on what their individual agencies priorities were, and there was little to no regard for what other entities were doing, or how they could approach the response and recovery in a unified effort.

In Oklahoma City, there was a command presence, and command direction from the onset. When Oklahoma City Fire Department arrived at the Murrah Building, a senior officer in the fire department immediately established command. While the command of the incident was initially chaotic, this was primarily because good Samaritans were interjecting themselves into the incident. In most major incidents, the general public does not realize that their efforts can cause issues with commanding an incident.

The Oklahoma City Fire Department commanded the incident until it was no longer safe to search for bodies. For a little more than a month, the fire department commanded the incident. This is not to say that the fire department did all of the work because they did not. The fire department organized and planned every detail, usually in cooperation with other agencies. Everything that was accomplished during this response was done so with the overarching goal of recovering all of the victims to be accomplished. In fact, each agency had specific tasks, but those tasks were either directed by, or requested and approved by, Incident Command (IC). In most instances, these goals were specific in nature, and each goal had a way of measuring success, they were attainable with the resources on scene, they were relevant to the overall mission, and except for the investigation, timelines were set to ensure ongoing progress. Even the FBI response was to a certain extent commanded by the fire department. While the FBI did the investigation on their own terms, they also coordinated and sought approval from the fire department to ensure that the FBI was not in the way of rescue and recovery operations.

2.5.2 Control

Controlling an incident helps to reduce the severity and reduce the lifecycle of a crisis. The act of controlling an incident involves leading (or guiding) the response, and in most instances, the recovery. To accomplish control, responsibility must be taken for the acts and achievements of all responders. Control involves maneuvering resources toward a desired result; in this case, the desired result was to save survivors and to recovers the bodies of victims without any injuries or death of responding personnel. Controlling an incident is done so that the response and recovery can keep on a specified course. This means that there must be a constant guidance and oversight so that the response and recovery is an operation that is organized and effective rather than haphazard.

Control can also be related to the accountability of personnel, and it is highly important for the safety of personnel. When responding to an emergency or incident, personnel accountability can usually involve two factors. The first factor of accountability is the number of personnel that are entering an area that is Immediately Dangerous to Life and Health (IDLH). Those commanding an incident need to make sure that they know exactly how many people are in a dangerous area, sometimes called a *Hot Zone*. Knowing the location of each of the first responders in a hot zone is extremely important. If conditions worsen or deteriorate, those in control need to ensure that all personnel are able to or already have evacuated the area. Essentially, the first part of accountability is to have a head count of who is in a danger zone (Tippett, 2013).

The second component of accountability is the physical location of all personnel operating on the incident scene, even those outside of the hot zone (Tippett, 2013). In the event of a catastrophic, or even potentially catastrophic, event (such as an explosion, an aftershock, a flash flood), it is critical that those controlling the incident to know the exact location of each, and every person involved in the response. By knowing where all the personnel are, it provides direction so that personnel can become a rescue priority if the unthinkable happens. If those controlling an incident know that personnel were on the second floor of a structure, they can go directly to that location, rather than wasting more time by searching the whole building.

An example of this is in order. Imagine if you will that rescue workers are working in a semi-collapsed building after an earthquake. As they progress through the building, they give regular reports of their location. While still in the building, a sudden aftershock causes the building to shift, thereby blocking any attempt to exit. The rescuers are pinned down to a specific area with no chance of self-extrication. Because control of the incident was exercised, and accountability efforts were in place, those rescuing the trapped personnel have a focus area of where in the building these individuals are trapped. They can send a Rapid Intervention Team (RIT) to their last-known location to start the search. Without control and accountability, the Rapid Intervention Team would have no idea where to start.

When we look at the Tokyo sarin attack response for control, we must, to a certain extent, make an assumption on the control of the incident. While this statement is speculative, based on lack of communication and the lack of cooperation, it is likely that accountability in the Tokyo incident was undertaken by each individual agency, if at all. The lack of a defined overall accountability is documented by Murakami (2000), and multiple accounts from victims describe that there were no specific zones set up to protect workers from areas that were identified as Immediately Dangerous to Life and Health (IDLH). This evidence also coincides with the accounts that there were numerous individuals who suffered secondary contamination. These eyewitness reports also confirm that there was no apparent accountability because anyone was allowed in the contaminated areas (Murakami, 2000). This lack of accountability put all individuals responding and in the general vicinity at risk.

In the Tokyo attack, there appeared to be no real control in place. Essentially, the right hand did not know what the left hand was doing. As has been described numerous times, each entity did what their specified job was, but there was no guidance for the overall response. Not only did each entity work autonomously, but in the case of the emergency medical response, there was no consideration of controlling where patients were transported to prevent overwhelming one or more hospitals (Murakami, 2000).

In looking at the Oklahoma City bombing, at the onset accountability was initially minimal at best, but that changed after the area was evacuated for a bomb scare. During this initial response, private citizens and some public safety officials were operating outside of the Incident Command System. It is important to note that in major and/or catastrophic incidents, it is not uncommon for untrained volunteers, off-duty first responders, and many others, to respond to the incident to see how they may be able to help. Unless the area has been secured prior to the incident, knowing who is present, where they are located, what they are doing, and the overall safety of the individuals is often not known during the initial stages.

After the area was evacuated for a bomb scare, control was strictly implemented (including accountability measures). As time went on, more accountability measures were implemented to ensure safety and an organized response. Some of the accountability measures included putting a perimeter fence in place and adding check-in points where an individual had to be identified before being allowed beyond the perimeter. Further accountability measures required a specific site identification card with pictures and credentialing, and a chain of command that had supervisors for each person working at the incident. Each supervisor could not have more than seven people working for them at a given time.

The incident was also controlled by priorities. Those controlling the response had an initial priority of life safety. They wanted to ensure that personnel did not become injured or killed while attempting to rescue individuals or recover bodies. This was effective because after gaining control of the incident, only relatively minor injuries and no deaths were reported. Prior to gaining control of the incident, a nurse was killed by falling debris (glass). There were reports that this nurse was climbing through debris and working outside of ICS when a piece of glass fell and killed her (Nordberg, 2010). It is very likely that had she been working within the ICS system, that she would have not been killed.

The next priority by those controlling the incident was to rescue living victims. There is a good possibility that the thought process of those in control was that they could not do anything for those that were already deceased. With this mentality, they would initially bypass the dead and put their focus on the living. While some may think this is cruel or disrespectful, if they did not bypass the dead, valuable time and resources could have been delayed when they could have been finding and responding to the living. As will be discussed in later chapters, as public safety workers, we must first focus on life safety.

As more resources arrived on scene at the Oklahoma City Bombing site, control was exercised to ensure the most effective use of these resources. Police officers who traditionally do not undertake rescues were effectively used to protect the victims' families from the media, and from others. As clergy and psychologists arrived on scene, they were sent to tend to the emotional needs of the victims' families, walking wounded, and in some cases, the severely wounded. As construction workers and equipment arrived on scene, they were controlled and given jobs based on the priority areas and their abilities. An extensive list of resources was managed and controlled throughout the entire response (The Final Report, 1996).

In comparing these two incidents, it is plain to see that having control of an incident is more conducive when an IMS method is utilized. While part of the problem with accountability in Tokyo was the lack of identification that the incidents were interconnected, it is also blatantly obvious that even after identifying this incident as a Weapons of Mass Destruction (WMD) attack, that accountability was lacking. No perimeters were set up, personnel were not screened, and first responders were not documented when they entered and/or left the area. An overarching control that set priorities or priority areas was not initiated, and accountability was lacking. Responders and hospital staff were contaminated and suffered injury, partially because of a lack of control at the incident. On the other hand, Oklahoma City did not micromanage the response, but they controlled nearly every aspect of it.

2.5.3 Cooperation

Cooperation is the process of working together to accomplish a mutual goal. Cooperation requires active participation and a unity between responding agencies and entities. It creates a synergy whereby a communal effort is put forward to provide the best possible response to an incident. Cooperation is working with outside agencies in an effort to make the agency more capable of completing a task. They essentially work on the same project rather than competing against each other.

As has been stated many times, cooperation in the Tokyo sarin attack was seriously lacking. To accomplish cooperation with other agencies and entities, there must first be an open line of communication and a willingness to work together. In Tokyo, the needed cooperation and communication was nonexistent. This led to and helped to create an extremely hectic and chaotic response. Looking at the big picture, the lack of cooperation and communication led to more confusion and uncertainty about what had happened (Murakami, 2000).

Because there was not any utilization of a full-encompassing IMS method, the response was a hodgepodge of (multiple) commanders from varying levels of government, each working toward their own goals. Even when command centers were in place, there was inefficiency, ineptitude, and ineffectiveness because of poor unit assignments that did not take advantage of each group's specialized training. Additionally, the multiple commands centers did not employ information sharing, cooperation, collaboration, and coordination, even after it was realized that it was one incident spread out over a large area.

These actions in the Tokyo incident were problematic, especially considering that a Chemical, Biological, Radiological, Nuclear, and/or Explosive (CBRNE) incident requires a mix of specialized responders. Often, specific specialized responders are needed to mitigate the agent or agents and prevent the spread of those agents to uncontaminated areas. Of course, to effectively undertake such an effort, police or some type of security force are needed to secure the area (and eventually investigate), firefighters are often needed to extricate victims and decontaminate them before letting them go or transferring them to a hospital, EMS (sometimes along with firefighters) are needed to administer prehospital care, and hospitals are needed to treat patients.

Murakami (2000) states that the death and injury toll could have been greatly reduced if the management of the incident did not have such a lethargic approach by the Tokyo government. Pangi (2002) first identified that there was no planning for such an event, and specialized training Chemical, Biological, Radiological, Nuclear, or Explosive (CBRNE) response was nonexistent at any level of government except for the military.

Information sharing was another issue. Because there was no coordination through information sharing between government entities, and relatively little information sharing between disciplines, the scope of the incident was not realized for over an hour after the attacks began. From the onset until the connection was made, each location that sarin was released was treated (by first responders) as separate incidents. According to Murakami's (2000), after completing interviews with a multitude of first responders and victims, it was found that each level of government worked independent of each other. In some instances, responders from the same level of government did not work in cooperation with their counterparts at the same level. Murakami's (2000) report stated, "There is no clear-cut chain of command… The effort of local units were extraordinary but the overall emergency network was useless" (p. 193). He went on to say that there needed to be more integration of resources.

In Oklahoma City, cooperation was a predominant factor in having such a successful response. From the onset of the incident, numerous agencies placed their mobile command centers next to the fire departments mobile command center. While each agency had tasks related to their discipline in public safety, they cooperated and collaborated with each other to ensure that nothing was missed, and they reduced duplication of efforts. Resources from nongovernmental organizations, including nonprofits, businesses, and organizations, local governments, state governments, and the federal government cooperated to ensure a holistic response and recovery. Whenever a gap was identified, many times an agency who was qualified to fill that gap would cooperate and offer to undertake the task (ODCEM, n.d.). This led to an all-encompassing approach that provided a superior process that sped the response and recovery process.

2.5.4 Collaboration

While collaboration is similar to cooperating, it is very different at the same time. Collaboration is the act of uniting with another agency (or agencies) in which they are not immediately connected with. Rather than being a helping hand to another agency, they become a stakeholder in successfully mitigating and ending the incident while working together to create the best possible outcome.

Collaboration, in an incident management sense, is more related to working together, sharing ideas, helping identify resources, and helping to plan the recovery and response to an incident. This is a synchronized and coordinated activity in which different agencies and disciplines continuously work side-by-side to develop and sustain the solution for the problem. It can be described as an orchestra with many different instruments coming together for the benefit of the end result.

In the Tokyo sarin attack, there was no collaboration. According to reports, those involved with this incident did not even take other agencies into consideration. Rather than working side-by-side, each agency worked independently of the others. As multiple people began to make their way to the subway platform and collapse or sit down waiting for help, only the EMS agency responded to those patients. Law enforcement did not help secure the perimeter, which led to more secondary contaminations. Approximately an hour after the incident started, a news crew took patients to a more distant hospital, they learned that the hospital had not been informed of the incident, and they did not even have a doctor available; they had to summon a doctor to treat these patients. Furthermore, the hospitals did not collaborate, so one hospital was severely overwhelmed while other hospitals saw few patients, or no patients at all (Murakami, 2000).

Essentially, all government levels were operating with little to no cooperation and collaboration between agencies. While it may sound as if this text is repeating itself, you cannot change the facts. Each individual agency did what they thought was best for their small portion of the overall incident, and they did not even consider what might be going on at other areas of the incident or with other agencies. Not only did the responding agencies work independently of each other, but they essentially worked in competition against other agencies. In one report by Pangi (2002), it was found that information was not shared for quite a long time; in one instance related to this attack, information was intentionally withheld for several months. Documentation found by Pangi (2002) revealed that the Self Defense Force (SDF) was still unwilling to share information about the attack with local police for several months, which played a role in slowing the investigation and prosecution of the perpetrators (Pangi, 2002).

Due to the lack of coordination in the Tokyo subway incident, one subway train was allowed to run the entire scheduled route to completion, and then begin the route again with the sarin puddle still on the train. Some commuters affected by the gas laid on the floor for over an hour before a passerby jumped into action (Murakami, 2000). While some of the issues can be blamed on the Japanese culture, it is ultimately first responder's that need to work together, even if it is different than what is normal in their culture.

In Oklahoma City, a collaborative effort was begun almost immediately. At the bombing site, first responders from multiple disciplines worked shoulder to shoulder meeting the needs of the incident. Local resources quickly began to create a unified front of how to best overcome the difficulties they were facing. From the onset, there was no competition between local agencies. They worked together to create the best possible outcome, even though the end of the incident would not be seen for over a month. As was mentioned previously, local and state agencies placed their mobile command centers next to the fire departments mobile command center. These agencies worked together to ensure that nothing was missed and that there was no duplication of efforts. They planned the response together, identifying resources and capabilities, and they worked together and discussed how to best cure each challenge they faced. These local resources were going beyond cooperating by providing resources, and they were collaborating by helping to identify the solutions (Final Report, 1996).

Local businesses and nonprofits collaborated as well. Heavy equipment companies collaborated among themselves and then collaborated with the Incident Commander (IC) and those planning the response. By doing this, the heavy equipment operators helped to ensure safe and effective operations. Local nonprofits went beyond cooperating and worked together with those utilizing the ICS method. Nongovernmental organizations not only helped to identify resources but also to offer solutions to completing the task at hand. These nongovernmental resources worked in concert with all other first responders.

If we look at the Family Care Center, there were multiple agencies that were working side by side rather than independently or in a way that there were clear-cut duties (Final Report, 1996). Both ends of the spectrum worked to provide care for the surviving family members. The nonprofits, pastors, law enforcement, the funeral directors, the Red Cross, and many more worked tirelessly to meet the needs of the effected families. The mission was more important than the individual organizations involved.

We also need to realize that the Emergency Operations Center (EOC) at the state level was activated in the first few minutes after the bombing. Equally impressive was that the Emergency Operations Center (EOC) was fully staffed within 25 minutes. It was staffed by those with the authority to make decisions from a multitude of disciplines. The Emergency Operations Center (EOC) incorporated and worked with the local responders (the boots on the ground), and they incorporated and worked with all levels of government. The Emergency Operations Center (EOC) staff members, all from differing disciplines and agencies, collaborated and worked together to ensure that every need was met at the bombing site and surrounding area (ODCEM, n.d.). The National Guard integrated their resources and collaborated with those in charge as well. Guard members accomplished work by providing security, working in the morgue, digging through evidence with the FBI, and undertaking many other jobs where their assistance was needed. They too collaborated with those leading the effort (Smith, 2010).

The federal response and integration of resources was also a collaborative effort. The federal government did not try to take over the incident, but rather they integrated all of their resources into the response. The Task Force Teams that were deployed to the Oklahoma City bombing site worked in unity with each other and with other agencies. K-9 teams were integrated with other rescuers, and they worked in a collaborative manner not only with these teams but also with those in charge of the incident. Military resources collaborated and worked side by side with other responding agencies and within the command structure. These resources integrated in a way that went beyond meeting the needs of the response, and they became part of the solution for the entire response.

2.5.5 Communications

Through looking at past disasters, we know that communications have always been an issue in any major or catastrophic incident. Unfortunately, communications will probably always be a weak link when managing an incident. While communication will likely be a problem for many years to come, those using IMS methods, such as ICS, should always strive to improve communications.

In a large or catastrophic incident, communications overload should be expected in most instances. In the planning stages for major incidents, state and local agencies have, and should continue to identify alternative methods for communicating. These communication mitigation measures would be best if they included methods to communicate locally and outside of the local area. It is also beneficial if communications can be established with nongovernmental agencies that may have limited knowledge about disaster communications.

In the Tokyo attack, communications between the responding agencies was essentially nonexistent. Interagency communications were fragmented at best, and in most cases, these communications were not even pursued. The various government agencies involved with the response did not cooperate or communicate with each other. According to reports from Pangi (2002), this ultimately continued throughout the duration of the incident. Each agency essentially operated within their own little bubble by failing to communicate effectively with other agencies.

While there was a breakdown in the communications equipment in the Tokyo incident, there is no mention in any of the resources about trying to mitigate this problem. One way that this could have been mitigated would have been sending someone with a radio to those hospitals to facilitate communications, or in sending runners to the hospital. In looking at the communications between EMS units and the hospitals, it was plain to see that the lack of communications was a contributing factor overwhelming the closest hospital, while most of the other hospitals had relatively few patients (Murakami, 2000).

Perhaps most disturbing part about communication in the Tokyo incident is that the hospitals that were treating patients did not learn from officials at the scene that this was a chemical attack. For three hours, St. Luke's Hospital sent researchers digging through the library trying to find answers. Additionally, doctors that did realize what was happening used the only communications method they could think of to contact hospitals, a fax machine. The confirmation that local hospitals needed did not come from a government agency, it came from the local news (Murakami, 2000).

In Oklahoma City, it was expected that there would be issues with communications during the preparedness phase while when planning for larger incidents. Those issues were in the process of being addressed, but a full resolution of identified deficiencies had not been fully implemented. Even though some mitigation measures had been implemented, this bombing was of such magnitude that it went beyond what was expected. It is likely that even if all proposed implementation measures were enacted, communications would still have been challenging (Manzi et al. 2002).

The After-Action Report ([AAR] n.d.) from the Oklahoma City bombing found multiple communications problems that occurred from the onset. The first communications failure in

Oklahoma City that was identified were with the phones. All phones, both land lines and cell phones, had issues that were described as intermittent to inoperable. For the most part, cell phones were inoperable because of the vast number of individuals attempting to make phone calls immediately following the bombing. There was also some infrastructure damage that occurred to cell phones due to the explosion, but it was not severe. Land lines were inoperable because of damage to the infrastructure caused by the bombing (ODCEM, n.d.).

When first responders arrived in the area, the only communications they had were their radios. As was expected, those with critical information, or urgently needing assistance, were overwhelming the only available radio frequencies at that time. With the amount of people being trapped or injured, these channels were immediately overwhelmed with attempted communications, and it was nearly impossible to effectively communicate (ODCEM, n.d.).

In an interview with Fire Engineering (1995), the Oklahoma City Fire Chief described how radio traffic was so heavy that he had difficulties in trying to find a break in radio traffic to give the Incident Commander (IC) situation updates. He had urgent information on the extent of the damage and where structural failures may collapse on first responders. The Chief described that even when radio silence was needed so that rescuers could listen for potential victims, it was extremely difficult to achieve (Fire Engineering, 1995).

Compounding the communications problem after the bombing, most communications between ambulances and the hospitals suffered a failure in communications, much like what occurred in Tokyo. The difference was that the Oklahoma City bombing communications issues were mainly rooted in hospital error. Because the emergency frequency had not been used in quite some time, many of the hospitals had turned the volume down or turned their radio off. In order to mitigate this problem, police officers were sent to each hospital to tell them to turn on or turn up their radios. If needed, it was planned that police or other personnel who had radios would relay information to the hospital to mitigate communication issues (Nordberg, 2010).

There was also the communications issue of each discipline having their own individual frequency. This caused interoperability problems when trying to communicate between discipline, unless there was face-to-face communications. The state had issues communicating with the locals, and the locals had interoperability issues communicating with federal response agencies. Communications between certain disciplines was nearly impossible; however, the state was able to communicate with federal resources via radio. Much of this was overcome by having face-to-face communications with mobile command centers that were brought in, who would then relay the information. This proved that mobile command centers that are placed within walking distance of each other helps to at least partially mitigate communications issues.

2.6 Conclusion

In looking at the response to the Aum Shinrikyo attacks on the Tokyo subway system and comparing it to the Oklahoma City Bombing, it is obvious that there needed to be a clear command and control in Tokyo to effectuate a positive response. It should also be apparent that integration of resources, interoperability, and open communications would have improved the response. The integration of resources and the use of an IMS method should have been a long-term goal long before the Tokyo incident ever happened.

Looking at the mistakes made in this subway attack, it is apparent that this was a critical incident that demanded a specialized response. Dissecting the response, it is noticeable that Chemical, Biological, Radiological, Nuclear, and Explosive (CBRNE) response standards were not met. Contaminated areas were not secured, and patients were allowed to leave the contaminated area without decontamination. Contaminated trains were allowed to keep moving while letting contaminated passengers board and leave the train, and there was evidently no command and control of any of the scenes. Add the lack of cooperation, coordination, and communication, and this response was a roadmap for disaster long before the incident ever occurred.

Instead of turning chaos into calmness, this response added to the chaos. Confusion ensued throughout not only the response but also through the entire investigation and even years beyond that day. Rather than removing the uncertainty, the Tokyo response added to the uncertainty, thereby substantially increasing it for many hours.

Through the use of an IMS method, these issues could have been mitigated, but only if the chosen IMS principles were practiced and used on a regular basis. In order to be effective, IMS methods need to become a response lifestyle, rather than using them for only major incidents. In a review of the incident, Pangi (2002) states that some method of IMS should be enacted in Tokyo, and throughout Japan. Had IMS principles and an IMS method been used, the outcome of the Tokyo sarin attack by Aum Shinrikyo would have likely been very different. It is likely that the death and injury toll likely would have been substantially less. Additionally, the incident likely would have been recognized as a terrorist attack utilizing a Weapon of Mass Destruction (WMD) substantially sooner. The earlier identification of this likely would have led to shutting down the train and securing the larger scene much faster. Secondary contaminations likely would have been reduced as well.

Utilizing an IMS method in the response mode would have been more effective because resources likely would have been identified, and specific tasks would have been assigned, based on the resource's abilities and equipment. There would not have been any duplication of resources, and personnel would have been more succinct operationally. If a Unified Command (UC) were set up, divisions could have been based on geographical areas. This would have allowed autonomy at a given site for the situation they faced, yet provided an integration of resources at a higher level.

Additionally, if an IMS method was integrated, public health authorities could have been brought into the mix, allowing them to know the number of patients and which hospitals were capable (and had room) for the type of injuries the patients suffered. Of course, this would have also provided clearly defined lines of who would be responsible for each aspect, and where strike teams and task forces were needed to improve the response. It also would have expedited the incident from a response into a recovery mode. This is based on the assumption that with a fully collaborative effort, ending the response mode would have occurred more quickly. Let us not forget that had an IMS method been used, the integration of the subway authority would have likely reduced the death and injury toll because that authority would have been informed sooner about the type of incident, thereby reducing the spread of the sarin contamination.

While there is no way to know for sure, it is believed that the Tokyo sarin attack and the Oklahoma City Bombing may have caused some countries around the world to rethink how they respond to major disasters. In other cases, major incidents that occurred in their country caused the responding agencies in other countries to create a system to better manage future incidents.

Chapter 2 Quiz

1. Incident management is
   (a) A way of understanding how to train first responders.
   (b) The skill of making decisions that will mitigate the negative effects of an emergency incident.
   (c) The skill of motivating people to follow your orders.
   (d) A method of making sure that all the paperwork is completed.

2. **True or False**: The utilization of a proven IMS method usually causes disasters to last longer.

3. The use of an Incident Management System (IMS) method helps to reduce the _____ of an incident.
   (a) Cooperation
   (b) Lifecycle
   (c) Collaboration
   (d) Communication

4. **True or False**: If an incident is managed, and resources are used in an efficient manner, it will usually reduce the negative effects of that incident.

5. What are the three common attributes of any emergency incident?

6. **True or False**: The Tokyo sarin incident was well managed, even without using an IMS method.

7. In the Tokyo incident, how did hospitals confirm that the method of attack was sarin gas?

8. **Yes or No**: Did the way the Tokyo incident was managed increase the number of patients that needed to be seen at the hospital?

9. What proven IMS method was used to manage the response in the Oklahoma City incident?

10. What mitigation measure was taken in the Oklahoma City incident when communication with some hospitals could not be made?

11. **True or False**: Placing mobile command centers in close proximity to each other at the Oklahoma City incident helped to facilitate communications, especially agencies that were using different frequencies.

12. A Family Assistance Center was set up for immediate family members at the Oklahoma City incident. Name three organizations that were included in providing comfort.

13. **True or False**: The military and the National Guard was under a different IMS method than the rest of the agencies on scene at the Oklahoma City Bombing site.

14 What agency was in charge of the criminal investigation of the Oklahoma City incident?
   (a) The Oklahoma State Police
   (b) The Oklahoma City Police Department
   (c) The FBI
   (d) The ATF

15 **True or False:** Using an IMS method, the Oklahoma City incident had fewer issues than the Tokyo incident which did not use an IMS method.

Self-Study

Final Report (1996). *Final Report: Alfred P. Murrah Federal Building Bombing April 19, 1995*. Stillwater, OK: International Fire Service Training Association.

Fire Engineering (1995). Report from Fire Chief. *Fire Engineering* 148: 10. Retrieved from https://www.fireengineering.com/articles/print/volume-148/issue-10/features/report-from-fire-chief.html#.

Manzi et al. (2002). Critical Information Flows in the Alfred P. Murrah Building Bombing: A Case Study. Chemical and Biological Arms Control Institute. Retrieved from https://www.ncjrs.gov/pdffiles1/Digitization/194411NCJRS.pdf.

Murakami (2000). *Underground: The Tokyo Gas Attack and the Japanese Psyche*. New York, NY.

Pangi (2002). Consequence management in the 1995 sarin attacks on the Japanese subway system. *Studies in Conflict & Terrorism* 25: 421–448.

Oklahoma City National Memorial and Museum (n.d.). Community response: Crisis management. Retrieved from https://oklahomacitynationalmemorial.org/wp-content/uploads/2015/03/okcnm-community-response-crisis-management.pdf.

Oklahoma Department of Civil Emergency Management (n.d.). After action report: Alfred P. Murrah Federal Building bombing 19 April 1995 in Oklahoma City, Oklahoma. Retrieved from https://www.ok.gov/OEM/documents/Bombing%20After%20Action%20Report.pdf.

Shariat et al. (1998). Summary of Reportable Injuries in Oklahoma: Oklahoma City Bombing Injuries. Oklahoma State Department of Health. Retrieved from https://www.ok.gov/health2/documents/OKC_Bombing.pdf.

Smith (2010). Guardmembers remember Oklahoma City bombing. *U.S. Army News* (25 October). Retrieved from https://www.army.mil/article/37587/guardmembers_remember_oklahoma.

3

Incident Management in Other Countries

The world as we have created it is a process of our thinking. It cannot be changed without changing our thinking.

Albert Einstein

All too often, we take for granted the incident management system (IMS) methods that began in the United States. Many believe that these methods that were developed in the United States are used around the world, or they would believe that they are not used anywhere except a few select countries. Either train of thought would be a major misconception. As we will see in this chapter, most countries have adapted an IMS method to manage incidents.

As a disclaimer, the information in this chapter was taken from various research rather than first-hand knowledge. Because the majority of the information came from the Internet, it is possible that some information in this chapter may be incomplete or incorrect. We apologize in advance for any misinformation.

While the current IMS system in the United States is the National Incident Management System (NIMS), most countries utilize a totally different IMS method. Some countries utilize the Incident Command System (ICS) component of NIMS, while others use similar bits and pieces of ICS, NIMS, or some other IMS method. Some countries have no standard IMS method across their nation, but rather leave it up to the provinces or the states what, if anything, should be used. In some instances, these foreign countries do not use an IMS method, or they only recently started using an IMS method. A basic review of these countries and the methods they use can help us understand the importance of the basic principles and concepts as they relate to managing an incident.

3.1 The United Nations

In 1999, the United Nations (UN) began the execution of the International Strategy for Disaster Reduction (ISDR or UNISDR). Member countries of United Nations were seeing the importance of disaster reduction and a coordinated response to disaster. For two years, the organization worked on a resolution to create disaster reduction while supporting coordination, collaboration, and cooperation among member countries that might respond to disasters. While prior versions were related to disaster reduction in the United Nations, most or all of the previous versions were recalled, and the new version was accepted in January 2002, four months after the September 11 attacks on the United States (UN General Assembly, 2002).

Much like what happened in the United States, the September 11 attacks created urgency for the United Nations General Assembly. There is no documentation to support this supposition, but there are many indicators that support this theory. While the International Strategy for Disaster Reduction (ISDR) was being discussed and planned for several years, it was not until four months after the September 11 attacks that the International Strategy for Disaster Reduction (ISDR) was passed. Because the United Nations is based in New York City, it would be fair to say that members of the organization saw the destruction, they saw how devastating it was to the community, and they could have even seen how the IMS method of ICS helped to organize the massive response at the World Trade Center. Perhaps they realized that arguing and nitpicking the smallest of items in the resolution was not productive and that the time for action was at hand. We were not in these meetings, so we really do not know. One thing that anyone in public safety can tell you is that after a major disaster, we typically see more laws, more spending, and more agreements being made in the first year than we have seen in the previous five years. Disasters typically force action from government entities.

Under resolution 56/195, the UN General Assembly mandated that the United Nations would be the focal point in this new system for the coordination of disaster reduction. Together, member countries would create a combined effort in disaster reduction activities. An organizational unit of the UN Secretariat was formed and was led by the UN Special Representative of the Secretary-General for Disaster Risk Reduction (SRSG).

Initially, they used the Hyogo Framework for disaster reduction. According to the United Nations (2005) this was officially called *The Hyogo Framework for Action 2005–2015: Building the Resilience of Nations and Communities to Disasters (HFA)*. It was essentially the first plan devised by the United Nations. It broke down the major areas of disaster reduction, and it explained (and detailed) the work that was required to reduce disaster losses. Many of those UN members and partners saw destruction after a major incident and realized that reducing disaster risks was critical for some countries to survive, and this was the initial way of doing so. The lengthy list of exactly who could benefit from disaster reduction included governments, international agencies, disaster experts, and many others. The Hyogo Framework helped to bring all the individuals and organizations into a common arrangement that would create better coordination when responding to a disaster. This framework also included language that would promote disaster resilience, especially in economically depressed countries. The Hyogo Framework outlined five priorities for action with a goal of significantly reducing disaster losses by 2015 through building resilience.

1. Ensure that disaster risk reduction is a national and a local priority with a strong institutional basis for implementation.
2. Identify, assess, and monitor disaster risks and enhance early warning.
3. Use knowledge, innovation, and education to build a culture of safety and resilience at all levels.
4. Reduce the underlying risk factors.
5. Strengthen disaster preparedness for effective response at all levels (UNISDR, 2005).

As part of strengthening disaster preparedness, it was suggested that the Incident Command System (ICS) should be used to ensure that all countries responding to a disaster were on the same page. This would also allow a higher level of integration of resources from other countries when a multinational response was needed.

In preparing for the ending of the Hyogo Framework, the UNISDR began working on a successor that would continue the strides that had been made in disaster reduction. The *Sendai Framework for Disaster Risk Reduction 2015–2030* is the successor instrument to the Hyogo Framework. It was adopted on 18 March 2015, at the World Conference on Disaster Risk

Reduction held in Sendai, Japan. The Sendai Framework also recommended that countries utilize, or continue to utilize, the ICS method to manage disasters.

While exact details were not found on when the United Nations began recommending the use of the ICS method, there were suggestions that 10 countries signed an agreement in 2003. Members of the Association of Southeast Asian Nations (ASEAN) adopted the use of ICS in that year, and it included the countries of Vietnam, Laos, Burma, Cambodia, Thailand, Singapore, Philippines, Malaysia, Indonesia, and Brunei. Additionally, the 133 members of the United Nations signed an agreement on the Hyogo Framework, and interestingly enough, numerous ICS classes were offered that same year (Cabag Jr., 2012).

In one advertisement from the Asian Disaster Preparedness Center ([ADPC], 2009), was revealed that a collaborative effort was being put forward to train countries about the ICS system. This collaboration included the US Department of Agriculture, Forest Service ([USFS], n.d.). According to the information, ICS was adapted to many countries including Bangladesh, Indonesia, Iran, Maldives, Myanmar, and Sri Lanka. This adaption was based on the structure of the countries national government while being cognizant of cultural considerations. This specific training offered *Train-the-Trainer ICS course curriculum* to key national agencies and trainers (ADPC, 2009). Train-the-Trainer courses provide the individual or agency with the knowledge and resources to become instructors and to teach the ICS system to others. According to the US Forest Service (n.d.), beyond the ASEAN, the agency had provided ICS trainings to Canada, Australia, Mexico, Bulgaria, Mongolia, and Sri Lanka by some time in 2007 (USFS, n.d.). Additionally, after the approval of the Sendai Framework, there were also a multitude of classes that were offered in ICS in 2015 and beyond.

3.2 Australia

In the 1980s, the Australian government created the Australian Inter-Service Incident Management System (AIIMS). This was the first iteration of an IMS system in Australia. It closely mimicked the National Interagency Incident Management System (NIIMS) system in use by the US Forest Service and that was based on the IMS method of ICS.

In 1989, the Australian government enacted legislation called the State Emergency and Rescue Management Act. The legislation mandated that an IMS system should be developed for all disciplines within public safety. The first implementation of Australian Inter-Service Incident Management System (AIIMS) method was sometime in 1991, and it implemented Incident Control System as the way to manage an incident. The Incident Control System was nearly identical to the USICS method, with one primary difference. The person in charge of an incident was not known as the Incident Commander (IC), but rather an Incident Controller (Turner, n.d.). In 2002, after hearing years of criticism that the system was fundamentally for the fire service, the Australian and New Zealand National Council for fire, emergency services, and land management undertook a revamping of the system to make it nonspecific to discipline type. While some changes were made, the system closely resembles ICS method used in the United States.

Sometime after the September 11 attacks, the US Fire Administration, in conjunction with United States Agency for International Development (USAID) began training Australians. This was done on-campus at the National Fire Academy to assist them in getting advanced training on the US version of ICS. Specific individuals were chosen to attend the National Fire Academy in Emmitsburg, Maryland. With the ICS training from the National Fire Academy, Australian first responders began to adapt the Australian Inter-Service Incident Management System (AIIMS) method to integrate with the ICS method. This ensured that whenever the

Australian personnel worked with other agencies from around the world, they would be able to work nearly seamlessly with each other.

The latest iteration of the Australian Inter-Service Incident Management System (AIIMS) method utilizing the Incident Control System is also compatible with the ICS method used in the United States. Much like the United States utilizes NIMS for integration of resources, and they use the ICS method to manage the incident, the Australians utilize AIIMS for the integration of resources and the Incident Control System to manage the incident. In recent years, the United States and Australia have seamlessly exchanged Incident Management Teams (IMTs) during eventful fire seasons. This integration from each other's country went smoothly, with relatively few problems.

Beyond the use of Incident Control System, another IMS method commonly used in the United States for healthcare facilities is the Hospital Incident Command System (HICS) to manage emergencies and disasters in hospitals. Similarly, the Australian healthcare system is closely mirrored to HICS, but it is called Hospital Incident Management System ([HIMS], n.d.). This IMS method allows hospitals to keep their basic hierarchy while still being able to integrate and be compliant with the larger IMS method for requesting external resources. In the case of Australia, the Hospital Incident Management System (HIMS) method integrates with the AIIMS method.

3.3 Bermuda

Bermuda mirrors the Incident Command System used by the United States. In fact, Bermuda's ICS system mirrors the US program so much, that the same number classification for levels of training is used. According to Bermuda's Authority in Workplace Safety and Health (n.d.), individuals can be trained on ICS-100, ICS-200, and ICS-300, although there was no mention of ICS-400 (Worksafe BDS, n.d.). These classes have very similar descriptions of the classes offered by the Emergency Management Institute (EMI) in the United States, and they are numbered much in the same way. While there is no concrete information on where the Bermuda Islands received their training, it could be speculated that they may have received their training from the US Forest Service. It is possible that this might be the reason the ICS method in Bermuda closely mimics the ICS method in the United States.

Gathering information from press releases, it appears as if the Incident Command System has been used in Bermuda since at least 2012. In 2013, off the Island of Bermuda, Exxon Oil, the Bermuda Department of Environmental Protection, and various other businesses (and agencies) participated in a full-scale exercise together. The exercise was a four-day exercise, and the command and control system used was the ICS method (Bermuda Department of Communication, 2013).

No information could be found as to whether Bermuda utilizes the HICS method to manage incidents within a hospital. If the HICS system is currently not used in Bermuda, there is a likelihood that it may be used in the future.

3.4 Burma/Myanmar

As was previously mentioned, in 2003, the Association of Southeast Asian Nations (ASEAN) signed an agreement with each other that they would implement the Incident Command System in each of their countries. This was done so that when one country needed the assistance of another, the method to manage the incident would be seamless in guiding resources from

the other ASEAN countries. In looking at the ICS method used in Burma/Myanmar, it looks identical to the IMS method created in the United States. This is in part because the United States Agency for International Development and the United States Forestry Service provided ICS training to Bermuda/Myanmar. By 30 June 2010, all training in Bermuda/Myanmar had been completed (UN Economic and Social Commission for Asia and the Pacific, 2010).

It appears as if Burma/Myanmar does not use the Hospital Emergency Incident Command System (HEICS) or Hospital Incident Command System (HICS). While there is no way to confirm it, no information was found that discussed the HEICS method or the HICS, or anything similar being used in this country after undertaking extensive search. For the sake of the citizens, we can only hope that this changes in the near future.

3.5 Bangladesh

The country of Bangladesh has been extremely proactive in disaster management and incident management, much to the surprise of many. They have implemented an IMS method that closely resembles NIMS, but it uses a different name and acronym. The Mutli-Agency Disaster Incident Management System (MADIMS) method is used widely in Bangladesh, and similar to its counterpart (NIMS), the ICS method is a key component of the system for managing incident operations.

To better understand just how comprehensive the use of the MADIMS method is, you only need to view the governments Standing Orders on Disaster ([SOD], 2010). Due to numerous disasters, Bangladesh is, and has been, active in disaster preparedness, so the use of the MADIMS method would seem to be a natural progression. The original framework was implemented in the country in 1997, and then updated in 2010. While the initial framework for disaster management (which included and their own IMS method) was implemented early on, the signing of the UN Hyogo Framework for Actions (HFA 2005–2015) inspired the government to go a step further. Based on Hyogo agreement, Bangladesh felt it was critical to update the Standing Orders on Disaster (SOD) and make sure that it was implemented in all sectors of the government, for all types of incidents (Government of the People's Republic of Bangladesh, 2010).

This update to the MADIMS method was driven by the admission of the Government of Bangladesh that their nation is vulnerable to disaster. By the government's own admission, "Bangladesh is one of the most disaster-prone countries in the world" (Government of the People's Republic of Bangladesh, 2010, p. iii). From a historical perspective, Bangladesh has suffered cyclones, landslides, tornados, floods, and other similar disasters that have affected the economy and caused loss of life. According to the SOD (2010), the country is extremely vulnerable to climate change, and according to the SOD climate change is a serious threat for food security and the sources of revenue for their citizens.

For the most part, the ICS method used by Bangladesh mirrors the method used by the United States and many other countries. The only slight difference that was found was the reporting method: that to say who reports to whom. What is unique about the Bangladeshi plan is that it is used by every level of government, including the military. Much like the ICS method used by most countries, all incidents are local and a request for outside assistance can be requested when local resources are overwhelmed (Government of the People's Republic of Bangladesh, 2010).

In taking a closer look at the Bangladesh IMS method of MADIMS, it could be said that it is probably one of the more developed IMS methods used in the world. It clearly compares to the NIMS system used in the United States. Based on how comprehensive the MADIMS method appears to be, it could be said that a lot of time, thought, and understanding went into properly developing this method of incident management.

In looking at the IMS method of Hospital Incident Command System (HICS), Bangladesh is a country that has received training and appears to be a leader in this area. As far back as 2007, the United States Agency for International Development began providing training in a program known as Hospital Preparedness for Emergencies (HOPE). The program was a four-day class that helped teach developing countries to understand and implement plans that assisted greater capacity to build resilience during disasters and emergencies. As a part of this training, those who attended were taught the Hospital Emergency Incident Command System (HEICS) so that they could better manage their hospital during confusing and chaotic incidents (Herbosa, 2007).

Keeping up with the changes to the IMS methods that were used, Bangladesh also received training on the new Hospital Incident Command System (HICS). In June 2014, a total of 29 healthcare workers from seven major healthcare facilities, attended a five-day training (ADPC, 2014a). From research conducted on the use of Hospital Incident Command System (HICS) method, Bangladesh continues to not only train on this IMS method but also to widely implement it as part of the standard in healthcare facilities across this nation.

3.6 Brunei

Like many countries, Brunei utilizes ICS method as a management framework to integrate personnel, equipment, procedures, facilities, and communications when responding to incidents. Prior to the implementation of ICS, the country saw many issues with multiagency responses that were quite similar to the problems faced in the United States prior to FIRESCOPE.

According to a research project based at the University of Brunei (2015), it was not uncommon for incidents in Brunei to have multiple Incident Commanders (IC's), who would at times argue or cause confusion. Sometimes this confusion was based on differing operating procedures. Because different disciplines had different organizational structures, this too added to the chaos and confusion. Additionally, information sharing was limited, so there was often a lack of thorough situational awareness. This too often led to more chaos and confusion, and it led to putting operational personnel at a higher risk (Soe et al. 2015).

Brunei also suffered from inadequate forms of communication hardware and incompatible communications frequencies. Much like the wildfires in California, various agencies would arrive on scene and would be unable to communicate with each other. They also each had their own vernacular, which did not coincide with each other. Even if these agencies did communicate with each other, there were issues with effectively communicating. They often lacked the ability or willingness to coordinated planning with each other. Essentially, every agency was doing what they felt was best, while ignoring the other agency(s). This was primarily because there were unclear lines of authority that identified who should be in charge of an incident (Soe et al. 2015).

After signing the ASEAN Agreement, the Government of Brunei decided to address the weaknesses they found in incident management. The Brunei National Disaster Management Center began working collaboratively with the US Forestry Department to create an IMS method that met the needs of the country in 2009. As with any IMS method, everyone must be on the same page for it to be effective, so the initial starting point of creating an IMS method for the Nation of Brunei was to organize resources. This led to the framework of

National Standard Operation Procedures (NaSOP). This further led to important factors in ICS including creating collaboration, reporting mechanisms, and common language as well as the terminology to be used in disaster management (Soe et al. 2015).

In 2011, Brunei National Disaster Management Center began conducting seminars, workshop, tabletop exercises, and simulation exercises in all four of Brunei's districts. The concepts were simple, by teaching and promoting the use of ICS, first responders will begin to grasp and use this method of incident management. This was a starting point, and according to a published report by the University of Brunei (2015), much progress has been made. The report also points out that there is more work to be done. This research paper points out that the initiative was not completely implemented to cover all incidents, and it does not involve nongovernmental organizations. The report puts part of the blame on the responding agencies by stating, "Silo mentality of the different government agencies hinders cooperation. In times of disasters, all agencies involved have to work together and should be able to take orders from the incident commander. Those in command of respective agencies prefer to use old, more familiar methods, rather than ICS" (Soe et al. 2015, p. 31). This is a common problem that occurs when changes happen.

Brunei also took steps early in the process to integrate healthcare facilities into disaster response. As early as 2005, the Brunei Ministry of Health had implemented a disaster management plan for the entire country, and all healthcare facilities were required to follow that guidance. In the disaster plan, learning and utilizing the Hospital Incident Command System (HICS) method was required.

The Brunei IMS method has one major difference from most systems currently in use. Whenever a disaster happens, whether in a hospital, a specific area, or even the whole country is affected, the National Disaster Management Center is the coordinating body for all disaster response. How this differs from most countries that have coordinating centers is that the Brunei Disaster Management Center guarantees that all aspects of managing a disaster are multiple ways. They include adherence to written policies, strategies, and the various practices implemented (and guided) by international, regional, and national drivers ("ASEAN Disaster Workshop in Thailand," n.d.).

As time went on, Brunei became more astute in Hospital Incident Command System (HICS), and they began to not only use it, but to intentionally test its limits to ensure that it was a reliable and resilient method of IMS. Soon, Brunei became a leader in hospital disaster preparedness in their region, and they began teaching hospital disaster management throughout the region. While this was typically done through Doctors without Borders (2018), the focus was on how Brunei prepares healthcare facilities for disaster management. The Hospital Incident Command System (HICS) is an important part of the teaching that they do throughout the ASEAN region.

3.7 Cambodia

In 2014, a law that outlined the larger picture of disaster management was submitted to the Cambodian government. In that law, there were provisions for the use of the Incident Command System. Relatively little could be found in research about the implementation of ICS before 2014, but we can assume that it was ongoing, based on the ASEAN agreement and the Hyogo Framework (An, 2014). Research into the use of Hospital Incident Command System (HICS) also did not yield any results.

3.8 Canada

The use of an IMS method in Canada revealed that the Incident Command System (ICS) is the primary method utilized. Some have the opinion that the system is used across the nation as a standardized approach to incident management (ICS Canada, n.d.), while others state that the use of ICS is left to each individual province (personal communications). Research revealed that Canadian's have made some slight adjustments to ICS to meet their needs, and the overall system appears to have widespread use.

An organization called ICS Canada (n.d.) claims that the first iteration of ICS in Canada happened in the Province of British Columbia during the mid-1990s. In 2002, the Canadian Interagency Forest Fire Centre (CIFFC), mandated that all provincial, territorial, and federal agencies should learn and utilize the CIFFC's version of ICS for wildland firefighting. This version of ICS included a standard doctrine and provided training materials for the wildland fire community across Canada. The mandate was directed at wildland firefighting operations, but the mandate did not include nonwildland firefighting organizations. Even without the mandate, many agencies who were not involved in wildland firefighting operations began to utilize this IMS method for daily response, and soon, the use of the Canadian ICS system increased (ICS Canada, n.d.).

When the CIFFC was reviewing potential updates to the wildland firefighting courses, it was realized that the Alberta Emergency Management Agency (AEMA) was looking to provide a single command and control system that incorporated an all-hazards approach that would meet their long-term provincial emergency management needs. These two separate entities with very different response areas cooperated and collaborated with each other to create the groundwork for the use of ICS in Canada (ICS, Canada). While it appears that not every province has decided to utilize the ICS method, the organization known as ICS Canada was actively recruiting provinces and organization to adopt this standard for incident management as of the writing of this book.

This brings us to the use of Hospital Incident Command System (HICS) in the country of Canada. Research revealed that the clear majority of hospitals in Canada have adopted the Hospital Incident Command System (HICS) method as a standard. Since 2006, the Ontario Hospital Association (2006) created a hospital emergency management toolkit, which included the call for implementation of the Hospital Incident Command System (HICS). From information found in research, it appears that this has become a voluntary standard across Canada, although it does appear as if there was some government urging.

3.9 China

Prior to May 2018, China had no nationwide emergency management agency or a nationwide IMS method in place. Until the formation of the new Chinese emergency management agency in 2018, the process of managing emergencies, planning for emergencies, and managing response and recovery efforts were handled by a hodgepodge of national, regional, and local governmental agencies. These agencies usually responded with no specific or defined way of handling incidents, and they usually did not integrate well when they were forced to work together. In a 2018 press release, it was stated that the new formation of the emergency management agency would improve the speed and efficiency of the Chinese response to disasters (Liqiang, 2018). No information was found on the use of a hospital IMS method in China, so it is assumed that there is none in place.

3.10 Germany

In Germany, their IMS method is known as Dienstvorschrift 100 (DV 100), Leadership and Command in Emergency Operations. The regulation of DV 100 was approved by the Board of Firefighting Affairs in Germany around the year of 2007. It is a regulation that was designed in a way that it was in harmony with the German Federal Constitution and allowed the states to be autonomous. DV 100 is a regulation that explains the command and control system for Germany, and it places the responsibility for emergency response to the 16 states within the boundaries of Germany. DV 100 was implemented in 1999 among the fire service, then later implemented by other national organizations. Since 1999, it appears to have become the single system utilized in Germany for the purposes of incident management (Dienstvorschrift, 2007).

Much like the ICS method developed in the United States, the roots of this method of IMS came from the fire service: it was later transformed to an all-hazards IMS method, for use in all types of incident. Also similar to the NIMS and its companion ICS, the DV 100 system is flexible and allows operational leaders to make basic decisions based on the circumstances evolving around them. The DV 100 system relies on motivated staff as a key to success, and much like NIMS, it does not micromanage operations. It allows operational flexibility to tactical crews so that they can achieve the mission's goals without being micromanaged. This IMS method is different because it identifies the disposition that the Incident Commander (IC) should possess, and the leadership styles that should be used. In most instances, it calls for the Cooperative Leadership, with an Authoritarian Leadership Style as needed. This regulation calls on the Incident Commander (IC) to be both confident and competent (Dienstvorschrift, 2007).

Confidence and competency are based on the protocols for an Incident Commander (IC). Requirements to serve as an Incident Commander (IC) include specific training and qualifications. To serve as an Incident Commander (IC) usually requires 24 months of training that helps ensure that they are competent in operational leadership. In most instances, it is also required that the candidate who wishes to become an Incident Commander (IC) must possess the minimum of a bachelor's degree, thereby putting an importance on higher education. Individuals are groomed and developed over a period of time to become specialists within their defined roles (Dienstvorschrift, 2007).

DV 100 has several mandates that require a noticeable involvement of elected officials. In this IMS method, elected officials are required to have a more hands-on role than the method used in the United States. In extended response and/or in major incidents, the overall command is held by the local or regional elected official (or authority). This might be a governor, a mayor, or some other political figure. To guarantee that the regulation is obeyed, they must oversee administrative-organization-component and the operative-tactical component. An administrative council is required by DV 100 so that quick decisions can be made to help manage major incidents. This council is tasked with focusing on political conditions, financial responsibilities, and other administrative tasks. This allows operation-tactical-component to focus on operations (Dienstvorschrift, 2007).

The Incident Commander (IC) is charged with managing the operational-tactical-component. They will use a continuous process of analyzing the situation, assessing the situation and deciding a plan of action, and implementing directives or orders. This is the German method for planning the response to an incident, where the United States uses the *Planning P* (Chapter 15). In assessing the situation phase, reconnaissance is heavily emphasized as an important factor to mitigating or ending the incident (Dienstvorschrift, 2007)

It should be noted that the arrangement of personnel commanded by the Incident Commander is similar to that of the ICS method. They have varying levels of responsibility and specific jobs. Technical and tactical considerations are handled by these individuals, based on their assigned position (Dienstvorschrift, 2007)

The synopsis of the DV 100 is a quick review; it is not a detailed analysis. While many portions of DV 100 are similar to NIMS and its companion IMS method, ICS, however, they are not exactly alike. An entire book could be written on the differences between the two, so it is strongly suggested that the reader do a more in-depth study on the German system to gain a better understanding.

While not very much information could be found, it appears that hospitals in Germany use the Hospital Incident Command System (HICS) to manage disasters and in-house emergencies. The lack of information could be related to security protocols, or it could be due to just a lack of information sharing. It appears that Germany began researching and preparing to use Hospital Incident Command System (HICS) since approximately 2005.

3.11 Haiti

Due to the geographical location, and the situation that Haiti is an economically depressed country, large disasters tend to affect this country considerably more negatively than most countries. Historically, the socioeconomic status in Haiti usually creates compounding and cascading effects when disasters strike. In this Caribbean country, even the most basic of disasters often leads to increased death tolls and severe devastation. These issues have amplified human suffering during many major incidents, and the country of Haiti is quite often reliant upon other countries for assistance when a disaster occurs. Because Haiti often relies on integrating outside resources, it is extremely important to have an IMS method. When the country decided to enact an IMS method as a basis for incident management, it was decided that the method used should be a system that was familiar to the countries, especially those most likely to respond to their request for assistance. Unfortunately, it would take a major disaster to move toward the standard use of an IMS method.

In January 2010, a 7.1 earthquake struck Haiti. While in the initial stages of the disaster, there was no estimate of deaths and injuries, but the country knew they needed outside assistance. It would later be realized that more than 222 500 killed, and the earthquake displaced over 300 000 people (United States Geological Survey, 2011). Haiti did not have an IMS method in place when the earthquake struck, which created additional challenges for those countries and agencies that came to assist them in their time of need. In the initial stages of the response, chaos, confusion, and uncertainty appeared to be the common theme, and it was ongoing for many days, even after outside resources arrived. The international response to this disaster was swift, and large contingencies of rescue workers began to arrive from a multitude of other nations within 24 hours.

With Haiti being a member of the Caribbean Disaster Emergency Management Agency (CDEMA), the first outside help to arrive was from neighboring Caribbean countries. In 2010, some of these countries were using an adapted form of the ICS method, but not all of them were on the same page. According to Fordyce, Sadiq, and Chikoto (2012), the country of Haiti had a new and inexperienced emergency management agency in place at the time that the earthquake struck. The fledgling agency was so new that it completely relied on help from regional and international countries for everything, even if it was the slightest disaster. As help arrived for the 2010 earthquake, it was quickly realized that incident management of such a large-scale disaster would be daunting (Fordyce et al. 2012).

As international help arrived, an IMS method (very similar to ICS) was used by most countries to help manage the incident. While it did not take long for CDEMA and other international response partners to initiate an IMS method (to help the people of Haiti), it was still a fractured response. This was partially because no IMS method was in place in Haiti. It should be noted that even though an IMS method was not nationally used, managing an incident of this magnitude would be difficult for even the most seasoned and adept user of an IMS method. While it was a monumental task, most resources were coordinated and managed to provide a more efficient and coordinated response and recovery. Those that responded and used an IMS method included member countries of the United Nations, and peacekeeping troops from the United Nations, and a multitude of international nonprofits. Additionally, numerous countries offered assistance from their military and other resources that they could loan. Thousands of nongovernmental agencies (beyond nonprofits) also jumped into the response and recovery and helped. Perhaps one of the most beneficial nongovernmental agencies was a multitude of ham radio operators, who facilitated better communications (Fordyce et al. 2012).

In the response mode, most of the resources were managed by the United Nations. Unfortunately, they were not the only organization coordinating efforts, and those managing portions of the response did not always integrate with the UN Command Center. This lack of unity in coordination added to the chaos, confusion, and uncertainty faced on the ground (Fordyce et al. 2012).

When the earthquake struck, it did not only damage private residences and businesses. The earthquake was so strong, and Haitian building codes were so weak, that many government buildings and hospitals were destroyed or damaged. According to an article in the *Academic Emergency Journal* (2010), every hospital in Haiti was damaged or destroyed. The first hospital to somewhat recover was marginally functional in two days, about the time that emergency medical relief was arriving from the United States. This original functioning hospital, and rapidly set up field hospitals, were quickly overwhelmed with the estimated 333 000 people who were seriously injured. As time went on, more hospitals and field hospitals were opened, but even then, there was great difficulty in meeting the needs of the Haitian people. In the initial stages of the response, there was no framework in place for managing this disaster, or the hospitals (Hausweld et al. 2010).

While the Government of Haiti had been working on disaster resilience, they were in the infant stages of creating response protocols when the earthquake struck. This massive earthquake killed over 50% of the government employees, which complicated the response and recovery even more. The United Nations also had a large presence in Haiti prior to the quake. They had been acting in a peacekeeping role, and they were coordinating humanitarian aid well before the earthquake. When the earthquake struck, they too were devastated, losing nearly half of their staff members, including the UN leadership that was on the ground in Haiti (Hausweld et al. 2010).

Almost immediately, the US government activated the National Disaster Medical System (NDMS). Teams who were specialized in providing for disaster medical needs were soon organized and dispatched to Haiti. An article on lessons learned by Grimm (n.d.) describes how the Baptist Child and Family Services medical team was sent to a hospital in Carrefour. This 50-bed hospital was credited seeing 300–400 patients per day for a period of several weeks directly after the earthquake struck. Grimm (n.d.) describes that when they arrived at Carrefour, the hospital was brutally overwhelmed, and was not capable of caring for all the patients in a timely manner. The hospital had no method of IMS to manage the incident, so this Strike Team that was sent to shore up this hospital provided a quick training on Hospital Incident Command System (HICS). Within a short period of time, the hospital began to be more efficient in treating patients with a high standard of care. With integration of resources, and through coordination and cooperation that was facilitated by using an IMS method, it was not long before a tent city was set up outside of the hospital for postoperative care (Grimm, n.d.). There is little doubt that the use of Hospital Incident Command System (HICS) was taught and utilized in many hospitals and field hospitals as more resources made their way into the country.

Much has changed in the way that disasters are managed in Haiti since the 2010 earthquake. While relatively little information could be found on the requirements of ICS when researching IMS methods in Haiti, there were several pieces of information that would lead almost every researcher to believe that the ICS method or IMS is in the late stages of implementation. It appears to have started when Community Emergency Response Team (CERT) trainers were sent to Haiti to help develop the private citizen into a basic responder in a disaster (Community Emergency Response Team, 2011). Even while individuals were still living in tent cities, they were learning what they could do to save lives in a disaster, including IMS principles. In the first training, over 400 Haitians, most of which were still living in a refugee camp, learned basic skills on how to manage and help in a disaster. Many even commented that if they had known these skills prior to the earthquake, they could have saved friends and family (Community Emergency Response Team, 2011).

Only two years after the earthquake, on 12 January 2012, a Mass Casualty Incident (MCI) occurred. A dump truck loaded with gravel apparently lost brakes in a crowded part of Port-au-Prince. Soon after the incident, the head of a local university hospital and the International Medical Corps, along with numerous nonprofits and local firefighters, began the process of creating an After-Action Review (AAR). In this incident, 26–30 people were killed and 56–57 injured after a truck lost control on one of the city's busiest streets. The After-Action Review states several times that there was a lack of knowledge and training related to the Incident Command System. The report called for the ICS method to be used nationwide (Donaldson, 2012).

A 2015 press release by the Office of US Foreign Disaster Assistance (USAID/OFDA) revealed that the Government of Haiti Directorate of Civil Protection (DCP) was in the process of introducing a new training program that will deliver ICS training to all first responders. The plan was focused on implementing a multiyear program with technical support from the US Forest Service and funding from USAID/OFDA (USAID Newsletter, 2015). While the ICS method was created in the United States, the Haitian government is adjusting the ICS method for their own specific use. The area that seems to have the most changes from the US ICS method is in the creation of protocols.

After the earthquake, the Caribbean Disaster Emergency Management Agency (CDEMA) decided to enact ICS as the preferred IMS method. The Caribbean Disaster Emergency Management Agency (CDEMA) decided to utilize the ICS method for all mutual aid calls. As part of this change, they put together protocols in the years of 2014–2015. As the consequences of the earthquake are still ongoing (eight years later), the Haitian government and the CDEMA have moved forward in initiating an IMS system to help them better manage disasters and everyday incidents (Hausweld et al. 2010).

3.12 India

India has also been a country that has also seen more than its fair share of disasters. In 2005 the National Disaster Management Authority (NDMA) of India implemented what they now call the Incident Response System (IRS). Incident Response System (IRS) provides guidance for disaster response in Section 6 of the Disaster Management Act. From the release of the document in 2005, local, state, and federal resources have been required to utilize this IMS method when responding to emergencies (NDMA, n.d.).

In taking a closer look at the IMS method used by India, it can be said that it closely mirrors the ICS component of the US system. The main purpose of the Disaster Management Guidelines is to identify the roles and responsibilities of different representatives and stakeholders at all levels of government. Much like NIMS, the Indian government has an overarching national management system that helps to facilitate coordination in multitiered responses which may include the national, state, and district levels of government. The Incident Response System (IRS) method also emphasizes the need for standardized and appropriate documentation of various activities so that improvements can be seen in planning, accountability, and incident analysis (NDMA, n.d.).

India also utilizes an IMS method for disasters or emergencies in hospitals. It appears that in 2002, in collaboration and cooperation with the UN Development Programme (UNDP), the Government of India implemented the Hospital Emergency Incident Command System (HEICS). The plan was to have all Indian hospitals in compliance with the disaster planning framework by 2008 (Government of India, 2002). While no concrete evidence was found, it can be assumed that hospitals in India updated the Hospital Incident Management System (HIMS) as new iterations of this method was released. This assumption is supported by Backer (n.d.) who stated that India planned to implement the 2014 iteration of the Hospital Incident Command System (HICS).

3.13 Indonesia

Indonesia is yet another country that has suffered many disasters, especially in the last 20 years. The use of the ICS method was initiated under the ASEAN agreement, and training was then supported by the US Government. The use of the ICS method is not complete in Indonesia (as of the writing of this book), but the use of this method in Indonesia has been gradually been gaining ground. Between 2004 and 2006 the US Department of State, Indonesian disaster/emergency managers were introduced to basic and then intermediate ICS courses through a series of workshops. These workshops were held in the region, although not initially in Indonesia. Later, in 2008, the US Forestry Service provided an ICS course specifically for Indonesian disaster/emergency managers in Jakarta (United States Agency for International Development [USAID], 2017).

Also, in 2008, the Indonesian government began to implement a system similar to the US version of the National Incident Management System (NIMS), but it was designed specifically for Indonesia. Much like the US version, the Indonesian version of NIMS utilized the ICS component as an important part of the NIMS method. While ICS can be used in a standalone capacity, it must be used if additional agencies are needed in a response. This has proven extremely beneficial for emergency incidents, drills, and exercises that have been undertaken by the Indonesian Government. After Action Reviews (AAR's) have identified only minor issues in the NIMS method and the ICS component, and these are addressed on a regular basis (USAID, 2017).

Since 2008, the Department of Justice's International Criminal Investigation and Technical Assistance Program (ICITAP) has conducted a series of basic, intermediate, and advanced ICS courses in Indonesia. These have been provided to at the national and provincial-level governments. While the principal recipients for ICS training have been the Indonesian National Police, it has also included disaster/emergency managers and other first responders. When requested, the US government has also provided several direct trainings (as well as technical support) on the ICS method to local or provincial agencies who wish to create or hone their skills in disaster management (USAID, 2017).

In 2008, the National Agency for Disaster Management in Indonesia issued Regulation 10. Regulation 10 requires the use of the ICS method for all response agencies. In addition, ICS became part of a mandated training with required competency for new disaster/emergency managers based on Regulation 10. In 2016, the Indonesian National Agency for Disaster Management updated Regulation 10 by creating Regulation 3. This provided a more substantial framework to guide the implementation of ICS training courses to local and provincial governments (USAID, 2017).

Between 2012 and 2015, the Indonesian government created an in-country ICS Steering Committee. The government also adapted ICS course materials to meet the needs of Indonesia and developed a training plan. With the assistance of several US agencies, they also arranged and provided ICS method train-the-trainer courses. In train-the-trainer courses, a variety of first responders from local and provincial agencies were taught how to deliver and teach ICS courses in their local area. Essentially, these individuals became master trainers for the ICS method. There was also consideration for the evaluation of the designation as a master trainer to ensure that there was an elevated level of competencies among those that deliver the courses. To ensure that this strategy was working, the use and evaluation of ICS was scheduled to be evaluated through conducting major disaster exercises (USAID, 2017).

Prior to master trainers, approximately 200 individuals in Indonesia had been trained per year since 2008. This has created a firm expansion in the use of ICS method in incident response. While the master trainer is in its preliminary stages, approximately 50 master trainers were certified in Indonesia by 2017. The organizations they represent include the disaster/emergency management agencies, the military, police, private industry (palm oil) organizations, forestry agencies (and organizations), and environmental agencies and organizations (USAID, 2017).

As of the writing of this book, officials in Indonesia are also considering mandating the implementation of the Hospital Incident Command System (HICS) through healthcare facilities in the country. This was brought about by the A 7.9 magnitude earthquake that struck the region in 2009 that caused mass confusion (Fuady, Pakasi, & Mansyur, 2011). While the considered mandate and implementation appears to be ongoing, it seems that Indonesia is moving in that direction, if they have not already completed the transition.

3.14 Iran

From a historical information, there is little known about when the ICS method (or some derivative of it) that was implemented in Iran. Reviewing an internet search, it appears as if the ICS method has been used in Iran for many years. It was found being used since 2010. From the flyers, discussions, and various other communications, it provides the impression that the ICS method was being used by numerous emergency response agencies prior to 2010. At fire conventions, disaster preparedness seminars, and various other venues, ICS for emergency incident response has been a topic for discussion for at least eight years, which gives the impression that the ICS method has been around Iran for quite some time.

Beyond the ICS method, it also appears that Iran has implemented a modified version of the Hospital Incident Command System (HICS). This method was modified based on the needs of Iranian Hospitals, which had different needs than more developed countries. While an overview of the HICS method will be more thoroughly discussed later in this book, the differences between the US version and Iranian version are slightly different, so it will be discussed in this chapter.

From a research paper by Djalali et al. (2015), it appears that Iran implemented the original version of Hospital Incident Command System (HICS) in, or around, 2006. It was later modified to better meet the needs of Iranian Hospitals in 2013. These modifications took place after a researcher discovered that there was a consensus among hospitals, and hospital staff, that Hospital Incident Command System (HICS) method was only moderately helpful in Iran. The vast majority of those surveyed felt that this method would be more useful in developed countries, but it did not exactly meet the needs of underdeveloped Iran. Having this information, a group of researchers began looking for modifications to make it more effective in Iran. These modifications were not done in a haphazard way. Potential changes were essentially voted on by an 11-member group that were all considered subject matter experts. These subject matter experts were used to review potential changes to the system (made by a researcher) and to vote whether these changes would improve Hospital Incident Command System (HICS) in Iran. In this Delphi study, if 85% or more of the subject matter experts agreed with what they survey proposed, then the change was accepted. If there was less than an 85% consensus, the proposal would either modify the suggestion or discard it from the study (Djalali et al. 2015).

Among several changes, the first was the addition of a Quality Control Officer to the Command group. The purpose of this new position is to assess and improve the performance of the Hospital Incident Command System (HICS) method in Iranian Healthcare Facilities. The Quality Control Officer was also put in place to recognize possible gaps, identify the related reasons, and then report them to the Incident Commander (IC). By doing so, they should be able to facilitate solving the problem or coming up with a mitigation measure (Djalali et al. 2015).

Security is another area where there were modifications in the Iranian model. In the Hospital Incident Command System (HICS) method, security was originally a second-level position. The security team would perform their duties under the direction of the hospital manager or general director (based on which name is used by the hospital). It was found that placing security in a third managerial level was not effectual in the Security Branch (of the Operations Section), the placement of security was changed to a new section on its own, reportable only to the Incident Commander (IC). This change was made because emergency department crowding is a major issue, especially during, and immediately following, a disaster. This new section will be charged with traffic control and crowd control. Additionally, hospital buildings are not exempt from disasters. After an event, a hospital might suffer structural and nonstructural damage. This change in security will also ensure that when there is structural damage, security can create and oversee the search and rescue teams that might be needed to rescue victims trapped in Iranian Hospitals. Adding security as a section, should result in better performance during times of chaos, confusion, and uncertainty, as well as remove a portion of the related stress from hospital administrators (Djalali et al. 2015).

To make this method more applicable to Iranian hospitals, the Infrastructure Branch was also moved to the Logistic Section. In an emergency incident, the Operations Section would need to manage both medical and nonmedical services, including what is considered technical/logistics services. By moving this function to the Logistics Section, it helps to facilitate a safe hospital strategy and frees up the Operations Section. This modification also allows the Operations Section to focus only on medical services, rather than potentially being distracted by nonmedical services as the facility manages a disaster (Djalali et al. 2015).

Another modification made to the Iranian version of Hospital Incident Command System (HICS) method was that the Business Continuity Branch was moved from the Operations Section to the Planning and Administration Sections. Business continuity is the capability of an organization to continue crucial functions during and after a disaster or emergency. Business continuity planning establishes risk management methods and procedures that will prevent or reduce interruptions to mission-critical services. Business continuity also restores full function to the organization as quickly and effortlessly as possible. In the case of Iranian hospitals, the researchers identified that the administrative offices will typically have responsibility for planning, and conducting hospital business, during normal operations. They deduced that moving this function in a disaster or emergency could cause more difficulties, so the decision was that HICS should not change who oversees business continuity in a disaster, and this was also the consensus of the subject matter experts (Djalali et al. 2015).

While these were not the only changes, all changes were made to be specific to Iranian hospitals. The ability for an IMS method to be customized provides evidence that shows the flexibility of the ICS method and the Hospital Incident Command System (HICS) method of incident management. It also shows that true IMS methods must be flexible in order to be adapted to other countries. It should be cautioned that making changes to any IMS method (at a local level) could lead to integration problems with other agencies. For this reason, significant changes should only be made at the federal level. Looking at the changes made to the Hospital Incident Command System (HICS) method in Iran, it was done using a group of subject matter experts, researchers, and even individuals who helped create the latest version, all of which came to a consensus. By doing so, they developed and modified Hospital Incident Command System (HICS) so that it is more feasible to implement in their country.

3.15   Iraq

Since the 2003 invasion of Iraq, the United States and its allies have been working to give Iraq the tools that are needed to improve social order. In doing so, they have helped Iraq to become a self-sufficient society that will be able to handle their own issues. Among the tools that were provided was the most up-to-date public safety training available in the world. Both Britain and the United States (as well as other nations and nongovernmental organizations) have provided ICS training to Iraqi public safety officials and nonprofits for many years. This initially appears to have begun in January of 2005 with law enforcement. By November of 2005, 836 law enforcement officers were trained in ICS (Center for Strategic and International Studies, 2006).

While no information could be found on any laws requiring an IMS method to be used in disasters and emergencies, it is clear that public safety agencies in Iraq have been receiving ICS for quite some time. Additionally, there were numerous advertisements hiring fire service trainers to teach ICS to the Iraqi fire service around the same time. This leads us to believe that public safety agencies in Iraq are widely using the ICS system. No additional information could be found on the use of Hospital Incident Command System (HICS) in Iraq; however, it is likely that hospitals were also trained in its use as a part of the structural rebuild and resilience building capacity of Iraq.

3.16   Japan

Japan is in a unique situation when it comes to utilizing an IMS method. Japan is similar to the United States in many ways, yet there are some stark differences as well. Much like the United States, there are three basic levels of government. In the United States there is primarily the federal level, the state level, and the local level (with a few exceptions such as tribal government).

In Japan, the three levels of government are national level, the prefecture level, and the local level. A myriad of laws, most of them dating back to the late 1940s, are the basis for disaster and emergency response, including the Disaster Relief Act of 1947, the Fire Services Act of 1948, and the Flood Control Act of 1949. These three laws helped to define emergency response in the nation of Japan (Nazarov, 2011). Only a small amount of information is available on the implementation of IMS methods used in Japan.

The Japanese government employs the Incident Command System (ICS), but due to the limited role of prefecture level of government, the ICS method appears to be relegated primarily to local governments. The local agency initiates the ICS method in the same manner as is done in the United States; however, it does not integrate into an overarching national incident management method that would help to integrate resources. From the limited information available, it appears that the Japanese government realizes that all incidents begin and end locally, and that local incident management is critical to mitigating the effects of the incident. The same basic principle is employed in the United States, and much like the United States, it appears as if the Japanese model of ICS makes the prefecture available to assist the local government when their resources are overwhelmed (Nazarov, 2011).

There is some question whether the ICS method used in Japan can expand similar to its US counterpart. Because the Japanese IMS method is not mandated nationwide, integrating outside resources such as nonprofit organizations, business, and other agencies that are not regularly involved in emergency response is problematic. A 2013 Crisis Response Journal article contended that due to the intermittent and pieced together approach of Japan's IMS method, there has been a breakdown in command, control, communication, collaboration, and coordination. Past disaster responses failed to integrate all resources under one command, and it led to a more confusing and less productive response. In many instances, nonprofits, businesses, and other nongovernmental resources acted independently of the command structure in place (Howitt et al. 2013). This helps provide insight that a haphazard implementation of an IMS method is equally as detrimental to a response as having no IMS method used.

It does appear as if Japan is working toward a national IMS method. Howitt et al. (2013) suggests that Japan was considering the ICS method and another overarching method to integrate resources; however, any strides in that area appear to have been without significant gains. A 2018 article in the *Journal of Contingencies and Crisis Management* revealed that while many strides have been made to improving information flow and chains of command, there are still obstacles that prevent it from being more functional. In fact, the authors even state that the largest barrier to an integrated IMS method for Japan is the sectionalism that is inherently built into the Japanese government (Okada & Ogura, 2014).

While Japan has seen issues in the ICS method that they use, they have also seen great success in customizing and utilizing a Japanese version of the Hospital Incident Command System (HICS). In fact, when they began to customize the American version, the University of Tokyo School of Medicine included the Emergency Medical Services Authority (EMSA) in the customization process. The reason that the Emergency Medical Services Authority (EMSA) was consulted is twofold. The first reason is that they are considered by many around the world to be the foremost authority in Hospital Incident Command System (HICS). The second reason was that they needed clarification on some issues to enable a better translation (Backer, 2016). In 2016, the University of Tokyo released the Hospital Incident Command System (HICS) guidebook (based on the 2014 revision) in Japanese, as well as guidance on Incident Planning Guides (IPGs) and the Incident Response Guide (IRG).

3.17 Maldives

Little was found historically about the legal mandates of implementing an IMS method in the Maldives. It was however obvious that the Maldives began utilizing an IMS method after signing the Hyogo Framework for disaster reduction. As was mentioned previously in this chapter, the Hyogo Framework brought all individuals and organizations into a common arrangement to create better coordination when responding to a disaster (Hyogo Framework, 2005).

As a part of strengthening disaster preparedness, it was suggested that the Incident Command System should be used to ensure that all countries respond to a disaster (Penuel, Statler, & Hagen, 2013). The Maldives was one of the 10 countries of ASEAN who signed an agreement to approve the adoption of the ICS system in 2003. Numerous ICS classes were offered that same year, and the Maldives sent representatives to these classes (Ramos, 2012).

There was a collaborative effort to train countries about the proper use of the ICS system. This collaboration included the US Department of Agriculture, Forest Service (USFS). There were minor to moderate adaptions made to the ICS system for each country, including the Maldives. This adaption was based on the structure of the countries national government while being cognizant of cultural considerations. This specific training offered *Train-the-Trainer ICS course curriculum* to key national agencies and trainers (Asian Disaster Preparedness Center, 2009).

In respect to hospital incident management, the Maldives appear to have a system in place. In the Ministry of Health's guidance on Health Emergency Operations Plan ([HEOP], 2018), the manual states that an (unspecified) IMS method will be used to manage disaster. Unfortunately, no information could be found on which IMS method would be used.

3.18 Malaysia

Malaysia is a country that has fully embraced the use of the ICS method. The ICS method used in Malaysia appears to be the same exact model that is used in the United States. Not only do they require that it be used on land, but they also require the use of ICS at sea, and in hospitals. While the ICS method is required in hospitals, only the ICS method is currently utilized rather than the Hospital Incident Command System (HICS). Recent calls for better disaster resilience in Malaysian hospitals has suggested that a modified version of the Hospital Incident Command System (HICS) be researched and implemented to meet the needs of the hospitals in this nation (Samsuddin et al. 2018). Historically, little was found about the laws requiring the use of ICS beyond the initial ASEAN agreement; however, research found a lot of information about ICS classes offered in Malaysia (ASEAN Disaster Workshop, n.d.).

3.19 Mexico

While information was limited on the use of an IMS method in Mexico, it was found that there were multiple agreements between Mexico and other North American Countries that points to the use of the ICS method. According to this information, along the border with the United States, whenever a joint response is needed, the ICS method will be used (Mexico–US Joint Contingency Plan, 2017). Looking deeper into the use of ICS in Mexico, it was found that numerous (bomberos) fire departments also utilize ICS to manage incidents. Urby (2014) states that ICS method is used in Mexico; however, there was often a lack of leadership or in some jurisdictions, there is only a partial implementation. Partial implementation and/or lack of leadership when using any IMS method will typically leave gaps when trying to integrate outside resources.

In researching the use of Hospital Incident Command System (HICS) in Mexico, there were no indicators that it is used in Mexico. This is not to say that the Hospital Incident Command System (HICS) is not used in Mexico; it just means that no information was found to confirm or deny that it was used.

3.20 New Zealand

The IMS system used by New Zealand is the Coordinated Incident Management System (CIMS) method. The CIMS method is very similar to the NIMS method used in the United States. This method was first introduced in 1998, and it is based on the same four basic tenets as the NIMS method, but they use different words. In New Zealand, those base tenets are identified as "Four-Rs." Those "Four Rs" are the following:

- Risk reduction
- Readiness
- Response
- Recovery (CIMS, 2014)

If we compare this to the NIMS method used in the United States, risk reduction is equivalent to mitigation, readiness is the equivalent of planning, and response and recovery are the same. Much like NIMS method, the Coordinated Incident Management System (CIMS, 2014) is flexible, uses common structures, roles, and responsibilities, it requires common terminology, and it is modular and scalable. This system is responsive to each community's needs, it fosters a coordination of response (among differing agencies), a coordination of resources, and integrates information management and communications. It also dictates a manageable span of control (number of people supervised), and it provides for the facilities needed for the response. It also is constantly evaluating the system to ensure that the Coordinated Incident Management System (CIMS) meets the ever-changing needs of first responders. One minor difference that was found from the NIMS method is that the Coordinated Incident Management System (CIMS) appears to put more of an emphasis on international compatibility with other IMS methods (CIMS, 2014).

While the Coordinated Incident Management System (CIMS, 2014) method is similar to the NIMS method in many ways, there are also some subtle differences. One definition in the Coordinated Incident Management System (CIMS) method defines coordination as being assisted by a defined command and control. According to this definition, command is vertical to a single agency, while control is horizontal to outside agencies. Another difference is in the general staff positions. Instead of four (or five if Intelligence and Investigation is enacted) general staff functions, the New Zealand IMS method has six general staff functions, and they identify the Incident Commander (IC) as *Control* instead of command. These six functions are the following:

- Intelligence
- Planning
- Operations
- Logistics
- Public Information Management
- Welfare (CIMS, 2014)

The Coordinated Incident Management System (CIMS, 2014) is also different in providing an Incident Management Team (IMT) who assists the Controller. While an Incident Management Team (IMT) is often used in the United States, they are typically used to fill key position in ICS. New Zealand Incident Management Teams (IMT) are typically technical experts who provide advice based on special knowledge and handling of comprehensive work. The Incident Management Team (IMT), when activated, could contain a Response Manager, a Technical Expert, or other individuals who have specific information for that specific type of incident. The Incident Management Team (IMT) can also include risk advisors. All of the Incident Management Team (IMT) in the Coordinated Incident Management System (CIMS) report directly to control (CIMS, 2014).

Another interesting concept in the Coordinated Incident Management System (CIMS, 2014) method is the integration of the community into any response. While most IMS methods suggest connecting with the community, Coordinated Incident Management System (CIMS) method goes one step further. In the Coordinated Incident Management System (CIMS) manual, it states

> Communities, organisations and businesses self-respond to emergencies, either as part of official pre-existing arrangements or on their own in a spontaneous or emergent manner. Response agencies need to accommodate, link with, support and coordinate community participation in response.
>
> Wherever possible, communities and the business sector should be appropriately incorporated in response coordination planning before incidents occur. Although CIMS is designed to apply to official response agencies, its principles can be applied at the community level where they form part of such pre-planned structures.
>
> CIMS 2014, p. 13

The Coordinated Incident Management System (CIMS, 2014) method also provides a color scheme for each general staff function being performed. This method has assigned pink for planning, dark blue for investigations, and purple for public information. This color coding would seem to be more helpful when trying to identify the functional-based individual you may need to interact with, or ask a question of, in an incident. The Coordinated Incident Management System (CIMS) method also provides a structure to identify the lines of authority between varying governmental agencies (CIMS, 2014), and while slightly different, it uses a planning process more like the German version of DV 100.

In looking at Coordinated Incident Management System (CIMS) using a holistic review, it more resembles the NIMS method of IMS, but it appears to be less complicated. While it may resemble NIMS, several stark differences have been revealed. Even with those differences, integration from Coordinated Incident Management System (CIMS) method to the NIMS method, or vice versa, would not be hard when an international incident occurs.

3.21 Palestine

The Palestinian government was in the process of creating an IMS method that fits their needs in 2017. According to information found in research, the IMS method they will eventually use will be similar to that of the ICS method used by the United States. While it will be similar to the ICS method used in the United States, some adjustments will be made to make it more relevant to Palestine and more effective for the style of government that this nation operates with. In a 2017 United Nations report, it stated

> All field actors need to be trained towards a Palestinian ICS, that has yet to be developed. Specific procedures need to be developed for how the Incident Commander will provide information up the chain to report on the situation on the ground and on how to request for assistance. During a major disaster, such as a significant earthquake, it is likely that resources will be scarce, and those dispatching resources to different areas will have to prioritize, based on the information coming from the ground. Any requests for international assistance will be based on information from the field and their ability to handle the situation.
>
> UNDP 2017, p. 46

Interestingly enough, it appears as if the use of the ICS method was introduced into the Palestinian Prison System in September of 2013 (United Nations Office on Drugs and Crime [UNODC], 2013). What makes this noteworthy is first responders (e.g. fire, Emergency Medical Services [EMS], law enforcement) are just now beginning to develop an IMS or ICS method, but the prison system has been utilizing an ICS method for over five years. While extensive research was done, there was no mention of a Hospital Incident Command System (HICS) method being in place in Palestine.

3.22 Philippine Islands

In 2010, a law was passed in the Philippines that required the Office of Civil Defense to create standard operating procedures for managing incidents. In Rule No. 7 of Republic Act 10121 (2010), the law mandated that the Office of Civil Defense was to establish ICS as part of the Philippines' on-scene disaster response system. In September 2012, the president of the Philippines signed Executive Order No. 82, which stated the activation of the ICS method during all human-induced crisis (National Disaster Response Plan [NDRP], 2016).

As part of setting the guidelines for the use of the ICS method, the Office of Civil Defense created the National Disaster Risk Reduction and Management Council (NDRRMC), formerly known as the National Disaster Coordinating Council (NDCC), is a working group of various government, nongovernment, civil sector, and private sector organizations of the Government of the Republic of the Philippines established by Republic Act 10121 of 2009. It is administered by the Office of Civil Defense under the Department of National Defense. The council is responsible for ensuring the protection and welfare of the people during disasters or emergencies (NDRP, 2016).

Additionally, it was found that the Hospital Incident Command System (HICS) is used in the Philippines. While the Hospital Emergency Incident Command System (HEICS) method was being tested in 2009, the University of East Ramon Magsaysay (UERM) Medical Center in Manila was severely damaged by Typhoon Ketsana. The typhoon devastated the hospital, including wiping out generators and filling the first floor of the facility with water. Patients had to be hurriedly taken to higher floors to prevent drowning. During the recovery process from the typhoon, they began to earnestly train on the Hospital Incident Command System (HICS) method, which they still implement today. While no legislation was found that mandated the use of the Hospital Incident Command System (HICS) method in the Philippines, it does appear that a majority, if not all, healthcare facilities are now trained and active in the use of this method (ADPC, 2014b).

3.23 Russia

Russia is a country that rarely shares all of the information about how they operate. It should be no surprise that only a small amount of information is known about how they manage incidents such as natural disasters and terrorist attacks. From research, it was found that the Russian government utilizes the military for the vast majority of emergency services. A military type of Command and Control system and also a Unified Emergency Prevention and Response State System (RSChS) as well as the Emergency Control Ministry (EMERCOM) are used in Russia.

After the 1988 Armenian earthquake and the Chernobyl disaster, a directive decision was made by the Presidium of the Supreme Council of Russian Socialist Soviet Republic. This directive essentially led to the formation of the Russian Rescue Corps, the predecessor of the current system: The Ministry for the Affairs of Civil Defence, Emergency Situations and Disaster Relief, Russian President Boris Yeltsin established the Ministry of the Russian Federation for Affairs for Civil Defence, Emergencies and Elimination of Consequences of Natural Disasters. After the formation, this became known as the Ministry of Emergency Situations. It was also called the Emergency Control Ministry or EMERCOM. In 2002, the Russian State Fire Service, which was originally the National Fire service, also became part of the Emergency Control Ministry (State Committee for Civil Defence, 2009). While local police, fire, and EMS handled daily problems, EMERCOM essentially became the equivalent of the Federal Emergency Management Agency (FEMA) in the United States. This reorganization was said to have affected incident management, but because only minimal information was provided, it is not certain how it affected incident management.

Roffey (2016) states that in 2004, Russian President Vladimir Putin approved the "Fundamentals of the Unified State Policy in the Field of Civil Defense for the Period up to 2010." This document created the groundwork and the jurisdiction for civil defense in the new political (and socioeconomic) conditions. In creating these fundamentals, objectives were defined, and guidance was given in how to implement the policy. In 2011, the Civil Defense Troops were changed into military rescue units and were melded into the already existing Search and Rescue Service within EMERCOM. Goals were identified, and plans were made for civil defense, with a completion date of 2020. These goals were to improve the current methods of protecting the Russian population, protecting resources and objects of cultural value against consequences due to military operations, and handling emergency situations due to natural and man-made disasters. Additional goals included developing services to be used for civil defense, preserving the existence of the Russian people during a time of war, cultivating the structures of education and training in the field of civil defense, and evolving international cooperation in the field of civil defense. In 2015, additional regular military defense units were earmarked to be available for handling natural disasters. A total of 12 000 soldiers from army units, railway troops, engineering and logistics troops, and airborne troops have been folded into the mix. Perhaps the most curious bit of information is that Russia was a signer and complied with the Hyogo Framework; however, there is no mention of what IMS method was used, or how they integrated with outside resources (Roffey, 2016). Due to the fact that the overall organization is military based, incidents are managed in a military style of leadership when it is a larger emergency than the local government can handle. It is still not known what IMS method is used by local governments, or if it is a mix of different IMS methods. No further information could be found on the IMS methods used in Russia, including the Hospital Incident Command System (HICS).

3.24 Singapore

The country of Singapore utilizes the ICS method to manage disasters and emergencies across multiple agencies. These agencies include EMS, emergency managers, the Singapore Disaster Response Force (SDRF), the Singapore Civil Defence Force (SCDF), and more. While little historical information was found, it is known that Singapore began using the ICS method even prior to the 2005 Association of Southeast Asian Nations (ASEAN) Agreement. It appears that the Singapore Civil Defence Force has used some form of an IMS method since at least 1998. The Singapore Civil Defence Force includes many public safety agencies (including fire and HazMat), and they cooperate with the vast majority of government organizations (Yung, 2012). Beyond this, no information was found except that Singapore was a signer of the Association of Southeast Asian Nations (ASEAN) and later the Hyogo Framework.

As was mentioned earlier in this book, necessity is the mother of invention, and this is (to a certain extent) what happened in Singapore. Between February and May 2003, an epidemic gripped Singapore with fear. A patient who had been visiting China contracted Severe Acute Respiratory Syndrome (SARS). When the patient returned to Singapore, this single patient created an epidemic that would infect 238 people. Thirty-three of those infected with the virus would end up dying (Chew, 2009). In response to the pandemic, some hospitals and physicians began looking for ways to manage future incidents such as this. Toward the end of 2003, Dr. Tham (2003) began delivering courses on the ICS method to hospitals, but this was not the Hospital Incident Command System (HICS). He had adapted the ICS method used by first responders to fit the needs of Singapore hospitals. What makes this interesting is that even though the Hospital Emergency Incident Command System (HEICS) was already developed and being used, and the US Center for Disease Control (CDC) was involved in containing this outbreak (Chew, 2009), curiously, it was not taught. There could be several different reasons why the Hospital Emergency Incident Command System (HEICS) was not used, including that it was not known in Singapore at that time and/or that it would not meet the needs of this outbreak. Even though no information could be found regarding the use of HICS in Singapore, it is a safe assumption based on other materials (Singapore training flyers, discussions about HICS, etc.) that it has been implemented.

3.25 United Kingdom

Fire and Rescue Service have been utilizing a modified version of the US ICS method since approximately 1999. While most countries around the world have based their modifications on the FIRESCOPE version of ICS, the United Kingdom chose to utilize and modify the ICS method developed by Chief Alan "Bruno" Brunacini. In fact, Brunacini generously donated his time to assist in creating the IMS method used in the United Kingdom (Arbuthnot, 2015).

According to Arbuthnot (2015), the need for an IMS method to manage operations in the United Kingdom was realized after several fire responses went horribly wrong. While numerous fires may have played a role, the Gillender Street fire in London during July 1991, and the Sun Valley poultry factory fire in Hereford, England, during September 1993 seemed to be the impetus. In both aforementioned fires, two firefighters died (in each one). Additionally, there had been questions about command and control issues in a government report, which also helped the push for solutions. Initially, the two main focuses for the UK ICS method were dynamic risk assessment and organizational structure on the operational (fire) ground. As the system started to develop, a third focus was added which revolved around command competency. After much debate and inquiry, a final IMS method was decided on, and in 1999, the

UK Fire Service released the *Incident Command* manual. Since then, there have been revisions, updating, and additions in 2002 and 2008 (Arbuthnot, 2015).

In the United Kingdom, the Metropolitan Police Service determined that an IMS method was needed after a police constable was killed in the Broadwater Farm riot that occurred in October of 1985. This event caused the Metropolitan Police to realize that a rank-based command system was not appropriate for rapidly changing events. This was (primarily) based on the premise that there was no specific individual definitively managing operations during the response to the riot, thereby causing chaos, confusion, and uncertainty (Future Learn, n.d.).

After reviewing this riot (and other incidents), it was determined that three essential roles were important in managing an incident. To mitigate problems with incident management, a few members from the Metropolitan Police created and implemented a new IMS framework called Gold–Silver–Bronze. This framework was also found to be useful in preplanned operations. Soon after its inception, it began to be utilized or integrated by other emergency services/public safety agencies (Future Learn, n.d.).

In this system, the Gold Commander is in complete control of organizational resources for an incident. The Gold Commander will be located somewhere other than the actual incident (usually in a control room) and will be in contact with their resources so they can receive regular updates. The specified control room is known as Gold Command, and this is where they will formulate the strategy. This remote command post was devised so that the person creating strategy has less of a chance of getting tunnel vision caused by being physically at an incident. Multiple Gold Commanders from different organizations (fire, police, EMS) will work together on a multiagency response in Gold Command and, if by chance they are not in the same geographical location, they will remain in continuous contact with each other by videoconference or telephone. This is not to insinuate that they will usually be in the same room, but rather they will typically be in the same geographical location, usually within walking distance of each other. Designated Gold Commanders from different agencies will periodically meet so that they can deliberate and articulate policies and improve working practices between their organizations. In these meetings, they will often preplan coordination for various types of incidents (Future Learn, n.d.).

The Silver Commander is responsible for managing tactics, and they are at, or near, the incident. It should be noted that the Silver Commander is not physically involved in dealing with the incident; they are strictly used for tactical planning. On larger incidents, they will typically be in a command vehicle or at an improvised command center with their counterparts from other disciplines. This command center will be identified as the Joint Emergency Services Control Centre (JESCC). After being given strategic direction by the Gold Commander, they will begin to formulate the tactics needed to meet the overall strategy for an incident. The tactical actions they develop will keep in mind the overarching strategy's that should be met, and plan accordingly. Once they develop these tactical plans, they then provide tactical direction to the Bronze Commander, or in cases where multiple commanders are used, the appropriate Bronze Commander (Future Learn, n.d.).

A Bronze Commander is in direct command of resources at the incident, and they ensure that their on-scene staff carries out the tactics and orders provided by the Silver Commander. In a multiagency response, a Commander or some other representative (from each required discipline) will take direction from their organization. If the incident covers a large geographical area, then multiple Bronze Commanders might be based on the geography of the incident. If an incident is extremely complex, each Bronze Commander can be given their own task (or responsibility) within the response to the overall incident. As an example, in a mass casualty incident, one Bronze Commander might oversee the taking statements, while another manages crime scene security, and yet another oversees survivor management (Future Learn, n.d.).

During the initial stages of an incident, the highest-ranking member of each discipline temporarily assumes the role of Silver or Bronze Commander. They will remain in that position until a more senior member or someone with specialized knowledge from their organization arrives on scene, then they will transfer command (Future Learn, n.d.). In personal communication, it was found out that police and medics do not utilize a formal ICS method, but these entities have created documents which mirror the fire service's architecture to a large extent. This became more of a priority since Joint Emergency Services Interoperability Principles (JESIP) (K. Arbuthnot, personal communication, 23 November 2018).

The Joint Emergency Services Interoperability Principles (JESIP) was a two-year coordinated research program that looked at improving integration of resources and improving coordination during a multiagency response. The primary focus was to improve response and recovery whenever a major multiagency incident occurred. This two-year shared effort, undertaken from 2012 to 2014, produced a joint doctrine that provides a framework that improves multiagency response by providing training and basic guidance. The program initiated one of the largest and most effective joint training initiatives, one that helped to integrate emergency services during a disaster in the United Kingdom (JESIP, n.d.).

A common misconception of the Gold–Silver–Bronze is that the organization in charge of an incident is dependent on the type of incident. Most people believe that the police are in charge of a criminal or terrorist incident, and that fire crews are in charge of fires and rescue type situations. This assumption is a misconception of the framework. Under the Civil Contingencies Act, these varying disciplines would coordinate the response to provide a seamless multiagency response, by utilizing Joint Emergency Services Interoperability Principles (JESIP) procedures and protocols. Each agency participating in the response has its own command structure. Representatives from each discipline will work together to provide a comprehensive and coordinated response (Future Learn, n.d.).

Comparing the UK system of IMS to the US counterpart, it is somewhat similar to the ICS method, however, multiple unnamed sources reported that the Gold–Silver–Bronze method has some issues. According to individuals who have worked in the field under this program, the issue arises when Gold Team members assert their authority to the boots on the ground, but the commands are contradictory to what responders on the ground think needs to happen in order to control the incident properly. While this view is subjective to the person, it also raises the question if a command staff that may be 30 miles away can make a better call than those that are right in front of the incident. In the United States, it is not uncommon for the Incident Commander (IC) to be off site; however, they receive detailed feedback from operations, and there are often joint planning meetings and daily evaluations. It is unclear if this occurs, these types of face-to-face meetings in the Gold–Silver–Bronze method. It is clear that Joint Emergency Services Interoperability Principles (JESIP) is similar to an Emergency Operations Center (EOC) that is used to supply resources to those in command of an incident.

When looking at the hospital IMS method used in the United Kingdom, hospitals integrate with public safety as part of the Gold–Silver–Bronze method, and they also have an internal IMS method for hospitals. The National Health System (NHS) in member countries of the United Kingdom (e.g. Wales, Scotland, England) provide the very specific Emergency Preparedness, Resilience and Response Framework that guides the hospital into a complete integration with public safety agencies. This framework appears to have been initially implemented in 2004, and it has undergone several modifications since then. In this framework, there are four levels of disaster. They are the following:

- *Level 1*: An incident that can be handled internally with the resources on hand.
- *Level 2*: Requires outside resources within the NHS region and liaison with local NHS office.

- *Level 3*: Requires outside resources within the NHS region and the regional NHS office will coordinate with commissioners at the tactical level.
- *Level 4*: An incident that requires NHS National Command and Control to support the response. Outside resources need from outside the region, and NHS is required to coordinate the NHS Response in collaboration with local commissioners at the tactical level (NHS England, 2015; NHS Scotland, 2015; NHS Tameside Hospital, 2016).

When a potential major incident is reported outside of the hospital, the facility will be notified. There are two levels of notification they may receive. The first is called a Major Incident Standby. In this notification, those activating the hospitals are unsure of the scope of the incident, or they do know the scope, and there is a potential that it may escalate to a larger incident. The standby classification allows hospitals more time to prepare staff and workers for receiving large quantities of patients or prepare for specialized procedures that may be needed such as decontamination, burn care, respiratory care for poison gases (NHS England, 2015; NHS Tameside Hospital, 2016).

The second level of notification is Major Incident Declared. This is an instance where it has been confirmed that a Major Incident has occurred, and the circumstances require the hospital to ready themselves for incoming patients. It is important to note that a Major Incident can be called without first having the hospital be on standby. The preparedness level of these hospitals also has redundant systems, which the Hospital Response Team are required to have (NHS England, 2015; NHS Tameside Hospital, 2016). As an example, the hospital is required to have the hardware, software, and the abilities to switch to a different means of communications, should their normal communications methods be damaged or knocked out.

Internally, the hospitals response to the incident is managed by the Hospital Control Team (HCT). Action Cards have a myriad of information that helps direct the Hospital Control Team (HCT) in what functions will be needed during a major incident based on the type of incident it is. Typically, a full set of Action Cards will be kept as a hard copy in the Major Incident Resource Pack and a full set of the Action Cards will be kept as a hard copy along with a copy of the Major Incident Policy in the office of the Director of Nursing as well as the Fire Safety/Emergency Planning Advisor, in an offsite location. Additionally, hard copies of these Action Plans are placed in each area that might be affected (e.g. Emergency Department, Burn Unit, Pediatrics) by a major incident and will be held available in all Departments and Wards that might be required to provide a response in the event of a major incident (NHS Scotland, 2015; NHS Tameside Hospital, 2016).

Action Cards are considered controlled documents, and changes to those cards usually require the approval by the Emergency Planning Resilience and Response Group. In most instances, the personnel identified in the Action Card as the lead of a given area is responsible for updating them and presenting the changes to the Emergency Planning Resilience and Response Group with any material changes (NHS Scotland, 2015; NHS Tameside Hospital, 2016).

The Actions Cards spell out who should be notified and in what order. Whenever a Major Incident alert is received, the Action Card identifies the notification process needed. Usually, the telephone switchboard operators and first on-call managers will be notified, and they will initiate staff call-out procedures. The Action Cards will suggest what actions the Hospital Control Team and Mobile Medical Team will need to take for that particular type of incident. After receiving notifications, the Action Cards will identify actions that will need to be taken in various areas and by various personnel, automatically. These actions may be for the Emergency Department, clinical staff as well as nonclinical staff (NHS Scotland, 2015; NHS Tameside Hospital, 2016).

Some of the actions that may be listed is how to maintain records on each patient, where they are located (updated regularly), and how they should be identified. Additionally, the Hospital Control Team (HCT) may determine that the hospital may need to accelerate discharge procedures to provide more beds for the influx of patients coming from a major incident. The Actions Cards may spell out the need to keep families informed, so part of the requirements of the Hospital Control Team (HCT) after a disaster is to provide a specific area for the patient's family, and to provide a liaison person to keep them informed. Another consideration that will be listed on the Action Cards is when a major incident may be part of criminal activity. In these incidents, the Hospital Control Team can direct personnel to enact already established protocols to protect and preserve potentially useful forensic evidence (NHS England, 2015; NHS Tameside Hospital, 2016).

In creating the Hospital Control Team (HCT), there are typically five main positions. These positions are the HCT Manager (usually the hospital's Chief Operating Officer), the Corporate Lead (usually the Chief Executive), the Nursing Officer (usually the Director of Nursing), Hospital Control Team (HCT) Administrator, and the Hospital Control Team (HCT) Documenting Officer. These five individuals work together to manage the incident in an organized manner. It should be recognized that the IMS method that is used is similar to the ICS method that is utilized by UK fire brigades, although it appears that it is not the same (NHS Tameside Hospital, 2016).

Much more could be written on the United Kingdom. This book only touches on the continual improvements that are seen in incident managements and IMS methods. Much like the United States, the United Kingdom is always trying to improve. It appears as if the key players in IMS methods are never satisfied with status quo, or that the systems they use are "good enough." This continual thriving to improve allows them to be one of the main leaders in the world for managing major incidents.

3.26 Vietnam

Somewhere around 2008, the World Bank (2017) began assisting the Republic of Vietnam to be more disaster resilient. This was done by beginning a Disaster Risk Management (DRM) project in the country. This reduced the negative effects of disaster from 2009 forward. In this project, significant upgrades were made to the infrastructure as well as the management of disasters (World Bank, 2017).

In 2015, the United States Agency for International Development (USAID) entered into a Memorandum of Understanding (MOU) with Republic of Vietnam, Ministry of Agricultural and Rural Development (MARD), to begin teaching classes in their country, to implement the Incident Command System. Since 2015, numerous classes have been taught in Vietnam (and other Asian countries). USAID has committed to providing assistance in training first responders in the country so that they could better manage disasters (United States Embassy in Vietnam, 2015). This was also done so that when other countries, including the United States, commit to help after a disaster in Vietnam, all resources can be integrated into the response seamlessly.

Reviewing research on hospital-specific IMS methods in the Republic of Vietnam, a 2011 report by the World Health Organization (WHO) found that all but one hospital in the country used Hospital Emergency Incident Command (HEICS). In looking at the report, the vast majority of the healthcare facilities had implemented most of the guidance on Hospital Emergency Incident Command System (HEICS), but not all parts of this IMS system were implemented (United States Embassy in Vietnam, 2015). While later information was not found (in further research), it would probably be safe to assume that these hospitals have implemented the Hospital Incident Command System (HICS) and that they have likely undergone improvements that build resilience over the past seven years.

3.27 Other International Uses

The use of the Incident Command System (ICS) method or National Incident Management System method (including ICS) appears to be used in numerous countries, and by numerous nonprofits and for-profit companies. In 2013, the Shell Oil Company released a report about the Incident Command System. In that report, they identified countries that utilized the ICS method, including Saudi Arabia, Ireland, and the International Maritime Organization to name a few. They also identified other similar IMS methods that would integrate with the ICS method including Australia (Australian Inter-Service Incident Management System), Brazil (ICS used by the fire department and civil defence of Rio de Janeiro), and other agencies and countries. Perhaps more interesting is that Shell Oil has regional teams around the world who use the ICS method as their system to manage chemical and oil-related disasters (Curd, 2013).

To some extent, you (the reader) may be wondering why this information was included in this book. The reasons are simple. First and foremost, to show that the rest of the world has also identified that IMS methods are beneficial in mitigating the effects of emergencies and disasters. The use of IMS methods around the world provides evidence that those utilizing these methods have a vested interest in creating calm from chaos, clarity from confusion, and predictability and confidence from uncertainty. The second, and perhaps biggest reason, is because it is good to get outside of our own little bubble that we operate in. One thing is for certain, this chapter definitely makes a partial case that some form of IMS method should be used to mitigate the effects of a disaster or emergency.

Real-Life Scenario

The Government of Burma (Myanmar) began implementing the ICS method in their country in 2003 and completed the full implementation in 2010. Over a three-month period in 2015, flood, storms, and landslides took a devastating toll on Burma, negatively affecting 12 of the 14 states and regions. Based on the protocols and training, the country was able to use the newly implemented disaster management system. Utilizing this new framework, those managing the incident were able to easily and successfully integrate outside resources as has never been seen before. Under the new protocols, the government realized the need for, and provided, transparent situation reports and expedited travel authorizations. This allowed them to provide extraordinary humanitarian relief in a very short amount of time. Some of the resources that were brought in included the Association of Southeast Asian Nations Coordinating Center for Humanitarian Assistance (AHA Center), the Myanmar Red Cross Society, and UN agencies (USAID, 2016).

From the onset of the response to the first floods, the Emergency Operations Center staff held a meeting with Vice President U Nyan Tun and local government officials. In this meeting(s), they developed a national recovery framework, and they utilized the ICS method to organize the recovery. After the incident was over, the Government of Burma was recognized for how successful the response and recovery went by the international humanitarian community. While the process worked extremely well, with the support of USAID and Japan International Cooperation Agency (JICA), undertook an After Action Review to help improve the quality and speed of disaster response in the country even more (USAID, 2016).

Chapter 3 Quiz

1 **True or False**: The United Nations has implemented their own IMS method that is different than any other country.

2 In trying to create disaster resilience, the United Nations has created two agreements that have included utilizing an IMS method. What were those agreements named?

3 What region of the world was the ASEAN agreement developed and signed?

4 What is the name of the Australian IMS method?

5 What is the name and acronym for the national system (similar to NIMS) in Bangladesh?

6 **True or False**: The country of Brunei utilized the US Forest Service to help improve IMS methods in their country.

7 Brunei has become an expert in Hospital Incident Command System. They began teaching this method throughout the region to other hospitals and other healthcare facilities. What organization usually sponsors Brunei led trainings?

8 What year did China begin looking at an ICS method to manage incidents?

9 What is the name of the German IMS method?

10 **True or False**: During the Haitian Earthquake in 2010, all of the resources were coordinated by the United Nations.

11 Iran modified the Hospital Incident Command System to meet the needs of Hospital by utilizing surveys and an ____ member group.
   (a) 11
   (b) 18
   (c) 8
   (d) 82

12 The largest barrier to implementing the ICS method in Japan is _____, which is inherently built into the way the government is set up.
   (a) Infrastructure
   (b) Sectionalism
   (c) Lack of bipartisanship
   (d) Their culture

13 **True or False**: Japan utilizes the Hospital Incident Command System.

14 What is the New Zealand method of IMS called?

15 Where was the ICS method first used in Palestine?

16 What method is used in the United Kingdom for IMS of emergency personnel in the field?

17 The UK Fire Brigades utilize an ICS method that is based on Fire Chief _____ method, rather than the ICS method that was created by FIRESCOPE.

18 **True or False**: Vietnam utilizes the Hospital Incident Command System (HICS).

Self-Study

Note: There are thousands of resources about what countries utilize for IMS, studies on the efficacy of these methods, how well these methods integrate with other methods, and much more. While this book could list hundreds of self-study resources, it would only take up space if these were not of interest to the specific reader. For that reason, you should research these issues yourself. You can search your school, college, or university library, or you can use a simple (or complex) Internet search. Simply look for the country you wish to research, then using that countries name (in a search box) and some key words, you will see some of the information available. Key search words might include any combination of:

Incident management, incident management system (IMS), incident command, incident command system, hospital incident command, hospital incident command system, hospital disaster, hospital disaster management, disaster management, command and control.

These are just a few of the search terms that could be used in conjunction with the country you wish to research. It is by no stretch of the imagination a comprehensive list of search term for IMS systems.

4

The Five Cs of Crisis (or Incident) Management

We've got to be judged by how we do in times of crisis.

Johnnie Cochran

If you were to conduct a literature review written on the control and management of emergency incidents, you would find common variables. While the military utilized C4 and C4-I as a way to manage troops, in a post-9/11 world, we now look at five reoccurring themes in order for the incident management system (IMS) method to be effective. These five reoccurring themes (or concepts) are critical to managing an incident. Those concepts are the following:

- Command
- Control
- Communications
- Coordination
- Cooperation

Not only are these Five Cs a reoccurring theme, but they also have a high level of importance in incident management. This holds true in all phases of incident management, including the preparedness mode, the response mode, the recovery mode, and the planning and mitigation mode of an incident. They are the basic concepts of managing any crisis, disaster, or emergency incident. These factors have a significant relevance and are underlying pieces that make incident management effective. The utility of each will be discussed in great detail.

4.1 Command

While no single element of the Five Cs has more importance than another (especially in terms of incident management), command is often seen as the cornerstone of any response. Without effective command, you cannot attain control of a crisis. Any incident has two separate, but very distinct, areas of command if they are to overcome the challenges of an incident: commanding resources and commanding the incident. When resources are commanded, an individual will need to direct those resources in the most effective way to mitigate the incident. Commanding an incident refers to identifying key areas of an incident that need addressed and taking the proper actions to control that incident so there is less chaos, confusion, and uncertainty. In almost every instance, effective command requires an authoritative presence.

When commanding resources if the individual in command of the incident does not have an authoritative presence and good leadership qualities, then their effectiveness will likely be diminished. This type of authoritative presence comes from experience and practice. They will need to have the knowledge, skills, and abilities for the incident they face. While they may not have been involved in every type of incident that may happen, the individual chosen should have a proven record of being a critical thinker and someone who has a proven record of being successful in a response.

Commanding an incident is often more difficult that commanding resources. The reason that it is more difficult is that the person commanding will need to collect data on the incident, utilize advanced critical thinking skills, and determine how to overcome the challenges that a community is facing. In many cases, the person that commands an incident will need a little bit of an ego in order to succeed. In order to have the upper hand against the challenges they face, they will have to have the attitude that they can beat the incident they are facing. As they fight the battle with the incident, they must either know, or believe, that they will be the victor. This is not only in the response mode of an incident but also the preparedness mode, the recovery mode, and the planning and mitigation mode.

In the preparedness mode, someone with an authoritative presence will need to guide what steps need to be taken to prepare for the most likely risks. They will direct their personnel on what they see as the most critical factors. While they will be listed as command, they will need to do more than tell personnel what to do; they (or someone on their staff) will need to be commanding the incident before it ever happens. In order to command the incident, they will need to identify what priorities will need to be addressed. To be successful, this will require situational awareness, and that situational awareness will usually need to be preplanned in order to be effective. While a more in-depth explanation of situational awareness will be discussed in Section 4.1.1, in essence situational awareness is having the most current information so that you can make an informed decision. Having the most current information allows those in command to set goals and objectives so that they can communicate with subordinates the importance that is placed on each type of risk. Command will process feedback from others, and in order to be in command of the incident, they will need to identify the resources that will be needed. They will command the process of how they should prepare, and how they will test the plans that have been made and communicated. In order to command an incident, even if it is in the preparedness mode, the person leading the effort will not only need to have a vision, but they must create buy-in (into their vision) so that others will be inspired to work with them in order to meet the mutual goals of planning.

In the response mode, one person would be, or should be, in command of resources. In instances where a Unified Command (UC) is needed, then a specific group of leaders will fill the role rather than a single person. Whether an individual is in command, or it is a Unified Command, all parties involved in command should gather information about the incident at hand (situational awareness) and determine how to manage the response to that specific incident. Using situational awareness, the person or persons in charge of an incident can create strategic goals and operational objectives, give orders and/or instructions, and assign tasks related to the goals and objectives so that they can command the incident. As part of an ongoing process of commanding the incident, they will work with key staff members to plan the next steps that will need to be taken, and what to do if those plans did not work as expected. While the person (or persons) in command may not personally do all of the work, they will make the final decisions in order to take command of the situation. If things go wrong, they will also face the backlash and repercussions.

In the recovery mode, the person in command will guide the recovery and, as necessary, they will delegate authority to ensure that the needs of the community come first. They will avoid getting tunnel vision, and they will have situational awareness. This situational awareness, especially in the recovery mode, will primarily come from the community, the media, social media, and those that work under their command. Those in command will make informed decisions based on that situational awareness. With help from their staff, they will identify what will help the community to be whole again in the quickest and most efficient manner. In order to command the incident, they will cooperatively work and liaison with other agencies, individuals, and entities to ensure that community needs are met.

In the planning and mitigation mode, the person in command will identify preparedness and mitigation priorities from After-Action Reviews, Risk Assessments, and other avenues of information. By having acute situational awareness of the most prevalent risks, and understanding the magnitudes of those risks, they can identify areas of a community that would benefit from mitigation and preparedness efforts. To be successful, they will use critical thinking to devise mitigation measures and take command. If done effectively, their efforts will take command of the risks by implementing mitigation measures or in making preparations for how to reduce the risks that the community may face.

4.1.1 Situational Awareness

Situational awareness is gained in two different ways: the first is from an initial impression or size-up of the situation. It does not matter if the agency is in the preparedness mode, the response mode, the recovery mode or the planning and mitigation mode of an incident, situational awareness is a critical part of managing an incident. Situational awareness is a basic concept of being able to command any crisis, disaster, or emergency incident. Having true situational awareness is critical in allowing command to make informed decisions.

When command is called to an incident, an initial size-up is needed. It is critical for the person in charge (usually called the Incident Commander [IC]) to undertake an assessment of the situation and the surrounding environment. This is usually referred to as *sizing up* the incident; however, this term actually refers to the size-up process.

Size-up is a mental evaluation of the situation. The initial size-up of the situation allows the person in command an opportunity to begin the process of controlling the emergency situation immediately. During the initial size-up, the person in command will immediately take control of the emergency and make an initial quick decision if more assistance will be necessary. Depending on the information gathered, the person in command may determine that more personnel and equipment is needed for the incident. This decision will be made based on the situation as they perceive it at that very moment.

The second part of situational awareness comes from ongoing intelligence and information gathering. Intelligence and information gathering are an important part of incident management. The person in command must have the most recent intelligence or information. This is extremely important in making informed decisions. The latest information will allow the person in command to form alternative strategies based on the changes in the incident. During the response to a large-scale crisis, they may want to appoint someone to gather information and intelligence on the evolving situation. This is especially true when the commander is overwhelmed and needs assistance in managing the crisis. The person appointed for this task would be assigned to collecting, evaluating, disseminating, and using incident data and intelligence. It is important to realize, they mode that the incident is in does not matter. Situational awareness with the latest information will promote improved and more efficient planning, response, recovery, and preparedness and mitigation measure.

4.1.1.1 Who Is in Command?

Individuals who are new to incident management game may mistakenly assume that the highest-ranking person will take command of an incident. In most instances, this assumption would be correct, but this is not always true. Beyond the previously mentioned attributes, a commander must also be knowledgeable about the type of incident they are commanding.

To put this in a more simplified explanation, you probably would not want to justify putting the police chief in charge of a mass casualty incident (such as a 12-car accident with injuries) just because the accident happened on "their road." The police chief would likely have little to no experience or knowledge in the intricacies of commanding a response and recovering from such an incident. The same holds true for a fire chief facing an armed gunman incident. In essence, this would be beyond their knowledge, experience, and abilities. By the same token, the fire chief may be exceptional in responding to fires and incidents that require extrication and medical care, but they may have limited knowledge in Hazardous Material (HazMat) response. In these cases, they will usually transfer the command function to someone with greater knowledge, experience, and abilities in that type of incident. This should hold true in all types of incident management, but unfortunately there are some that let their ego get in the way.

In looking at the HazMat incident, let us look at a more likely scenario that would involve a semitruck involved in an accident, and it is leaking a hazardous material. It might be decided that the highest-ranking HazMat Officer or even possibly the most knowledgeable and experienced individual would assume command of the incident, even if they were not an officer. This would be done to bring about a quick and positive outcome. In some instances, the most experienced person should take command, but even if they do not take command, they should be right there with the person in command, providing technical assistance to advise them of the best way to move forward.

It should be realized that in a viable agency, it is not about the number of bars or stars that the individual has on their collar or lapel: it is about mitigating the incident by choosing the person best suited to make order out of the chaos, predictability out of the uncertainty, and who can create simplicity out of complexity. Commanding an incident should not be about pride or ego. The only ego that is needed at an incident is the mindset that those in command can take control of the incident and bring it under control.

It should be realized that commanding an incident is more than just barking orders. In order to effectively command an incident, the leader must be inspiring and encouraging. If the person in command is encouraging and inspiring, they will empower their followers and motivate them. Few people want to follow orders from someone that does not motivate or inspire them (Carter, 2008).

The person in command should also love what they do. If the commander loves what they do, that quality of the response to the incident will be abundantly apparent to those for whom they are commanding (Alyn, 2011). They must lead by example, rather than being a leader that does one thing, and then says another (Compton, 2010). They should also be someone who displays a positive attitude, even when things are going to hell in a handbasket (Northouse, 2009). They should be a person that shows humility by thinking less of themselves and more about the mission and the individuals that they are commanding. The person leading must also have integrity and honesty, and they must treat those that they are commanding with respect (Brady & Woodward, 2008; Northouse, 2009).

The person leading a charge may be an authoritative figure, but that does not mean that they need to be an authoritarian. They must display an air of confidence, and they must be decisive (Crawford, 2010). By being decisive, it might mean they do not make the best decision every time, but if they do make a wrong decision, they will be honest enough and have enough integrity that they admit their mistake (Smeby, 2013). They must truly care about the welfare and safety of those that they command, and they must place that safety as an extremely high priority. Finally, in order to be effective in commanding an incident, they will have a vision, and they will get others to see their vision and then get them to "buy in" to that vision (Sargent, 2006; Kouzes & Posner, 2012).

The character of the person in command will transfer to the response and the responders. Those working under their command will see the true character of a bonafide commander, and in most instances, they follow the lead of that person. At this point, you may be thinking that these are leadership skills: They are! In order to be able to manage an incident, you have to be a good leader. A leader commands the situation rather than letting that situation command them or their agency.

4.2 Control

Controlling an incident is a critical part of ensuring that the lifecycle of the incident is shortened. Many may think that in order to control an incident, they must be able to micromanage all aspects. This could not be further from the truth. In all modes of an incident, planning, response, recovery, as well as preparedness and mitigation, it is important to find people that you can trust to delegate authority to and let them do their job. Some of the most successful people in history have eluded to this.

Lee Iacocca, a well-known executive at Ford Motor Company and Chrysler said, "I hire smart people and get out of their way." Similarly, the well-known founder of Apple Computers, Steve Jobs said, "It doesn't make sense to hire smart people and then tell them what to do; we hire smart people, so they can tell us what to do." This type of mentality is needed when controlling an incident, especially large incidents that require a lot of resources. As Peter Drucker once said, "Tell people what you need, NOT what to do."

In order to control an incident, especially larger incidents and disasters, the person in command will need to delegate authority. This holds true in all modes of an incident. The commander will have to separate organization levels using an IMS method. In doing so, personnel who are exceptional at their specific job can be delegated authority and use their expertise to control the incident. To do this, the commander will need to identify someone who has the skills, knowledge, and abilities, as well as someone who is efficient at managing resources, while still providing tactical direction in the incident. In the Incident Command System (ICS) method, this would be the Operations Section Chief (OSC).

In order to control the incident, the person in command will also need to provide logistical support to those that are undertaking tactical actions. This logistical support will be in the way of facilities, services, and supplies. It will include identifying the services and support that will be needed to control the incident, as well as meeting the requested needs. This is important to managing a large incident because personnel need food, water, transportation, supplies, equipment, places to sleep, and much more. They may need radios, fuel, maintenance for equipment, medical assistance, and even more. By supplying the needs of first responders, the person in logistics helps to control the incident. They can plan for how they will obtain the supplies in planning; they can deploy the supplies during the response and recovery; and they can mitigate by securing the supplies that might be needed through the signing of agreements and making purchases. In the ICS method, the person in charge of controlling logistics during the response and recovery phase is known as the Logistics Section Chief (LSC); however, the Logistics Section Chief (LSC) for an incident will not always be the same person that works in logistics during the planning, preparedness, and mitigation modes of IMS.

Controlling an incident also requires a large amount of planning. This includes planning before an incident, during an incident, and after an incident. During the response and recovery phase, an individual will be appointed to be in charge of planning, and they will be identified as the Planning Section Chief (PSC). These individuals will provide planning services for both the current situation, and future activities. In the planning, preparedness, and mitigation modes of IMS, this part of controlling an incident may be undertaken by a person or group of people to help make sure that every effort is made to plan for the worst but hope for the best. As will be described later in this book, there are specific ways that planning must be accomplished in order to control the incident.

Another way to control an incident rather than letting it control you is financially. We all need to remember that money does not grow on trees, and that we will be held accountable for our spending. Many organizations fail to realize that in the planning, preparedness, and mitigation modes of IMS, actions can be taken to reduce chaos and confusion in the finances of the response and recovery mode. In major events, someone will have to pay the bill for the response and recovery, and possibly even for preparedness and mitigation efforts. It does not matter if it is a local, state, or federal government footing the bill, or even if it is a foundation, an insurance company, or some other form of donation, these people will want to see documentation about how the money was spent. We need to remember the golden rule of financial support: he who has the gold makes the rules. If specific planning work is not done prior to an incident, then when an incident occurs, it will be difficult if not impossible to control the costs. Part of the planning to control finances will include how to provide cost assessments and how to properly document expenditures. Part of the preparedness and mitigation for funding the response will be to create methods of recording the time that personnel are working (including mutual aid personnel) and a method for procurement control. Procurement control is undertaken so that not just anyone can buy something and charge it to the incident. While it is necessary to support the incident, at some point, the managing of claims for the incident will need to be rectified. We will have to pay for what we used. In the ICS method (during the response and recovery modes), this will be accomplished by the Finance (and Administration) Section Chief (FSC).

Another way of controlling the incident is based on managing the public perception. With the increase of social media, and the advent of the 24-hour news cycle, it has become increasingly important for public safety agencies to provide public information. In the ICS method, this duty is performed by the Public Information Officer (PIO).

There have been instances when no information came from a responding agency, and social media and/or the media began to speculate what was happening, or what had happened. If these media channels decide to put out negative coverage, or even wrong information, it can undermine the response and recovery mode. Imagine if you will that you are just minutes away from issuing a press release to shelter in place to protect people from a mass shooter, and someone on social media tells everyone to get out of the area. Imagine if you will, a wildfire is encroaching on an escape route and the media is telling people to evacuate in a way that would unknowingly put the population at a higher risk. To manage public information in an effort to assist the controlling an incident, it is most likely Public Information Officers (PIO's) will need to do the heavy lifting during the planning and the preparedness and mitigation modes. In the planning mode, they will need to set up as many automations as possible to send press releases. Also, in the planning stages, protocols and policies can be made to ensure only information that should be released actually is released and that there is a priority on public information. Rather than enter each e-mail address, or in entering individual fax number, they can preplan a group (or groups) that send out press releases to multiple numbers and addresses in just seconds. In the planning and mitigation mode, they can begin to build relationships with their social media partners and the media. By doing so, they help to mitigate the potential for contradictory or misinformation to be released. This helps to better gain control of the incident.

A way of controlling the incident is also to provide a safe operating environment. In the ICS method, this is a task that is relegated to the Safety Officer (SOFR). Having an unsafe environment can, and does, prevent those involved with the incident from working in a fast and efficient manner. Additionally, if personnel feel safe, and they are safe, they will have less stress while working in an inherently dangerous situation. Should personnel become injured while operating at an incident, then other personnel will need to remove themselves from actively working on the incident and make the injured member their top priority at that moment. In the response and recovery mode, safety must be constantly monitored, and a culture of safety should be instilled in all personnel. In the planning mode, protocols and policies should be made that address safety issues, and in the preparedness and mitigation mode, equipment, training, and other safety mitigation measures can be reviewed. As an example, if preparing and mitigating safety for a pandemic, there may be the purchase of advanced Personal Protective Equipment (PPE), and training on how to use it. After learning how to use it, then a mitigation measure would be to identify how to decontaminate that equipment and the personnel that wore it. By doing so, you create a safer environment for personnel, and you reduce the chance of infecting more people.

Finally, in order to control an incident, it is important to ensure that all assisting and cooperating agencies have their needs met and that they are used in an efficient and effective manner. In the ICS method of IMS, this is typically done by the Liaison Officer (LOFR). If an assisting or cooperation does not have their needs met, such as food, water, sleeping accommodations, medical needs, then they cannot operate at optimum levels, and it will slow the response and recovery. Not taking care of their needs, especially when they are part of the solution, would be inhumane. Additionally, utilizing them in an efficient and effective manner helps to ensure that there is no duplication of services and that the person in charge is not waiting for specific resources when that same resource is already working on the incident. As an example, if there is a fire crew that has all licensed EMT-Ps (Paramedics) or Fire Medics, and they are only blocking roads to prevent entry, yet more paramedics are needed desperately at the triage area, it does little to control the incident. It is also an inefficient use of resources. If the Liaison Officer (LOFR) notifies the person in charge that these individuals are the resource that is needed, and they are trained to the level needed, then a single law enforcement officer can block the road while five paramedics go to work at triage. In the response and recovery mode, it is extremely important that the Liaison Officer (LOFR) gets to know the assisting and cooperating agencies and that they act as a voice for these agencies, including their safety and proper utilization. In the planning mode, protocols and procedures can be initiated that promotes diversity and the humanitarian and efficient use of these agencies. In the planning and mitigation mode, mitigation measures could be put into place that identifies all of the skillsets that these resources may have. Creating agreements with how these resources will be housed and fed can also be a mitigation measure, as well as enhancing and simplifying reporting procedures.

As can be seen, controlling the incident is part of managing an incident, and a large part of IMS. While controlling a major incident or emergency may be overwhelming, it is possible to control an incident if IMS is implemented properly. The first time you manage a major or extremely complex incident and positively control all of the major outcomes, you walk away with the sense that you can tackle the world.

4.3 Communications

As was mentioned in Chapter 1, the 1960s and 1970s was riddled with public safety challenges. Personnel faced many challenges, especially when attempting to manage a major incident. One key area that had ongoing issues was in communications (Federal Emergency Management Agency [FEMA]). Terminology between multidisciplinary, and in some cases multijurisdictional, agencies were often difficult and presented huge challenges. This often led to fractured and disorganized tactical operations during an incident. An investigation into these communication issues found that it usually had two primary causes:

(1) differing terminologies between different agencies or
(2) a lack of compatibility (interoperability) between agencies due to hardware and frequency differences

4.4 Responder Communication Problems

4.4.1 Terminology

During an incident, agencies could usually communicate within their own agency in an effective manner. When a crisis expanded, and outside agencies were called in, problems began to become apparent. Each individual response agency had its own self-evolved professional vocabulary (or lingo) in many instances. This lingo included the use of acronyms and undescriptive words such as "10-codes" to convey a message, whether it be via radio or through face-to-face communications. This was especially problematic when responding agencies came from different geographical areas; one agency's use of a term, acronym, or other dialect would be in direct contrast with another. This sometimes put operational responders at risk over a lack of communications, or miscommunications.

4.4.2 Interoperability

Joint communications and interoperability during this time period was essentially nonexistent. In many larger incidents, communication system incompatibilities were unmistakable and could be seen by even the most junior personnel that were responding to multiagency incidents. In later years, even where sophisticated modern communications systems were in place, there were major incompatibilities. In most instances, these incompatibilities and agency (or jurisdictional)-specific problems were centered around interoperability and interagency communications. For the most part, the systems available were technologically incompatible. There were reports that during some incidents, first responders were able to physically see each other from a distance, yet they had no method for interagency communication capabilities. This led to a slower response which in turn endangered the public and first responders simply because of a lack of interoperable communications. Over the last 50–60 years, the use of communications and communications devices have evolved substantially. This is not to say that there are no communications problems anymore, it means that it is much better than it used to be.

4.4.3 Current Communications Facilitation

Clear and concise communications at an emergency incident can mean the difference between life and death. In order to facilitate effective communications, a variety of methods are generally used in the response and recovery mode. Such methods include but are by not limited to:

- Face-to-face communications
- Radio communications

- Telephone (land line or cellular) communications
- Written communications including paper and e-mail
- Electronic messaging system
- High-tech communications systems such as satellite and other space-based systems

Communications between personnel is a two-way process by definition: there must be both a sender and a receiver for the information being relayed, regardless of the method used. The process of communicating has three very distinct parts: sending, receiving, and validation (or confirmation). The sender transmits the information to the receiver, who in turn processes the information. The receiver of the information should then repeat that information back to the sender to validate (or confirm) they understand the sender's message. This confirmation should hold true even when face-to-face communication via a runner is utilized.

The validation process is substantially more important when any type of technology (radio, social media, computer, etc.) is used to facilitate that communication. In the absence of face-to-face discussion where the sender can gain visual verification (body language and facial expressions), it is critical to make sure that the receiver has understood the meaning and context of the message. By confirming the information, the receiver confirms that the communication process was completed.

Anyone who has worked in public safety has heard a validation to a message be similar to "Copy that, '10-4'," or some other insufficient validation phrase. When hearing this type of response, the sender of the message may assume that the message was received, but it could be misunderstood, or the person who answered with an inadequate validation may not even be the person that the message was intended for.

An example of the potential hazards that lack of validation might cause could be summed up in a 9-1-1 call, where a sender says that they need the police, but then immediately hangs up. The receiver has no idea why the caller needs the police. Even if the dispatcher can locate the call, he or she has no idea of the nature or severity of the incident. It could be a barking-dog complaint or an armed robbery in progress. The response from law enforcement to these two contrary examples would be markedly different. The lack of validation regarding the exact nature of the call and a lack of communication could put lives at risk.

Similarly, an event might include heavy rains and a dam that is nearing capacity. The person in charge decides that the pressure must be relieved immediately to prevent a failure of the dam. The communications sent, might be to tell the person at the dam, "release water," and the response is "O.K.." Twenty minutes goes by, and the person at the dam contacts the person in charge and states, "The water is a foot from the top, shouldn't we start dumping water?" A confirmation of this order may have averted this type of incident before it became critical.

Downstream, the same thing could have happened. An order to someone in charge of evacuation could have advised them to evacuate a wider area, and the person sending the message hears "Copy that" come over the radio. The sender assumes that this validation response was directed at their transmission. After the dam fails, it is found that people died because the response was not directed at the order from the commander, but rather response to a deputy advising that a bridge was out, and he would have to go a different route.

Many of the communications that occur during a significant crisis are potentially critical in nature. In real-world incidents as well as in scenarios and exercises, poor responses are most often related to failures in communication. During the initial period of amassing resources for a response, communications are frequently confusing due to the high volume of radio traffic between dispatch, initial responding agencies, command, and arriving units. Communicating with other agencies such as utilities and nonlocal responders creates additional challenges and greater chaos. It is imperative that all communications be clear and concise to assure that the response forces and command personnel are able to better categorize and then prioritize this information. This helps to ensure that the information was received as presented.

Interoperability among agencies has come a long way since the 1960s, and it has substantially evolved since September 11, but there is still much work to be done, especially at the local level. Many at the local level become complacent in communications, and in major incidents, this complacency can cause disastrous results. Another issue that arises is the inability of local level governments to integrate communications among first responders. This holds especially true when varying agencies utilize different bandwidths for their primary communication.

4.5 Integrated Responder Communications

In the years since 11 September 2001, and the terrorist events of that day, the terms "interoperability" and "integrated communications" have become buzzwords in public safety. An amazing amount of commercial and technical solutions have been introduced and are being developed by both public and private companies, and response agencies, in an effort to solve public safety communications problems. Many of those developments are based solely on technological solutions, but technological solutions will not solve all the communications problems that exist in public safety. Public safety leaders must also create and enact policies that support interoperability and clear communications, and then enforce them to ensure it becomes a culture within their own agency.

An awareness of the available communications systems and frequencies, combined with an understanding of the requirements of the incident, will guide those managing a crisis in how to communicate. This means that those in charge of an incident must research what is available in their area and preplan how they are going to communicate with other agencies. In doing so, they will help to bring order out of chaos because they will be able to communicate more effectively. One major consideration that has implications for all emergency personnel is the use of "codes" during radio transmissions in the process of communicating. Whenever possible, all radio communications between agencies and all portions of their organization should be given in clear and simple language. Of course, in day-to-day operations, there are some instances, where this is not possible; however, the use of clear and simple language will relieve many of the misunderstandings that have occurred during crisis incidents (Illinois Terrorism Task Force, 2011).

It was briefly mentioned that there are times when codes are acceptable. Anyone who has been involved in public safety knows that things are not always textbook incidents. Instances when information transmitted in plain language may contribute to the destabilization of the situation, and put people at risk, may require the use of codes. Case in point: in an incident that happened several years ago, a law enforcement agency had just switched to plain language. The implementation was haphazard. While several officers were dealing with a mentally unstable man (due to drug use), an officer responding to the air broadcast the transmission, "Do they still need my assistance, or do they have that nut-case in custody?" Hearing the word "nut-case" across the radio, the man went ballistic. Before it was over, eight officers were assaulted, and three officers were injured requiring hospitalization.

In some instances, the use of codes should be acceptable, providing that all of the responders (and dispatch) know and understand those codes. Once again, those codes should only be used when it could lead to the destabilization of the incident. In the mentally unstable scenario above, if the word "nut-case" was replaced by the numbers commonly used by law enforcement of "5150" or the cryptogram of "Code purple" used in other areas, then there is a good chance the situation would not have escalated. While the use of codes is not usually acceptable in incident management, it could be contended that destabilization of an incident has the exact opposite effect of managing an incident. That is why using other than plain language should only be used sparingly and only be used as absolutely necessary.

4.6 Creating a Communications Unit for Responders

In a larger crisis, it might be advisable to create a Communications Unit. The Communications Unit is responsible for developing plans for the effective use and utilization of crisis communications equipment and facilities. This includes

- Installation and testing of communications equipment
- Supervision of a communications center
- Distribution of communications equipment to crisis personnel
- Maintenance and repair of all communications equipment (FEMA, 2014)

Collectively, these activities are known as creating a Communications Plan. A simple way of documenting and organizing these activities is to use the readily available ICS Form-205, the Incident Radio Communications Plan, which is found on numerous websites. It is in your best interest to print copies of this form prior to needing it as Internet connection may not be available in a major crisis. The ICS Form-205 will be discussed later in this book along with other common forms used in incident management.

When created, and implemented, the Communications Unit's major responsibilities are effective communications planning for the incident and the managers working the command elements of the incident. This holds especially true when involved in a multiagency or multijurisdictional incidents. When managing a major incident, it is critical to

- Determine the needed radio nets for command staff, tactical staff, support staff, and air units
- Plan the use of, and establish, interagency frequency assignments
- Ensure interoperability
- Expand the use of all allocated communications, including setting up on-site telephone and public address equipment as well as providing any required off-incident communication links
- Avoid codes of any type (numeric or alpha based) as codes should never be used for radio communications in large-scale incident communications

A leader should be appointed to the Communications Unit who has knowledge in communications and the intricacies involved so as to avoid errors (FEMA, 2014). By appointing a knowledgeable leader, there will likely be a reduction in the number of communications problems in the response to the large-scale event.

4.7 Radio Networks

Radio networks are an important aspect when it comes to a multiagency response such as a large-scale incident. A radio network is set up to ensure that no single frequency is overloaded when responding to a major incident. If the Incident Commander (IC), their staff, and all tactical responders are all operating on the same frequency, the frequency will become so busy that it will slow operations based on a lack of information flow.

Imagine trying to pour five gallons of water through a small kitchen funnel. In no time at all, the water begins to fill the funnel and it will overflow the sides of the funnel. However, if multiple funnels all the same size are used, then more fluids can transfer quicker because one funnel is not overloaded. This is what can happen if there are not multiple networks to carry the large volume of fluid radio traffic.

By dividing (and in some cases subdividing) the communications into radio networks (or "nets"), there will be less confusion and better communications (FEMA, 2014). The most common way to divide communications in a crisis are

- *Command net*: Command net links are used to connect the person in command, his/her staff, and all supervisors on one frequency. In extremely large incidents, there may be separate communications channels for each separate general staff position, such as Logistics, Planning, and Operations as well as other major functions.
- *Tactical net*: Tactical nets are established to connect local and nonlocal agencies, departments, geographical areas, or specific functions responsible for a specific portion of the tactical response. In larger incidents, there may be multiple tactical nets. The number of nets needed is based on how large and/or complex the incident has become.
- *Support net*: Support net is used for logistical and nontactical functions. Support nets may be established primarily to handle changes in resource status, but also to convey logistical requests and other nontactical functions. These can include cooling stations, food services, medical, communications, sheltering, and other support for an incident.
- *Ground-to-air net*: Ground-to-air net is used to coordinate ground-to-air traffic. These nets are used to give orders to, or receive information from, air units. The Ground-to-Air Net can either be a specific tactical frequency or it may be a designated separate frequency. It is acceptable to use regular tactical nets for ground-to-air nets.
- *Air-to-air net*: Air-to-air nets will normally be predesignated and assigned for use at the incident by local aviation authorities. This is typically a way for air support to communicate with each other. This is important in identifying where they are located (to avoid collisions) and to tactically identify the specific locations of their last tactical response completed (such as a retardant or water drop on a fire).

Communications among responding agencies does a lot to help mitigate the crisis at hand, and it is an important part of an effective crisis response management strategy. While communications among responders is important, they are only the tip of the iceberg when it comes to communications.

4.8 Stakeholder Communications

Communication among responders is vital to help effectively mitigate a crisis and implement an effectual incident response strategy, but responders are not the only individuals that require communications. The Incident Commander (IC) also has the responsibility to coordinate or designate their staff to coordinate with key people and officials (known as stakeholders) from all concerned agencies, businesses, and organizations. Stakeholders might be able to assist in the response, or they might require information or assistance, but all stakeholders have a personal interest in a positive outcome for a wide variety of reasons. Some stakeholders are often, but not exclusively, part of the Emergency Operations Center (EOC); for this reason, each will be covered individually.

When communicating with stakeholders, be mindful that everyone does not communicate in the same way. Portions of the population may use English as a Second Language (ESL), have disabilities that affect communication such as those who are deaf, those with Traumatic Brain Injury (TBI), or individuals with cognitive or psychological disorders. Additionally, older individuals may not be able to understand technology or be able to see an informational crawl on a television (Warnick, 2015). Successful crisis communications need to encompass all possible populations not just the average person. FEMA (2016) call this a whole-community approach.

Every effort should be made to make communications with stakeholders all-inclusive. While the Emergency Alerting System (EAS) reaches the vast majority of populations, FEMA, (2014) states that up to 25% of the population need accommodations when it comes to emergency alerting. Recent advances using the Integrated Public Alert and Warning System (IPAWS) has likely helped to reach 90% of the population, but this still leaves 10% that are not receiving possible life-saving crisis communications (FEMA, 2014).

This is just one portion of the overall picture that is crisis or disaster communications. As we look through some of the stakeholders that might need to integrate into our communications, consider how you might effectively communicate with these individuals in your response area. This thought process will also help you think outside the box while knee-deep in an incident. Some of the stakeholder examples later are traditional resources; the vast majority of them are nontraditional resources. Nontraditional resources are frequently overlooked during an incident, but they can be a significant asset during an incident response, especially in large-scale incidents.

4.8.1 Government Stakeholders

Local, state, and federal officials (or tribal governments) are stakeholders in your community. Each can play a critical role in mitigating the effects of an incident, and open lines of communication should be utilized to communicate with them. These stakeholders have a vested interest in ensuring that the incident is resolved quickly, and with a positive outcome. While government stakeholders are concerned about the individual and the community as a whole, they are also concerned about the short and long-term economic effects that a given incident might have on a community, a state, or the nation. Being able to communicate with these individuals and/or agencies during a disaster or major emergency should be a priority.

Establishing a point of contact and gathering important information prior to an incident will save a considerable amount of valuable time, and it can help diminish the probability of failures. Communicating with federal, state, and local elected officials can also be beneficial in expediting approvals for additional resources and financial support. A simple call to a congressman or senator who is not afraid to put some pressure on the Department of Homeland Security, the Department of Justice, FEMA, or some other entity may yield substantially quicker results. The same principles hold true at the state and local levels.

At the federal level, most understand that FEMA, the Federal Bureau of Investigation (FBI), and/or the Bureau of Alcohol, Tobacco, Firearms and Explosives (BATF) may become involved in helping to manage the incident. There are a substantial number of other agencies within the federal government that can assist in managing an incident by providing valuable resources and specific expertise, which in turn could help hasten the end of the incident. This can include agencies such as the Center for Disease Control (CDC), Department of Energy, the US Department of Agriculture, the Department of Defense, and many others. Under the Department of Homeland Security, there are 22 listed agencies that are tasked with assisting local communities during a Presidentially Declared Disaster (PDD).

At the state and local levels, there are also numerous resources beyond the state or local police and the state or local emergency management (or Homeland Security Agency) that can aid in the event of a major incident. The state or local health department, the Department of Transportation, the National Guard, and others can be called upon to help mitigate an incident. Additionally, local governments (except for some in extremely rural areas) often have street departments, water departments, and community centers. For example, a local street department may be able to block off roads with barricades, or they may be able to deliver sand which could be used to soak up or divert/contain a spill. Depending on where you live, they may also have their own utilities (gas and electricity), government-based Internet service, or similar services. It is important to be aware of all levels of available resources so that you can make best use of the services they provide.

The time to identify these resources is prior to a major incident, not in the middle of one. By cultivating these relationships with these resources prior to an event, you help grow a relationship that will create quicker communication, and a desire to help you however they can when the big one happens. Then, when you are in the middle of an incident, you will have a champion in your corner who will be willing to fight for everything available to assist you.

4.8.2 Media Stakeholders

A stakeholder that can be especially important to the outcome of an incident is the media. The media can be an Incident Commanders (ICs) best friend, or they can be their worst enemy. In a major crisis, the Incident Commander (IC) can assign the task of dealing with the media to a Public Information Officer (PIO), or in the case of a multiagency response, the formation of a Joint Information Center (JIC) could be activated. If the Incident Commander (IC) has delegated authority for communicating with the media to a Public Information Officer (PIO) or a Joint Information Center (JIC), the person in charge of the entire incident (usually the Incident Commander [IC]) still needs to have the final say on the type and content of information prior to the release of information.

One key factor that is critical in providing information to the media is to only provide the facts. Speculation that is released to the media can, and usually does, undermine public confidence severely, especially if it is later found out to be wrong. Even when giving the facts, certain information may need to be withheld based on confidential information, or in instances where it would compromise an investigation.

It is important to realize at the onset that the person in charge needs to be exceptionally careful not to become so focused on the tactical response that they overlook the benefits that the media can provide. The most popular method of using the media in their role as a stakeholder is in providing information and direction to the general public. For instance, if a disaster requires evacuations, the media could provide evacuation orders and route information. The media can also be used to identify the location of local emergency shelters, or where to get a hot meal. Similarly, the media could also be used to inform the public if a perpetrator is on the loose, or if a child is lost in a specific area.

The media is a powerful resource that should be substantially engaged, unless it may compromise the operation, the security of that operation, or create an unneeded panic. In many cases, lack of information can create a panic as well. When you do not communicate with the public (including through the media), people's imaginations can, and usually do, run wild. What may seem like a medium-sized incident to you, can be a huge incident to the public. Of course, there is also the rumor mill that will likely start if there is no information being released, and in no time at all, the rumors will become facts in the minds of those spreading rumors. This is but another reason that it is usually in your best interest to keep the media informed.

It is also important to realize that if the media is not included and/or updated on the response and regularly briefed, they could get inaccurate information from bystanders who might want their 15-minutes of fame. Worse yet, this bystander who has no training in what is going on could suddenly become an expert in their own mind, and report that the responders are not doing a proper job. In the case of an unscrupulous news organization, there is also a possibility that they could potentially fabricate, or even misrepresent, a story describing what they believe happened, or is happening. This could potentially cause added complications and panic among the general public. Those who work in public safety only need to watch the news to see what an integrated element the media has become in relationship to a major incident. The media can show up at any scene which they have determined might be newsworthy. Whether it is a car accident, structure fire, or terrorist incident, it is possible that a newspaper or television station could show up on your scene and be your best friend or your worst nightmare. It is in your best interest to use the media to your advantage when managing any type of incident, rather than potentially letting them undermine the management of the incident.

4.8.3 Social Media

Social media is perhaps the newest trend in incident communications, and its use is so widespread that it should not be disregarded. A Congressional Research Service (CRS) report by Lindsay (2011) found that social media is useful before, during, and after a disaster. Social media can be used for communicating preparedness and readiness information before an incident ever occurs, and it can be used in communicating instructions to the community during an impending/current incident. Social media can also be used to communicate notifications of trainings and drills, so the community realizes that it is not a real incident.

Social media can be useful in all modes of an incident as well. During the planning phase, social media can be used to get the pulse of the community when it comes to the community priorities. Social media can also help to find what might be useful or needed in planning for the elderly or disabled. In the response mode, if social media communication is already in place, directions can be given over social media, and can request help from the community. During recovery efforts, social media can be used to direct the community to resources for feeding, where to file claims, where there are long-term shelters, and much more. As part of preparedness and mitigation, social media can be the mitigation measure to communicate with stakeholders, and plans can be made on how to more effectively utilize social media (Lindsay, 2011).

Social media can also be useful in communicating calls for assistance. These calls for assistance can come from citizens, or they can come from people active on the scene of the incident. If most methods of communication are overwhelmed or incapacitated, a person that needs assistance can put out a call for help using social media. If a sandbagging operation is being started in order to hold back the river, and first responders need the community to help, then a call can go out over social media for volunteers. If social media is used and monitored properly, it can also help to provide better situational awareness to those in charge of the incident.

An example of potential uses of social media might be local officials sending out an alert stating that everyone should take shelter due to an impending tornado. Once the tornado has passed, the agency could provide the all clear and/or identify areas to stay clear of, what usable travel routes were available, the locations of disaster shelters, what medical facilities were affected or unaffected, and when and where food, water, and other essentials are available. If those in charge wanted quick information, they could ask citizens to provide feedback regarding the damages in their area including

- Blocked roads
- Infrastructure damage (downed powerlines, damaged roads, etc.)
- Injuries
- Deaths
- Immediate or critical needs (food, water, medicine, etc.)

The use of social media can then become intelligence for a damage assessment, critical needs, and other important situational awareness information that could help guide response. During the recovery phase, social media can be used to identify points of distribution (POD), the available materials that have been donated, and how to go about obtaining donated materials and assistance. All of this can be listed on social media and it could be used to reach the general population. Another advantage of social media is that when you list information on social media, many people will share that information, which in turn amplifies the number of people you reach.

4.8.4 Local Utility Companies

Communication with utility companies can be vital. This is another stakeholder who could provide vital services during an emergency response. Utility companies might need to turn off electricity, shut down gas lines, or potentially they may even be needed to restore utilities during or after an incident. In major incidents that require long-term recovery, the utilities can (in some instances) route temporary electricity and gas to areas critical for promoting recovery. This could include recovery centers, feeding centers, sheltering facilities, and other crucial locations.

When we think of utilities, we typically think of gas, electricity, and water. Other utility companies to be considered are the local Internet provider, the local phone company, and while not traditionally identified as a utility, the cell phone service providers that service an area. These utilities can be extremely important to those that need to communicate with others outside of a major incident, especially those in the command structure and local medical facilities. Utility companies are especially important during large-scale disasters such as ice storms, blizzards, earthquakes, or hurricanes. In order to utilize these utility company stakeholders, you will need some way of communicating with them; this holds especially true if an Emergency Operations Center (EOC) has not been activated, the utility company is not represented in the Emergency Operations Center (EOC) or if standard methods of communication (e.g. phone and Internet) are nonoperational.

It is important to note that most utility companies have standing mutual aid agreements nationwide. They work together under an IMS method to restore utilities in a fast and efficient manner. In the United States, the most common method for managing utility company response and recovery is the ICS method. Because utilities are part of the infrastructure of a community, those in charge of the incident can communicate the highest priorities for restoration, which in turn will help restore the infrastructure more quickly to support an incident and to support an expedient incident recovery.

4.8.5 Local Businesses

Local businesses are stakeholders that can also provide significant resources including food, shelter, heavy equipment, trucking, tools, transportation, and various other necessities. Most local businesses are willing to assist in an incident out of the goodness of their heart and/or because of civic responsibility. Depending on the incident, businesses can also have an economic interest in assisting. Regardless of a business's motivation, they have a vested interest to ensure that response and recovery of a crisis is mitigated quickly to lessen the economic impact on the business and their community. When resources are difficult to find in the initial stages of a crisis, many local businesses will have resources on-hand that may take hours, or even days, to obtain through other channels.

For example, imagine a large fire involving a HazMat spill. Local businesses could be incorporated into the response to provide sand and the dump trucks to deliver it. Another local business could provide heavy equipment to make trenches, dams, or pits, which might assist in damming, diverting, and/or diking the spill. If the crisis involved long-term operations, other local businesses might provide food, water, portable restrooms, tools, and specialized equipment that may be needed. Many times, those in command of an incident overlook communicating with these available resources in their community. They also either fail to realize, or make assumptions about, the willingness of these businesses to contribute to the response and recovery.

By opening lines of communication in the preparedness, planning, and mitigation mode of an incident, public safety can ensure that additional help is available in the response and recovery mode. In many towns and cities throughout the United States, resources for response and recovery will be available in your own neighborhood, if you will only cultivate a relationship with them. By engaging in communications with these local businesses, you create more options for when an incident does occur, and your response can be part of the economic recovery for the community.

4.8.6 Civic Organization and Advocacy Groups

Civic organizations and advocacy groups are stakeholders that can provide substantial assistance during major incidents. As with government stakeholders, communications with these organizations and groups should be initiated long before their services are required or needed. Specifically, advocacy groups (e.g. organizations for hearing impaired, sight impaired, disabled, disabled veterans, elderly) intimately know the needs of those they represent. Not only do they understand the specific needs, but they also know the most appropriate methods of communication. This is because of their daily interaction with the individuals they represent. In some instances, they may already have systems in place and the technical expertise about the best ways to contact, and communicate with, their constituents.

By utilizing ongoing communications with civic organizations in the planning, preparedness, and mitigation modes, those that will command a response to a disaster can better understand how to respond and recover from an incident. This can, and usually will, reduce response and recovery times, create less confusion, and increase inclusivity when communicating with large, diverse populations.

As an example, civic organizations such as the local Parent Teacher Association (PTA) may assist first responders in identifying which students may be home alone during an emergency. With their unique knowledge, they could potentially be enlisted in helping evacuate those children. Similarly, the local hearing-impaired advocacy organization may have a TTY phone tree already in place. By communicating with, and utilizing this resource, they may be able to send mass notices to the individuals they represent. With a brief phone call, the organization could send a TTY message to their constituents that provided information on what action should be taken (e.g. evacuate, shelter in place), the location of hearing-impaired (friendly) shelters, the location of food and water Points of Distribution (POD), or even the location of victim assistance centers.

Predisaster identification and communication with these advocacy groups is a key factor in integrating them into a local incident management system. A study by the National Organization on Disabilities (2007) called *Special Needs Assessment of Katrina Evacuees* (SNAKE) reiterates the need for communication with advocacy groups and organizations prior to a large-scale incident. Post-Hurricane Katrina surveys of 26 advocacy organizations found that 85.7% of those surveyed said they did not know how to link with the local emergency management system prior to Hurricane Katrina making landfall. The report found that in preparing for disaster, the New Orleans Emergency Management Agency (NOEMA) failed in communicating and collaborating with various agencies to determine the risk of elderly and disabled prior to the event. Access and Functional Needs (AFN) advocacy organizations had never been contacted by New Orleans Emergency Management Agency (NOEMA), nor did most of these organizations know how to contact this agency. Treaster (2005) found that if New Orleans Emergency Management Agency (NOEMA) had been involved with advocacy groups, there likely would have been less people left behind and fewer individuals killed by Hurricane Katrina. This contention is also supported by other researchers, and it paints a dismal picture of just how bad public safety failed (White & Whoriskey, 2005; Brodie et al. 2006; Evacuteer, 2011).

Postdisaster research by the Center for Disease Control (CDC) found that deaths during the Hurricane Katrina disaster disproportionately affected the Access and Functional Needs (AFN) population, especially the elderly (Benson, 2007). The CDC report revealed that 71% of those that died in Hurricane Katrina were over the age of 60, and 47% were over the age of 75 (Benson, 2007, p. 1). At the time Hurricane Katrina made landfall, the elderly comprised only 15% of the population in New Orleans, yet they accounted for more than four times that percentage in the death toll. While there is no way to tally the amount of disabled and elderly lives that may have been saved by integrating communication between public safety and advocacy groups, it can be assumed that there would have been a substantial reduction of deaths based on the statistics.

By opening lines of communication with these stakeholders prior to ever needing them, you will create good will. These civic organizations and advocacy groups will usually be appreciative that you have thought about them, and their constituents. In most instances, they will bend over backwards to try to be part of the solution, rather than being part of the problem.

4.8.7 Houses of Worship

Local houses of worship are also stakeholders that can play an important role in a major incident. Houses of worship typically have moderate to large buildings with sufficient square footage that can be utilized to house and care for individuals in short-term durations. They are often a cross section of the community they serve, and they usually have individuals with specializations that might be beneficial during a crisis, especially if they have been trained to Community Emergency Response Team (CERT) standards.

By communicating with houses of worship in the planning, as well as the preparedness and mitigation modes, information can be gathered that will usually be beneficial during the response and recovery mode. You can learn how many active members they have, and how many might be able to help you, to help their neighbors. As an example, let us suppose a severe winter storm moves into a rural area, making the local interstate slick. Local first responders receive, and systematically begin to respond to, hundreds of calls for cars sliding off the roadway, some with injuries. The emergency manager, law enforcement, ambulances, and the fire department start to become overwhelmed. They are overwhelmed by the ongoing emergency response and subsequent injuries, not to mention trying to identify what to do with all the stranded motorists. The Emergency Manager could communicate with local houses of worship near the slide-offs and ask if they can temporarily house, and feed, the stranded individuals until the interstate is travelable again. The local house of worship could agree, and in short order, an army of 30 volunteers set up cots and makes food while local first responders and volunteers begin to deliver people to the faith-based establishment. As time goes on, some travelers realize they left their life-critical medicines in their car. A member of this house of worship is a pharmacy technician. This pharmacy technician calls their boss, the pharmacist, and a doctor on call at the local clinic, to arrange for emergency medicines to be filled. The local pharmacist, or a trained volunteer, delivers the medicines to the church, perhaps even by snowmobile. The church members make sure that all those stranded have their needs met, allowing local first responders to continue to work through the night. In the morning, after the interstate becomes passable, the church members provide the use of a phone for the victims to call tow trucks and/or provides them transportation from the makeshift shelter to wherever the individual needs to go. By initiating communications with the House of Worship long before the incident, the person or persons in charge has provided alternative means of mitigating the situation.

4.8.8 Volunteer Organizations

Another stakeholder to keep in mind and that can provide a substantial amount of help are volunteer organizations. In all modes of a disaster, these organizations can be a key player and an invaluable resource, providing those in charge will effectively communicate with them. The sheer number of resources these organizations can bring to the table is phenomenal, and some believe that they can bring even more resources to the table than FEMA. In most instances, volunteer organizations can be divided into three levels:

- Local (or tribal) volunteer organizations
- State level Voluntary Organizations Active in Disaster (VOADs)
- National Voluntary Organizations Active in Disaster (NVOADs)

Local (or tribal) volunteer organization are those organizations that you can turn to during all modes of an incident. While your agency is in the planning stages, you can utilize these local organizations to better identify what they can do and what their areas of expertise are. You can identify the resources they have, and the resources they have available to them from their parent organization (if they are a local chapter), or from other resources they may have. During the preparedness and mitigation modes of an incident, you can train these volunteers, especially in the effective use of the IMS method your agency uses, and you can create protocols and procedures so that when they are needed, it is a smooth and integrated response and recovery.

In the recovery mode, the state-level and national-level Voluntary Organizations Active in Disaster (VOADs) can bring a multitude of individuals and physical resources to assist with the recovery. Through a statewide and nationwide network of interconnected volunteer organizations, everything from personal items for the victims, to major equipment, can be obtained. This allows those in charge of the incident to overcome many of their challenges with trained personnel that only need direction and very little supervision. It is important to note that in the United States, all Voluntary Organizations Active in Disaster (VOADs) require their members to be trained in the National Incident Management System (NIMS) method and the ICS companion method as their IMS method.

As an example of how they might be applied using the same winter storm scenario as above, those in charge of the incident could communicate with (and use) voluntary organizations to do much of the same work as the houses of worship did, and then some. The local American Red Cross or Salvation Army could set up shelters, cots, feeding, stations, and provide for the basic necessities of the stranded individuals. If needed, additional support could be requested from the state and national level of the organization. Concurrently, the local Community Emergency Response Team (CERT) organized under the emergency management agency could provide medical care, medicines, and ensure that local access and functional needs (disabled, elderly, etc.) individuals have their needs met. The Community Emergency Response Team (CERT) may have a group of individuals who have snowmobiles or four-wheel drive vehicles that are capable of checking vehicles that have slid off the roadway. As they check them, they tag them as searched so that first responders do not spend extra time checking them. After marking the car with a flag, they notify the local HAM radio operator providing all the necessary information (car location, number of victims saved, and where they are taking them), and they transport individuals to the shelter. They could also set up signs on the interstate that point travelers to the emergency storm shelter. In doing so, they help alleviate and mitigate the problem of ongoing accidents. These actions would relieve some of the stress on first responders, permitting them to respond to more serious accidents while improving response times to the nonlife-threatening accidents. This allows those in charge of the response to focus on the more urgent issues, while these other stakeholders take care of the secondary issues of those that are stranded.

4.9 Communications Wrap-up

There are literally thousands of stakeholders that were not mentioned in the previous paragraphs who would be able to assist. While not every single one will be used on every incident, there may be incidents where communication with these individuals and groups will help you to better manage the incident. It is in the best interest of local first responders to preidentify traditional and nontraditional stakeholders (and other resources), and to communicate with them prior to an incident so that when they are needed, their capabilities, their willingness to help, their understanding of incident management, and their role in incident management, as well as various other aspects are already in place. The list of those that need communicated with should be updated annually (at a minimum), with bi-annual updates being preferred so that when an incident does occur, a potentially valuable resource that could play a part in managing an incident is not forgotten.

4.10 Cooperation and Coordination

Cooperation and coordination are so strongly integrated in an incident management context, they are difficult to separate. Cooperation is participation toward a common purpose with no expectation of shared work. Coordination involves making decisions and working together for a common goal. This usually begins in the preparedness and mitigation phase of incident management (Kapucu & Garayev, 2011). In essence, coordination speaks to shared work, while cooperation speaks more to the willingness to work together.

As a leader of a public safety agency, one of the highest priorities should be establishing a system to expand regional coordination and cooperation prior to an incident. Regional coordination allows the local first responder agency to diversify costs, pool resources, share risk, increase their overall return on investment, and usually promotes cooperation. To facilitate effective incident management, local agencies should have written and signed mutual aid agreements within their region as well as Memorandums of Understanding (MOUs) with the private sector, nongovernmental organization (NGOs), and private individuals. They should also develop intergovernmental agreements between various governments and governmental agencies, including multiple agencies that are under the purview of the same local government (e.g. street department, mass transit, fire department, law enforcement, and health department).

Unfortunately, there are some jurisdictions that only have a verbal arrangement, a handshake agreement, or other nonformal agreement. This leaves the leadership of the agency at great risk, especially if legal action is brought against that jurisdiction based on their response to an incident. By showing that the agency had taken the necessary steps (in writing) to plan for, respond to, recover from, and prepare and mitigate for an incident, it provides a higher level of protection from litigation. It also allows government officials to see, and know, you are protecting your community. Additionally, it is important to remember that public safety is a dangerous profession, and that we could lose our life on any incident. If you make a handshake or verbal agreement, will your replacement know who and what agreements you have? With a paper trail, they will have a much easier time of getting up and running.

Agencies that should be included in collaborative efforts are federal agencies (e.g. FEMA, DHS, ATF, FBI, and CDC), state agencies (e.g. Department of Health, State Fire Marshal's Office, State Emergency Management Agency, and Department of Aging), local and/or tribal agencies (e.g. street department, water department, parks, and recreation), and other governmental agencies that may be needed in the response and recovery of an incident. For example, each of the nine fire departments in a county supplies its own fire and Emergency Medical Services (EMS) needs; however, each could also take responsibility for one type of technical response such as:

- HazMat Team
- High angle rescue team
- Trench rescue team
- Collapse rescue team
- Emergency lighting (with light trailers)
- Decontamination units
- Swift water rescue team
- Heavy vehicle/machinery extrication team
- Confined space rescue team

Each department would take responsibility for gathering the equipment and specialized training for their area of responsibility. This coordination is important in today's world. Due to dwindling public safety budgets, and a reduction in personnel, it might be cost prohibitive for any single agency to assume the expense of equipping and training for all types of technical rescue. If there was enough collaborative relationship in place, first responders could potentially participate in another agency's technical rescue team, however, that would need to be decided among those jurisdictions.

A group of local law enforcement agencies in a single county or separately a six-county law enforcement agency could potentially put together a regional SWAT Team to address instances that include hostage standoffs, drug raids, fugitive warrants (that involved violent or extremely dangerous criminals), and similar incidents. In creating a regional SWAT Team, each agency may take responsibility for different portions of the SWAT Teams' needs such as night vision, an armored assault vehicle, rappelling equipment, forcible entry tools, and more. Another option is to devise how much each agency will contribute annually, so the regional team could purchase the needed supplies. Either way, these agencies are collaborating for a common goal while diversifying costs, pooling resources, sharing risk, and increasing their return on investment.

Similarly, we can promote nonbinding formal agreement between an agency and another organization. For example, a local emergency management agency may have a Memorandum of Understanding (MOU) with the local portable toilet company. In that agreement, the company agrees to provide sanitary services on extended incidents. By the same token, the local police department may have a Memorandum of Understanding (MOU) with a local advocacy group who deals with abused spouses, or victims of human trafficking. The Memorandum of Understanding (MOU) creates a partnership and provides the reason why the partnership in that area is beneficial. While these concepts will be discussed more in-depth in the Chapter 5, this brief overview should help you understand the basic concept.

In order to bring about a successful response and recovery from an incident, especially a large-scale or especially complex incident, cooperation and coordination is a critical factor. If outside resources are brought in, and they do not coordinate or cooperate with each other, then more chaos will ensue in almost every instance. Whether you are responding to another's jurisdiction, or they are responding to your jurisdiction, no matter if it involves different disciplines (police, fire, EMS, etc.), or just the same types of agencies, cooperation and coordination matters can enhance or even derail the effectiveness of the overall incident management. This is an extremely important factor to remember. We have all heard the saying that there is no "I" in team. Without cooperation and coordination, an effective and efficient response will be nothing more than a pipe dream.

4.11 Cooperation and Coordination in the State of Illinois

The state of Illinois implemented a program known as Mutual Aid Box Alarm System (MABAS). Presently, this is one of the most progressive models of regional coordination and coordination in the United States. MABAS in cooperation and coordination with the Illinois Office of the State Fire Marshal, the Illinois Department of Public Health (EMS Division), and the Illinois Fire Chiefs Association created a partnership with the Illinois Emergency Management Agency (IEMA). In this agreement, they establish a statewide, nondiscriminatory mutual aid response system for fire, EMS, and specialized incident operational teams (MABAS, n.d.). MABAS has a defined resource response plan to supply resources within the state, when the Governor orders a Declaration of Disaster. This statewide mutual aid association can quickly provide emergency resources to communities that are overwhelmed, or do not have the resources available locally.

As of the writing of this book, MABAS has mutual aid agreements with 1175 of the state's 1246 fire departments and the association has agreements with state, federal, and local agencies to provide mutual aid, usually in the form of equipment, during times of large-scale incidents (MABAS, n.d.). The system has even expanded beyond the state of Illinois borders and is working with agencies in other states.

MABAS is a unique program because every agency that participates has signed the same mutual aid agreement, and these agencies are able to work together almost seamlessly on any emergency scene. First and foremost, it is a requirement of MABAS that all public safety use the NIMS method of incident management along with the companion IMS method of ICS. All MABAS agencies operate on common radio frequencies, utilizing at least one radio per member agency that was provided by MABAS. Additionally, assets are strategically placed around the state at participating fire departments, saving local first responder agencies a considerable amount of money while promoting local training and maintenance. MABAS assets in the state of Illinois include

- 25 Decontamination Units
- 7 Warehouse Units (includes most tools needed in a disaster)
- 82 Light Towers
- 11 Cascade Trucks (for filling SCUBA and SCBA)
- 68 Expedient (Command) Shelters with heating ventilation and air conditioning (HVAC, holds 20 first responders)
- 70 ATVs (with medical and Urban Search and Rescue [USAR] capabilities)
- 8 Mobile Ventilation Units
- 3 Tent Cities (Sleeps 200. Includes cots, HVAC, generator and accessories)
- 1 Cross Contamination Trailer (for large scenario blood borne pathogens)

- 17 Boats and trailers (with one rigid hull boat and one inflatable boat)
- 5 Mobile Units (to support Incident Command)
- 69 IMTA Team trailers (carrying forms, clipboards, and other supplies to support an Incident Management Team [IMT])
- 1 Mass Casualty Unit (including 100 backboards, large quantity EMS supplies and morgue supplies)
- 1 EMS Triage/Treatment Unit (Advanced Life Support capable)
- 1 Logistics Support Unit (Food, water, and PPE to support 500 first responders for a period of 72 hours)
- 1 Mechanics Vehicle (with fuel trailer)
- 2 Heavy Duty Utility Trucks
- 1 Urban Search and Rescue Bus

The State of Illinois also has a similar organization for Law Enforcement and Homeland Security known as the Illinois Law Enforcement Alarm System (ILEAS). Much like MABAS, ILEAS provides for the needs of local law enforcement through Homeland Security Grant programs and purchasing equipment for regional placement. Some equipment that ILEAS has purchased includes

- 26 000 APR respirators
- $15 million of SWAT equipment
- Large specialty vehicles such as armored transport/response vehicles and mobile command posts
- 1000 mobile digital computers (laptops for squad cars)
- 1000 interoperable public safety radios
- 6000 radiation detectors (ILEAS, n.d.)

Most likely due to reasons of national security, information is available but limited for ILEAS. For more on the cooperative effort within these two regional collaborative efforts (MABAS and ILEAS), you should visit their websites and glean the information that may assist you in your specific discipline.

4.12 Private Sector Cooperation and Coordination

While some of what is being written may seem redundant, it is important for you to realize that all of the resources are interconnected at almost every corner of incident management. When we look at the private sector, businesses such as heavy equipment, bus, and Medi-transit should be included in not only Memorandums of Understanding (MOUs), but every effort should be made to forge good relationships which promote cooperation and coordination. Cooperation and coordination may extend to consulting, electricity needs, building supplies, and other companies with valuable equipment. Prior to, and during times of disaster (or emergency), advanced cooperation and coordination should extend to houses of worship, the Salvation Army, and advocacy groups (e.g. Centers for Independent Living, Senior Centers, Disabled American Veterans, disability groups). Much like the need to communicate with these agencies, it is also extremely helpful to forge agreements, so they are on the same page as your agency.

It should be noted that even though the American Red Cross is federally mandated to respond to domestic disasters within the United States, local coordination and coordination should be forged in advance of a need. Nothing helps to facilitate mutual aid more than direct communication, discussing the resources that are available, and starting a coordinated and collaborative friendship with the same goals in mind.

4.13 Strengthening Intelligence/Information Sharing with Coordination and Cooperation

Coordination should also focus on intelligence and information sharing that facilitates the proper and timely dissemination of information. By sharing information across multijurisdictional and multidisciplinary responders, a common operating picture incorporates nonprofits, the private sector, and citizens (as deemed necessary). Intelligence is derived by gathering, analyzing, and fusing relevant information from a wide range of sources on a continual basis.

Since the September 11 attacks, various means have been employed to share information and to gather information. One method that has been successful in mitigating criminal activity and terrorist attacks is the use of *Fusion Centers*. Fusion Centers are two or more agencies that combine to create the method (or function) of intelligence gathering and dissemination. They serve as focal points that gather, receive, analyze, and share relevant information about threats between federal government agencies, state agencies, local, tribal, and territorial agencies, and private sector partners. Besides sharing potential threats, Fusion Centers also relay information about identified vulnerabilities and consequence assessments. The input from threat reporting is a method of preparedness and response that can help to manage an incident. While not the typical form of incident management, it is a form of incident management that could be described as a mitigation measure. One of the best methods of managing a disaster or an incident is to prevent it from ever happening. Fusion Center information sharing can do just that.

An example of how this coordination and cooperation creates a beginning for incident management might include a local fire department who notices 10 sets of missing fire gear, while at the same time, a large heavy rescue truck is stolen from an unmanned volunteer fire department 20 or more miles away. Both incidents are reported to authorities. Around the same timeframe, two men in a rental truck buy a large amount of ammonium nitrate from a fertilizer plant more than 80 miles away from the fire department, and a gun shop 200 miles from the fertilizer plant reports the theft of 12 semiautomatic weapons and over 10 000 rounds of ammunition. Each of these are separate incidents and each occurred in a different county. If this information was not shared through a Fusion Center, then analyzed and disseminated, a domestic or foreign terrorist group could have attacked a critical part of infrastructure causing significant death and destruction, but because the shared evidence put this all together, it led to an arrest. This is why Fusion Centers are critical in coordination and cooperation.

In some locations, law enforcement, fire, and EMS do not work together on a day-to-day basis. While they may cross paths and work with a few people here and a few people there, they do not integrate a large portion of their agency into a single response, except for major incidents. Still, a viable way of managing an incident is to integrate various disciplines into the same response. Even so, there are many ways that different disciplines can work together and assist each other.

An example of the importance of coordination and information sharing could include a tornado incident that has caused severe damage to a local jurisdiction. This tornado left a path of destruction covering over 20 square miles. The resources of the fire department and local EMS will likely not be enough to identify exactly where assistance is needed, especially not in a short amount of time. The intelligence provided by the law enforcement and the local emergency management agency who are collaborating can help the person in charge to make informed decisions by collecting intelligence from other agencies. As the fire department, ambulances, and law enforcement responds, the person in charge can receive intelligence and information from each of those agencies. This will help them to quickly determine the damage and what other resources might be needed, as well as manage other aspects of the disaster.

Most modern-day disaster incidents are dependent upon real-time and accurate situational awareness from both the point of the incident and ongoing emergency response activities. Strengthened information sharing and coordination capabilities helps to enable a more accurate situational awareness and allows the development of a real-time common operating picture. These efforts can take place at local, tribal, regional, state, and/or federal levels.

A prime example of this would be the National Crime Information Center (NCIC) program. The National Crime Information Center (NCIC) system began in 1967 and has been a collaborative tool for law enforcement agencies since that time (Federal Bureau of Investigation, n.d.). Much like the National Crime Information Center (NCIC), technology has been emerging in all public safety fields with a significant amount of that technology having a goal of increasing coordination. While the technology can bring large amounts of data into one place, it takes a person to identify what information is important to the community, and how it affects local first responders and the community being served. The police officer, the homeland security professional, and the emergency manager must build cooperative and collaborative relationships, forge agreements, and allow the new technologies to assist them in planning, preparedness, mitigation measures, managing incidents they respond to, and in recovering from these large-scale incidents that may happen.

4.14 Cooperation and Coordination During an Active Incident

While collaborative preparedness sets the stage for a more successful response, coordination in the most chaotic part of an incident is also critical. Certain jobs and/or tasks that are assigned during an incident demand cooperation and coordination in order to properly manage an incident. Working together, and making decisions together, will usually provide a more comprehensive tactical response; a response that effectively utilizes assets where they can do the most good. While certain tasks all but demand cooperation and coordination, they will only be briefly discussed in this chapter as examples to show the importance of utilizing those attributes in specific jobs in managing an entire incident.

4.14.1 Joint Information Center – Cooperation and Coordination

In the case of operations that are conducted under Unified Command (UC) or that include multiple agencies responding, a Joint Information Center (JIC) is usually needed, implemented, and utilized. A Joint Information Center (JIC) replaces the singular Public Information Officer (PIO) that was initially appointed by an Incident Commander (IC). A Joint Information Center (JIC) coordinates all information released to the media and the public about the incident and, in most instances, uses a singular spokesperson unless the members of the Joint Information Center (JIC) determine that a spokesperson from a specific agency should present for the agency they represent.

The Joint Information Center (JIC) will typically be positioned in a location away from the command post. Public Information Officers (PIOs) from most of the agencies, or at least the major agencies involved will discuss, and collaboratively work together, in order to develop what message or messages will be presented to the media and the public. They will jointly cooperate with each other to create a time schedule for news conferences, press releases, and the release of information. This will keep the media and the public informed while helping to instill confidence. They will hold strategic meetings, and they will make decisions on which information can be released by virtue of committee.

While this may seem like an unimportant task, it is critical to the overall response. The use and the proper management of the Joint Information Center (JIC) is a cooperative effort to ensure that all involved agencies have their needs, protocols, and operating strategies met. An example of how this could be important might be a crime scene that involved the use of explosives and nobody claiming responsibility. Fire department and EMS personnel may have no problem with the release of all the information they know, while law enforcement may want to keep certain details hidden from the public in an effort to determine which tips were credible, or even perhaps for later use to potentially trap the perpetrator in a lie. If not for a Joint Information Center (JIC), potentially detrimental information or even specific details that should not be spoken about, could be shared and cause more issues.

Another relative example might be the need of an evacuation of an area due to wildland fires. Fire and EMS may provide a message that everyone needs to evacuate in the given geographical area, while law enforcement may block roads or shut down highways that lead to the fire. Law enforcement can collaboratively provide the best routes to take in order to prevent congestion. If the fire switches direction, inadvertently putting the citizens themselves in harm's way, all of the agencies that are coordinating and cooperating can change the evacuation routes based on the dynamics and fluid changes in the incident. By coordinating and cooperating among the disciplines and agencies, they can send a unified comprehensive message of the fluid changes in the incident rather than sending separate and potentially conflicting messages that may cause more confusion and endanger the citizens they serve.

The primary advantage of implementing a Joint Information Center (JIC) is that there will be a unified and all-inclusive message. Because the Public Information Officers (PIOs) of multiple lead agencies cooperate and coordinate to provide that unified message, it serves in facilitating better incident management.

As was mentioned in previous paragraphs, it may be especially important to gauge the information released in incidents with active crime scenes. This holds especially true with incidents of suspected acts of terrorism. For that reason, singular Public Information Officers (PIOs) on incidents that will likely utilize a Joint Information Center (JIC) should be cautious from the onset about the information that is released. By waiting to ensure that no potential investigative information is released, they create more cohesiveness with response partners, and they do not undermine the potential for a unified voice. In many instances, it is best to have a collaborative mindset from the beginning, and release only the most obvious details. This is working in a coordinating and cooperative manner, even before other entities have arrived on location.

Another important and even critical part of cooperation and coordination with the Joint Information Center (JIC) is the process for which all information that is intended to be released. Any information being considered for release should first be mutually agreed upon and cleared through the Joint Information Center (JIC), and then approved by the Unified Command ([UC] or their designated spokesperson). This helps to ensure that information that might be incorrect, or that is sensitive in nature, will not be provided to the media or the public. While the media and the general public may desire cooperation and coordination from the Joint Information Center (JIC) or a Public Information Officer (PIO), it is important to note that your coordinated and cooperative responsibilities do no lie with the media or the public. Your collaborative partners are the agencies that have responded to the incident, and it is important to make sure that you do not undermine that help.

Sometimes, especially in larger incidents, the media and the public will want to hear from differing agencies, rather than a single spokesperson, with the hope of gaining more information. They will even sometimes request to hear from the lead agency. There are times when the Public Information Officer (PIO) who is the face of the incident may need to bring that agency or agencies into briefings. An example of this may be a passenger train accident that involved multiple federal, state, and local agencies. In this scenario, a Public Information Officer (PIOs) has, for the most part, been the single spokesperson. It may be determined that a briefing, or even part of a briefing, should include a spokesperson from the National Transportation Safety Board (NTSB) to explain the steps being taken. In times when it is decided to provide these individuals, the Joint Information Center (JIC) as well as the Unified Command (UC) should be involved in the message. The individual should be briefed on what information should, and should not, be released, and it should be a coordinated and cooperative statement that satisfies all response partners. This will promote a cooperative and coordinated message, and it will send a unified message. It also helps prevent the release sensitive information.

The importance of a unified message cannot be understated. Whether a singular Public Information Officer (PIO) is the spokesperson, or if different agencies are providing different parts of an overall briefing, it is imperative that there are no discrepancies that could undermine the response or public confidence. Even seemingly insignificant conflicting information coming from those involved in an incident may cause a loss of confidence, and the handling of an incident may come into question and/or heighten anxiety. In the release of information, cooperation and coordination can help manage an incident, or totally undermine it completely.

Imagine a press conference where a local agency, such as a local Sherriff or Fire Chief, stands up and says, "Well, it appears to me that there were two explosive devices that caused this building to come down." After investigation, it was determined that the explosion was caused by a gas leak and ignited by the flip of a light switch. The general public will lose trust in any information this individual presents in the future. At the same time, the conspiracy theorists will say that it was explosive devices, and that "Big Brother" was just trying to cover up the real cause of the incident. A secondary negative may be that the public might lose confidence in the local first responders. All of these issues can occur because of a few seconds when someone blurted out inaccurate information that was based on conjecture, rather than facts.

When dealing with the media and the public, we should all keep in the forefront of our minds what the late 16th President Abraham Lincoln is quoted with saying: "It is better to be thought a fool and be silent, than to open your mouth and remove all doubt." In this case, it is better to coordinate a unified message, and instill confidence, than to have the public, the media, and possibly first responders to lose trust in the important work that is being done.

4.14.2 Liaison Officer – Cooperation and Coordination

Incidents that are multijurisdictional in nature will usually require a Liaison Officer (LOFR) for the coordination of agencies responding to an incident. The Liaison Officer (LOFR) is the point of contact for incident response personnel. Their primary job is to coordinate and cooperate with assisting agencies. Only one Liaison Officer (LOFR) is assigned for each incident; however, on very large incidents, the use of Assistant Liaison Officers may be required to facilitate all the outside agencies involved. Any agency participating in the incident by directly contributing tactical resources is known as an assisting agency; this includes fire, police, or public works equipment that has been sent to another jurisdiction. An agency that supports the incident, or that supplies assistance other than tactical resources, would be considered a cooperating agency. Coordinating agencies might include the American Red Cross, Salvation Army, or a utility company. On some law enforcement incidents, a fire agency may not send fire equipment but may supply representatives for coordination purposes. In this case, the fire agency would be considered a cooperating agency rather than an assisting agency.

As the point of contact for outside agencies, the Liaison Officer (LOFR) and their assistants maintain a list of assisting and cooperating agencies and their respective representatives. They assist in establishing and coordinating interagency contacts as well as collaborating with supporting agencies with incident status reports. The Liaison Officer (LOFR) monitors crisis operations to identify current or potential interorganizational problems, participates in planning meetings, and provides the current resource status (including limitations and capabilities) of assisting agency resources.

Being the Liaison Officer (LOFR) and using cooperation and coordination does not stop until the incident is completely over. In the ICS method, as the incident is coming to a close, or agencies are being demobilized, the Liaison Officer (LOFR) will assist in filling out the ICS Form 221-Demobilization Plan. In assisting with this, the Liaison Officer (LOFR) will provide agency-specific demobilization information and requirements for each agency. They play an important role in coordinating and cooperating with these agencies. In the demobilization, some agencies may have union agreements that may be different from other agencies or departments; thus, the Liaison Officer (LOFR) would ensure that those agreements are addressed and that it takes the appropriate priority. It is only through the hard work and the cooperation and coordination of this Liaison Officer (LOFR) with the responding agencies that it is ensured that all requirements are met and that there is a full integration of resources.

4.14.3 Agency Representative(s) – Cooperation and Coordination

When outside agencies respond to an incident, responding agencies should try to provide an agency representative (AREP). This is especially true in extremely large multijurisdictional incidents. Agency Representatives (AREPs) are provided to assist in coordination and cooperation efforts. Agency Representatives (AREPs) are usually not directly related to a tactical assignment, but rather their job is to work with the Liaison Officer (LOFR) to assist in promoting cooperation and coordination. An Agency Representative (AREP) is typically an individual assigned to an incident who has the delegated authority to make decisions on matters affecting that particular agency's participation at the incident. Agency Representatives (AREPs) report to, and coordinate with, the Liaison Officer (LOFR); if there is none, they report directly to the Incident Commander (IC) or Unified Command (UC).

It cannot be emphasized enough that Agency Representatives (AREPs) roles are clearly defined prior to being deployed. It is important that they have decision-making authority for their respective agencies. Agency Representatives (AREPs) should not be used in a Unified Command (UC) structure unless they are qualified by their agency as an Incident Commander (IC). This is because a lack of qualification and/or authority can delay implementing critical actions needed in a major incident. If certain actions are delayed, it could ultimately have a negative impact on an incident, thereby increasing the likelihood of a regrettable outcome. Considering this is negating the purpose of incident management method, it is an imperative point that they have the authority to act. Having authority can be, and usually is, critical to every incident.

Agency Representatives (AREPs) responsibilities include ensuring that all agency resources that are used have been properly recorded at the incident by coordinating and cooperating with the Liaison Officer (LOFR). They should obtain briefing information from the Liaison Officer (LOFR) or the Incident Commander (IC), and they should inform assisting or cooperating agency personnel that the position for that particular agency has been filled. Agency representatives should attend briefings and planning meetings as required by the Liaison Officer (LOFR) and coordinate and cooperate with others in the planning meeting to identify the best use of their resources. They should also provide input on the use of agency resources unless technical specialists are assigned from their agency.

One very important role of the Agency Representative (AREP) is to ensure the well-being of agency personnel assigned to the incident and to advise the Liaison Officer (LOFR) of any special agency needs or requirements. This will take someone willing to work in a cooperative and coordinating manner in order to assist in management of the incident. Agency Representatives (AREPs) also should be the go-between for their home agency and those in charge of commanding the incident. In some instances, they may need to play the diplomat so that their agency can cooperate and coordinate even more.

4.14.4 Non-command Cooperation and Coordination

Coordination between agencies is not only relegated to the command function of an incident. Cooperation and coordination must also be integrated into the tactical and operational efforts as well. Coordination between ground resources and air assets occurs every time that airplanes, helicopters, drones, or other air assets are used. Coordination should also occur whenever an incident or emergency covers a geographical area that spans multiple jurisdictions. Two of the most prevalent occurrences of coordination during tactical operations are when a *Strike Team* or a *Task Force* which will be better explained in later chapters.

When multiple units are assigned to complete the same tactic together, it is known as a *Strike Team*. Strike Teams are a set number of resources *of the same kind and type* operating under a designated leader. These resources communicate and coordinate together to provide a known capability, and they are typically a highly effective unit. An example of a Strike Team would be five different patrol officers, all from differing agencies, working together to storm an active shooter in a mall under the direction of one team leader. They are coordinating and cooperating to meet the need of the incident.

Real-Life Incident

In 2011, the Mississippi and Ohio Rivers began to rise due to heavy snow melting in the upper Midwest and an especially wet spring. This flooding caused billions of dollars of damage. One city that was affected was the southernmost town in the state of Illinois. Cairo, Illinois, is located where the Ohio River and the Mississippi River meets, and these two rivers doubled the threat for this small community.

During the flood, an Incident Management Team (IMT) was brought in to help local officials manage the incident. This team supplemented or were placed in the command structure to help facilitate better management of the incident. This team was brought in at the Incident Commander's (IC) request. As planning meetings were taking place, one Agency Representative (AREP) was asked if their agency could perform a specific operation. The Agency Representative (AREP) asked to be excused from the meeting and stepped out to make a call. When he came back, he said his agency had permission to perform the operation. Shortly thereafter, the same Agency Representative (AREP) was asked if their agency could perform another operation after the first was completed. As he began to excuse himself to make another call, one of the General Staff members confronted him and said, "When you call, you tell the person on the other end of the line to get their butt down here! We can't keep waiting for you to get permission, this situation is changing by the minute." Within a few hours, the agency sent a new Agency Representative (AREP) who had the authority to act.

While there is no documentation of this story, it was shared in an ICS 400 training class during August 2011 in Shelbyville, Illinois. The instructor of the class was part of the Incident Management Team (IMT) that was deployed, and he described how that Agency Representative (AREP) and the agency was beginning to slow down planning a response before being replaced.

A Task Force can be defined as any combination of resources put together to accomplish a specific mission, and usually include different disciplines and areas of expertise. Perhaps one of the most apparent areas of coordination and cooperation at the tactical level of an incident, is when there is a call for a *Task Force*. A Task Force has a designated leader who operates utilizing common communications but *uses differing resources* to accomplish an objective. This allows multiple and/or diverse key resource elements to be managed under one individual's supervision.

Perhaps one of the best examples to help understand the utility, the cooperation, and the coordination involved with a Task Force would be the 2007 Super Bowl in Miami. As might be imagined, post-9/11 security for a Super Bowl event is extremely complex, yet the same principles were used to manage all public safety at this event. Strike Teams were set up at strategic locations in and around the stadium to address such issues as fire, EMS, and general security issues. There were also specific Task Force Teams being positioned in preassigned locations so that there could be a quick response to a specialized problem (Nuñez, 2009).

One specific Task Force was assigned to the Explosive Ordinance Disposal (EOD) Team, who were part of the larger contingent identified as the Bomb Management Team. The Bomb Management Team was a multiagency and multijurisdictional team tasked with Improvised Explosive Devices (IEDs) and Chemical, Biological, Radiological, Nuclear, and Explosive (CBRNE) protection and response. This multidisciplinary/multijurisdictional team was tasked with all-hazards canine sweeps, vehicle and cargo inspections, air monitoring, and postal inspection. Another group of Task Force Teams formed was the Joint Hazard Response Teams (JHRTs). Their primary purpose was to detect, identify, and neutralize any explosives found. In each of these teams, there was a mix of responders from varying agencies including

- An FBI Bomb Technician
- An ATF Agent
- A Miami-Dade Fire Department HazMat Technician (or an FBI Hazardous Material Response Unit member)
- A HazMat Technician from the 44th Civil Support Team
- A canine team
- A Department of Energy Radiological Assistance Technician

This diverse set of JHRT Task Force team members was typically led by a Miami-Dade Bomb Technician, a local resource (Nuñez, 2009). In order to have a unified response, all agencies and individuals had to cooperate and coordinate together to meet the end goal. While there was not a terrorist attack or a major incident that occurred, the cooperation and coordination of all involved help to ensure the safety of others by utilizing an IMS method and principles. Much like the example of driving a car, the more you collaborate, the easier it gets, and the more natural it becomes.

4.15 Conclusion

As this chapter comes to a close, it is important to note that the Five-Cs are an integral part of incident management. These principles are interwoven into the now mandatory ICS method, which is a component and the NIMS method used in the United States. Unfortunately, in most classes on the ICS component of the NIMS method of incident management, instruction is given as to what personnel should do, however, they do not tell you why you should do it that way. They explain how it should be done, but only touch on the principles that drive these IMS methods.

Most first responders have inquisitive minds and are critical thinkers. If they are treated as robots by being told what to do, it is highly likely they will not comprehend why they do it in a certain way. By understanding the underlying principles, they will likely have a better grasp on why they do it in a certain way and be more apt to do so.

As this book moves forward, you will see, Command, Control, Communications, Coordination, and Cooperation will be interlaced into incident management. Without the Five-Cs, there will be more chaos, confusion, and uncertainty; with the Five-Cs there will be an advanced level of calm, understanding, and confidence. Each individual, and each agency, needs to be a part of the solution in order to facilitate a positive outcome. The underpinnings that were described in this chapter are purposely built into every effect IMS method for a reason. They are key elements of managing an incident. As you see these key underpinnings integrated into a response, it will all make more sense to you why things are done in this way.

Real-Life Situation

Hurricane Katrina decimated many fire departments, EMS providers (including EMS squads), and emergency management agencies on the Gulf Coast in 2005. A nationwide nonprofit known as Helping Our Own was contacted by someone within the NIMS structure, and asked if they could help provide equipment that might help put these agencies back in service.

Helping Our Own was uniquely positioned to undertake this task because they collected surplus fire and EMS equipment, then refurbished it, and donated it to fire departments and EMS squads in the United States. Within minutes, Helping Our Own put out a call for assistance and within the first 24-hours had over US$2 million dollars of equipment donated. Along with the donated equipment, a fire department on the West Coast offered two trucks with firefighters to drive them to help accomplish this task. This created a task force that would supply equipment to Hancock County, Mississippi, on the Gulf Coast.

As the West Coast trucks were making their way to the Michigan headquarters of Helping Our Own, the organizations box truck and semi began picking up equipment. When the organizations trucks came in, the equipment was sorted, and serviceable equipment was packaged to be loaded on the delivery truck and equipment that was not serviceable was set aside to be refurbished. Some workers began refurbishing equipment on 12-hour shifts, while other workers loaded and unloaded trucks, and sorted the equipment.

The Task Force Leader would send the semi out to pick up more equipment, and the West Coast and organizations box trucks were loaded with equipment going to Hancock County, Mississippi. Over the period of four weeks, seven fire departments were put back into service, including delivering multiple fire apparatus, a mobile command unit, and millions of dollars of fire and EMS equipment.

The use of a Task Force allowed the organization to focus on the task at hand, while coordinating with their Group Supervisor, and ESF-4 (Firefighting) contact. Because the Task Force used different resources that cooperated and coordinated together, using command, control, and communication, the task force was able to focus on their task and provide a lot of equipment in a short amount of time. It was critical that the Task Force leader communicated with the Group Supervisor to determine what equipment was needed at each location, and what priority equipment was needed, and where it was needed.

Chapter 4 Quiz

1 Which of the following are included in the Five-Cs?
- Command
- Coding
- Comparison

- Control
- Communication
- Compassion
- Coordination
- Coordination
- Configuration
- Contamination

2. Are any of the 5-Cs more important than another?

3. What are the two most common communications failures in incident management?

4. What are six ways of communicating during an incident?

5. What are the four responsibilities of a communications unit?

6. **True or False**: Social media is not an important part of communications.

7. **True or False**: Coordination involves making decisions and working together for a common goal and usually begins in the preparedness phase of crisis management.

8. In the _____ _____, businesses such as heavy equipment, bus, and Medi-transit should be included in Memorandums of Understanding (MOUs). Consulting, electric, building supply, and other companies with valuable equipment or responsibilities during times of disaster or emergency should also be considered.

9. **True or False**: Advocacy groups such as veteran groups, hearing impaired organizations, and other access and functional needs organizations are stakeholders that should be included in cooperation and coordination efforts.

10. **True or False**: When initiating and operating a Joint Information Center (JIC) only one person makes final decisions, so coordination and cooperation are not as important.

11. **True or False**: A Liaison Officers primary job is to coordinate and cooperate with outside resource agencies.

12. It is important that the Agency Representative (AREP) has the _____ to act and make decisions.

13. A Strike Team is

14. A Task Force is

Self-Study

Information on Zoo and Aquarium all Hazards Preparedness (ZAHP). Retrieved online at https://zahp.aza.org.

Edison Electric Institute-Mutual Assistance. Located online at http://www.eei.org/issuesandpolicy/electricreliability/mutualassistance/Pages/default.aspx.

Department of Homeland Security Brief-Mutual Aid Agreements. Retrieved online at https://www.hsdl.org/?view&did=765527.

Emergency Management Magazine (4 August 2010). *Effective disaster management strategies in the 21st Century*. Retrieved from http://www.govtech.com/em/disaster/Effective-Disaster-Management-Strategies.html.

Key search terms: Any variation of the main words command, control, communication, cooperation, coordination with one of the subwords, in disaster, in an emergency, in a terrorist attack, incident management. You can also focus on main words found in this chapter such as Mutual Aid Agreements, Memorandum of Understanding, Liaison Officer, Agency Representative, MABAS, ILEAS, etc.

5

The National Incident Management System (NIMS)

As everyone in Louisiana knows, there was often no communication or coordination between the state and federal government in the aftermath of Hurricanes Katrina and Rita.

Bobby Jindal

Before introducing you to the National Incident Management System (NIMS) method, there are some important things to realize. The first is that this chapter is a synopsis of what one individual believes are important aspects of the NIMS method. It is not, nor should it be considered a full overview of NIMS. An entire book could be written, and somewhat has been, about the NIMS method. It is also important that the NIMS method integrates and overlaps with the Incident Command System (ICS) method and vice versa. While NIMS is the overarching method for managing an incident, ICS is more localized and focuses on managing the incident with the resources on hand. ICS is a companion to the IMS method of NIMS, however it can also be used independently, depending on the situation; although this is not recommended. This chapter specifically talks about NIMS, and the factors that make this system successful.

The NIMS method is mandated under Homeland Security Presidential Directive (HSPD) Number Five (HSPD-5) and Homeland Security Presidential Directive (HSPD) Number Eight (HSPD-8). The NIMS method has been the core doctrine for managing incidents in the United States. The NIMS principle utilizes and creates an integrating command and control system that spells out communications, coordination, and cooperation so that outside resources can be integrated seamlessly. In most instances, the foundation of NIMS doctrine creates a flawless integration of resources, even if agencies have never worked together before, including agencies from a different country. The NIMS method integrates these resources so well because it is comprehensive, flexible, and adaptable. When issues arise with NIMS, the vast majority of these problems are caused by user error.

The NIMS method is an all-inclusive, national approach to incident management. It is appropriate for all jurisdictional levels and is functional for disciplines and interdisciplinary responses. It is intended to be valid across a full variety of potential incidents, hazards, and impacts. The functionality of NIMS is not impacted by size, location, or complexity of an incident. It is also designed in such a way as to improve coordination and cooperation between public and private entities in a variety of incident management activities, and it provides a well-established distinctive standard for overall incident management.

Before going into what the NIMS method is, perhaps we should discuss what the NIMS method is not. The NIMS method is not a response plan, and it should not be used as one. It should not be saved and only used in only major emergencies, major disasters, acts of terrorism, or during only large-scale incidents. The NIMS method should be used whenever an emergency incident occurs. The NIMS method should also not be considered a communications plan, even though it guides us in how to communicate better. It is not an incident or organization flow chart, and it is not designed to be more useful to any one discipline or type of incident response personnel. It is not a rigid and unforgiving framework, but rather a flexible and lenient framework that facilitates ease of use. Finally, it is not a structure that commands an incident, although it does use the ICS method as companion for its command and control function.

The NIMS method is designed to provide a dependable nationwide framework and technique that permits government entities at all levels (federal, state, tribal, and local), the private sector, and Nongovernmental Organizations (NGOs) to work together effortlessly. While most individuals think that working together is in the response phase, the NIMS method is based on the modes of preparing for an incident, preventing and incident, responding to an incident, recovering from an incident, and providing mitigation efforts that will lessen the effects of virtually every type of incident.

5.1 NIMS Method Guiding Principles

No matter what the incident, there should be a management of priorities. In order, these priorities should be to first save lives, then stabilize the incident, and then focus on protecting property and the environment and sometimes the acronym of Life, Incident, Property, Environment (LIPE) is used. To achieve these priorities, incident personnel will implement the NIMS method, utilizing the principles of flexibility, standardization, and unity of effort. By implementing these principles, everyone is on the same page, with the same overarching priorities.

5.1.1 Flexibility

The NIMS method has components that allow it to be flexible to the point that it is adaptable to any situation. It does not matter if it is a planned special event, a mundane local incident, or if it is an incident involving interstate mutual aid and/or federal assistance, all players can be on the same page. The NIMS method allows coordination of efforts in a fast and efficient way. The flexibility built into the NIMS method allows it to be scalable and, therefore, applicable for incidents that vary widely in terms of hazard, geography, demographics, climate, cultural, and organizational authorities. Flexibility will be discussed more in-depth later in this chapter.

Using the analogy of the previously used car, imagine this flexibility as the ability to use the steering wheel to guide your direction of travel, rather than just going the direction the wheels are pointed. By using the steering wheel, you can avoid potholes in the road, or you can go a different direction, so you do not risk having an accident. If need be, you can put the car in four-wheel drive and plow through the mud and snow to get to your destination. This analogy somewhat represents what flexible means in this context.

5.1.2 Standardization

Standardization is crucial to interoperability among varying agencies and organizations during incident response and recovery. Managing an incident (of any size) requires coordination and standardization among first responder personnel. The NIMS method provides standardized structures that improve integration, coordination, collaboration, and communication between all jurisdictions and disciplines.

Agencies that have implemented the common facets and framework of the NIMS method are able to work together almost seamlessly. This promotes consistency among a multitude of agencies and organizations that are involved with diverse aspects of an incident. The NIMS method provides and promotes standard terminology. This terminology expands past vernacular of simple verbal responses and moves into even the most complex part of an incident of ordering the resources that are needed.

Using other than the standard terminology, the resources ordered could be apples, and oranges might be delivered. By using standard terminology, including very specific resource typing and credentialing, those in charge of the incident get exactly what they order. This fosters more effective communication and understanding among those agencies and organizations that responded to the incident.

FEMA provides several tools for ensuring that those in charge of an incident get exactly what they need. The first tool is the Resource Typing Library which is found at https://rtlt.preptoolkit.fema.gov/Public. This website not only identifies physical resources and their qualifications but also personnel resources along with the requirements to be that type of human resource as well. Additional tools to facilitate effective mutual aid can be found at https://www.fema.gov/resource-management-mutual-aid. It would be nearly impossible to list all resource typing and qualifications in one book, so it is best for those interested in typing and qualifications to use FEMA's tools.

The NIMS method of IMS defines standard organizational structures that are used to improve integration and connectivity. The NIMS method defines standard practices that facilitate allowing incident personnel who have never worked together before to effectively integrate with each other. The NIMS method also helps to create a unity among the various organizations involved, and to create the mindset of "us versus the incident" rather than an "us versus another organization."

If we look at the analogy of the car again, we should focus on the brake and the accelerator. If the location of the brake and the accelerator changed from car to car, we likely would not know which way to coordinate our legs until we sat in the vehicle. Even then, our responses would be slowed because we would have to think about the way each vehicle was set up. Comparing that to the standardization of NIMS, we can see that it is standardized with the accelerator on the right and the brake on the left. If we immerse our organizations and agencies in the mindset of the NIMS method, then we won't have to think about which pedal is the brake and which is the accelerator, we can get from "Point A" to "Point B" without thinking about it.

5.1.3 Unity of Effort

Unity of effort is the act of organizing the energies between multiple organizations or agencies and harnessing those energies so that all of them are working toward the same objective. By harnessing these energies in a unified manner and effort, it prevents organizations from working on tasks that are not a priority, and it effectively reduces instances of duplication of effort. Unity of effort in the NIMS method of incident management means that all agencies and organizations are organized to achieve a shared common objective(s) and/or goal(s). This combined and unified effort enables agencies and organizations with no previous working relationship, to support each other while still maintaining their own lines of authority.

Going back to our analogy of an automobile, if our legs did not work in a unified manner, then it could be possible that neither leg or foot would ever touch the accelerator or brake, and we would go nowhere. By the same token, if both legs pushed all the way down on the gas and the brake at the same time, then it would either kill the effort of moving forward, or it would sit there and spin its tires. If the legs worked intermittently together, we would not be able to get to our destination as quickly. If both legs and feet work in a unity of effort, then they would work together in a unified effort to reach the final destination.

5.2 Key Terms and Definitions

Part of standardization of the NIMS method is standardization of terms. Rather than reinventing the wheel, this book has taken these key terms and definitions from the Third Edition of the *National Incident Command System* book released by the Federal Emergency Management Agency (2017a). Paraphrasing these terms would likely somewhat change the meaning, so these are directly quoted.

Area Command: When very complex incidents, or multiple concurrent smaller incidents, require the establishment of multiple ICS organizations, an Area Command can be established to oversee their management and prioritize scarce resources among the incidents. Due to the scope of incidents involving Area Commands and the likelihood of cross-jurisdictional operations, Area Commands are frequently established as Unified Area Commands, working under the same principles as a Unified Command (UC).

Authority Having Jurisdiction: The Authority Having Jurisdiction (AHJ) is an entity that can create and administer processes to qualify, certify and credential personnel for incident-related positions. AHJs include state, tribal, or federal government departments and agencies, training commissions, NGOs, or companies, as well as local organizations such as police, fire, public health, or public works departments.

Emergency Operations Center (EOC): An Emergency Operations Center (EOC) is a facility from which staff provide information management, resource allocation and tracking, and/or advanced planning support to personnel on scene or at other EOCs (e.g. a state center supporting a local center).

Incident Commander: The Incident Commander (IC) is the individual responsible for on-scene incident activities, including developing incident objectives and ordering and releasing resources. The Incident Commander has overall authority and responsibility for conducting incident operations.

Multiagency Coordination Group (MAC): MAC Groups, sometimes called policy groups, typically consist of agency administrators or executives from organizations or their designers. MAC Groups provide policy guidance to incident personnel, support resource prioritization and allocation, and enable decision-making among elected and appointed officials and senior executives in other organizations as well as those directly responsible for incident management.

Unified Command: When more than one agency has incident jurisdiction, or when incidents cross political jurisdictions, the use of Unified Command enables multiple organizations to perform the functions of the Incident Commander jointly. Each participating partner maintains authority, responsibility, and accountability for its personnel and other resources while jointly managing and directing incident activities through the establishment of a common set of incident objectives, strategies, and a single Incident Action Plan (IAP) (FEMA, 2017a, pp. 4–5).

Regardless of the incident's cause, size, location, or complexity, the NIMS method provides a framework that works. It does not matter if this response is a single agency response, a multiagency response, or a multijurisdictional response, the NIMS method provides a framework for emergencies, natural disasters, or terrorism response that promotes the Five-Cs (Chapter 4.1) of incident management. Organizations that have integrated the NIMS method into their planning, response, recovery, and preparedness and mitigation efforts (and incident management structure) have an advantage against the incident they face. They can be called to an incident at a moment's notice, and they already know and understand the procedures and protocols for any incident.

Because of the way the NIMS method is set up, they will understand the expectations for equipment and personnel, and they will be able to know if the exact equipment needed is available, rather than send something that may not work. NIMS provides cohesiveness in the planning, response, and recovery modes that allows differing entities to easily integrate with each other. If conditions require it, these differing agencies can even establish a Unified Command to meet the needs of the incident. What really makes the NIMS method the best method for managing incidents in the United States is that it is comprehensive, flexible, and adaptable.

5.3 Understanding Comprehensive, Flexible, and Adaptable

When talking about IMS methods, we often hear the words comprehensive, flexible, and adaptable. This holds especially true when we hear or talk about the ICS component of NIMS, and the NIMS method. Many who have attended any ICS method classes have heard these words. The instructors and students alike typically take these words with a grain of salt, and then move on to the next slide. Even if great detail was given by the instructors to help students better understand these concepts, some first responders would only block it out. When these individuals miss the details, they fail to realize that these are the underlying factors that make NIMS work, and work well. They are so interconnected that in most instances each describes the other, and they have multiple connections. They are intertwined into throughout NIMS.

The mentality of blocking out or giving little authority to these factors can be somewhat understandable. After all, they are sitting in a classroom studying about an IMS method when they really want to be doing something else. Still, by not knowing and understanding these concepts, it is paramount to having a vehicle and not knowing whether it is a car, truck, or a semi. You would not know whether it ran on gas, diesel, or racing fuel, and you would not know the towing capacity or the weight capacity. These important factors can tell us what vehicle we are driving, and what capabilities this vehicle has. When we do not know the particulars of an IMS method, we also do not know what the system will do, nor do we know the abilities of the system. For that reason, we should look at what type of vehicle we have with the NIMS method, which should help us to understand what it will do.

5.3.1 Comprehensive

Many have heard that the NIMS method and other IMS methods are comprehensive. What exactly does comprehensive mean in the context of incident management? When we discuss comprehensive, we refer to how the framework incorporates all aspects into a singular system. It is structured in such a way that it provides a complete system (or framework) to deal with any part of managing an incident. Whether you are a single response agency on a minor medical call, or if it is a nationwide response with over 300 outside agencies at a catastrophic incident, the structure of the NIMS method can, and does, support incident management. The framework for the NIMS method utilizes best practices as well as new and existing structures to create calm out of chaos, simplicity out of complexity, and predictability out of uncertainty.

In the initial response to an incident, the framework of ICS, or in the case of healthcare facilities Hospital Incident Command System (HICS) method, has been developed to assist in the management of the incident. If the incident ends without becoming more complex, then the framework does not expand. NIMS is a uniform structure that is flexible enough to be a workable method of managing incident across a complete range of potential incidents, no matter how difficult the incident may be. Through nearly 50 years of trial and error with the ICS method, the NIMS method has been built on the best practices of those that worked through the intricacies, and the recognized problems of incident management.

The NIMS method is comprehensive because it is usable nationwide, and it is a systematic approach to incident management. It not only includes the existing structure of the Incident Command System or Hospital Incident Command System (HICS) but also integrates Multiagency Coordination Systems (MACS), Emergency Operations Centers (EOCs), and Joint Information Systems (JISs). It incorporates preparedness beliefs and philosophies for virtually every type of risk or threat. The NIMS method also provides critical values that support a common operating picture, based on situational awareness that helps to make more informed decision. The NIMS method also supports interoperability of communications and the management of information. It has become a tool that is built on standardized resource management activities that not only facilitates coordination but also empowers coordination among similar and dissimilar response agencies. It does not matter whether the response and recovery are typical, or atypical, whether it is based with only public safety, or it incorporates a multitude of non-governmental organizations. NIMS is so comprehensive that it facilitates integration across a full gamut of agencies and/or organizations.

Due to the ability to be scalable, it is used for all sizes of incidents which make it even more comprehensive. It does not matter whether a small- or large-scale response is required, and it does not matter whether if the incident occurs in a hospital, a trench, a bank, or on a highway. The NIMS method is also a system that can be active and constantly changing with changes in any incident, which helps promote ongoing incident management. It is comprehensive before, during, and after an incident. As an incident becomes more complex, the NIMS methods is so comprehensive that it facilitates multiagency disciplines (law enforcement, fire, EMS, non-profits, etc.) and it allows outside resources to efficiently and effectively work together as the incident becomes more critical. The NIMS method is designed to allow comprehensive multiagency coordination to be accomplished through a wide range of elements. Some of these elements include facilities, equipment, personnel, procedures, and communications. Emergency Operations Centers (EOCs) and Multiagency Coordination (MAC) groups are just two examples of the coordination elements that makes NIMS comprehensive.

While there is no such thing as a perfect incident management system, NIMS is so comprehensive that some believe that it is the closest thing to a perfect incident management system that the emergency response industry has ever seen. Even better, it is improved on a regular basis because of the constant research and development that is federally required, and which was designed to improve the system.

In connecting the adjective comprehensive to the previous example of the vehicle, NIMS is the complete package. It is the body of the vehicle, the wheels, the drive train, the charging system, the braking system and more. NIMS is what we need to get from one location to another in a safe and efficient manner.

5.3.2 Flexible

The flexibility of an incident management system is also part of the comprehensive nature of NIMS. The core facilitating factor of the NIMS method that makes it flexible is the pre-existing structure of the companion method of ICS. With the ICS method, a single agency can utilize the command system on smaller incidents to have a single agency manage a small crew of less than seven personnel, or the command system can utilize a large contingency of individuals to manage thousand or tens of thousands of personnel. Some of these larger-scale response and recovery efforts have been seen after incidents such as Hurricanes Katrina, Rita, and Michael as well as numerous wildfires.

The NIMS method is flexible enough that the function of public information can change from a singular Public Information Officer (PIO) with a single individual managing information to a Joint Information Center (JIC) with multiple response agencies providing expertise. A Joint Information Center (JIC) can include as many Public Information Officers (PIOs) from different agencies as is needed. Their only goal is to create and release important information with a unified message and coordinated information to inform the media, and the general public as needed. The NIMS method framework is so flexible that it can incorporate a virtual Joint Information Center (JIC), whereby public relations companies and those enroute (or possibly delayed individuals due to commitments) can participate via the phone or Internet to ensure that the best interests of the public, the agency, and the incident are considered by all involved individuals.

The NIMS method promotes flexible communication and information sharing to ensure that information for situational awareness is free flowing. It creates flexibility by ensuring communication problems of the past (e.g. local vernacular, the use of 10 codes, different methods of reporting) do not hamper the management of an incident by mandating and using plain language. It is flexible because of the standardized forms for each specific task, and the availability of technological equipment to overcome differing radio frequencies and coding.

While there are many reasons why the NIMS method can be considered flexible, perhaps the core reasons are the paradigms that the system addresses and holds in high regard. Those paradigms include the Incident Command System (ICS), the Multiagency Coordination (MAC) systems, the Emergency Operations Center (EOC), and the Joint Information Systems (JIS) that have been developed to support incident management at the site of an incident.

When comparing the adjective flexible to the previously mentioned example of a vehicle, we could say that the engine runs on whatever fuel is needed. Similar to military vehicles that are multifuel, the engine can run on diesel, kerosene, heating oil, or jet fuel. The NIMS method is like these engines, because it is flexible enough that we can run it using the fuel that is in abundance for the job at hand.

5.3.3 Adaptable

When we speak of how the NIMS method is adaptable, we also need to realize that adaptable is interconnected with the concepts of comprehensive and flexible. By being adaptable, we need to realize that it can be used for almost any purpose. Whether it is an incident that is large and complex such as an earthquake or a hurricane, or managing a disaster at a hospital, or it can be as simple as daily managing a public safety agency. The NIMS method has the capability to be an effective tool for all types of incident management, and many command structures. As has been mentioned before, one of the primary reasons for this is the existing framework of ICS as a foundation for incident management, but there is also the standardization of qualifications, equipment, and other resources.

The NIMS method is adaptable to the extent that a singular agency can handle a child who fell off their bike and scratched their knee to a drug overdose that will include local law enforcement, an EMS agency, and the local fire department. It is adaptable enough to handle a small street or block festival, or a major sporting event such as the World Series or the Super Bowl. It is adaptable enough to handle a small grass fire in the local city park or as large as a 40-square mile wildland fire that brings in responders from other countries such as Australia or New Guinea.

Whether it is a single engine plane crash, or a large crash such as a jetliner, or even similar to the crash of the Space Shuttle Columbia, this incident management system can meet the needs at hand.

The NIMS method of IMS allows the framework to adapt to any fluid incident. The system is designed in such a way that if the principles are known, understood, and the basic tenets are followed (by all), any agency or individual who might be able to provide resources can be utilized as long as they can follow directions. While it is preferred that they have been trained on the NIMS method and the ICS method, it is not necessary. Granted, having untrained individuals adds another layer of complexity, and while it is not preferred, the NIMS Framework is so adaptable that it is possible.

In comparing the adjective of adaptable to the example of a vehicle, it is not car, a truck, or a semi, it is a combination of whatever we need it to be. It is like we have and wave magic wand, and it is a motorcycle for a short trip without packages. If it is a short-range trip with only a few packages, presto; it is now a car that can help us to transport a few small packages. If it is a trip that involves more packages than the car can handle, presto; it is now a truck that can include a moderate level of packages. Should it be needed, we can change that vehicle into a semi that is hauling two trailers, adding 70 000 pounds of packages.

5.4 NIMS Components

When the NIMS method was first introduced, many mistakenly thought that NIMS was just an expansion of the ICS method. Time has revealed to the vast majority that the NIMS method is much more than initially thought. It is also much more than just an organizational chart that we see pinned on a wall, or on a computer screen. NIMS is a consistent, nationwide, systematic approach for managing an incident of any magnitude. Part of its success is because it uses a compilation of components that integrate into the overall system rather than be used independently of each other. Those components include the following:

- Preparedness
- Communications and Information Management
- Resource Management
- Command and Management
- Ongoing Management and Maintenance
- Supporting Technologies

These components of NIMS were not designed to stand alone; they were designed to integrate and work together. While each component is not dependent of the other, each one enhances and supports the other.

5.5 The Importance of Preparedness with NIMS

Many believe that they should focus on what to do if (and when) an incident actually occurs. By looking at incident management in this short-sided manner, they fail to realize the importance of preplanning various aspects of incident management that can help to mitigate the negative effects. If we wait until an incident actually occurs before we begin to see what resources are available, and how to initiate contact with these resources, then we are already in a difficult situation and trying to bail ourselves out. If we already have preplanned who, what, when, and where available resources are and where they will come from, plus have agreements in writing, then we are ahead of the game.

Essentially, the idea is to integrate the basic concepts that are taught in emergency management. These concepts better facilitate and create an enhanced form of incident management. It does not matter if you are a law enforcement agency, a fire agency, an EMS agency, an Emergency Management Agency (EMA), or a volunteer organization, incorporating planning and mitigation into incident will help to enable a more integrated and effective response and recovery. It will also make incident management easier and more user friendly.

While emergency management agencies are typically the agency to collect, make agreements, and preplan for a major incident, nothing prevents other local agencies from doing the same. This holds especially true in rural areas of the United States, where there are few paid first responders and even fewer emergency managers. Local fire departments, police departments, EMS squads, and various other first responders can (and should) identify potential resources that may be able to help them. It does not matter if it is a small- or large-scale incident, by initiating contact and developing a rapport with them, you increase your ability to manage an incident, while decreasing the chances that an incident will manage the agency.

Through creating bonds with other agencies, you will know who to contact at 4 a.m. when you need to use school buses and school bus drivers to evacuate residents from a deadly Hazmat spill on the nearby interstate. You may also be able to quickly incorporate the local equestrian trail riding association, the local ATV group, and the local Unmanned Arial Vehicle (UAV) club to begin grid searches in the national forest in an effort to find a missing child or an adult with dementia. Preplanning can, and does, help facilitate a broader spectrum of integrated incident management as well as create a bond that will enable making that 4 a.m. call.

Arguably, one of the most effective tools in emergency incident management preparedness is the *Cycle of Preparedness* (DHS, 2013b). The primary reason for the Cycle of Preparedness is critical to incident management is that it creates a 360° view of a specific issue prior to an emergency response. This 360° view can then be used to respond and recover from the same or even similar incidents with greater efficiency.

5.5.1 Cycle of Preparedness as a Part of NIMS Incident Management

Preparedness involves more than guessing what will happen and then trying to preplan what to do in a disaster. The type of comprehensive preparedness that will be needed to be prepared for nearly any type of incident is typically accomplished utilizing the Cycle of Preparedness as identified in the NIMS method training materials (FEMA, 2017a). The Cycle of Preparedness is an endless cycle of planning, organizing, training, equipping, exercising, reevaluating, and improving to bring about a successful response, including coordination, during an incident response (DHS, 2013b).

In utilizing the Cycle of Preparedness, the first responder agency should identify a moderately plausible scenario. The most plausible scenarios can be identified by using the principles in the Comprehensive Preparedness Guide (CPG) 101: Developing and Maintaining Emergency Operations Plans (EOPs) and Comprehensive Preparedness Guide (CPG) 201: Threat and Hazard Identification and Risk Assessment (THIRA) Guide. These will guide you through the process of Threat Identification, Hazard Identification, and Risk Assessments (THIRAs). These are critical to preparedness, because they identify the highest risks within a jurisdiction. After completing a Threat Identification, Hazard Identification, and Risk Assessments (THIRAs), the agency will know the types of incidents that would most likely affect their community, based on the unique factors specific to their jurisdiction (DHS, 2013b).

In most instances, when choosing a scenario based on Threat Identification, Hazard Identification, and Risk Assessments (THIRAs), it is important to begin with only a slightly challenging scenario. You should save more complex scenarios for future Cycle of Preparedness settings, after personnel are more proficient and accustomed to this process. Once the scenario is chosen, then the response should be tested to Homeland Security Exercise and Evaluation Program (HSEEP) standards (DHS, 2013b).

While the following explanation may be an oversimplified version of the process, the parent agency would plan for a potential event, organize their resources, train the responding agencies on the actions they should take (and how they will be dispatched and incorporated), and identify equipment not already on hand. If the needed equipment is not owned by the primary agency, the agency would then try to identify if it is located within another agency that would be a response partner, or if the equipment would need to be purchased. They then undertake the exercise process, evaluate, or reevaluate their preparedness plans, and improve their preparedness efforts. This would be a continuous process for each risk that the response area faces (DHS, 2013b). In most instances, a jurisdiction will cover one risk to test, then move on to another risk. By doing this, they can become better prepared to respond more effectively.

5.5.2 NIMS Drills and Exercises to Support Preparedness

It should first be recognized that drills and exercises fall under preparedness, planning, and mitigation modes of incident management. This underlines the importance of drilling and exercising as a way to mitigate the negative effects of an incident and to reduce chaos, uncertainty, and complexity. By undertaking drills and exercises in the proper manner, your agency, your jurisdiction, and/or your organizations can overcome common problems that may arise and help to build resiliency.

There are seven types of drills and exercises that can help to evaluate and improve resiliency (DHS, 2013b). Exercises are meant to be building blocks, where each type of drill helps better prepare the agency to take on the next level toward improvement. The previous completed drills and exercises each lay a stronger foundation for future drills and a real-life incident. The seven types of drills that are usually used include

- Seminars
- Workshops
- Tabletop Exercises (TTX)
- Games
- Drills
- Functional Exercises (FE)
- Full-Scale Exercises ([FSEs] DHS, 2013b)

Each individual drill or exercise should be used to identify issues, make corrections, or adjustments to make a real-life response and/or recovery stronger, then move on to the next level of testing. When utilizing the Cycle of Preparedness, the agency should start with the least labor intensive (a seminar), make any necessary adjustments, then move to the next (a workshop) type of exercise. This process continues until all seven types of drills are completed and tested in each area of a potential response. Unfortunately, many agencies skip some types of drills and exercises along the way. When they fail to undertake all parts of the process, they often miss opportunities to better hone their skills, primarily because they fail to build foundational strength.

Using these drills and exercises as building blocks helps to clarify the roles and responsibilities of the primary agency, and all responding agencies. It also helps to improve interagency cooperation with other responding agencies and/or organizations. Certain drills and exercises can be used to help identify gaps in resources and coordination. Drills and exercises with other responding agencies and organizations help to improve and enhance performance among all participating agencies. Additionally, all types of exercise can serve as a way to identify areas where improvement may be needed, thereby automatically improving resiliency (DHS, 2013b).

Each type of drill in the NIMS method plays a critical and unique role in increasing preparedness and resiliency. In order to help you better understand how these exercises can contribute to incident management, as well as understand the role it plays in preparedness and resiliency, each will be briefly explained.

5.5.3 Seminar

A seminar is best described as an informal discussion. It is designed to provide participants with new or revised plans, policies, or procedures. As a discussion-based exercise, seminars are a valuable resource for those agencies who are developing or making major changes to existing plans or procedures, or they are introducing a new plan that has not been in place. Seminars can also be beneficial when trying to evaluate the capabilities of other potential responding agencies. In providing a seminar, it allows those that will be involved with a response, and those who will be assisting in incident management, to review the new or revised plan or procedure, and to clarify or understand their roles and capabilities. It can also provide expectations as to who will respond to that type of incident, with which equipment, and the standard procedures they should be familiar with if such an incident should happen (DHS, 2013b).

5.5.4 Workshops

A workshop is similar to a seminar, but it is part of the designing process for a specific response. Workshops are mostly used to determine the schedule for a training and exercise plan, or in identifying training schedules of an individual agency exercises which will support the overall larger exercise. The purpose of the workshop is not to have only first responders present, but to bring elected and/or appointed officials together with response agencies (and organizations) so that those officials can identify and set exercise program priorities. Workshops are also used to develop a multiyear schedule for exercise activities, and the planning and funding of entwining each individual agency training activities to meet the needs of those priorities (DHS, 2013b).

This type of workshop ensures that the entire community exercise and all the planned initiatives are coordinated. These workshops also help to reduce the duplication of efforts through identifying which agencies are responsible for specific training and equipment needs. Workshops provide an efficient use of resources rather than a duplication of effort, it avoids the unneeded overextending of agencies and personnel, and it maximizes the value of training and exercise funding (DHS, 2013b).

5.5.5 Tabletop Exercise (TTX)

Tabletop Exercises (TTX) are designed to promote dialogue based on the many issues related to the imaginary emergency being simulated. Tabletop Exercises (TTXs) can be used to develop awareness, corroborate plans and procedures, and to practice ideas on the mitigation of the imaginary incident (DHS, 2013b). It is also used to evaluate the types of systems required to lead the prevention of, protection from, mitigation of, response to, and recovery from a specific incident. Many see this as a game that is played, to see how they can overcome chaos and complexity.

Tabletop Exercises (TTXs) are generally focused on creating an understanding of incident response to a given scenario, identifying strengths and weaknesses, and to assist in changing perceptions and/or preconceived notions (DHS, 2013b). During Tabletop Exercises (TTXs), the players are encouraged to have in-depth discussions that collaboratively review areas of concern and to work on mitigating any gaps or issues that may arise. Through the active involvement of participants and the issues they have identified, the players can recommend revisions to current policies, procedures, and plans which will help to mitigate any of the identified issues (DHS, 2013b).

Tabletop Exercises (TTXs) can range from basic to complex (DHS, 2013b). In basic Tabletop Exercises (TTXs) such as a Facilitated Discussion, the scenario is offered and stays the same. It presents the known facets of this scenario and then provides those participating in the Tabletop Exercises (TTXs) may be told where they are in the scenario. Those involved apply their knowledge and skills to a list of complications that are offered by the facilitator. Those complications are then discussed as a group, actions are agreed upon, and then documented. After the exercise is over, those decisions will be analyzed to determine if better or different actions might have been a better choice in that specific scenario (DHS, 2013b).

In Tabletop Exercises (TTXs) that are more comprehensive in nature, players may receive prescribed messages at specific times, or tasks that need completed during the exercise from the facilitator (DHS, 2013b). To make it more challenging and/or more realistic, facilitators will typically announce various problems that the players will need to deal with while in the middle of the virtual response. While it is common to only introduce a singular complication at a time, there may be instances where the problem may be a cascading event. These complications can be introduced via a hand-written or typed message, a simulated telephone call, videotape, or any other means that may enhance the play. In this type of exercise, the participants will discuss the issues raised, problem by problem. In some instances, the participants will explain their recommendations for the proposed action by referencing research, books, established authorities, plans, and/or procedures that supports their proposed actions. The participant's actions are often incorporated into the exercise and become part of the scenario as it progresses (DHS, 2013b).

The significant benefits of Tabletop Exercises (TTXs) are that participants are making decisions in a nonemergent environment. They can evaluate actions when lives and property are not on the line, and they can (in most instances) identify how to take corrective action before it is an actual event.

5.5.6 Games

A game usually involves two or more teams that typically compete against each other (DHS, 2013b). A game uses set rules and can use data from previous incidents and/or procedures designed to mimic a hypothetical situation, or past real-life situation. Games allow players to see the significance and the consequences of their decisions and actions in a nonemergent manner. These games are valuable tools that can help to validate plans and procedures and these types of games can assist in evaluating the needed resources for an incident (DHS, 2013b).

Based on the exercise design and the objectives of each specific game, decision-making may be either unhurried and cautious, or it may be rapid (potentially even with timed responses) and more stressful (DHS, 2013b). The informal format of a game allows the facilitator to present "what if" questions that can help participants use critical thinking, which also usually promotes thinking "outside of the box." This type of critical thinking helps to create a deeper understanding of what an incident may entail. By utilizing these unusual or surprising questions, it increases the benefits of the exercise by allowing them to dig deeper and to review and come up with solutions they may not have previously thought of.

The design of the game can be prescripted or decided as the game progresses. Each of these methods have benefits, and the design should be constructed based on whether this is a new and unusual scenario, or if it is a scenario that the participants should already be familiar with. The key factor to remember when conducting games is that the players should identify key points in critical decision-making. The evaluation of these games should hinge on quality. Improving quality will improve a response when it occurs.

5.5.7 Drills

A drill is an operational based exercise. It is a coordinated and controlled action that is typically used to confirm a specific function, or capability, in a single organization. Drills are commonly used to offer training on new equipment, confirm procedures, or they are used to practice and/or maintain existing skills (DHS, 2013b). In a local fire department, a drill may entail practicing foam firefighting, or the use of a new (or old) set of extrication tools and rescue jacks. In an emergency management agency, a drill may incorporate activating an Emergency Operations Center (EOC), or it may include setting up an emergency shelter or decontamination system. In law enforcement, drills may include the response to an active shooter, deployment for civil unrest, or a manhunt. For nonprofits, drills can be used to set up shelters, provide feeding of multitudes, or setting up a donation center. Of course, there are many other types of scenarios that could be drills, but no matter what the scenario, a drill should always test the capability of the agency.

Drills can also be utilized to understand if plans can be accomplished as intended, to identify shortcomings in planning and policies and to calculate whether more training is needed on a specific issue (DHS, 2013b). These drills are often used to reinforce certain specific methods of responding and to highlight and to engrain best practices into a first responders mind. While a singular drill is useful as a stand-alone tool, a series of consecutive drills can be even more helpful in preparing a singular agency for a collaborative effort with other agencies (DHS, 2013b). This holds especially true if a Full-Scale Exercise is planned, or if it is likely these other agencies will be working together if an incident occurs.

Prior to the drill, the facilitator and the designers should create clearly defined plans, procedures, and protocols for the community, and for the drill. There should be no conflict between the everyday plans and procedures, and the drill itself. It is also important to ensure that all personnel are aware of those strategies and trained in the existing plans, processes, and procedures that should be used in the drill.

5.5.8 Functional Exercises (FEs)

Functional Exercises (FEs) are designed to confirm and assess agencies capability (DHS, 2013b). These can include multiple functions and/or subfunctions that may be needed, or inter-reliant sets of functions in a disaster or emergency response plan. Functional Exercises (FEs) are typically undertaken to test existing plans, policies, procedures, and to evaluate the capabilities of staff members involved in management, or even the incident management end of an incident (DHS, 2013b). A Functional Exercise (FE) helps to identify where improvements can be made in the command, control, communication, cooperation, and coordination functions of an incident.

The situation being exercised is presented through an exercise scenario (DHS, 2013b). As the Functional Exercise (FE) progresses, informational updates should be provided which allows the command and management staff to create decision-making activities. A Functional Exercise (FE) is presented in a realistic manner, and the setting for the exercise should be presented as if it is actually the incident, and the subsequent events around the incident are unfolding as the exercise progresses (DHS, 2013b). While the scenario should be realistic, it should be noted that the actions of personnel and equipment are almost always pretend or simulated actions.

Controllers of the scenario will normally use a Master Scenario Events List (MSEL) to ensure that the participants actions remain within the determined parameters as well as ensure that the crucial exercise objectives are experienced (DHS, 2013b). As an added benefit, a Simulation Cell (SimCell) can be added to Functional Exercises (FEs). A SimCell is an adaptable process that is helpful in controlling the direction of the exercise. It allows the participants to simulate interaction with a wide assortment of (currently) noninvolved officials and organizations by injecting elements (such as a sent message or directive) from those that are not present. These SimCells allow virtual interaction without the individuals and/or organizations being present for the exercise (DHS, 2013b).

5.5.9 Full-Scale Exercises (FSEs)

FSEs tend to be the most multifaceted, and they usually utilize more personnel and equipment than any other type of drill or exercise. A Full-Scale Exercise (FSE) is a multiagency, multi-jurisdictional, multidiscipline exercise which helps to evaluate and authenticate many, if not all, phases of the Cycle of Preparedness (DHS, 2013b). This evaluation and authentication allow the agencies involved with the exercise to find the strengths and weaknesses within the current plan so that improvements can be made and later implemented. Full-Scale Exercises (FSEs) usually incorporate a large diverse group of players, all of which operate under a singular IMS method (usually the ICS method). The exercise will typically involve a variety of public safety agencies, and various organizations, and most of those organization being from different jurisdictions, and simulated victims (DHS, 2013b).

In the United States, Full-Scale Exercises (FSEs) often include many participants who all operate under the ICS method, while potentially utilizing a Unified Command, and incorporating NIMS principles into the exercise (DHS, 2013b). Full-Scale exercises not only can serve to test how tactical operations play out but also how other agencies work together within the chosen IMS method. By setting up the Full-Scale Exercises (FSEs) IMS system as expected in a real incident, then all functionality of the response can be tested and evaluated.

The facilitators of a Full-Scale Exercise (FSE) will present the events by providing an exercise scenario (DHS, 2013b). As the Full-Scale Exercise (FSE) progresses, event updates will be provided which will help to guide the operational activity. Full-Scale Exercises (FSEs) are simulated as if it were a live real-time environment, and a Full-Scale Exercises (FSEs) can, and usually does, create extremely stressful situations. These stressful environments are part of the design, primarily because this exercise is planned to impersonate the real event (DHS, 2013b).

From the beginning of the exercise, personnel and various other resources are dispatched and deployed to the simulated scene. In order to reduce the risk of an accident or other problem, the act of dispatching and deploying is usually done from a nearby staging area. While the crews and resources are staged in an area, the dispatch times are prearranged to the mock the actual times it would take for a resource to arrive on scene in real-life situation. Upon arrival, the personnel and equipment proceed as if the Full-Scale Exercises (FSEs) was a real incident.

As the Full-Scale Exercise (FSE) progresses, complex but plausible problems are interjected to make the exercise more realistic and credible. To overcome these problems, personnel will need to use fast paced critical thinking and problem-solving skills to overcome these issues with an effective solution. The way the problem was solved is documented, and it allows the primary agency to review and determine if changes are needed for the protocol or procedure.

Real-Life Scenario

After years of planning and holding various drills and exercises, the Christian County (Missouri) Emergency Management Agency decided to undertake a Full-Scale Exercise in 1992. Approximately 14 agencies were involved from law enforcement, fire departments, EMS squads, and ambulance services. The scenario being exercised was a bus accident that involved traffic control, extrication, and medical care for approximately 30 patients. Prior to the exercise, safety issues were addressed, and all participants were told that a real emergency would be identified as a "Coca-Cola" emergency, to match with the company's slogan of the time, which was "the real thing." Several first responders laughed, while other less-experienced first responders felt this type of precaution was unnecessary. The scenario was set up in such a way that it tested school and tour bus accident response and recovery plans.

When first responders were dispatched, they found individuals that had various simulated injuries from minor to critical, and participants were arranged so they were under seats and in various positions that mimicked a bus accident. Approximately two-thirds of the way through the scenario, one of the individuals who had repositioned themselves several times, found that they were legitimately trapped between several seats. As soon as this real safety issue was realized, a "Coca Cola emergency" was transmitted across all channels, and real-life extrication began. In the end the individual was extricated with no injuries, however, all participants learned the importance of identifying a real emergency in an exercise.

Less than a year after this full-scale exercise, the incidents that were practiced occurred. The first incident was early in 1993, involving a tour bus that was taking individuals to the nearby destination of Branson, Missouri. The improvements that were made from the exercise helped to reduce the loss of life and promote better interagency cooperation. In the fall of 1993, there was also a head-on collision between a school bus and a grain truck on a curvy rural road in Christian county. Between the lessons learned from the tour bus crash and the full-scale exercise, this response went off with virtually no problems.

A Full-Scale Exercise (FSE) requires the highest level of support, and the location where the Full-Scale Exercise (FSE) will be held is usually required to be quite large so that it can facilitate the large number of participants that will be involved. When choosing a site for a Full-Scale Exercise (FSE), the planners will need to remember that the Full-Scale Exercises (FSEs), and all activities related to it, will require close monitoring of safety and logistics. The highest priority and the most serious consideration should be given to the safety of all involved. When conducting a Full-Scale Exercise (FSE), a sufficient amount of safety officers should be incorporated to closely monitor the exercise. Individuals who will be participating in the exercise should also be briefed that they are also responsible for ensuring safety, and that they should point out safety issues to the Safety Officers (SOFR's) if they recognize them. These safety issues should include issues found prior to, during, and after the exercise. Full-Scale Exercises (FSEs) usually have synchronized and instantaneous parts of an incident going on at the same time. While these multifaceted events permit the response capabilities to be tested, it also puts the participants at a substantially greater risk. For the entirety of the Full-Scale Exercise (FSE), there should also be a preassigned method to identify that an actual emergency or safety issue has arisen. If a real-life emergency occurs during a Full-Scale Exercise (FSE), the exercise should stop immediately, and the entire exercise should shut down until the situation is mitigated.

5.5.9.1 Use of Nonresponse Personnel

The use of nonresponse personnel (actors) from the general public is an excellent way to add realism to drills and exercises and to provide an opportunity to simulate victim care. This is especially important in Full-Scale Exercises (FSEs). Recruitment of actors can be from local colleges and universities, medical and nursing schools, drama clubs, theaters, civic groups, emergency response academies, and federal and state military units.

Consideration should be given to utilizing volunteer actors from within the *Access and Functional Needs (AFNs)* community as well. This is an important facet so that you can provide first responders the opportunity to practice meeting the needs of this portion of the population during emergency operations. Often, Access and Functional Needs (AFN) populations are forgotten or intentionally not used in drills and exercises. Subsequently, the Access and Functional Needs (AFN) population is often underserved or misunderstood during an actual emergency.

Many researchers elude to the fact that emergency services organizations need to better understand preparedness for Access and Functional Needs (AFN) individuals and populations (Davis & Mincin, 2005; National Organization on Disabilities [NOD], 2007; Risoe, Schlegelmilch, & Paturas, 2013). Due to the irregularity of emergencies, Incident Commanders (IC's) and emergency planners must take a *Whole Community* (holistic) approach while taking into consideration the rapid onset and unexpected nature of emergencies.

A Whole Community approach is a framework that helps agencies across the United States to integrate the community into their response and recovery. When unusual or uncommon devastating events occur, Access and Functional Needs (AFN) and first responders tend to have preparedness plans that are less developed: this is even more true when dealing with emergencies that are less common (NOD, 2007; Parr, 1987; Risoe et al. 2013).

Planning for Access and Functional Needs (AFN) individuals (and populations) are often forgotten or ignored by the first responder community, producing an appearance that first responders are insensitive to the Access and Functional Needs (AFN) community (Litman, 2006; NOD, 2007). Because these individuals and the population (as a whole) are underserved, they nearly always account for a higher rate of death and suffering in a disaster. For this reason, it is in the individual agency's best interest to undertake preparedness measures that are inclusive. One method of doing this is to incorporate them into drills and exercises. It is suggested that the amount of Access and Functional Needs (AFN) individuals incorporated into a drill or exercise should be approximately 20% because it is close to the percentage of the overall population they represent.

In most scenarios, children and those that use English as a Second Language (ESL) should also be included in drills and exercises. This will assist the planners with better understanding and dealing with children, and those who do not fluently speak English. By doing so, it will help the agency to address these gaps or any issues before it is needed in an emergent situation.

5.6 NIMS Method of Resource Management: Preparedness

In public safety, we often hear the phrase "resource management," yet do we really know what that means? The novice to IMS methods may think that resource management means telling a resource to go here and do that, and another resource to go there and do something different. In looking at this train of thought about resource management we must ask yourselves "is that really managing your resources"? Absolutely not!

In the terms of an IMS method such as NIMS, resource management is the act of effectively and efficiently managing all of the resources in such a way that there is no duplication of efforts, and there are no gaps in the response and recovery (FEMA, 2017a). Resource management

in the planning mode identifies what resources are available within your agency and mutual aid agencies, while the preparedness and mitigation mode help to identify what resources are available, qualified, and typed, and what resources need purchased or further developed.

The term "resources" are used as a broader term, and they may include personnel resources, equipment, nondurable goods (e.g. water, paper, printer ink), expert human resources, information technology (IT) resources, local, state, and federal resources, and much more. Resources are technically anything that you may need in preparedness, response, recovery and in planning and mitigation.

The function of resource management, in the NIMS method of IMS, provides the ability of many different organizations and agencies to collaborate and coordinate so that they can methodically and thoroughly manage resources. These resources will include personnel, specialized teams (e.g. swift water rescue, SWAT, HazMat, Bomb Squads), facilities, equipment, and supplies.

An important part of ensuring that you are ready to meet the demands of a disaster, terrorist attack, or a major emergency is to take preparedness efforts. All communities and individual agencies should work together before an incident occurs to prepare for the day that horrible day that this incident occurs. To do this, they should develop plans for identifying, managing, estimating, allocating, ordering, deploying, and demobilizing resources (FEMA, 2017a).

Part of the preparedness process involves identifying resource requirements based on the threats and vulnerabilities that are unique to each community and/or agency (FEMA, 2017a). Perhaps one of the most difficult parts of planning involves identifying different strategies to acquire the needed resources that may, or will, be needed in a major event. Some tactics that may help planners to be better prepared might include stockpiling resources, establishing Mutual Aid Agreements (MAAs), developing Memorandum of Understandings (MOUs) and/or contracts, and identifying and making contacts with nontraditional vendors (FEMA, 2017a). These will be discussed later in this chapter.

Whenever undertaking the planning process, resource planners should consider the availability of the supplies that will be needed in the event that a worst-case scenario occurs. In some cases, and with some items, stockpiling may be a consideration or even an avenue they might undertake. Of course, this would be dependent on many factors, such as availability during an incident, the shelf life, the likely need of the item, and a whole host of other considerations that will be unique to your community.

Stockpiling should be given the proper weight when planning. Considerations that may affect your decision might involve the shelf life and the durability of the item. Is it an item that will need replaced annually or is it a durable good that will last 10-years or more? Is it an item that is always readily available or do you sometimes have to wait for it to be manufactured? If a major disaster occurs, will other communities be relying on the same assets or vendors?

After Hurricane Katrina made landfall, there was a severe shortage of firefighting Personal Protective Equipment (PPE). Many sets of gear and Self-Contained Breathing Apparatus (SCBA) were contaminated or destroyed in at least three states: Mississippi, Alabama, and Louisiana. Because a large portion of protective clothing is custom-made to fit the individual, this presented a major problem. Within the first three days after the Hurricane, all stockpiled firefighting protective clothing was sold out, and some manufacturers began to run their production lines around the clock. During this shortage, many agencies donated their surplus new and used protective clothing to affected fire departments. While not the best solution, it filled the gap, at least temporarily. Even so, the new manufacturing of this clothing took years to meet the need.

In making the decision about whether to stockpile or not, resource planners need to know whether a sufficient number of items are on hand at all times, or if they can be produced quickly enough to meet needs of the community. If there is "Just-in time" delivery, or the item needs to be ordered, then manufactured while the agency waits, then it is delivered. As an example, there is a relatively small amount of new ambulances that are ready for delivery today. In fact, if a major event such as Hurricane Katrina occurred, most ambulances would take months or possibly even years to deliver. While financial constraints might be an issue, the community may consider having used, or even new, ready reserves in place as a planning measure.

Beyond taking care of the needs of first responders, we must also look at the needs of the public. FEMA may take up to 72 hours to have boots on the ground, and in an extreme disaster that covers multiple states (or perhaps an entire state), it would be safe to assume that it may take longer to arrive. In the planning mode of resource management, stockpiling of nonperishable food and water might be an avenue that the community might want to pursue. Of course, the resource planners should also incorporate protocols for protecting these resources as well as handling and distributing procedures donated resources.

For incidents that may need surge capacity, planning often includes prepositioning resources, and where stockpiles should be located. Resource planning should anticipate conditions that require restocking supplies when inventories reach a predetermined minimum. As an example, the local fire department, police department, EMS agency, public works agency, and the Emergency Management Agency could each stockpile six pallets of water. In order to help ensure freshness, they could use this water for events such as festivals, structure fires, and when a pallet is near being used, they could order another pallet. By the same token, they could make arrangements for multiple local grocery stores to keep a certain amount of bottled water on-hand, reserved for the agency in the event a disaster occurs.

If stockpiling is done, certain inventory control measures should be put into place (FEMA, 2017a). In some communities, a computer program would be the best solution, however smaller, less affluent, agencies and organizations can also do this with a pen and paper. No matter what the method of tracking stockpiled inventory, basic information should be kept on these items, which include

- Name of the item
- Other names for the resource
- Current condition as related to readiness
- Expiration date
- Status: The resource's current condition or readiness state.
- Resource type definition
- Readiness to assist Mutual Aid
- Location including latitude/longitude and United States National Grid coordinates to ensure interoperability.
- Point of Contact
- The owner
- Manufacturer/Model (Equipment Only)
- Purchase, lease, rental, or maintenance agreements or other financial agreements associated with the resource.
- Certifications associated with the resource.
- Deployment Information including
- Minimum Lead Time (in hours)
- Maximum time the resource can be deployed
- Any restrictions on the resource
- Reimbursement process for this resource
- Release and return instructions

- Any information regarding actions necessary to maintain the usability of the resource.
- Other unique information (FEMA, 2017a)

By inventorying all available resources, including personnel, it reduces the potential of double-counting any resource, including personnel and/or equipment (FEMA, 2017a). While inventorying and sharing resources are important, it should also be realized that their primary response area should come first. An agency should not make all of their resources (that they have available) eligible to be deployed. They must also consider the needs of the community should something happen while they are deployed.

Because of financial constraints, the vast majority of local governments and various organizations (and/or jurisdictions) do not have the financial resources to purchase and maintain all of the physical resources necessary to effectively respond to every potential threat and hazard (FEMA, 2017a). In order to effectively manage physical resources, the agencies will usually leverage the physical resources from other jurisdictions. They also need to prearrange these resources through Mutual Aid Agreements (MAAs) with other agencies and/or initiate a Memorandum of Understanding (MOU) to utilize the resources from the private sector. By engaging private sector resources, the local entity will need to involve volunteer organizations, businesses, other public safety agencies, and encourage these entities to further develop these Mutual Aid Agreements (MAAs) and Memorandum of Understanding (MOU). In order to understand NIMS as an IMS method, we should first take a close look at Mutual Aid Agreements (MAAs) and a Memorandum of Understanding (MOU). These agreements are part of the NIMS method, and it helps prepare for an incident as well as manage our resources before, during, and after a response (FEMA, 2017a).

5.6.1 Which Agreement Should Be Used?

Utilizing the proper agreement for the appropriate resource is important. A simple way of determining which type of agreement should be signed is by identifying the type of help that the entity will receive. You must ask yourself if you are trading resources so that the assistance can be beneficial to your agency and to their agency, or if the agreement is just beneficial to only your agency in the form of a resource, or resources.

An agreement that is beneficial to both agencies is a reciprocal agreement. A reciprocal agreement should utilize a Mutual Aid Agreement (MAA), because it is mutually beneficial. When "Agency-A" has a disaster or a major emergency, they can call "Agency-B," and they will respond with the needed resources that they have available. By the same token, when "Agency-B" has a disaster or major emergency, then "Agency-A" will respond with their available resources. It is mutually beneficial to both agencies, so a Mutual Aid Agreement (MAA) would be the proper agreement.

If the agreement is only beneficial for an agency, and it is not a trade-off in physical resources, then the agreement should be a Memorandum of Understanding (MOU). Memorandum's of Understanding (MOUs) are often used to link nonprofits and businesses with response agencies. In some cases, the Memorandum of Understanding (MOU) my spell out a set amount paid for the services that may be provided, or it may only discuss how they will make this company or organization financially whole for the work they perform.

As an example, a city or county Emergency Management Agency (EMA) may sign an agreement with a boat dealer to provide boats for flooding evacuation. They may also sign a Memorandum of Understanding (MOU) for tree removal services after a storm with local companies. Local hardware stores can potentially be engaged to sign a Memorandum of Understanding (MOU) to stay open later after a disaster occurs so that first responders and the public can purchase what is needed to help mitigate further damage. Additionally, nonprofits could sign a Memorandum of Understanding (MOU) to provide feeding services, management of donations, sheltering, and much more.

Essentially, the Memorandum of Understanding (MOU) is for services and supplies that may be needed by the agency while a Mutual Aid Agreement is an even, or at least near even, trade off. This is not to say that a Memorandum of Understanding (MOU) does not have financial gain for the signers; it means that the help an agency will receive is not in the form of a public safety entity.

5.6.2 What is a Mutual Aid Agreement (MAA)?

Mutual Aid Agreements (MAAs) are mutually beneficial agreements between agencies, organizations, and jurisdictions that provide a prearranged and quick method of assistance. Mutual Aid Agreements (MAAs) can be specific being only on one type of resource such as personnel, equipment, materials, or other associated services, or they may be broad spectrum, including all of the resources that they maintain. The key purpose of the Mutual Aid Agreement (MAA) is to facilitate a speedy, short-term placement of emergency support services prior to, during, and after an incident. Many have the misconception that a Mutual Aid Agreement (MAA) requires both entities to provide mutual aid, but it does not. A Mutual Aid Agreement (MAA) is only a tool that can be used when an agency immediate resources is overwhelmed ("Best Practice," n.d.). This agreement covers a full range of functional areas, better described as tasks based on the function they perform. These can include

- Animals/Veterinary Services
- Administrative Support
- Continuity of Service
- Colleges and Universities
- Coroner/Mortuary Services
- Critical Incident Stress Management
- Building Inspectors and Architects
- Damage Assessment
- Decontamination
- Emergency Managers
- Emergency Medical Services
- Evacuation
- Financial
- Firefighting
- Hazardous Materials
- Law Enforcement
- Logistical Support
- Mass Care Shelters
- Military Support
- Medical Support
- Private Sector Support
- Public Information Officers
- Public Works
- Radio Communications Support
- Schools
- Transportation/Buses

- Traffic and Access Control
- Search and Rescue ("Best Practice," n.d., p. 2)

Those that need to make agreements for mutual aid can include two, three, or even substantially more response agencies, private organizations, hospitals, public utilities, governments, as well as almost any type of organization that can provide needed resources during an emergency or a disaster. It is important to remember that these agreements may vary substantially based on the support needs of each jurisdiction and/or organization, however, in order to be a mutual aid agreement, it must benefit all involved in some way ("Best Practice," n.d.).

The beauty of a Mutual Aid Agreement (MAA) is that as the resources at the incident become overwhelmed, those on-scene or the Emergency Operations Center (EOC) can request the next level of mutual aid. While some levels of mutual aid may require specific paperwork to be signed such as disaster declaration or a resolution by a government official before any resources can be dispatched, the mechanisms are in place to facilitate the exact assistance that may be needed ("Best Practice," n.d.).

There are various types of Mutual Aid Agreements (MAAs), and they are each based on the unique circumstances that an agency or combined agencies face ("Best Practice," n.d.). In some instances, especially rural areas who have limited personnel, there is sometimes Automatic Mutual Aid (AMA) agreements. Some Automatic Mutual Aid (AMA) agreements even have time and day restrictions, while others are specific to the type of incident. These agreements automatically dispatch Automatic Mutual Aid (AMA) partners and the needed resources without incident-specific approvals ("Best Practice," n.d.).

As an example, if two rural volunteer fire departments are often short of personnel from 8 a.m. to 5 p.m., Monday through Friday, then they may initiate an Automatic Mutual Aid agreement with another fire department to respond during those hours, so the two combined departments can provide the appropriate resources. Other automatic agreements may include the automatic dispatch of a resource within a city or a county if a particular incident happens. These could be incidents such as a sawmill fire, a multicar pile-up on an interstate, and various other incidents where it is already known that local resources will be overwhelmed.

Another type of Automatic Mutual Aid (AMA) agreement can be forged between local private agencies and governmental agencies. As an example, in areas where the local ambulance is not government-based, but rather is hospital or independently based, an Automatic Mutual Aid (AMA) can be made between the local ambulance company and the local fire department for the ambulance to respond to all structure fires. There can also be agreements between law enforcement and the ambulance company to stage nearby whenever a raid or SWAT action is taking place.

Another example may be an Automatic Mutual Aid (AMA) agreement might be between law enforcement agencies. In the event of a mass shooting, a manhunt, a riot, or some other type of incident, automatic aid from neighboring police departments, the county sheriff's department, and potentially even the state police might be initiated. The type of Automatic Mutual Aid (AMA) should match the unique needs of what the community may need in certain circumstances.

Another type of agreement is the Local Mutual Aid (LMA) agreement that typically involves agencies that are based within your geographic location. Local Mutual Aid (LMA) agreements are typically between neighboring jurisdictions and/or organizations. This type of mutual aid is not automatically dispatched, and it is not normally requested unless the initial response agency is overwhelmed and/or realizes that they will need assistance ("Best Practice," n.d.).

In some instances, entire counties of fire departments have formed mutual aid associations and are committed to assisting each other when requested. These mutual aid agreements are usually based on all the resources and equipment that the agency has, and they will typically cover a larger geographic area than an Automatic Mutual Aid (AMA) agreement. Local Mutual Aid (LMA) agreements are typically based with agencies within a 5–10-mile radius, however, in rural jurisdictions, the distance could be substantially further.

There are also Regional Mutual Aid (RMA) agreements ("Best Practice," n.d.). Most states have taken the step to divide their state into regional subdivisions. A Regional Mutual Aid (RMA) agreement would be in one region of a subdivided state. These types of agreements are typically sponsored by a council of government (or similar organization) and are made between multiple jurisdictions. Regional Mutual Aid (RMA) agreements are best utilized for when a small geographical area of the region has responded to an incident or disaster, and they have been on the scene for an extended amount of time and are needing relief ("Best Practice," n.d.). Because the region is nearby to other jurisdictions, these types of agreements do not provide additional support when a large disaster occurs.

In cases such as a large disaster, many regional agencies will usually be dealing with responding within their own area, rather than trying to help another agency. It should also be noted that in some instances, the governing body (the council of government) may need to pass a resolution in order to create and/or activate this type of agreement ("Best Practice," n.d.).

Intrastate (Statewide) Mutual Aid (IMA) agreements are often coordinated through a state entity such as the state emergency management agency or state homeland security division ("Best Practice," n.d.). These types of agreements typically incorporate both state and local government resources to increase preparedness effort statewide. In some instances, nongovernmental resources are incorporated into these preparedness efforts. Some states require that there must be a state disaster declaration, signed by the governor, in order to initiate this type of mutual aid ("Best Practice," n.d.).

The state of Illinois uses Intrastate Mutual Aid (IMA) agreement and preposition resources at fire and EMS agencies around the state. The Mutual Aid Box Alarm System (MABAS) organization, discussed in Chapter 4, is independent of the state government, however, they have signed agreements with essentially every fire and EMS agency within the state. This Intrastate Mutual Aid (IMA) agreement, while not state run, works in coordination and cooperation with the Illinois Emergency Management Agency (IEMA) and has an Intrastate Mutual Aid Agreement (IMA) with MABAS. The state of Illinois also has a program that includes law enforcement mutual aid under ILEAS, but those agreements are done utilizing a Memorandum of Understanding (MOU).

Interstate Agreements are for out-of-state assistance that can be utilized through the Emergency Management Assistance Compact (EMAC). The Emergency Management Assistance Compact was ratified in 1996 by the United States Congress under PL 104-321 (EMAC Act, 1996). It was signed and is a law in all 50 states, the District of Columbia, Puerto Rico, Guam, and the Virgin Islands (EMAC, n.d.). In the Emergency Management Assistance Compact (EMAC), there are 12 functional areas that include the disciplines of

- Animal Health Emergency
- Emergency Medical Services
- Fire and Hazardous Materials
- Human Services
- Incident Management
- Law Enforcement
- Mass Care
- Public Health and Medical
- National Guard

- Public Works
- Search and Rescue
- Telecommunicator Emergency Response (EMAC Act, 1996)

Another interstate agreement to recognize is between power companies. The Edison Electric Institute (EEI, n.d.) has gathered the vast majority of electric power industry into an ongoing mutual aid agreement. Utilizing the NIMS method as the overarching IMS method for organizing resources, the electric power industry prepositions resources when it is known that a disaster is incoming, and they respond after a disaster (whether known prior or not) creates power line damage. These mutual agreements have seen great success after Hurricane Sandy, Hurricane Michael, and many other disasters (EEI, n.d.).

While these are the main Mutual Aid Agreements (MAA) agreements utilized under the NIMS method, they are not the only ones. There are a multitude of other Mutual Aid Agreements that help to ensure that the needed resources are available when the unthinkable happens. In fact, it is so prevalent in planning for disaster that many zoos, animal rescue organizations, and faith-based organization have signed agreements with similar organizations.

5.6.3 What is a Memorandum of Understanding (MOU)?

A Memorandum of Understanding (MOU) is a written agreement made between two different types of entities (e.g. agencies, organizations, and businesses) to provide some type of resource, or resources, to assist in achieving a certain goal. This understanding is not an agreement that is a mutually beneficial agreement whereby both parties assist each other. It may however have a financial component to it. Unlike a contract, the Memorandum of Understanding (MOU) is not a binding agreement (in most states), and there is usually no legal obligation that requires either entity to perform. Essentially a Memorandum of Understanding (MOU) is written to show the intent of how one business or organization may be able to help the other entity when that help is requested, and how they intend to help.

An example of this might be that an Emergency Management Agency (EMA) or the county sheriff's department may sign a Memorandum of Understanding (MOU) with a local towing company. In this Memorandum of Understanding (MOU), they might specify that when activated during a mass evacuation, the towing company would remove disabled cars from the evacuation route in order to keep traffic moving.

5.6.4 Identifying and Typing Resources[1]

As previously mentioned, resource typing is stipulating what is required to be a specific resource, and then categorizing that resource by specifying the minimum capability that it can provide. By defining the type of resource and the capabilities of that resource, it allows the NIMS method to establish and promote the all-important common language. Typing resources helps to promote common language by defining the minimum capabilities for personnel, teams, facilities, equipment, and supplies. Resource typing enables communities to plan for, request, and have confidence that the resources they receive have the capabilities they requested.

FEMA leads the development and maintenance of resource typing. Resource typing includes all resources that can be shared on a local, regional, intrastate, interstate, and a national scale. Jurisdictions can use these definitions to categorize their own local assets as well as for ordering

---

1 This heading is for advanced IMS students.

resources when an incident occurs. When identifying which resources to type, FEMA selects potential resources that are

- Commonly used
- Sharable
- Can be deployed across jurisdictional boundaries
- Can be identified by the following characteristics:
  - *Capability*: The core capability (for which the resource is most useful)
  - *Category*: The function for which a resource would be most useful (e.g. firefighting, law enforcement, health, and medical)
  - *Kind*: A broader characterization of the resource, such as equipment, personnel, teams, facilities, and supplies
  - *Type*: A resource's minimum level of capability to perform the given function (FEMA, 2017a)

In typing a resource, FEMA uses a quantitative method for determining a resource's type, depending on the service it can provide and the mission envisioned. As an example, a mobile kitchen unit is typed according to the number of meals it can serve. Dump trucks are typed according to the number of cubic yards they can haul (per load). Water tenders (for hauling firefighting water) are typed by the gallons of water they can haul and the gallons of water (per minute) they can pump. In some instances, such as bomb squad/explosive team, the kind is based on their level of Personal Protective Equipment (PPE) and the level of training.

It is important to remember that typing of resources is done by a number system, with the lower numbers being more skilled or more capable than higher numbers. As an example, anything that is listed as a Type 1 resource will have better capabilities than a Type 2 resource of the same-named resource. Type 2 resources will have better capabilities than a Type 3 resource of the same name.

While this is an oversimplified explanation, the level of capability for equipment is typically based on size, power, and capacity of that equipment. When it comes to typing personnel, whether an individual or a team, they are typically typed by experience and qualifications. The typing of basic capabilities ensures that you order what you need, however, it is important to note that when higher capabilities are ordered, it limits how many of those types of resources may be available. Ordering more skilled resources will, in most instances, increase the distance from the scene and thereby increase the time it will take the resource to arrive. For this reason, it is in your best interest to only order the resources that are needed rather than ordering above the level or capacity that you need to manage the incident.

In order to be a viable resource with the NIMS method, the resource must be able to be identified, inventoried, and tracked so that there can be a determination if that specific resource is available. These typed resources also must be beneficial in incident management, incident support, and/or coordination of the companion ICS method (and/or in Emergency Operations Centers [EOCs]). The resource must also be capable of being compatible, and they must be able to integrate and deploy using the common systems in place for resource ordering, managing, and tracking.

5.7 NIMS Qualifying, Certifying, and Credentialing Personnel[1]

Qualifying, certifying, and credentialing are an important part of having individuals and teams that are able to assist while responding and recovering from an incident or emergency. Those in charge of an incident, or those actually ordering resources, order what they need. They do not want to order apples, and when the shipment arrives, they find they were shipped oranges. This same analogy holds true for personnel and teams. Nobody wants to order a ladder truck and receive a brush truck, and nobody wants to order a crowd control team only to get a single security guard. This is why qualifying, certifying, and credentialing are critical in the NIMS method of incident management.

Qualifying, certifying, and credentialing is undertaken at the local level and led by the Authority Having Jurisdiction (AHJ). It should be a priority for the Authority Having Jurisdiction (AHJ) to ensure that personnel deploying have the knowledge, skills experience, training, and the capabilities to accomplish the duties for the roles that have been assigned to them. By ensuring that the standards required are met, the jurisdiction helps to ensure that their personnel are prepared to perform their incident responsibilities based on a nationwide standard.

It is important to note that these standards are based on the previously mentioned qualifications, certifications, and credentialing, but what do those terms really mean? When we look at qualifications, we must realize that it is a process of ensuring that the individual or team will be useful. Personnel must meet the minimum established criteria in regard to training, experience, physical fitness, medical fitness, and capability in order to fill the specific position for which they may be used. While we should never discriminate based on disability, if someone is not medically or physically capable of performing a task, they become a liability rather than an asset. If a person just received a certification and has relatively little experience, then they will be slower and more methodical than the person who has 10 years' experience. Much like the person who only started driving the car, they will think more about what they are doing rather than just do it out of instinct and experience.

Certification (and recertification) is an act or process of providing an individual or a team with an official document that attests to a status or level of achievement. Certification can come from the Agency Having Jurisdiction (AHJ), or a recognized training entity that is a third party in the process. These third parties could be a government entity (e.g. state or federal), an organization (such as the National Fire Protection Association [NFPA], the National Registry of Emergency Medical Technicians [NREMT], etc.), a college or university, or a for-profit business. Certifications should recognize and confirm that the individual has met established criteria that falls in line with national standards. It is up to the Authority Having Jurisdiction (AHJ) to ensure that the individual and/or the team keep their certifications current and in good standing.

The third and final part of the readiness process is credentialing. Credentialing is when the individual or team is provided some method of identifying the individual and/or team that authenticates and verifies their qualifications. This is usually some form of an identification card. It is important to note that these identification cards should be a picture form of an identification card to prevent those who are not qualified from taking or stealing a badge to gain entry in a potentially secure are. While credentialing includes issuing an identification card from the Authority Having Jurisdiction (AHJ), there will usually be a separate incident-specific badging process when they arrive on scene. This incident-specific badging process will typically include verifying the person's identity, reviewing their qualifications, and reviewing their deployment authorization.

At the local level, the standard for each type of personnel should meet or exceed the standard set forth in the Resource Typing Library Tool (RTLT). The Resource Typing Library Tool (RTLT) is an online collection of the NIMS method resource typing definitions and job titles or position qualifications. The Resource Typing Library Tool (RTLT) can be found online at https://rtlt.preptoolkit.fema.gov/Public. Once on the home page of the website, you can do a search based on the type of resource, the agencies discipline, their core capability, and various key words that can help identify the resource.

By identifying nationally standardized criteria and minimum qualifications, the NIMS method provides a consistent minimum standard for response personnel based on their qualifications and their credentials. Additionally, in the NIMS method, been Position Task Books (PTBs) have been developed. These are a basic tool that describe the minimum proficiencies, activities, and responsibilities necessary to be qualified for a position. Position Task Books (PTBs) provide the basis for a qualification, certification, and credentialing process that has become an established standard nationwide with the NIMS method. Because the NIMS method uses the qualification, certification, and credentialing process as a performance-based approach, it enables communities from across the nation to plan for, request, and be assured that personnel from other organizations through mutual aid agreements.

5.8 NIMS Method of Resource Management Response and Recovery[1]

Managing personnel, specialized teams, equipment, and supplies are an elaborate part of response resource management, yet these critical details are often overlooked. It is important to realize that the companion IMS systems that integrate with the NIMS method are critical for managing resources in the response mode. While the ICS method is the most common method utilized in response, we also need to realize that healthcare facilities utilize the companion method of Hospital Incident Command System (HICS). Both companion methods leave the effective management of resources in the hands of the Logistics Section Chief (LSC). This will be discussed in greater detail in Chapter 7.

The 2017 NIMS guidance identifies six major areas for resource management in the response mode. These six major factors for effective resource management are

- Identify resource(s) needed for the incident
- Order and acquire the resource(s)
- Mobilize the resource(s)
- Track and report resource(s)
- Demobilize and reimburse resource(s)
- Restock resource(s) in an incident (FEMA, 2017a)

5.8.1 Identify the Resource[1]

In order to manage resources, you must first identify what resources are needed. Initially, this will be undertaken in the first few minutes of the incident, but as additional resources are brought in, and the command structure is expanded, this responsibility will be a coordinative and cooperative decision that will be made during the planning process for incident response. In doing so, the Command Staff and General Staff will identify what equipment is needed, where it is needed, when it is needed, and who will be receiving or using it. While some resources will only perform one type of task, others may be useful for multiple tasks, and the ability to perform multiple types of tasks may be more advantageous than single task resources. Whenever there is a decision to be made between the two, careful considerations should be explored so that the appropriate resource is identified and effectively utilized.

5.8.2 Order and Acquire the Resource[1]

When it is identified what resource or resources are needed, the Logistics Section Chief (LSC) will need to order the resource, with an understanding of the amount of time it will take the resource to be physically acquired and to be put into commission. If the ICS method has not been fully expanded, the person handling logistics (usually the Incident Commander [IC]) will manage this responsibility.

While undertaking the planning process, it is extremely important to remember the lead-time. If a resource is coming from a long distance, it could take days to physically acquire that resource. If an ordered resource is a human resource, then consideration must also be given to the rest and feeding that may be required enroute and before putting them into service. As an example, during the 2017 Gatlinburg TN wildfire, human resources came from all over the country. If human and equipment resources were ordered from Colorado, Utah, or Nevada, then considerations for needed down-time while traveling, and prior to being deployed in the incident, so that they could be at their best.

5.8.3 Mobilize the Resource[1]

Mobilizing is the process and procedures used for activating, assembling, and transporting all resources that have been requested to support an incident. In this process, you will plan transportation and logistics needs based on the incident's response priorities and equipment requirements. To ensure timely arrival of the necessary equipment, you will need to determine the priority, the lead time, and the importance that the resource has on the incident. Varying decisions may need to be made, depending on the urgency of need, and whether the resource is just delivered, or if the resource will arrive with personnel assigned to it.

If it is equipment only, then it will need to be decided if it can be shipped overnight (if it is small enough) via US Postal Service, FedEx, or UPS, or if the item will need to be transported by truck. If being brought by truck, will standard delivery time be sufficient, or should it be expedited? Can you wait four days, or should team drivers be utilized to get it on scene in two or less days? If equipment that comes with personnel, will it arrive in the timeframe needed if they drive the equipment (if it is even highway drivable), or will it need to be shipped in an expedited manner and the personnel will come by bus or plane? Of course, there are numerous questions that could be asked about mobilization, but these were provided to help you to think through the process. It is important to realize that you will need to use critical thinking to ensure that you choose the best option while being fiscally responsible.

5.8.4 Track and Report Resources[1]

It is important that the resource is tracked from the leaving destination, and that regular reports be provided until it arrives. In tracking, the continual and ongoing tracking of the exact location of the resource is critical. On a continual basis, the staff will need updates in preparing to receive resources, to ensure safety and security of the equipment and personnel, and to ensure that the resource will be efficiently used. This is important to ensure coordination and movement of equipment to the appropriate area after arrival.

5.8.5 Demobilize and Reimburse the Resource[1]

After the resource has arrived and been pressed into service, it is extremely important to only keep that resource as long as it is needed, and/or return the resource on the time schedule that it was allotted. Ensuring that you demobilize the resource quickly and efficiently is critical in guaranteeing that the equipment will be available for its home base, or another assignment, as soon as possible. Demobilization should include decontamination, disposal, repair, and restocking as is needed and required. This demobilization includes both dispensable and durable resources.

It is also important to ensure that while the resource is in service for your incident, that you have in place a way to track the costs and provide repayment in a timely manner. This includes all incident expenses, including contractors, equipment, transportation services, and any other costs affiliated with the incident. In some instances, your entity may need to reimburse from their own coffers, while in other incidents, this may be covered by a state or federal entity. No matter who is paying for the resource, tracking costs is important to ensure that everybody is made whole for their services, and that there was no wasting of money.

5.8.6 Restock Resource(s) in an Incident[1]

While restocking of outside resources should be ongoing, once an incident has been completed, and outside resources have been made whole, you must also consider making your agency or community whole again as well. This can often be accomplished by utilizing a resource inventory system, or equipment checklist to assess the availability of equipment and supplies. In order to make the community whole, you will need to restock at least pre-incident levels of preparedness for the resource. You should also consider what lessons you learned from the incident and incorporate them into community preparedness. You could and probably should procure the original items that had to be restocked and keep them on-hand, as well as any additional resources that may be needed to be prepared for future events.

5.9 NIMS Multiagency Coordination Systems[1]

Multiagency Coordination Systems (MACS), not to be confused with Multiagency Coordination (MAC) Groups, is an all-encompassing identification of the four Command and Coordination systems in the NIMS method of incident management. The Multiagency Coordination Systems (MACS) include the Incident Command System (ICS), Emergency Operations Centers (EOCs), Multiagency Coordination (MAC) Groups, and the Joint Information Systems (JIS).

When it is realized that an incident has become more complex, Multiagency Coordination Systems (MACS) become progressively more important. Because coordination and cooperation are key concepts to managing an incident, the Multiagency Coordination Systems (MACS) provides the avenue that creates a more enhanced coordinated and cooperative response, which then creates enhanced response and recovery efforts. The ICS method is a standardized, on-scene, all-hazard incident management concept that will be discussed at great length in the following sections. Except for the ICS method of incident management, the Multiagency Coordination Systems (MACS), including the Emergency Operations Centers (EOCs), Multiagency Coordination (MAC) Groups and the Joint Information Systems (JIS) will be better explained below.

5.9.1 Emergency Operations Centers (EOCs)[1]

At the tribal, local, state, and federal levels Emergency Operations Centers (EOCs) are an important element in emergency management. Emergency Operations Centers (EOCs) are physical locations where staff from multiple agencies come together to address imminent threats and hazards. The primary function is to provide coordinated support to an Incident Commander (IC), on-scene personnel, and/or other Emergency Operations Centers (EOCs) around the nation. Emergency Operations Centers (EOCs) can take many forms and may be located at fixed locations, temporary facilities, or they may be a virtual structure with staff contributing remotely (FEMA, 2017a).

The purpose, authorities, and composition Emergency Operations Center (EOC) can differ substantially, but the individuals that work in an Emergency Operations Center (EOC) will typically exchange information, support incident decision-making, coordinate resources, and communicate with personnel on scene and at other Emergency Operations Centers (EOCs). The personnel at a local Emergency Operations Centers (EOCs) can and usually do support staff at an Incident Command Post (ICP), various field personnel that are not associated with the response but rather working on recovery (such as debris removal, performing muck out, or managing a long-term shelter), or staff in another Emergency Operations Center (EOC) such as in a widespread disaster, communicating resources available for the local Emergency Operations Center ([EOC] FEMA, 2017a).

Emergency Operations Center (EOC) staff can help reduce the stress, chaos, confusion, and uncertainty from the Incident Management Team (IMT) by taking control of certain operations, such as emergency shelters or Points of Distribution (POD), and they allow those involved in the response to focus on the immediate issues while they handle underlying issues. When a specific on-scene incident command is not established, such as in an ice storm, staff in Emergency Operations Centers (EOCs) can and often do direct tactical operations.

A good example of no incident command in place might be that in an ice storm. The Emergency Operations Center (EOC) may be activated to prioritize and facilitate check-the-well-being calls, emergency slide-offs, activating additional support teams, directing power company resources, and more. While calls for assistance will typically come into the local 911 center, the Emergency Operations Center (EOC) is capable of handling deploying resources for overflow calls that local public safety cannot handle in a timely manner. These are just a few ways that an Emergency Operations Center (EOC) can be utilized.

It does not matter if it is a virtual or physical Emergency Operations Center (EOC), the primary functions of staff member are to collect, analyze, and share information, provide the needed resources and to fill requests (including allocation and tracking). They are also responsible for coordinating plans and determining. On occasions, they may need to also provide policy direction and coordination of resources.

In some instances, various agencies, organizations, and departments have their own specific operations centers. While these operation centers will often have a representative, or some form of contact that integrate with other Emergency Operations Centers (EOCs), these organization-specific operations centers differ from a multidisciplinary Emergency Operations Center (EOC). The entity specific operations centers have been named Departmental Operations Center (DOC), and it is important not to confuse them with a multidisciplinary operations center. Departmental Operations Center (DOC) staff members will typically coordinate only their entity's activities while they communicate with other organizations and Emergency Operations Centers (EOCs). Departmental Operations Centers (DOCs) will focus specifically on their operations and are usually more concerned about increasing cooperation and guiding a response within their own agency, thereby providing a more coordinated response.

An example of a Departmental Operations Center (DOC) could be various electric companies. In instances where there have been major power disruptions, electric companies will sometimes open their own Departmental Operations Center (DOC) so that they can coordinate efforts. In many of these centers, they will set up IMS by enacting the ICS method of incident management, or something that closely resembles it. Agency representatives will interact with the Incident Commander (IC) or Emergency Operations Center (EOC) who will set the priorities for their response. The Departmental Operations Center (DOC) will typically provide a liaison working in the local and/or state Emergency Operations Center (EOC) as well. It should be noted that Departmental Operations Centers (DOCs) are popular in health departments, electric companies, universities, and some EMS agencies.

5.9.1.1 Which Stakeholders Should Be Represented in the EOC?[2]

Bringing representatives from various stakeholders together in an Emergency Operations Center (EOC) heightens the concept of unity of effort. This is because it enables multiple agencies from differing disciplines to share information, provide legal and/or policy guidance to those working on-scene, plan for contingencies, deploy resources proficiently, and to generally support the mission at hand.

The purpose of an Emergency Operations Center (EOC) is to integrate different stakeholders together with the primary goal of supporting the incident. It is important that the Emergency Operations Center (EOC) staff represent different disciplines; also known as multidisciplinary. Emergency Operations Center (EOC) would be less effective without a multidisciplinary approach because each varying discipline can offer a wide variety of resources, all in one place, with an individual that understands and knows their own resources, and they can offer and commit them for the response. While the make-up of Emergency Operations Center (EOC) disciplines will be unique to the geographical location, many Emergency Operations Centers (EOCs) will use a combination of emergency management, the fire service, law enforcement, EMS, public works, power companies, public health, and more.

It is important to note that the time to determine who should be represented in an Emergency Operations Center (EOC) staff is not after an incident occurs, but rather during the emergency operations planning process. When identifying what disciplines should be represented in the Emergency Operations Center (EOC), some things to consider is who has authority and/or responsibility for core services, what resources and information might a stakeholder have (or have access to), and what expertise and/or relationships does a stakeholder have. How the Emergency Operations Center (EOC) is organized may also depend on the type and complexity of the incident.

It does not matter which stakeholders are represented, the Emergency Operations Center (EOC) will always be overseen by elected and/or appointed officials. These officials usually include governors, tribal leaders, mayors, or city managers. While these elected and/or appointed officials may be present in the Emergency Operations Center (EOC), it is more common for them to be at an alternate location. While they may be overseeing the Emergency Operations Center (EOC) from a distance, they will typically make decisions and provide policy and input on the legal aspects. They will also identify priorities on issues such as emergency declarations, large-scale evacuations, types or availability of emergency funding, sign waivers to ordinances and regulations, and decide where limited resources should be used.

5.9.1.2 EOC Organizational Structure and Management[2]

Because the Emergency Operations Center (EOC) is an aspect of the NIMS method, and because the Emergency Operations Center (EOC) must integrate with the NIMS method, there are certain factors that must be met to promote the integration of an Emergency Operations Center (EOC). One function that should be met is the modular management characteristic. In the NIMS method, modular management is when leaders are responsible for the functions of positions that are not staffed. This management characteristic also applies to Emergency Operations Centers (EOCs).

In an Emergency Operations Center (EOC), the individual in charge, usually the Emergency Operations Center (EOC) Director, is responsible for the duties of other Emergency Operations Center (EOC) team members not present until those positions are staffed. If the position is not staffed, the Emergency Operations Center (EOC) Director will continue to fill that position until it is filled or the Emergency Operations Center (EOC) is demobilized.

---

2 This title signifies highly technical information for advanced students and practitioners.

In managing the Emergency Operations Center (EOC), most locations use the ICS method, or some derivative of it, to manage the Emergency Operations Center (EOC) staff. The reason this practice is common revolves around familiarity that most stakeholders have with the ICS structure. Another reason for choosing the ICS method is that it helps to align the Emergency Operations Center (EOC) with the on-scene incident organization. In some Emergency Operations Centers (EOCs), they will use the standard ICS organizational structure but will modify some of the titles to create an ICS-like organizational structure to reduce confusion by separating their functions (in name) from their on-scene counterparts. While the make-up of the organizational structure is up to each individual Emergency Operations Center (EOC), they should always ensure that integration with the NIMS method is a top priority.

To learn more about how an Emergency Operations Center should be structured and managed, you can take multiple online independent study courses on Emergency Operations Centers (EOCs). These interactive web-based courses are free, and available from the Emergency Management Institute (EMI).

5.9.1.3 EOC Activation and Deactivation[2]

Emergency Operations Centers (EOCs) can be activated for a multitude of reasons, however, the activation is typically based on a threat, or the expectation of an event or events. They can also be activated in response to a large or evolving incident. Some examples of a threat can include rumors of civil unrest due to an officer involved shooting. It could also be a threat that outlaw bikers are traveling through the community and/or are planning to camp in (or near) your community. A threat could be a wildland fire that is encroaching on a community, or it could be that a tornado warning has been issued.

When it comes to events or expected events, it might be a winter blizzard, a potentially deadly heat wave, or a solar flare that risks damaging the power grid and electronics. An event can also be a large festival that may need coordination, or a huge outdoor concert that may need outside support. When it comes to an incident-based activation, it could be almost anything that public safety agencies might respond to.

While each community and their elected and appoint officials must determine what circumstances should activate the Emergency Operations Center (EOC), there have been trends seen nationwide regarding what should constitute the activation of an Emergency Operations Center (EOC). These activation trends include

- Two or more jurisdictions are involved in an incident;
- It is realized that the incident could expand rapidly, suffer cascading effects, or require additional resources;
- Past similar incidents required activation;
- The EOC director or official directs that the EOC be activated;
- An incident is imminent (e.g. hurricane warnings, tornado warnings, elevated threat levels)
- The Emergency Operations Plan specifies that the EOC should be activated for an event;
- The community will likely suffer significant impacts to the people, or economy (FEMA, 2017a).

It is not uncommon for Emergency Operations Centers (EOCs) to have multiple levels of activation. This allows the local or state Emergency Operations Center (EOC) to scale the response to the needs of the incident. It also allows them to gauge the level of support and coordination that is appropriate for each individual incident.

There are typically three levels that pertain to activation of an Emergency Operations Center (EOC). A Level-1 activation is a full activation of the Emergency Operations Center (EOC).

Level-2 Activation is an enhanced steady-state or partial activation. A Level-3 Activation is normal operations (or steady-state).

5.9.1.3.1 Level-1 Activation[2]

During a Level-1 Activation, every member of the Emergency Operations Center (EOC) staff is activated. This almost always includes all major assisting agencies that have been identified in the Emergency Operations Plan (EOP). These individuals are typically in a specific building (or portion of a building), but in some cases, staff members can participate virtually.

While they may be able to participate virtually, this is not usually the best option because of the potential for the interruption of services that support virtual participation. The purpose of a Level-1 activation would be to support the response to a major incident or to support operations to a credible threat.

5.9.1.3.2 Level-2 Activation[2]

During a Level-2 activation, only certain members of the Emergency Operations Center (EOC) staff are activated. This is done so they can monitor a credible threat, risk, or hazard. During this type of activation, the staff member can also be physically in the Emergency Operations Center (EOC) or they can participate virtually.

While it is usually preferred that they have a physical presence, a virtual presence is usually more accepted in a Level-2 activation than a Level-1 Activation. Unless there is a high likelihood of an interruption of services, a virtual presence (and work) by an agency can be acceptable if the director is agreeable. Emergency Operations Center (EOC) Level-2 Activations typically support the response of a new and/or potentially evolving incident.

5.9.1.3.3 Level-3 Activation[2]

During normal operations (steady-state), Emergency Operations Center (EOC) personnel maintain an operational readiness so that they can quickly activate the Emergency Operations Center (EOC) if needed. They stay in a steady-state by monitoring and assessing potential threats and hazards.

In the steady-state the Emergency Operations Center (EOC) will conduct routine and ongoing coordination with other departments and agencies so that when activated, they already have relationship and contacts in place. During their down-time in the steady state, they will develop and potentially implement plans, undertake training and exercises, as well as maintaining facilities and equipment within the Emergency Operations Center (EOC).

5.9.1.4 What Triggers Levels of Activation?[2]

When we look at the activation of an Emergency Operations Center (EOC), it is best to have preset conditions that automatically trigger a higher level of activation. These are usually predetermined in the Emergency Operations Plans (EOPs) based on the hazards or threats being faced. It is important to note that most Emergency Operations Centers (EOCs) increase staff as the incident they are involved with grows in size, scope, and complexity. As an incident grows in the field, it usually requires additional support and coordination from the Emergency Operations Center (EOC).

At their discretion, the Emergency Operations Center (EOC) Director can activate additional staff. This is typically done to involve more disciplines, mobilize additional resources, inform the public, address media inquiries, involve senior elected (and appointed) officials, and to potentially have the proper staff in place to request outside assistance. Whenever outside assistance is requested, it is important to remember that the resource chosen matches the typing of the incident in the NIMS Resource Typing Library using the Resource Typing Library Tool.

5.9.1.4.1 Deactivation[2]

Deactivation is the process of demobilizing or disbanding the Emergency Operations Center (EOC). In most instances, the Emergency Operations Center (EOC) Director will begin the deactivation of staff as the circumstances allow. When deactivation is complete, the Emergency Operations Center (EOC) returns to its normal steady-state condition.

The process of deactivation typically occurs when the incident no longer needs the support and coordination that has been provided by the Emergency Operations Center (EOC), or whenever those functions can be managed by an individual stakeholder. Deactivation can be a slow process, such as deactivating one or two functions at a time, or it can be done by deactivating the entire Emergency Operations Center (EOC) at the same time. How the Emergency Operations Center (EOC) is deactivated will typically be based upon the needs of the incident they were supporting. When the Emergency Operations Center (EOC) staff is completing the demobilization, they should transfer any ongoing incident support and/or recovery efforts before fully deactivating.

An important part of completing the mission is to undertake an After-Action Review (AAR). After-Action Reviews (AARs) are usually done so that it can be determined what went right, what went wrong, and what needs improved upon. This will allow the Emergency Operations Center (EOC) to identify areas of improvement and to discuss ways to overcome any shortfalls that were identified during the activation.

5.9.1.5 Multiagency Coordination (MAC) Group[2]

A Multiagency Coordination (MAC) group, when activated, is a collection of individuals who deliver guidance to an incident by providing coordinated decision-making and resource allocation among the cooperating agencies. Multiagency Coordination (MAC) groups will usually create the priorities amid an incident, match the appropriate agency policies to the response, and provide strategic guidance (and direction) to support incident management activities. They will not micromanage a response, but they will usually provide overall direction (FEMA, 2017a).

As an incident becomes more complex, multiagency coordination and the need for additional resources become increasingly important. While the Emergency Operations Centers (EOCs) will typically support on-scene incident command, another method is sometimes needed to support those who are working and managing the incident directly. On-scene resources using ICS method and an activated Emergency Operations Center (EOC) sometimes need guidance on priorities and policy. This is usually done by creating a Multiagency Coordination (MAC) group (FEMA, 2017a).

Multiagency Coordination (MAC) groups are often referred to as policy groups. While this may be the name some call the Multiagency Coordination (MAC) group, technically they should not be called a policy group. The reason they should not be called anything other than the Multiagency Coordination (MAC) group (or MAC Group) is that we want to continually support the common terminology needs of the NIMS method. Any terminology that we use which is not the proper NIMS method name can add complications to the incident because common language is not being used (FEMA, 2017a).

Those involved with a Multiagency Coordination (MAC) groups can come from any discipline. These disciplines might include emergency management, public health, critical infrastructure, or even the private sector. These groups can also be from any governmental level (e.g. local, state, tribal, or federal) or they may be a group comprised of a singular, or multiple, private businesses.

The Multiagency Coordination (MAC) group is part of the off-site incident management structure in the NIMS method. The Multiagency Coordination (MAC) group typically consists of representatives from stakeholder agencies or organizations. This group is quite often composed of senior officials such as agency administrators, executives, or their designees. While this is not a hard rule, the individuals that are chosen for this group should have authority to make decisions and commit resources, or they should have specialized knowledge in policy, protocols, procedures, or the jurisdictional laws. When we talk about having a specialization in law, it usually goes beyond state law and focuses on local administrative law and ordinances that apply to that jurisdiction.

In some jurisdictions at the governmental level, local administrative law or some governmental policy may require approval (and/or guidance) from the Multiagency Coordination (MAC) group. This required approval will typically be to authorize additional resources and/or provide guidance to Emergency Operations Center (EOC) personnel and/or incident command personnel (FEMA, 2017a).

The composition or mix of the Multiagency Coordination (MAC) group is also important. Sometimes membership is obvious, especially if it is an Agency Administrator or the Chief Elected Official. Organizations directly affected, and whose resources are committed to the incident, should most likely be represented. Sometimes Multiagency Coordination (MAC) group members are not as obvious, and additional coordination should be considered.

Specific community and civic organizations might enhance the capabilities of a Multiagency Coordination (MAC) group. These organizations should consider including local chambers of commerce, local volunteer organizations (e.g. American Red Cross, Salvation Army), or other organizations that have specialized expertise or knowledge. Often, these organizations will not have physical resources or funds to contribute to their response because their organization suffers from financial constraints. Even if they may not be able to contribute physical resources, the influence they have in the community may be equally critical to ensuring a more productive response and recovery. These relationships often come with political influence or technical expertise which can help to improve the success of incident response and recovery. Multiagency Coordination (MAC) group typically make decision through a consensus vote of the entire group.

The goal of multiagency coordination is to prioritize and organize the need for critical resources, especially when those critical resources are difficult to obtain. By creating a Multiagency Coordinating (MAC) Group, those operating under the NIMS method of incident management can assist with the coordination of the response, especially in operations. A coordinated effort should consist of a combination of agreed-upon foundations of the areas in which the Multiagency Coordination (MAC) can coordinate for the incident. Typically, areas that are under the purview of this group include personnel, procedures, protocols, business practices, and communications systems and/or methods.

The Multiagency Coordinating (MAC) group is usually established and organized to make cooperative multiagency decisions. Multiagency Coordinating (MAC) groups can be national, regional, or local management group. Using a Multiagency Coordinating (MAC) group will usually result in better interagency planning, coordination, and operational leadership for incidents. The group is a vital management apparatus for strategic coordination. They help to ensure incident resources are resourcefully and properly managed in a cost-effective manner. During a process that sees significant competition for resources during the response to a major incident, Multiagency Coordinating (MAC) group can help relieve some of the coordination requirements by making decisions about the prioritization and allocation of limited resources.

At a local level, some communities place the physical Multiagency Coordination (MAC) groups at (or in close proximity) to the Emergency Operations Centers (EOCs). This is done so the Multiagency Coordination (MAC) group can authorize additional resources, approve emergency authorities, and provide guidance on emerging issues face to face.

While face-to-face meetings are preferred, if communications avenues are still operable, the Multiagency Coordination (MAC) group can function virtually, using a multitude of technologies.

It is important to remember that elected and appointed officials are key players in incident management. The reason is that they are responsible for the protection and well-being of their constituents as well as the overall effectiveness of all ongoing incident management efforts. Because these elected and appointed individuals operate at the policy level of incident management, they will usually provide guidance regarding priorities and strategies for incident response and recovery.

The Incident Command (IC) staff and the Emergency Operations Center (EOC) are responsible for keeping elected and appointed officials informed about the situation, resource needs, and other pertinent information. By creating effective communication between these incident personnel and policy-level officials, it creates a trust between these officials and responders. This communication should help to ensure that all leaders have the information they need to make informed decisions. Multiagency Coordination (MAC) groups provide a way to integrate and organize policy-level officials which in turn enhances the unity of effort at the highest levels.

It is important to remember that when Multiagency Coordination (MAC) is formed, they may need administrative and/or logistical support. Administrative support can be provided by the Emergency Operations Center (EOC), while in other instances, separate staff members may be needed to help the Multiagency Coordination (MAC) group by meeting its logistical and documentation needs. Administrative support for the Multiagency Coordination (MAC) group might include taking care of tracking critical resources, helping to provide better situational awareness or status, and investigative information.

5.9.1.6 Joint Information Center[2]

Joint Information Centers (JICs) are another structure that supports command, coordination, and cooperation in the NIMS method of incident management. Joint Information Centers (JICs) work together to integrate incident-related information and public affairs into a unified group. This group will provide a consistent, coordinated, accurate, accessible, timely, and complete information. This information is not only for the public but also it provides stakeholders with the most consistent and coordinated information while incident operations are ongoing.

Joint Information Centers (JICs) have the capability and the responsibility for operating transversely with all Multiagency Coordination Systems (MACS). Joint Information Centers (JICs) supports these systems by specifically overcoming information management issues in the NIMS method of incident management. The way in which they are organized promotes better coordination and cooperation. Tasks affiliated with the Joint Information Centers (JICs) include

- Creating and issuing coordinated messages
- Recommending and managing public information tactics and strategies
- Providing guidance on information issues that could affect incident management
- Rumors control and inaccurate information that could undermine public confidence

These issues can totally undermine a response and recovery, so the Joint Information Centers (JICs) are a coordinated effort that allows the Incident Commander (IC) or Unified Command (UC), the Emergency Operations Center (EOC) director, and the Multiagency Coordination (MAC) group to do their jobs with little to no worry about public information.

As an incident gets larger, and more resources are delivered to the incident, they may need to expand from a single Public Information Officer (PIO) to a Joint Information Center (JIC). To ensure coordination of public information during incidents that involve multiple agencies and/or jurisdictions, the use a Joint Information Center (JIC) helps to share the load of information verification, coordination, and dissemination of accurate, accessible, and timely information.

In the early stages of an incident, the Public Information Officer (PIO) will typically discuss with the Incident Commander (IC) or the Unified Command (UC) of when a Joint Information Center (JIC) should be activated. While the authority to open a Joint Information Center (JIC), the singular Public Information Officer (PIO), or the lead Public Information Officer (PIO) if more than one is present, can make recommendations about when to open one.

The Joint Information Center (JIC) will be activated only at the direction of the Incident Commander (IC) or Unified Command (UC). When initiated, they will be organized at either a predetermined location, or at a location that is specific to the incident. While the Joint Information Center (JIC) will usually be a physical location, away from Incident Commander (IC) or Unified Command (UC), the Emergency Operations Center (EOC) director, and the Multiagency Coordination (MAC) group, they should be in close proximity to all. In the NIMS method of incident management, the Joint Information System allows federal, state, tribal, or local governments to deliver a unified message.

5.10 Conclusion

While this has been a brief overview of the NIMS method of incident management, even the newest student of IMS methods can see how each function intertwines with the others. Moreover, these functions support and enhance each other with minimal effort. The NIMS method organizes chaos, reduces or removes complexity, and provides certainty in how to respond to an incident. This is important because chaos, uncertainty, and complexity play a role in every emergency incident that first responders are involved with.

Sometimes the issues faced in an emergency incident are relatively minor and simplistic, and other times these issues are extremely significant and extremely complex. The NIMS method is simple to use and understand, it can assist with taking even the most complex incident and subdivide that incident into manageable tasks. It does so by spreading the work across multiple individuals and systems.

One shining example of how complex an incident can become would be to look at Hurricane Katrina. Even 11-years after the incident, the NIMS method, including the integration the ICS method, is still being used in the reconstruction of Louisiana, Mississippi, and Alabama. This is but one example of how complex, how long-lasting, and how important an emergency incident management system is, especially if we want to provide holistic solutions.

One thing that is often forgotten is that the NIMS method is not only for response. The NIMS method of incident management should be recognized and used in all modes of an emergency and incident management, and it should be used across all disciplines and jurisdictions. There are many that use the NIMS method in response, then terminate it as the incident moves into recovery. NIMS should be used in all aspects of an emergency, including planning, response, recovery, and preparedness and mitigation.

Planning is perhaps one of the most overlooked areas. What many fail to realize is that planning is critical to managing an incident. Without undertaking planning steps to respond to (and manage) an incident, we essentially fly by the seat of our pants. Planning can make the difference between knowing what you should do, and guessing about what might work, especially is the cycle of preparedness that was utilized in the planning process.

Another area commonly overlooked is preparedness and mitigation. By not preparing, we have not established what resources will likely be needed and what tactical and operational strategies work and do not work. If we do not undertake preparedness and mitigation efforts, we may even spend wasted time looking for specific resources that probably would have already been known if we had prepared for, and tested, our capabilities. It is important to learn from our past experiences, which is also part of preparedness.

Past lessons learned should also guide us in mitigation measures. Mitigation measures help to provide community resilience, increasing the ability to manage the incident because it did not impact the community as hard as it could have. While most remember to implement the NIMS method and its companion ICS, some fail to realize that they should also be used in the recovery efforts. In many instances, when the response is over, they terminate Incident Command Posts (ICPs), rather than leaving the structure in place for the recovery.

Unless we are proficient in the use of these tools, and we practice them on a regular basis, we have little hope of fully integrating them into a complex response that incorporates agencies from across our state, region, or nation. Not only should we be prepared to accept the assistance from others, but we need to plan and prepare to assist another jurisdiction when they need us in a complex emergency. We may be called to a major earthquake, tsunami, terrorist attack, or hurricane. If we are not fully cognizant of the NIMS method and how it works, there is a possibility that a single agency could cause more confusion than they provide help.

Being knowledgeable and proficient in all facets of the ICS and NIMS method is critical to ensuring a rapid organized response that overcomes chaos, uncertainty, and complexity. Frankly, having well-trained, knowledgeable, and fully capable resources is what will be, and should be, expected of all responding agencies. When resources arrive and the have the knowledge, skills, and experience, and they can integrate into the response as if they were part of the original entity, they become an asset. Without this integration and skillset, that resource will likely be nightmare for the Command Staff.

A final thought for this chapter relates to making your voice heard. If you use the NIMS method or the ICS method, you should be vocal about suggested future changes. Proposed changes to the NIMS and ICS method are always presented to the public for comment. Organizations and individuals should always ensure that they review any proposed changes and provide feedback based on experience. By becoming part of the process, you ensure that potential changes will be beneficial to you and your agency. You should always review any and all proposed changes and make a determination based on how the changes will affect you and your agency. If those changes will not be beneficial, you and your agency have the right, and some would say the duty to voice your opinion. Be careful not to be the old timer that says, "We ain't never done it that way before!" Public safety and the challenges we face change daily, and if we are not identifying and changing to meet the needs, then we are falling behind. If changes to the NIMS and ICS methods will improve your efforts, then you should voice your opinion, the same as if it will not be beneficial. You can be an active participant of the research and development of the NIMS and ICS method, and you should be.

Real-Life Incident

In 2018, in a rural area of Tennessee county, a manhunt was underway for a suspect. The charges were murder, and the suspect began his crime in one county, then fled to another county, and allegedly committed another murder. During this manhunt, multiple issues arose that could have been overcome by using NIMS and ICS.

One of the first issues to arise was with coordination and cooperation. While the Sheriff in the first county notified the second county about the incident, neither county coordinated or cooperated with the other. Reports from multiple sources who were involved with the manhunt identified the main problem was a "pissing contest" between the two jurisdictions. Federal, state, regional, and local resources were brought in to assist, however, the expertise of these agencies in tracking a suspect was largely ignored by at least one county Sheriff.

Once during this seven-day ordeal, information and intelligence was not properly vetted, which skewed situational awareness and led to the deployment of resources away from a grid search. This led to resources not being effectively used, and frustration among resources.

The consensus of most resources was that command and control appeared to be fractured. Additional issues arose when responding resources were not provided with food and water while holding a perimeter for 12 or more hours. Communications issues arose at all levels, even within the same agency. One state agency that responded with personnel failed to listen to their people that were assigned to, and familiar with, the area. Communications were also broken when a Sheriff from a totally different county (who was an avid animal hunter) made suggestions on the typical psychology of how animals avoid hunters, which were reported to have been ignored.

Out of frustration, the Sheriff from two counties away decided to freelance, and to use his own suggestions in order to capture the suspect. He went to one of several areas that he suggested, and he found a deer-hunting stand in a tree. He climbed up to the deer stand and waited. Within 30 minutes, the suspect walked down the path where the Sheriff was located, and the suspect was captured.

This incident has been troubling to many law enforcement officers who were present. From multiple sources (who asked for anonymity), they stated that it appeared as if no planning efforts were undertaken prior to, or during, the incident. They identified that command and control was essentially handled by one person in each county. The span of control (number of people one person can supervise) was double and sometimes triple what it should have been to be effective. Logistically, there were multiple failures with taking care of the resources that responded, especially with feeding, hydrating, and housing them. While many agencies provided agency representatives, it appeared as if liaising with these organizations was somewhat ignored, and the Incident Commander (IC) of at least one agency only cared about boots on the ground. The response was so fractured, that it led to frustration by many, and freelancing from at least one person, if not more. Perhaps the most telling sign of this incident is that neither county appears to want to learn from their mistakes. No After-Action Review (AAR) was scheduled or undertaken to identify problems and to take corrective action.

Chapter 5 Quiz

1. What are the three themes that are synonymous with any emergency incident, especially in the beginning stages?

2. What will happen if the lifecycle of an emergency incident is not managed?

3. In managing an incident, what plays an important part in enjoying a successful outcome?

4. NIMS is
    (a) Comprehensive, Flexible, and Adaptable
    (b) Incomplete, Flexible, and Malleable
    (c) Comprehensive, Rigid, and Adaptable
    (d) Comprehensive, Flexible, Rigid

5. As you gain further knowledge and experience in incident management, it will become
    (a) Harder to understand

(b) Increasingly less natural to use
(c) Increasingly more natural to utilize
(d) None of the above

6. **True or False**: NIMS is a consistent, nationwide, systematic approach for managing an incident of any magnitude.

7. What are the five components of NIMS?

8. **True or False**: Preparedness is not an important part of incident management.

9. What functions are included in the cycle of preparedness?

10. **True or False**: Even if you use the cycle of preparedness, it will have no effect on coordination when working with other nearby agencies during incident response.

11. What are the seven types of drills and exercises that can be used to make sure that an emergency event is well planned for?

12. **True or False**: When conducting a full-scale exercise, you should make sure to include approximately 20% of access and functional needs (disabled and elderly) individuals as actors to gain a better understanding of their needs during an incident.

13. **True or False**: Resource management is basically telling people where to go and what to do.

14. A _____ is mutually beneficial to both parties.
    (a) Mutual Aid Agreement (MAA)
    (b) Memorandum of Understanding (MOU)

15. A _____ is an agreement that spells out what one party intends to do in the event that their services are needed. Even though it may involve the exchange of money, as a resource it is beneficial only to the agency.
    (a) Mutual Aid Agreement (MAA)
    (b) Memorandum of Understanding (MOU)

16. What four characteristics does FEMA use to type resources?
    (a) Capability
    (b) Category
    (c) Type
    (d) Kind

17. **True or False**: FEMA is responsible for NIMS qualifying, certifying, and credentialing personnel.

18. What three functions are part of the Joint Information Systems (JIS)?

19. **True or False**: When initiating and operating a Joint Information Center (JIC) only one person makes final decisions, so coordination and cooperation are not as important.

20 True or False: When changes to the NIMS method are proposed, and public comments are open, it does not matter what your comment will be, the powers that be will do what they want.

Self-Study

National Incident Management System (3rd ed.). Retrieved from https://www.fema.gov/media-library/assets/documents/148019.

NIMS Guidelines for Credentialing of Personnel. Retrieved from coordinated responses. https://www.fema.gov/resource-management-mutual-aid.

NIMS Intelligence and Investigations Function Guidance and Field Operations Guide. https://www.fema.gov/nims-doctrine-supporting-guides-tools.

NIMS Resource Center. https://www.fema.gov/national-incident-management-system.

NIMS Training Program. https://www.fema.gov/training-0.

Comprehensive Preparedness Guide (CPG) 101: Developing and Maintaining Emergency Operations Plans (2nd ed.). http://www.fema.gov/plan.

CPG 201, Threat and Hazard Identification and Risk Assessment Guide (2nd ed.). http://www.fema.gov/threat-and-hazard-identification-and-risk-assessment.

Emergency Management Assistance Compact (EMAC). http://www.emacweb.org.

Incident Resource Inventory System (IRIS). https://www.fema.gov/resource-management-mutual-aid.

National Emergency Communications Plan (NECP). https://www.dhs.gov/national-emergency-communications-plan.

National Information Exchange Model (NIEM). https://www.niem.gov.

National Planning Frameworks. http://www.fema.gov/national-planning-frameworks.

National Preparedness Goal. http://www.fema.gov/national-preparedness-goal.

National Preparedness System. http://www.fema.gov/national-preparedness-system.

National Wildfire Coordinating Group (NWCG). http://www.nwcg.gov.

Resource Management and Mutual Aid Guidance. https://www.fema.gov/resource-management-mutual-aid.

Resource Typing Library Tool (RTLT). https://www.fema.gov/resource-management-mutual-aid.

United States Coast Guard (USCG). http://www.uscg.mil.

Using Social Media for Enhanced Situational Awareness and Decision Support. https://www.dhs.gov/publication/using-social-media-enhanced-situational-awarenessdecision-support.

6

An Overview of the Incident Command System

If you invest the time earlier to create structure and process around communication, planning, and goal-setting, you can prevent missteps before they occur.

Christine Tsai

Public safety in the United States has become larger, faster, and better at what they do over the past several decades. Technological advances in recent years have led to vast improvements not only in effective emergency response but also in the way that first responders do their job. In many cases, empirical research has guided the modern-day first responder in how to do their job more effectively and efficiently.

Modern-day first responders have seen technological advances in all fields during the past 60 years. Technological advances have changed the dynamic, and what used to take centuries to improve, an advance now only takes a few years. The world of public safety has not changed at all in some ways, and yet in other ways, technology has helped it to grow by leaps and bounds. Because Incident Command System (ICS) has improved, it too could be a technological advancement.

Some universal truths in public safety never change. A firefighter will tell you that a fire can only burn up, over, then down unless, of course, it is "helped" by someone or something. Another universal truth is that "all bleeding will stop, eventually, if left unattended." The problem with that analogy should be focused on how it stops. Will it stop because it clotted, or will it stop because the person bled out? In any case, these truths remain as true today, as they were one hundred, or even one thousand, years ago.

If we go back several hundred years, we are told that Benjamin Franklin is credited as being the first American Fire Chief. It makes you wonder what Ben Franklin would say if he was to take a tour of a modern-day fire department. Just like when he was a Chief, he would still see fire burn up, over, and then down, but a lot has changed since his era. What would he think of the advancements in apparatus, hoses, hydrants, thermal imaging cameras, Self-Contained Breathing Apparatus (SCBA), and a whole host of other technological advances that have come about since his tenure as fire chief?

Similarly, some of the most famous lawmen who helped tame the Wild West would probably find modern-day technological advances amazing. We could only imagine what Wyatt Earp or Wild Bill Hickok would think of the advances that have been seen in law enforcement. Imagine their amazement in seeing the modern-day semiautomatic pistol, eavesdropping devices, modern-day radios, bullet-proof vests, Kevlar helmets, computerized record checks, and other such technological advances.

Emergency Incident Management Systems: Fundamentals and Applications, Second Edition.
Mark S. Warnick and Louis N. Molino Sr.
© 2020 John Wiley & Sons, Inc. Published 2020 by John Wiley & Sons, Inc.
Companion Website: www.wiley.com/go/Warnick/EIMS_2e

The 1970s, we saw the creation of Emergency Medical Services (EMS). This important part of public safety was initially based on the tactics employed by field medics on the battlefield in Vietnam. Current EMS systems provide early treatment and quick transport to medical facilities, often saving much time. In the years prior to EMS (as we know it today), local morticians were usually the one to transported injured and ill people to the hospital. Sometimes they would use the same hearse that would eventually carry their patient to their final resting place. In rural areas, the hospital may have been many miles away and taken more than an hour.

Today, transport to a specialized facility several hundreds of miles away can only take 20 or 30 minutes by helicopter or fixed-wing aircraft that are designed to facilitate such transfers. Simple bandages and tourniquets have been replaced by advanced life-support ambulances staffed by Emergency Medical Technicians (EMTs) many of whom are trained to the paramedic level. The advanced equipment on ambulances can keep a patient alive in many scenarios, something that was not possible 50 or more years ago. The crew of an ambulance can now provide early intervention and treatment, thereby saving more lives. In the past, many of these individuals would have a slim chance of survival, or even a complete recovery. Modern medicine, as a whole, relies on a system of step-by-step interventions that start in the prehospital setting.

Technical advances have helped public safety to work smarter, protect personnel better, and refine procedures to maximize effectiveness. As a community and a profession, most personnel place a high value on education, training, and real-world expertise. The training of responders on the street, as well as in the executive suite, can literally be important in life and death in many instances.

As a professional community, public safety has come to realize that organizing personnel and equipment on an incident begins almost immediately. Without this organization, there will be chaos. An Incident Management System (IMS) method is of great importance during large-scale incidents as well as everyday responses because it provides an organizing mechanism. Under many different laws and policies, the federal government mandates the use of the National Incident Management System (NIMS) method, including the ICS method to control an incident. The use of ICS was initially mandated on any incident involving a hazardous material (HazMat). This mandate came in 1986 with the adoption of 29 CFR 191 0.120 by the Department of Labor's Occupational Safety and Health Administration (OSHA). As was mentioned in chapter five, in 2002, the use of the NIMS method with ICS for on-scene management was mandated with Homeland Security Presidential Directive five and eight (HSPD-5 and HSPD-8), for any response agency that took federal funds.

6.1 Taking Control with ICS

Using ICS should start with the beginning of the response. The initial size-up of the situation allows the Incident Commander (IC) an opportunity to begin the process of controlling the emergency situation immediately. With a simple form of communication, the Incident Commander (IC) will effectively inform dispatch they have arrived on scene, and that they are taking control of the incident. At the same time, they announce to all other personnel that they are the Incident Commander (IC). This lets everyone know who is in charge of this incident, at this time.

Usually, and frequently because of protocol, the officer in charge of the incident is the first officer arriving in the first emergency vehicle. It is important to note that a well-trained officer will have begun preparation enroute to the incident by gathering as much information as possible about the situation from dispatchers. They often will create a mental picture of the area around the incident. Depending on the amount of information provided, the officer in charge of an apparatus may begin to assign tasks prior to ever arriving on scene.

During the initial size-up, the officer will decide if more assistance will be necessary, depending on manpower and equipment needs as they perceive them at that time. This decision should also be based on several factors such as availability resources. As an example, the Incident Commander at a fire scene may need to determine if more water is needed, if there is enough air supply, what rehab requirements might be, if a Rapid Intervention Team (RIT) is available and or needed, as well as any other incident needs. In extreme weather conditions, more manpower may be needed in order to rotate crews and form specialty teams. They may make the decision to request mutual aid to accommodate this rotation of personnel. The officer has the option to call more resources, of any type, based on their perceptions of the incident and what the needs of the incident are.

As more personnel arrive on the scene, they will report to the Incident Commander (IC) and ask for direction. The Incident Commander (IC) is the one who will most likely be making a number of decisions in a short period of time. Early in the response the Incident Commander (IC) will need to evaluate the incident and decide whether they have the training, skills, and experience necessary for the situation. If they determine that they do not have the skillset for the response, they should call in an officer of higher rank, or even someone that is not an officer but who has more expertise on the type of situation they are facing. If they have the needed capabilities and experience, they should be left in place. When the person with more capabilities and knowledge of this type of incident arrives, the initial Incident Commander (IC) will transfer command to that individual. This will be done in person and announced over the radio. In most systems, when this transfer of command occurs, the first-in officer, the initial Incident Commander (IC), will usually become the Operations Section Chief (OSC) and begin directing the tactical operations as needed, with the higher-ranking or more qualified officer assuming the role of Incident Commander (IC).

One of the very early actions that should be taken by an Incident Commander (IC) is to appoint a Safety Officer (SOFR). If a Safety Officer (SOFR) is not appointed, then the Incident Commander (IC) will personally assume those duties and be responsible for any liability with safety issues. If the Incident Commander (IC) appoints a Safety Officer (SOFR), that person works directly for, and directly reports to, the Incident Commander (IC) as a part of the Command Staff. The appointing of a Safety Officer (SOFR) usually is dependent on the size and scope of the call. The more complex or larger a call is, the greater the importance that a Safety Officer (SOFR) be appointed to ensure the health and welfare of all responding personnel. An incident Safety Officer (SOFR) has a unique capability of stopping operations without going through the chain of command first. If the Safety Officer (SOFR) sees an act that they believe will endanger the health and welfare of any personnel, they have a duty to act. While the Incident Safety Officer (SOFR) has the ability to override the Incident Commander (IC), they should use extreme caution to not abuse that authority. Like all positions in IMS, the Safety Officers (SOFRs) position can be filled upon arrival at the incident, or the position can be filled later as more personnel arrive. In most instances, it will be the best option to appoint a Safety Officer (SOFR) in the beginning of an incident rather than later in an incident.

It is the sole decision of the Incident Commander (IC) to determine if they can manage all aspects of an incident on their own, or if they need to delegate tasks to others. Once the Incident Commander (IC) determines that they need assistance in managing an incident, the Incident Commander (IC) begins to fill the other Command Staff and General Staff positions as needed for the incident. Once these positions are filled and other critical objectives are met, the planning process begins. Many emergency responders call these meetings "huddles," as this is where the strategies for the long haul are developed on both a strategic and a tactical level.

If we use a structure fire for an example, usually one firefighter (or officer) will be assigned to choose a crew and attack the fire (operations), while another officer or firefighter will be assigned to equipment (logistics). In smaller incidents, it is unlikely that other sections of the IMS method (such as planning or finance) will need to be filled in a simple scenario, so ultimately these duties will be assumed by the Incident Commander (IC). In larger or more complex incidents, planning and finance will likely be needed.

In the initial stages of an incident, or in smaller incidents, the Incident Action Plan (IAP) is decided and transmitted orally. It will be verbalized to the staff as they arrive, and as they are assigned to the IMS position that they will be assuming. It is important to note that some incidents grow rapidly, or they become long-duration incidents. While some agencies vary on how many hours personnel will be on scene constitute a long-duration event, an incident that will last 12 hours or more should be the maximum. Some agencies use 6 hours, while others chose 8, 10, or 12 hours to signify a long-duration event.

Whenever it is realized that an incident will be a long-duration event, the Incident Commander (IC) should begin to divide the day into operational periods. On extended responses, operational periods are usually from 8 to 12 hours in length. This duration can be changed at the discretion of the Incident Commander (IC) based on the needs of the incident. In some instances, operational periods will be only during daylight hours, while other incidents may require around-the-clock operations. While some in the emergency response community will tell you that you have to use 12-hour operational periods, this may not be the best avenue, depending on the circumstances surrounding the incident. The Incident Commander (IC) will need to evaluate what timeframe that personnel could work while being effective and safe. This will be based on the unique incident they are facing and how physically and/or mentally taxing the incident is to personnel.

Many of the procedures used in setting up Incident Command has become second nature to members of the emergency services community, especially in the United States. The mandates directed under HSPD-5 have significantly reduced the number of responders who do not understand the ICS method and how it works. In extreme situations, such as terrorist activities, or major disasters, Untrained Spontaneous Volunteers (USVs) and others may become involved who are not fully trained in the procedures and concepts of the ICS method, which typically causes more chaos. Untrained Spontaneous Volunteers (USVs) can pose a major problem in disaster response and recovery, especially if they have no training and do not want to take direction. The chaos they can add is capable of partially undermining if they are not properly identified and managed.

In most instances, volunteer organizations in the United States who regularly respond to major disasters are trained on the concepts and principles of the ICS method, and they work within the NIMS method of incident management. In order to better organize and manage volunteers and volunteer organizations responding to disasters, the National Voluntary Organizations Active in Disaster (NVOAD) was formed. NVOAD provides a structure that promotes substantially more effective services to those affected by disaster through providing organized volunteers. One key issue that NVOAD members address is credentialing (including verified training) and the management of volunteers (Sauer et al. 2014).

Sometimes spontaneous volunteers can create chaos even if properly identified and managed properly. Depending on the scope of the disaster and the number of volunteers, the resources needed for a response could increase significantly because of Untrained Spontaneous Volunteers (USVs). When these individuals arrive at a disaster, the Incident Commander (IC) will need to determine if volunteers should be turned away or utilized. If these volunteers are utilized, ICS may need to expand to ensure that these volunteers are managed in accordance with response objectives. Additional considerations will need to be addressed to ensure the feeding, housing, transportation, and needs of these volunteers.

If you want to better understand just how disruptive and chaotic Untrained Spontaneous Volunteers can be, speak with past Incident Commanders (ICs) who have managed a major disaster. In Hurricane Katrina, the Joplin Tornado, and other similar events, Incident Commanders (ICs) Command Staff, General Staff, and even basic personnel will be more than happy to explain how much chaos they can create. The common theme found among those managing or involved in these major disasters was that in most instances, albeit not all instances, these volunteers added to the chaos and confusion, and they are frequently referred to as "The disaster after the disaster." For this reason, each incident should be evaluated on a case-by-case basis as to whether volunteers would be beneficial or problematic. Management of spontaneous volunteers should be included in the preparedness mode, as well as the preparedness and mitigation mode so that there are plans to manage them in an incident, especially when Untrained Spontaneous Volunteers (USVs) are likely to show up at a specific type of incident.

Beyond volunteer assets that can be utilized to assist, various other resources are available and used regularly. Incidents that are too large or complex for a single agency to manage will often rely on other local public safety agencies to assist. This is particularly true in the case of smaller, rural agencies. In these instances, the value of mutual aid becomes evident. One agency may need to rely on its nearest neighboring agency to supplement manpower or equipment. First responders as a whole are a proud group of people, but they tend to know when they need help, and they are almost always ready to help and lend a hand when called on.

On significantly larger incidents requiring massive mutual aid, responders can create more chaos just in the sheer number of personnel converging on a scene. If you take into consideration the varying types of equipment and resources, and that each department has its own chain of command, the potential for more chaos is present. This can be a dangerous situation and must be rectified. If all agencies are using the same IMS method, the management becomes substantially easier, and chaos and uncertainty will be reduced.

On an everyday "normal emergency response," depending on the situation, more than one discipline may be required to fulfill the needs of the incident. Going back to the example of a structure fire, while the fire department is fighting the fire, there may be a need for law enforcement to handle the traffic and the crowds. EMS personnel will most likely be on the scene to tend to any injured victims, and for any personnel that might get injured fighting the fire. There is also the possibility that the power company, the gas company, and/or the water company may be needed. The response to even day-to-day incidents can be made more effective with proper use and implementation of an effective IMS method.

During the initial period of amassing the resources that will be needed for the response, communications are often confusing. This is often due to the high volume of radio traffic within each department and between departments. The matter of communicating with other disciplines creates even more challenges and if not managed properly, it can cause more chaos, confusion, and uncertainty. Through utilizing IMS methods, these issues can be substantially reduced, especially if these issues are addressed in the planning mode.

Interoperability among agencies has come a long way since September 11, yet there is still much to be done. The importance of interoperability cannot be stressed enough. Lack of interoperable communications can put response personnel at great risk, and this is yet another reason an IMS method is needed for the managing of incidents.

Perhaps the biggest reason that an IMS method needs to be in place, even at the smallest of day-to-day emergency events, is responder safety. The Incident Commander (IC) has little chance of maintaining an effective span of control (number of people that a person can effective supervise) unless they delegate authority. While the Incident Commander (IC) is responsible for the overall incident, the Command Staff and the General Staff take responsibility for jobs within the incident and manage the personnel that are active in their area of the response.

In our simple structure fire example, the Incident Commander (IC) will need to make many decisions in a short amount of time. These decisions might include where to place equipment, if search and rescue of victims might be needed, who will perform search and rescue operations, where to start the initial attack, how many firefighters to send in, and anticipating what additional resources might be needed. If the fire continues to escalate, there may be danger of collapse or exposure involvement, and they must decide when to remove personnel from the fire. In looking at this incident, it is not practical to make decisions and attempt to design a system to help accomplish this attack in the heat of the battle. During an incident is not the time to come up with a means of organizing this situation.

There are many tasks to accomplish before this incident comes to a close, many of which may need to happen quickly, and even simultaneously. Having a good organizational method or system can make the difference in how the incident plays out. An effective IMS method is a system for both the day-to-day events and "The Big One."

Relevant Research

The investigation *A Failure of Initiative*, undertaken by Select Bipartisan Committee to Investigate the Preparation for, and Response to, Hurricane Katrina, looked into the failures during Hurricane Katrina's preparedness and response mode. In the Command and Control chapter of the report, the committees' research found that there were major issues with command and control during the response phase. Many of the command and control problems evolved from the destruction of government buildings, communications systems, and generators. Essentially, large parts of their infrastructure was totally inoperable. In some areas, there was enough infrastructure, but there was no unity of command or communication between differing agencies.

The report found that while operating at the New Orleans Superdome, the New Orleans Police Department thought the National Guard was in charge, and the National Guard thought that the New Orleans Police were in charge. An Incident Command Post (ICP) was never established, and the commanders from either agency never met face to face. This was confirmed by Federal Emergency Management Agency's (FEMA's) Field Coordinating Officer (FCO), who stated that there was no Unified Command (UC) and nobody in charge.

Similarly, 7000–8000 people gathered at the cloverleaf (overpass) where both the National Guard and State Police were present. When the State Police and National Guard were separately asked who was in charge, neither knew. The report found that there were failures in unity of command, unity of effort, and communications at all levels of government; local, state, and federal. This lack of command, control and communications created more chaos and more frustration for the victims of this massive disaster (Select Bipartisan Committee, 2006).

6.2 Common Components of Incident Management Systems

An effective Incident Management System (IMS) provides for a solid management structure and creates a logical system for conducting on-site operations in a manner that is both effective and safe for all people involved. Keeping with the theme of the three common characteristics synonymous with almost every critical incident (chaos, complexity, and uncertainty), the overall objective of an incident management system is to produce order where there was chaos, predictability and stability where there was uncertainty, and to provide simplicity out of what was complexity.

When properly used, the ideal IMS method will be applicable to even small-scale daily operational activities in all areas of public safety regardless of which discipline is using the system. The nature of the specific service or activity will have no particular bearing on how the system is employed at the rudimentary level. An IMS method should seek to manage, organize, categorize, and simplify any activity at the strategic level rather than the tactical level. In some cases, an IMS method may have applicability even in areas that are not directly involved in the public safety arena. This can include such daily operations of public work entities and/or other tasks where several people or agencies will need to work in unison, yet separately, to achieve a specific stated goal.

An example of this might include a construction crew that hit a gas line. In order to bring this type of incident under control, there may need to be a gas line repair crew, fire crews, public works, street crews, and others, who may need to be involved. They will need to work together to bring about a successful operation that keeps the public safe and repairs the gas leak and potentially street repairs as quickly and efficiently as possible.

6.2.1 The ICS Component of NIMS

As an effective IMS, the ICS component of NIMS will help to bring about order to any type of complex situation, if used properly. This is even truer when we look at chaotic events, which public safety agencies are accustomed to dealing with, on a daily basis. These tasks can be accomplished because the ICS method is a standardized operational structure with common terminology that fully integrates with the NIMS method. These common elements provide a useful yet flexible management tool, thereby allowing the public safety agency to concentrate on both strategic and tactical objectives. They can undertake operational work while providing for the highest level of safety and accountability.

The true value of using the ICS method on a day-to-day basis, especially for small-scale or common incidents, is that it can rapidly be expanded and adapt to any situation regardless of its scope or size. As the most popular IMS method used over the past 35 years, ICS is particularly adaptable to incidents involving multijurisdictional (cities, states, or even national scope) or multidisciplinary responses (police, fire, EMS, public works, and private industry). When the ICS method is properly employed by all responding agencies in a community, it has been proven to be extremely effective in controlling the short- and long-term consequences of any type of disaster. It is effective for both natural disasters and incidents and man-made disasters and incidents.

Utilizing the ICS method and integrating it with the NIMS method provides the framework that facilitates a flexible, yet rapidly deployable and expandable (as well as collapsible) organizational format. This format supports organizational and tactical functions that need to be performed with almost any type of incident. The basic concepts of the ICS method will not change because differing or a substantial number of agencies have responded to an incident. The specific needs of an incident will also rarely have a significant impact on the base concepts of the method. When using the ICS method some of the basic operating requirements that make it flexible and expandable include that the method can be used for

- Single discipline, single jurisdiction, and single agency response.
    - Car fire (Fire Department response)
    - Heart attack (EMS response)
    - Breaking and entering (Police Department response)

- Single jurisdiction responsibility with multiagency involvement
  - Dwelling fire with entrapment (Fire, Police, and EMS response)
  - Bank robbery with a person shot (Police and EMS response)
  - Building collapse with entrapment (Fire, Police, EMS, and Public Works response)
- Multijurisdictional responsibility with multiagency and multidisciplinary response.
  - Coastal storms that causes flooding in multiple cities or counties (local, regional, and or state-level responses with local city, county, and state jurisdiction)
  - Major bombing (local, state, and federal response and both state and federal jurisdictions)
  - Terrorist attack (nationwide response as well as a multitude of jurisdictions)

The system requirements that allow the ICS method to be effective include reasons such as the organizational structure that can adapt to any emergency or incident, and it is applicable, adaptable, interoperable, and acceptable for all types of agencies. It can easily expand in a rapid manner (from an initial response into a major incident), and it is capable of reduce its size just as easily. Of course, decreasing the organizational structure would be done to meet the organizational needs as the incident sees a decrease in scope and complexity. Additionally, the ICS method has common elements in organization, terminology, and procedure, which reduces confusion and increases compatibility. It can be implemented in a manner that includes an agency's procedures and operational systems, and it is simple enough that it is easy to understand. The ICS method is always relevant because it has a research and development system in place. This gives it the ability to adapt over time and it makes it constantly current to the ever-changing needs of public safety. It also integrates into the overall framework and concept of the NIMS method.

A special note of caution should be given to those that live and respond in the United States. If you respond within the United States, you should use ICS (supported by NIMS) as the operating platform for emergency response, as mandated by the federal government. Using another operating system, or even modifying the ICS and/or NIMS platform, may cause integration and interoperability issues as well as make your jurisdiction noncompliant with the NIMS requirement. This noncompliance could lead to the withdrawal of federal funding (FEMA, 2004b, 2015d; McEntire, 2007). While there are no documented cases (at this point), noncompliance could at some point in the future affect disaster funding for a local jurisdiction. Those who choose to implement another IMS method, or that choose to modify ICS and/or NIMS, should proceed with extreme caution. They should understand that there may be consequences for those actions.

6.3 ICS and NIMS-Differences and Commonalities

ICS is the core command and control mechanism in the United States. As an incident requires response from multiple local agencies, it is important to have an effective cross-jurisdictional, multijurisdictional, and multiagency coordination method that is used as a common processes and system. The ICS method provides that flexible, but consistent, fundamental instrument for coordinated, cooperative, and collaborative incident management tool. It does not matter if it is an incident where additional resources are required and provided from within a diverse group of organizations within a single jurisdiction, or whether these resources are made up by outside the jurisdiction. It does not matter if it is for a smaller local issue or a more complex incident with national implications.

The ICS method is a widely applicable incident management system that was created to allow the effective and resourceful management of an incident by integrating a combination of facilities, equipment, personnel, procedures, and communications into a singular operating system while utilizing a common organizational structure. The ICS method is a critical type of incident management that is recognized in a uniform yet standard format. The attractiveness of ICS as an IMS method is that it permits incident managers to identify the key concerns connected to the incident without neglecting any component of the command system, even when operating under austere conditions. In the overall scope of things, the ICS method is the most advantageous method for managing an incident.

Similarly, the NIMS method is a comprehensive, national approach (to incident management) that is applicable across all territories, regions, states, and local levels and across all disciplines, even nongovernmental organizations. It is suitable for a comprehensive assortment of potential incidents, hazards, and impacts, regardless of the size, location or complexity of the incident. It is designed in such a manner as to enhance the coordination and cooperation between public, private, and government organization. It does so by providing a cooperative system that uses an assortment of incident management actions, which organizes incident management in a singular framework. It helps to facilitate increased coordination, cooperation, and collaboration from nontraditional resources, or resources that are needed but not available locally.

Due to resource typing and interoperable communications, NIMS provides an almost seamless integration of resources into a local, regional, or even national response. By all entities utilizing the NIMS method, resources should be able to effectively integrate with a minimal amount of effort, primarily because everyone is on the same page. This increases communication, cooperation, and collaboration. Because ICS is used as the basis of command and control, and everyone is trained in ICS, resources from across the nation are able to come into a local incident and be part of the solution rather than part of the problem.

As was mentioned in chapter five, the broad application of the NIMS method allows the foundation for efficient and effective emergency response. Entities that have integrated the NIMS method into their planning and incident management structure can arrive at an incident from one side of the United States to the other, with little notice, and still understand the procedures and protocols governing the response. They will also know the expectations for equipment and personnel. NIMS provides commonality in preparedness and response efforts that allow diverse entities to readily integrate and, if necessary, establish unified command during an incident.

It is important to remember what the NIMS method is not. NIMS is not a response plan, and it is not only used in large-scale incidents. NIMS principles should be used locally, on a daily basis. NIMS is not relevant to only "certain" emergency management or incident response personnel; it is relevant to anyone that may respond to an incident. NIMS also is not an expansion of the ICS method, but rather a separate system that provides better communication, collaboration, and cooperation between response agencies. This includes the usual response personnel and those that are not normally involved in response, such as houses of worship, business, and industry.

6.4 Incident Management System and NIMS Integration

In order for any Incident Management System to fulfill the above operating requirements, several commonalities must be effectively in place during the design, implementation, and use of the system. Additionally, a constant system to update that IMS method must be built into the system to assure that it remains effective. This effectiveness refers to both operational methods and technologies that may change within an agency.

While many of the issues that will be discussed in this section have been spoken about previously in this book, it is important to realize that there is so much overlap in the foundation between the ICS method and the NIMS method that they are intertwined extensively. In the

ICS method, as has been mentioned previously, there are some common elements that must be in place to make it effective. The common elements of the ICS method are

- Common terminology
- Modular organization
- Integrated communications
- Consolidated incident action plans
- Manageable span of control
- Predesignated incident facilities
- Comprehensive resource management

6.4.1 Common Terminology

When working together in a multiagency response, one of more frustrating aspects of trying to work together is the fact that we all speak our own industry lingo. The only real commonality, in many instances, is that we speak English. Each of the major public-safety disciplines can appear to have their own dialect. Some agencies still use codes and other abbreviations. An example of utilizing a specialized jargon would be the 10-Code system that many law-enforcement agencies still use. This system dates to the days of tube-set radios.

Further compounding these factors are the multitudes of jargon dialects that often tend to be regional in nature. Sometimes these dialects can be traced to subregions or even local areas. The reasons for the multitude of variations are generally societal in nature. Many times, it is based on the makeup of the emergency services community and their geographic roots. Many of the terms tend to be holdovers from previous years. If we look at names for a fire hydrant, we can see that depending on where you go in the United States, it could be called a fireplug, a fire hydrant, a pump, a johnnie pump, or it can be called a hydrant. It is easy to see that if common language is not used, it could slow the response and add confusion.

One area in common terminology that has led to some heated discussions is the terms "tanker" or "tender." These terms are often used in the fire service. Many Midwest fire departments identify a tanker as a truck that carries 1200 gal of water to support firefighting operations. According to the NIMS method resource typing, this is a water tender. Multiple types of tenders are identified in NIMS resource typing. There is the previously mentioned water tender, the foam tender (used for foam firefighting), a fuel tender (for delivering fuel to airborne or ground equipment), a helicopter tender (supports helicopters with fuel and support equipment), and a water truck (for supplying potable water). NIMS resource typing identifies a tanker as an aircraft to deliver firefighting retardant (usually mixed with water). These airborne tankers are divided into four levels:

- Type 1 Tanker holds 3000 gal of fire retardant
- Type 2 Tanker holds at least 1800 gal of fire retardant
- Type 3 Tanker holds at least 600 gal of fire retardant
- Type 4 Tanker holds at least 100 gal of fire retardant

With the Midwest and (some) East Coast fire departments use the term "tanker," while at the same time West Coast fire departments call the same unit a tender, it is easy to see how lack of a common language can lead to confusion in a major incident where east meets west. Many West Coast firefighters may think that an Air Tanker was requested, when in fact they were asking for an apparatus that could refill their engines or wildland firefighting apparatus.

The potential for confusion from not using common terminology is ever-present. With common terminology and resource typing, not only can confusion be reduced but also the timely delivery of resources is increased. It is also fair to say common terminology can reduce costs associated with an incident by not wasting time and money on the wrong resource. This essentially leads to a more organized response.

The previous examples represent a fraction of the overall problem. Real problems in terminology can occur when terms are used in radically differing manners by different services. This language can make a life-and-death environment even more hazardous by creating misunderstandings. These misunderstandings could cause individuals or agencies to act in a manner inconsistent with safety and counterintuitive to the tactical objectives of the incident. At best, this could lead to needlessly endangering lives, and at worst, a Line of Duty Death (LODD).

An anecdotal real-life example was the response to a shooting call with victims suffering gunshot wounds. As is common, a police and EMS response was ordered by the dispatcher, simultaneously. In most operating systems, the EMS component of the response would intentionally stage in a safe area while law enforcement secured the scene. EMS would wait for an "all clear" signal from the police to ensure that they were not placed in any jeopardy from the perpetrator while doing their duty. This is a common practice because EMS staff generally have no defensive capabilities, and that is not their job. In one particular incident, a police officer told the responding EMS crew that the scene was "secure." The EMS crew took that to mean that the perpetrator was no longer a threat in the area. However, that was not the case. In the "police speak" of that geographical area, the term "secure" referred to the surrounding the perimeter with officers to the prohibit anyone from entering or exiting without the police knowing. The perpetrator in this scenario was still in the area and still armed and dangerous. The EMS crew assumed that the term "secure" meant safe. This was not the case. Luckily, the police had direct radio contact with the EMS unit and soon ordered it to a safe position. It was not until after the perpetrator was apprehended, and the victims were transported off scene that the true nature of the "miscommunication" was revealed. As result of the incident, the agencies involved made adjustments to communications by ensuring the use of common terminology and they provided training for their respective personnel.

It is far beyond the scope of any IMS method or system to change culture that is deeply entrenched. Verbal communication problems can be overcome by the use of plain English and resource typing mandated in the ICS and NIMS method. Interoperability should be viewed in a scope that is far beyond the technical aspect of having communication systems that are on the same frequency. The ICS method relies on common terminology so that verbal interoperability at all levels are assured, no matter what vernacular is in use at a local level. When responding to an incident, if we do not speak the same language, then the response will be less effective and more time-consuming. If we do not speak the same language, more lives will be at risk and there will be more chaos and confusion. The use of ICS and NIMS include the practice of common terminology to avoid turmoil and mistakes in communications, as well as providing for simplified and exact resource ordering.

This is not to say that agencies should always abandon their codes and specialized lingo. There are times and places when codes are used and should be used. It is critical that these codes and geographically based vernacular should not be used in a multiagency or multijurisdictional response. A prime example of when codes should be used happened in the 1980s in a Massachusetts Police Department. The department was sent to an address for a man acting erratic and irrational. When they arrived on scene, they found a mentally unstable man that was angry, and he was lashing out at everyone, even the police. One officer took the lead. He was slowly but systematically calming the man down. Within five minutes, the man was fairly calm, and officers were now able to stand next to him and talk. As they were attempting to talk the man into voluntarily going to be mentally evaluated, another officer said something over the radio that changed everything in an instant. Instead of using the code for a mentally unstable man, the officer on the radio said, "Did you get the nut case settled down, or do you need me to still come over and help arrest him?" Needless to say, the mentally unstable man became immediately enraged, and a battle ensued. In the end, it took multiple officers to subdue the man, and several of them received battle scars in the process.

In most instances, plain English should be used, however first responders can keep codes in their toolbox for instances such as the Massachusetts incident. Common reasons for using codes would be to identify if somebody standing next to you has a warrant, or they are a violent offender. Another reason for using codes can be for when a coroner is needed, when an investigator or arson investigator is needed but you do not want the listening public to hear this, and other similar situations. While these instances are rare, they must be considered.

6.4.2 Modular Organization

In the beginning of an incident, the Incident Commander (IC) is charged with taking responsibilities of personally handling all aspects of the incident. They are responsible for all functions, including safety, public information, logistics, operations, finance, and administration, and more. The modular design of the ICS method allows them to essentially have various functions in a box, much like the meals in a box that can be purchased. By simply opening the box they need help with, they gain the support they need.

As an example, when the Incident Commander (IC) assumes control of an incident, they may see that the incident is too fluid or dynamic for them to handle safety, so they might appoint a Safety Officer (SOFR). When the person that will fill that function is appointed, they automatically know what they need to do, because there is an imaginary proverbial box they will open. In this imaginary box is all the information on the tasks they should perform, and how they should do their job. The same holds true for any function, and the Command and General Staff already know what is expected, what their tasks should be, how those tasks should be performed, and their powers and abilities in the position they are chosen for. By simply appointing an individual to undertake a function, the Incident Commander (IC) has delegated authority to another individual who will take care of all of the duties in that module for that function. Any time that another position is added to the organizational chart, it will almost add another module to the overall module design.

It is important to note that the modular design does not only cover functions, but facilities and personnel as well. When an incident occurs, certain facilities may need to be set up. Because of the modular design of the ICS method, the Incident Commander (IC) or the Unified Command (UC) can order facilities and know that they will get what is needed. The same holds true for specialized crews. When they receive a resource, they do not need to ask a hundred questions to understand their capabilities. Because of the modular design of both the ICS and NIMS method, they know that they received the resource in the imaginary box. It will be the complete package.

6.4.3 Integrated Communications

As has been mentioned previously in this book, communications are an important factor in managing an incident. In the ICS method, incident communications are enabled through the development of a communications plan. The use of a common communications plan that include interoperable communications allow all parties involved to interconnect. In many instances, especially in major disasters where communications infrastructure has been damage, this is no easy task.

Thanks to the technological advances that have been seen over the last 30 years, there is a multitude of ways that we can communicate, especially when major disasters occur. No longer do we need to rely on only radio communication. By utilizing effective operable and interoperable voice, video, and information systems, interoperable communications have seen advances that were never thought of 40 years ago. While most local agencies will rely on radios, text messages, cell phones, and e-mail, state and federal resources usually have additional communications methods that promote interoperable communications. When these resources arrive on scene, they will typically identify and document disaster emergency communication and information systems capabilities, requirements, solutions, and mitigation strategies that are critical to the mission and the incident.

There are many benefits of creating interoperable communications. With interoperable communications, those managing the incident can create and develop effective command and control communications frameworks, thereby increasing the resiliency of the response. Interoperable communications also support the coordination of secure, yet accessible (to personnel), communications for a full gamut of resources from various governmental organizations, including federal, state, tribal, and local agencies. Integrated communications also allow the Incident Commander (IC) (and their Command and General Staff) to create a better information sharing system, which will in turn provide a greater situational awareness about the incident.

6.4.4 Consolidated Incident Action Plans

Integrating all of the moving parts of an incident can become a monumental task, particularly on larger incidents. It can be difficult to keep everyone on the same page, and if not for the consolidation of the Incident Action Plan, it would be possible and even likely that some individuals would inadvertently veer from the overall plan. Utilizing the planning process that is built into the ICS method, the coordination of strategic goals, tactical objectives, and support activities are consolidated into a seamless task.

Command and General Staff, along with the occasional other expert (as needed), meet during every operational period to plan the response. This prevents a duplication of efforts and ensure that everyone involved is going the same direction. This becomes extremely important when outside resources are integrated into the response. If not for the consolidated planning, federally delivered resources could have one goal and/or objective, while state-delivered resources had another goal and/or objective, and local resources have even another goal and/or objective that is totally different than the other agencies. By consolidating planning, everyone is on the same page, moving in the same direction, and enhancing the work that others or doing.

6.4.5 Manageable Span of Control

Within ICS, the span of control refers to the number of people a supervisor can effectively manage. In the 2008 version of the ICS method, it was stated that supervisors should manage between three to seven subordinates five being ideal. There was also an exception that provided guidance for law enforcement operations in a large-scale disaster. The span of control for law enforcement was from 8 to 10 subordinates that might be optimal. The number of people a supervisor can manage, or the span of control, is critical to effective and efficient incident management.

Because ICS and NIMS in the United States is constantly being updated, there was a change in the span of control in the 2017 version. In the new version, the span of control is adjusted to provide more flexibility. In this new guidance, the optimal span of control is still a supervisor to five subordinates, but it also gives discretion to incident personnel. The new version allows incident personnel use their best judgment to determine the actual distribution of subordinates to supervisors. This should be based on aspects such as supervisory experience, past experience, and the type of the work that is being supervised. This applies to ICS at the scene of an incident and at the Emergency Operations Center (EOC).

As an example, after a flooding event, muck out is an important, and usually first step, to restoring a structure back to normal. In larger structures, one experienced supervisor might be competent enough to manage 12–15 people. By the same token, an experienced public works supervisor, with an experienced crew, might be able to manage 14 workers, while clearing roads with heavy equipment, chain saws, tractors, and more. In the newest version of the ICS method and the NIMS method, this is now allowable.

6.4.6 Predesignated Incident Facilities

Various types of operational support facilities are established in the vicinity of an incident to accomplish a variety of purposes. Typical designated facilities include Incident Command Posts (ICPs), Bases, Camps, Staging Areas, Mass Casualty Triage Areas, Helibase, Helispots, Emergency Operations Centers (EOCs), Joint Information Centers (JICs), Joint Operating Center (JOC), and other facilities as is required.

These predesignated incident facilities relate to the modular design of the ICS method, and it allows facilities to be set up rapidly while still ensuring that it meets all functionality standards. When an incident first occurs, the Incident Command Post (ICP) may be in a police car, a firetruck, or an ambulance. As time goes on, the Incident Command Post (ICP) may transition into a mobile command center or even a building that has all of the prerequisites to be an Incident Command Post (ICP). As outside resources are ordered, it may be determined that a base camp is needed, so holding true to the modular design, a base camp can be ordered, a location selected, and the base camp will be set up to minimum standards.

6.4.7 Comprehensive Resource Management

Comprehensive resource management, as has been described in Chapter 5, is the process of categorizing resources, ordering resources, dispatching resources, tracking resources, recovering resources, and demobilizing resources. Comprehensive resource management ensures that resources are always identified and tracked to increase accountability and safety. The principles that apply for the NIMS method in regard to comprehensive resource management also apply to the ICS method of incident management.

6.5 Conclusion

As this chapter ends, we need to remember that public safety is an ever-changing field. On a regular basis, we see public safety technological advances that are making leaps and bounds. We need to realize that in the last 50 years there has been a revolution in technology. While some in public safety may resist technological advances, others embrace them. As first responders, it is imperative that these new advances are tested before putting them in service.

While not the latest greatest electronic gadget, the ICS method integrated with the NIMS method can still be considered a technological wonder. ICS and NIMS provide a system for taking control of a situation, for managing it, and for integrating the resources that are needed. Because there is ongoing research and development, these IMS methods keep up to date with the latest advancement in IMS, and it provides a technological advantage to those who become proficient in their use. These methods put everyone who uses them on the same playing field, playing the same game, and providing the right plays at the right time. Just like a football team has common terminology for specific plays, NIMS has common terminology for specific equipment. Using these common components as a way to communicate nationwide with others provides clarity and a coordinated supply chain that supports the quick deployment of exactly what is needed, when it is needed.

While the NIMS method and the ICS method are different in some ways, they integrate with each other and provide a more comprehensive approach to incident management. The NIMS method enhances the power of the ICS method by providing for the needs of the incident. Moreover, it allows outside responders and individuals with different vernaculars, specialties, governmental backgrounds, and disciplines to work together for the good of a community. In the end, this saves lives, property, the economic status of a community, and the environment.

As we move forward into this book, we will take a closer look at every aspect of the ICS method, as well as how it integrates with the NIMS method. The best way to become more proficient at using a system is to study every aspect of it, then apply it. Through repetition, it will become second nature.

Real-Life Incident

On 6 January 2017, at the Fort Lauderdale-Hollywood International Airport, an armed gunman shot 11 people, killing 5 of them. According to an After-Action Review of the incident, the ICS and NIMS were not properly implemented. This led to more chaos and confusion, especially when a second active shooter scare occurred (Broward County Sheriff's Office, 2017).

Numerous deficiencies were found. In the planning stages, there were no considerations for mass care, insufficient evacuation transportation, what to do when security and TSA left their posts, where witnesses should be taken, no clear communication plans for additional resources, and planning and exercises that did not meet core capabilities. Additionally, there were problems with coordination and cooperation of key airport personnel.

At the response level, it was identified that differing disciplines initially set up different Incident Command Posts (ICPs), a lack of coordination with the airport in decision-making, and the Incident Commander (IC) was not known by most people on scene. There were also issues with resource management, the coordination of a Unified Command (when activated), the Mobile Command Post was placed too close to the incident, the self-deployment of law enforcement officers, and staging areas for outside resources were not communicated to those resources.

Other issues identified included lack of coordination and cooperation with the Federal Bureau of Investigation (FBI), undercover officers wearing masks who could have been mistaken for a perpetrator, abandoned TSA checkpoints, a lack of credentialing for outside resources, and because officers abandoned their cars in the road, exit routes that were completely blocked.

Perhaps one of the most valuable lessons learned was when a second scare of another active shooter was broadcast. While the scare was unfounded, law enforcement evacuated everyone from the airport. This left people standing outside of the terminals in 80 °F heat from 4 p.m. to 1 a.m. with no food or water. This included children, elderly individuals, and others with Access and Functional needs. During the evacuation, it was noted that those with Access and Functional Needs took too long to evacuate because of insufficient planning.

These are just a few of the many issues that were identified by the Broward County Sheriff's Office (2017). In looking at the report, it was clear that a lack of planning lead to unnecessary (major) problems. An important part of the ICS method and the NIMS method is to implement the cycle of preparedness and thoroughly test Emergency Operations Plans (EOPs) on a regular basis.

Chapter 6 Quiz

1. **True or False**: Using ICS is a way to begin taking immediate control of an incident.

2. What Command Staff position should (in most instances) be appointed early in an incident?

3. How long is a typical operational period?

4. What does NVOAD an acronym for?
   (a) National Voluntary Organizations Active in Disaster
   (b) Not Victims of American Disasters
   (c) New Volunteers Ordering Accommodation Devices
   (d) National Volunteers Active in Disaster

5. **True or False**: In the initial period of amassing resources for an incident, communications will usually flow smoothly.

6. _____ is the most common IMS systems used in the United States, and it uses the companion method of _____?

7. _____ usually cause more chaos and confusion.

8. The _____ method provides for a flexible but consistent fundamental command and control instrument for coordinated and collaborative incident management.

9. **True or False**: According to the report *A Failure of Initiative*, because there was a Unified Command in place the Superdome shelter was chaotic

10. _____ is designed to enhance coordination and cooperation between the public sector, the private sector, and government entities.

11. **True or False**: NIMS uses resource typing and interoperable communications to provide a near seamless response.

12. **True or False**: In the United States, public safety entities that do not use ICS and NIMS may lose federal funding and be denied federal grants.

13. **True or False**: Cooperation and coordination involves making decisions and working together for a common goal and usually begins in the preparedness phase of crisis management

14. **True or False**: Utilizing the NIMS method and integrating it with the ICS method provides the framework that facilitates a flexible, yet rapidly deployable and expandable (as well as collapsible), organizational format.

15. **True or False**: With common terminology and resource typing not only can confusion be reduced but also the delivery of resources is decreased.

Self-Study

National Incident Management System (3rd ed.). Retrieved from https://www.fema.gov/media-library/assets/documents/148019.

NIMS Intelligence and Investigations Function Guidance and Field Operations Guide. https://www.fema.gov/nims-doctrine-supporting-guides-tools.

NIMS Resource Center. https://www.fema.gov/national-incident-management-system.

NIMS Training Programs. https://www.fema.gov/training-0.

Incident Resource Inventory System (IRIS). https://www.fema.gov/resource-management-mutual-aid.

National Emergency Communications Plan (NECP). https://www.dhs.gov/national-emergency-communications-plan.

National Planning Frameworks. http://www.fema.gov/national-planning-frameworks.

Fort Lauderdale-Hollywood International Airport (FLL). Active shooter/mass evacuation critical incident: critical incident report. Retrieved from http://www.trbas.com/media/media/acrobat/2017-10/94837910-10093411.pdf.

A failure of initiative. Bipartisan Committee (2006). Retrieved from https://www.nrc.gov/docs/ML1209/ML12093A081.pdf.

7

Command Staff, General Staff, and Their Functions

Surround yourself with the best people you can find, delegate authority, and don't interfere as long as the policy you've decided upon is being carried out.

Ronald Regan

The Command Staff designations in the Incident Command System (ICS) method refer to a standard set of major positions and functional units. These positions have been predesignated for the ICS method of incident management, and they are created with a modular design. Once activated, the Command Staff knows what their responsibilities and tasks are, and they carry them out with minimal direction.

7.1 Incident Commander (IC)

The structure of the ICS method enables the Command position (or Incident Commander [IC]) to focus on strategies that help manage the incident and it allows other positions to fulfill the tactical needs of the incident. Doing so ensures that the strategic goals set by the Incident Commander are met. The Command position may be undertaken by a single individual, an Incident Commander (IC), or if the circumstances necessitate, it may be handled by multiple individuals from differing agencies or disciplines. When multiple individuals are utilized, this is known as a Unified Command (UC).

A singular Incident Commander (IC) is typically used when the incident falls within the Incident Commanders (ICs) scope of knowledge, and the use of outside resources are not needed. An Incident Commander (IC) may direct a single-agency response or in some instances a multiagency response if they have expertise in the other agencies capabilities and the tasks they will undertake. As an example, if the incident is a motor vehicle incident on the highway, a single Incident Commander (IC) would likely be able to manage fire department extrication and firefighting, Emergency Medical Services (EMS) response and tactics, as well as the diversion of traffic by law enforcement.

The Incident Commander (IC), at their choice, may choose to personally manage all Command Staff Functions. Command Staff functions include Safety Officer (SOFR), Liaison Officer (LOFR), and Public Information Officer (PIO). While this method of filling all Command and Command Staff functions are ill-advised, it is possible, and to a certain extent viable on extremely small and/or simplistic incidents. The point that needs to be remembered when undertaking this method is that only one person is managing the personnel, the intricacies of the incident, safety, liaising with other agencies, and dealing with any media that may appear. By the Incident Commander (IC) assuming responsibility for these tasks, it can be surmised that they assume all legal liability. While they are still legally liable after appointing individuals for command staff, appointing others to fill these positions reduces the possibility of litigation based on the negligence of the Incident Commander (IC).

If the Incident Commander (IC) does decide to implement any or all of the Command Staff, they are allowing the Command Staff to focus on specific issues while the Incident Commander (IC) is more focused on the management of the overall incident. Due to the modular design and flexibility of the ICS method, the Incident Commander (IC) can expand (or contract) the incident as they see fit, and as they determine that circumstances dictate.

As an example, the Incident Commander (IC) may arrive with their crew on the scene of a small grass fire on a back-country road. At this point, they feel that they can handle all Command and Command Staff aspects of this incident on their own. If a gust of wind comes and more than triples the size of the fire from ¼ of an acre to 2 acres, the Incident Commander (IC) may determine that a Safety Officer (SOFR) is needed to ensure that all firefighters are not involved in high-risk tactics or that they are unaware of deteriorating conditions. As the fire grows, the Incident Commander (IC) may determine that they need to call for more assistance from neighboring fire departments. If the Incident Commander (IC) is extremely busy managing the crews that are already on scene, they may decide to appoint a Liaison Officer (LOFR) to ensure that there is successful integration of the outside resources with the crews that are already on scene. The Liaison Officer (LOFR) can also help to ensure that no single individual that reports to the Incident Commander (IC) has a span of control that exceeds the ICS methods thresholds. If the incident progresses, then the Incident Commander (IC) may want to appoint a Public Information Officer (PIO) to manage the media or to send out evacuation orders if the incident requires it.

In some agencies, the Command and Command Staff are already predesignated, saving time and confusion when an incident occurs. Additionally, a common practice in most fire agencies is to have a designated Safety Officer (SOFR) on each apparatus that responds, or a specific Safety Officer (SOFR) that is not assigned to an apparatus, but rather a station or an entire department. The same holds true for the Public Information Officer (PIO) and the Liaison Officer (LOFR); although a preappointed Liaison Officer (LOFR) is less common.

On the other hand, if an incident expands to the point that it is beyond the area of expertise of one person, then the IC may choose to pursue a Unified Command. An example of this may be an improvised explosive device is detonated and a multiagency response is required. This multiagency response may include HazMat teams, explosives teams, EMS crews, law enforcement investigations teams, fire crews, extrication crews, and other potential specialty personnel. For most Incident Commanders (ICs), the combination of these very different specialties will be outside of the Incident Commanders (ICs) scope of knowledge. Therefore, a Unified Command should, at the very least, be considered and most likely enacted. The same would hold true for areas that cover large geographical areas and/or are highly complex.

7.2 Unified Command

An especially collaborative effort for managing an incident is when a Unified Command structure is implemented. A Unified Command (UC) can be simply explained as a group of Incident Commanders from the numerous jurisdictions and/or agencies that cooperatively work together to form a single command structure. Unified Command (UC) is extremely beneficial for managing incidents that cross jurisdictional boundaries and that utilizes varying disciplines.

In creating a Unified Command (UC), each individual organization maintains its own authority, responsibility, and accountability; however, designated individuals will jointly determine incident goals, objectives, and strategies with their counterparts from other agencies. It allows the command module of ICS to make mutually agreed-upon decisions which, in turn, maximizes resources.

In changing from the initial Incident Commander (IC) to the more collaborative Unified Command (UC), the ICS 201 form (readily available on the Internet) should be prepared by the current Incident Commander (IC). The ICS 201 form, when filled out properly, provides the incoming members of the Unified Command (UC) with the basic information regarding the incident situation and the resources allotted. The information provided in the ICS 201 form helps to facilitate collaboration and cooperation. It provides the objectives and the Incident Action Plan (IAP) currently in place for the response. While the Unified Command (UC) begins to create a joint strategy, the Incident Action Plan (IAP) remains in force and continues as developed until the end of the operational period (the time set to accomplish specific goals) or a new formal Incident Action Plan (IAP) is completed. This is done to prevent a stoppage or delay of tactical operations and helps to create a smoother beginning to the extremely cooperative and coordinated process of managing an incident by utilizing Unified Command (UC). When implementing a Unified Command (UC) three main premises must always apply:

- The incident will function under a single, coordinated and collaborative Incident Action Plan (IAP).
- The Operations Section Chief (OSC) will have responsibility for implementing the Incident Action Plan (IAP).
- One Incident Command Post (ICP) will be established (per incident).

During multiagency incidents with a Unified Command (UC), Incident Action Plans (IAPs) should always be written, and almost never verbal plans. Incident Action Plan (IAP) meetings are set to identify and define specific (SMART) objectives which will be discussed later in this book. In an Incident Action Plan (IAP) meeting, representatives from all involved primary agencies are gathered to discuss goals and to determine the actions that should be taken. The results that come from the discussion within this command meeting will decide goals and objective for tactical operations for the next (not the current) operational period. In order to do this, the individuals that make up the Unified Command (UC) collaboratively establish resource requirements, identify resource assignments, determine immediate resource availability, and identify resources not on scene that can be obtained. The end result of the planning process is an Incident Action Plan (IAP) that addresses multijurisdictional and/or multiagency priorities and provides tactical operations and resource assignments for unified and collaborative effort.

The use of Unified Command (UC) typically improves information flow and cooperation and coordination between multiple agencies. The use of a Unified Command (UC) typically results in fewer conflicts, and it helps to provide a 360° view of each agency's capabilities that would potentially not be realized under a singular command. In most instances, a Unified Command will reduce conflict among various agencies, disciplines, and jurisdictions. This is often the agencies involved have a greater understanding of the priorities and strategies being utilized as well as a better understanding of the rationale behind decisions. Key benefits of using a Unified Command include

- A shared understanding of priorities and restrictions
- A single set of incident objectives
- Collaborative strategies
- Improved internal and external information flow

- Less duplication of efforts
- Better resource utilization

Command meetings, which occur every operational period, provide agency representatives with an opportunity to discuss and to agree or disagree on important issues prior to the task of developing the Incident Action Plan (IAP). The agenda for the meeting usually includes stating the priorities and objectives; presenting jurisdictional limitations, concerns, and restrictions; developing a collective set of incident objectives; and establishing, and agreeing on, acceptable priorities. While this is just one section of ongoing cooperation and coordination, it may be one of the more important areas because it will set the stage for the overall response to the incident.

7.3 Command Staff

By default, the highest-ranking individual in the first-arriving response team functions in the Incident Commanders (ICs) position until they are relieved of this position by someone of higher rank and/or ability. It is important to note that when a person of higher rank arrives, it should not be assumed that they will automatically take command of the incident. In fact, some higher-ranking individuals may opt to allow the individual who initially assumed command to remain so that they can gain more knowledge, skills, and abilities. Depending on the mentality of the higher-ranking officer, this could be considered a teaching moment. The higher-ranking officer may also leave the person that assumed command in place so that they can initially tend to support issues for the incident, to gather intelligence, so that they can inform elected officials, and a whole host of other reasons. Another reason they may leave the lower ranking officer as the Incident Commander (IC) may be related to the lower ranking officer has more knowledge or expertise in responding to that incident. No matter what the reason, it is important to realize that just because someone is a higher rank does not mean that they are always the best person to manage the incident as the Incident Commander (IC).

As can be seen in the hierarchal organizational chart (Figure 7.1), the Incident Commander is in charge, with the three additional functions that create the Command Staff. The Command Staff includes the Incident Commander (IC) as well as the Safety Officer (SOFR), the Public Information Officer (PIO), and the Liaison Officer (LOFR). These Command Staff positions are essential for the successful outcome of an incident. It does not matter whether these functions are undertaken by the Incident Commander (IC) themselves or if these functions are carried out by individuals appointed by the Incident Commander (IC), however, these three functions must be fulfilled and managed in all incidents. If these functions are not staffed by different individuals, the responsibility falls back to the Incident Commander (IC).

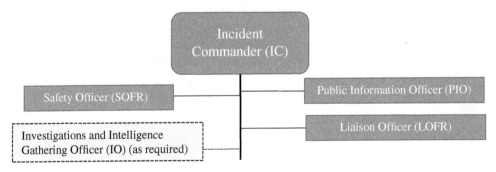

Figure 7.1 Command Staff.

In public safety, we are often told to do something, but not why we are doing something, or why we should do it in a specific way. This mindset often causes frustration. This book was designed to answer those questions. The Incident Commander (IC) needs to ensure that Command Staff functions are covered for numerous reasons.

As an example, the way an incident is managed could be questioned in a court of law, especially if there is an investigation into the response. This investigation, whether true or not, may focus on the actions taken, potential civil rights violations, or a magnitude of other issues that can arise during an incident response.

Another reason these functions are mandated is to ensure the safety of personnel and the public, which should be a high priority to anyone managing an incident. If the function of identifying, protecting, or mitigating safety issues is not addressed, the worst possible thing that could ever happen to an Incident Commander (IC) would be an unnecessary death.

When it comes to liaison, you should remember that in Chapter 1, many failures in the past have occurred because of lack of communication, coordination, and cooperation with assisting agencies. By ensuring that the Incident Commander (IC) and/or the Command Staff are liaising with other agencies, it provides a clear line of communication that promotes cooperation and coordination of all resources.

Similarly, public information can be a key aspect in managing an incident, as well as potentially shoring up or undermining the response and recovery. If public information is not handled properly, it can lead to litigation, calls for an investigation, and various other issues that can put the actions of an Incident Commander (IC) under a microscope.

While these are just a few of the potential reasons that these functions are mandated, it is plain to see that these Command Positions can play a critical role in protecting your community and yourself when you are an Incident Commander (IC). In a large-scale incident, the potential for managing an incident without making some kind of mistake is extremely hard. Whenever an Incident Commander (IC) activates the Command Staff, they reduce the chances of liability because they now have individuals that are not only helping to manage the incident, but they are also present to provide input and ideas that may prevent a catastrophe.

7.3.1 Safety Officer (SOFR) Function

The specific function of a Safety Officer (SOFR) is to develop and recommend procedures for ensuring the personal safety of all incident personnel. The Safety Officer (SOFR) evaluates and tries to identify situations that might become hazardous or unsafe. Only one Safety Officer (SOFR) is assigned for a specific incident, and this person has direct reporting responsibility to the Incident Commander (IC). If an incident expands, or if the incident covers a large geographical area, or if the incident becomes too complex, it may be too complex for a singular individual to manage all safety aspects.

In instances when the inability to manage all safety aspects occurs, the Safety Officer (SOFR) may appoint and utilize assistants. These assistants will report directly to the incident Safety Officer (SOFR). An Assistant Safety Officer is often representative of outside agencies or jurisdictions involved in the incident; however, this is not always the case. Anyone with the proper qualifications may be appointed to the Safety Officer (SOFR) position or as an assistant. In most instances, these qualifications are predetermined by the jurisdiction and based on either NFPA 1521 or based on FEMA's Incident Safety Officers (SOFR) Task Book.

Safety Officer (SOFR) assistants may have more specific responsibilities such as air operations, hazardous materials, or other specialty areas, or they might even be assigned to a specific task or geographical area. The Safety Officer (SOFR) should have an in-depth understanding of all tactical operations that may be taken during an incident, so they can help to identify and prevent issues before they become a problem. The officer need not be a certified responder in every technical area represented on the incident scene, but they should have a good working knowledge of public safety. This is because on a large-scale incident, it would be nearly impossible to have the expertise for every tactic that might be undertaken. We only need to look at incidents such as the Hurricane Katrina or the events surrounding 9/11 to realize how hard it would be to have that kind of extensive knowledge.

In order to be effective while managing an incident, the Safety Officer (SOFR) and their assistants must be authorized to stop any action that the individual, or the group, the Safety Officer (SOFR) determines is unsafe without going to the Incident Commander (IC) for permission. By the same token, any time a Safety Officer (SOFR) or their assistant stops dangerous operations, they must report this action to Command at once. Shutting down a portion of an operation may have a cascading effect that disrupts the overall incident operations and tactical objectives. It should be noted that while the Safety Officer (SOFR) has the ability and responsibility to stop any unsafe actions before reporting the action to the Incident Commander (IC), this authority should not be abused.

7.3.2 Public Information Officer (PIO)

The Public Information Officer (PIO) is responsible for developing and releasing information about the incident to the news media, to incident personnel, the general population, and to other appropriate agencies and organizations. Usually, only one of the Public Information Officer (PIO) will be assigned for each incident; however, multijurisdictional incidents may have multiple the Public Information Officers (PIOs) operating at the Joint Information Center (JIC), but typically only one will be the face of the incident. The Public Information Officer (PIO) may have multiple assistants as necessary. Those assistants may represent assisting agencies or jurisdictions. The Public Information Officers (PIOs) functions are not only limited to the release of information to the press. He or she may also assist all members of the Command and General Staff in obtaining information needed by staff members which will enable them to complete their respective functions in their role in the overall incident.

In the case of operations conducted under Unified Command, the previously mentioned Joint Information Center (JIC) will likely be needed. A Joint Information Center (JIC) uses the Public Information Officers (PIOs) from various agencies to coordinate all information released about the incident to the media and other entities. The use of a Joint Information Center (JIC) is especially vital in incidents with active crime scenes as well as cases of suspected terrorism. In most instances, when a Joint Information Center (JIC) is implemented, it is critical that all information to be released is cleared through the Joint Information Center (JIC) via staff meeting that clearly lays out what information could, and should be, released. This can only be accomplished if using experienced individuals that have been trained, and are well versed, in the job of the Public Information Officer (PIO).

In order to instill confidence in the public, all information to be released from the Joint Information Center (JIC), and by various entities involved in a response, should be approved by the Joint Information Center (JIC) first, and then approved by the Incident Commander (IC) or the Unified Command (UC). In major incidents, it is imperative that there is a unified message without conflicting information. Conflicting information will usually cause a loss of confidence from the public, reduce confidence in the handling of an incident, and heighten anxiety.

Depending on the scope of the incident, and underlying factors, the Public Information Officers (PIOs) may need to create units or appoint certain individuals to handle specific tasks. In our communication-enriched environment that we see today, information flows constantly from a multitude of sources. This can lead to rumors that could undermine a response, or even rumors that could potentially be used later in litigation or a criminal case. Additionally, social media is an area where a good social media strategy by a Public Information Officer (PIO) might enhance a response. If an individual, or unit, is implemented, then the Public Information Officer (PIO) and their team can get ahead of the situation and stop rumors quickly while using social media to their advantage.

We can learn many lessons from other agencies. An October 2016 exercise might help you to better understand the importance of rumor control. During the planning stages of an Active Shooter exercise in rural Moweaqua, Illinois, it was brought up that the rumor mill may start up seeing squad cars parked as if they just responded to a local country church. The team formed a plan to help mitigate the potential effects of the rumor mill. The strategy used was to post the event on social media for at least three weeks prior, and to engage other local people in the conversation, and why this was being done. This created a buzz where many in the town were talking about how wonderful the police force was, and how they worked hard to be prepared. Even though there was a buzz already started (and promoted) for weeks before the training exercise, a Public Information Officer (PIO) was appointed for the day of the event, and they were responsible for marking the roads with signs that announced a public safety training event was ongoing. For six hours, this church was used for training purposes and multiple scenarios, and not one negative post was found on social media. There was no sign of the characteristically negative scuttlebutt of some complainers was not even heard in town. Of course, there was probably someone that disagreed with the exercise, however, because the Public Information Officer (PIO) did their job right, even if the complainers tried to make an issue of it, others would have provided the facts presented by the Public Information Officer (PIO). Of course, it never hurts to make a good impression on the public.

If we look at the previous scenario and try to imagine what might have happened if the Public Information Officer (PIO) had not done their job, the scenario could have been quite different. Someone driving by could have seen the squad cars parked at the church as if they had just responded, and they could have called (or texted) someone in the rumor mill. This is where chaos begins, but it grows exponentially in a short time.

We have all seen, or played, the game telephone. It starts out with the first person whispering "Mary had a little lamb" or some other phrase in someone's ear. By the time it gets to the fifth or sixth person, it changed to Mary's lamb is at the concert. Much like the game telephone, by the third or fourth time it was repeated, it could have been a mass suicide, or a mass shooting with 12 people dead. Traffic on this rarely used rural road could have turned into a parade of cars that blocked incoming vehicles during the training. It also could have caused a cavalcade of 911 calls, and the news media rushing to the scene. By the time the Public Information Officer (PIO) received the information that the rumor mill had started, irreparable damage could have been done to the church, and the law enforcement agency.

In the public's eyes, perception often becomes reality, and some people do not want confused by facts or evidence, especially when the negative stories being disseminated seems to take on a life of its own. When it comes to public information, it is best to be proactive to prevent these types of stories from taking over your response. Even worse, these rumors and stories could begin to break down the important relationship with their community.

7.3.3 Liaison Officer (LOFR)

Incidents that are multijurisdictional and/or multidiscipline in nature often require a Liaison Officer (LOFR). The Liaison Officer (LOFR) is critical in helping to coordinate outside resources that may respond to a crisis. The Liaison Officer (LOFR) is the point of contact for emergency response personnel. The Liaison Officer's (LOFR's) primary job is to communicate, cooperate, coordinate, and collaborate with assisting agencies so that these resources can integrate into the response nearly seamlessly. Only one Liaison Officer (LOFR) is assigned for each incident; however, similar to the Public Information Officer (PIO) and the Safety Officer (SOFR), the use of Assistant Liaison Officer's (LOFR's) may be required. Any agency participating in the crisis by directly contributing tactical resources is known as an assisting agency; this includes fire, police, or public works equipment sent to another jurisdiction.

An agency that supports the incident or supplies assistance other than tactical resources would be considered a cooperating agency. Cooperating agencies can be organizations such as the American Red Cross, the Salvation Army, Volunteer Organizations Active in Disaster (VOAD), or even a utility company. On some law enforcement incidents, such as a massive manhunt, a fire agency may supply experts to help assist with coordination of resources, or in helping to manage ICS. In these cases, the fire agency would be considered a cooperating agency rather than an assisting agency.

As the point of contact for outside agencies, the Liaison Officer (LOFR), and their assistants, maintain a list of assisting and cooperating agencies and their respective agency representatives. They assist in establishing and coordinating interagency contacts as well as collaborating with supporting agencies with incident status reports. The Liaison Officer (LOFR) monitors crisis operations to identify current or potential inter-organizational problems. The Liaison Officer (LOFR) also participates in planning meetings and provides the current resource status (including limitations and capability) of assisting agency resources. In essence, they are an advocate, and to a certain extent, a sponsor for these assisting and cooperating agencies.

The Liaison Officer (LOFR) will be the point of contact for all responding agencies that arrive on the scene. This holds especially true if the resource was not given a specific tactical task when ordered. Some agencies may have specific union agreements, or they may have special considerations that may be different from other agencies. In situations such as these, the Liaison Officer (LOFR) would ensure that those agreements are addressed.

As an incident begins to wind down, or if during extended operations when a resource needs to be released, the Liaison Officer (LOFR) will assist with the demobilization. As the demobilization process begins, Liaison Officer (LOFR) will attend the demobilization meeting(s). They will assist with filling out ICS Form 221-Demobilization Plan as it is developed. This will usually be done by the Liaison Officer (LOFR) providing agency-specific demobilization information and requirements and priorities of the resource. After the demobilization meeting and finalization of the process, the Liaison Officer (LOFR) will supply cooperating and assisting agencies with demobilization information. This information should be shared with these agencies at least one operational period before the planned demobilization. The Liaison Officer (LOFR) will be responsible for recording any issues that occur with demobilization file all records with the Documentation Unit. Information that they will provide should include the phone number of the resource so that they can be contacted, if needed, later on.

7.3.4 Investigations and Intelligence Gathering Officer (IO) Alternative Placement

In the 2017 iteration of the National Incident Management System (NIMS) method with the ICS companion method of incident management, it was determined that there were instances when Investigations and Intelligence Gathering (I/I) should have more direct contact with the Incident Commander (IC). While these instances are rare, and they should be decided on a case-by-case basis, the option for Investigations and Intelligence Gathering (I/I) to be a Command Staff position is an option for alternative placement.

The determination to place Investigations and Intelligence Gathering (I/I) within a Command Staff position may be best suited for incidents where there is little to no need for tactical information to support the response, and/or the Investigations and Intelligence Gathering (I/I) is handling classified intelligence that is being kept on a "Need to know" basis. The placement of Investigations and Intelligence Gathering (I/I) in the Command Staff position would also be viable when supporting Agency Representatives are providing real-time information to the Command Staff and playing an active role in guiding the response.

An example of this may be an incident that is a pandemic and the Investigations and Intelligence Gathering Officer (IO) would be directing the Incident Commander (IC) in how to best prevent the further spread of the pandemic. Real-time information from the Investigations and Intelligence Gathering (I/I), especially if they were a public health entity, would guide where and how to restrict travel, enact quarantines, provide medicines, and a whole host of other guidance. When the Investigations and Intelligence Gathering (I/I) requires immediate access to the Incident Commander (IC) and other Command and General Staff, especially when that information is mission-critical, then placement of Investigations and Intelligence Gathering (I/I) in the Command Staff position may be the best location.

Another example may be when an organized terrorist attack has taken place, and more are expected in the near future. The Investigations and Intelligence Gathering Officer (IO) would help guide the response based on intelligence that their staff has gathered. Of course, this is speculative as there is no evidence that the placement of Investigations and Intelligence Gathering in the Command Staff position has happened at this point and time.

The placement of Investigations and Intelligence Gathering will be discussed more in-depth in Investigations and Intelligence Gathering (I/I) chapter. It is however important to note that there are areas in which Investigations and Intelligence Gathering can be placed: Command Staff position; General Staff position; within the Operations Section (OPS); and within the Planning Section.

7.4 General Staff

As an incident expands, the Incident Commander (IC) may determine that they need assistance in the *General Staff* position. The type of assistance they will need is completely up to the Incident Commander (IC), and because the ICS component of NIMS is flexible, they can determine when they will need assistance, and what function(s) they will need to activate. They may need help in managing operational issues (*Operations Section*), and/or they may need help with tactical and operational planning (*Planning Section*), and/or managing logistical issues (*Logistics Section*), and/or administration and finance tracking of the incident (*Finance/Administration Section*), or they may decide that an Investigations and Intelligence Gathering (I/I) Section may be needed. The Incident Commander (IC) can implement each of these positions individually (as they are needed), or they can implement them all at the same time. It is important to note that no functions in the General Staff positions should be combined. The reason for this is that it can be confusing to handle multiple General Staff functions at the same time, and there will likely be important issues that may be missed during the process.

The Incident Commander (IC) can expand or collapse each function as they see fit. The decision to activate or deactivate a General Staff section is based on an increase or decline in active operations and the support needs of expanded operations. As the incident becomes too burdensome or complicated for the Incident Commander (IC) to handle on their own, General Staff positions can be activated to assist in managing the incident. By the same token, as an incident that has utilized various or all General Staff positions begins to be under control, the

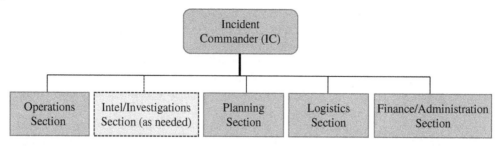

Figure 7.2 General Staff.

Incident Commander (IC) can take on the responsibility of the General Staff member and let that member return to their normal duties.

In Figure 7.2, you will notice that there is a direct line between the Incident Commander (IC) and the various General Staff functions. In the General Staff, the individual who supervises a section is known as a *Section Chief*. The Section Chief manages and directs only those under their specific chain of command, and they report only to the Incident Commander (IC) unless there is a safety issue. If there is a safety issue, they may be required to interact with the Safety Officer (SOFR) and cease operations without the initial involvement of the Incident Commander (IC). This was done so that in the event that there is a safety concern, personnel can be immediately taken out of harm's way while the Incident Commander, the Safety Officer (SOFR), and the appropriate General Staff Section Chiefs determine the safest way to proceed.

Following the principles of span of control, it is possible that if the Section Chiefs use the span of control to the maximum, one individual could conceptually manage seven *branches*, who could manage seven *divisions/groups*, who could manage seven *strike teams, task forces, single resources,* or *units*. These terms and the structure of the hierarchal organization will be discussed in greater detail later in this book. Under the normal hierarchy, it would not be impossible for a Section Chief to manage over 2500 people. Conceptually, with all General Staff in place utilizing the maximum span of control, it is possible to manage over 10 000 first responders. While the exact numbers are not known, we saw the management of thousands of individuals and agencies during the response and recovery for Hurricane Katrina. If additional personnel are needed, each General Staff Section Chief could appoint up to seven Deputy Section Chiefs, and they each could manage seven branches, thereby increasing the number of personnel that could be integrated into the response, with each Deputy Section Chief reporting to the Section Chief. This would still allow a manageable span of control while increasing personnel exponentially (Figure 7.3).

7.4.1 Operations Section Chief (OSC)

In most instances, the General Staff position of the Operations Section Chief (OSC) will use the most manpower. The Operations Section Chief (OSC) is responsible for all field operational elements directly related to the primary mission. The Operations Section (OPS) encompasses all of the tactical aspects of an incident. After the Incident Commander (IC) makes the initial appointment of the Operations Section Chief (OSC), he or she will then provide a briefing to them that should include

- The current incident objectives
- Recommended strategies to manage the incident
- The standing and status of all current tactical assignments
- The current organizational structure

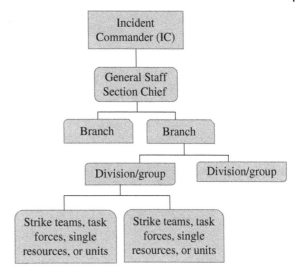

Figure 7.3 Hierarchal Structure.

- The location of resources, and their assignments
- The current resource ordering process and point(s) of contact
- The location of Staging Areas and the resources assigned to them (Emergency Management Institute [EMI], 2012a)

This Incident Commander (IC) should present this information to the new Operations Section Chief (OSC) both verbally and in writing. This is done using both forms of communication to ensure that there is no miscommunication and to provide documentation. For this briefing, the ICS form 201-Incident Briefing form should be utilized. The use of ICS form 201, as well as other ICS forms, can be an important aspect of managing and documenting an incident, especially at the operational level.

ICS forms help to ensure that a whole community approach is used to manage the incident, and they provide guidance to help ensure that nothing is missed. These forms also document the actions proposed, and taken, which can be extremely useful in evaluating a response, and in protecting Command and General Staff from liability should litigation be initiated.

Additionally, the Operations Section Chief (OSC) will ensure that ICS forms specific to their job will be maintained. Common ICS forms that are maintained by the Operations Section, or that may be required to work collaboratively with other sections include

- ICS form 201-Incident Briefing
- ICS form 202-Incident Objectives
- ICS form 203-Organization Assignment List
- ICS form 204-Assignment List
- ICS form 205-Communications Plan
- ICS form 205A-Communications List
- ICS form 206-Medical Plan
- ICS form 207-Incident Organizational Chart
- ICS form 208-Site Safety and Health Plan (ICS Form 208)
- ICS form 209-Incident Summary Status
- ICS form 213-General Message
- ICS form 214-Unit Logs for all Operations Section Personnel
- ICS form 215-Operational Planning Worksheet
- ICS form 215A-Hazard Risk Analysis

- ICS form 221-Demobilization Check-Out
- ICS form 225-Incident Personnel Performance Evaluation (EMI, 2012a)

The ICS form 201-Incident Briefing, and all other forms will be discussed more in-depth as we proceed through the upcoming chapters. When it is expected that an incident will last an extended period, an Incident Action Plan (IAP) will be needed. The Operations Section Chief (OSC) will initiate and oversee organizational duties based on the Incident Action Plan (IAP), and they will administer and manage its implementation. They will also manage the preparation of operational plans, any requests for operational support, as well as being responsible for the release of resources. At times, they may be required to make quick (and on-the-spot) changes to the Incident Action Plan (IAP) as required for safety or tactical reasons. Should they decide that a change of plan is needed, they should report those changes to Incident Commander (IC) as soon as possible. The Operations Section Chief (OSC) is also responsible for briefing and assigning personnel under the hierarchal control of operations as well as maintaining the unit Unit/Activity Log (ICS form 214). Operations can be labor intensive, and the Operations Section Chief (OSC) can appoint Deputy Operations Chief(s) to assist with the administration of their duties (EMI, 2012a).

When it is expected that an incident will last an extended period, an Incident Action Plan (IAP) will be needed. The Operations Section Chief (OSC) will initiate and oversee organizational duties based on the Incident Action Plan (IAP), and they will administer and manage its implementation. They will also manage the preparation of operational plans, any requests for operational support, as well as being responsible for the release of resources. At times, they may be required to make quick (and on-the-spot) changes to the Incident Action Plan (IAP) as required for safety or tactical reasons. Should they decide that a change of plan is needed, they should report those changes to Incident Commander (IC) as soon as possible. The Operations Section Chief (OSC) is also responsible for briefing and assigning personnel under the hierarchal control of operations as well as maintaining the unit Unit/Activity Log (ICS form 214). Operations can be labor intensive, and the Operations Section Chief (OSC) can appoint Deputy Operations Chief(s) to assist with the administration of their duties (EMI, 2012a).

In the overall response, the Operations Section Chief (OSC) is supported by the members of the Command Staff and General Staff. This position is filled as soon as it is practical and should be first considered at the onset of an incident when (or shortly after) the first operational units arrive at an incident. Often this position is assumed by the first arriving tactical officer when their upper-level chain of command arrives at the emergency incident (EMI, 2012a).

The Operations Section Chief (OSC) should ensure that a special focus (especially during the onset of an incident) should be placed on the span of control in respect to tactical operations. During the initial onset of an incident, the atmosphere is usually extremely chaotic: An inexperienced Operations Section Chief (OSC) may inadvertently overlook small details, such as the span of control, because of the chaos they are facing. Exceeding a span of control initially may seem like a minor detail; however, a mistake in this area could cause major problems as the incident expands. This is not to say that span of control is not equally important throughout an incident, because it is. This is noted because most mistakes in span of control occur during the initial onset of an emergency incident, when the incident is most disorganized, and failure to manage span of control early in the incident could lead to even greater problems as the incident expands.

The Operations Section Chief (OSC) has the responsibility to safeguard and work toward optimal operational efficiency to ensure the safety of all first responders. They will also be responsible for establishing the timing of an operational period, as well as creating Staging Area plans and the demobilization of these Staging Areas when they are no longer needed. A critical aspect of the Operational Section Chief (OSC) is to attend Operations Briefings, then in the next operational period, they are responsible for assigning personnel as was agreed upon in the completed and approved Incident Action Plan (IAP). They are also responsible for briefing the Staging Area Manager on types and numbers of resources that should be sustained at each Staging Area, and they should brief tactical leaders (Branches, Divisions/Groups, Task Force/Strike-Team Leaders) on various issues. These issues include

- The ordering processes
- Protective equipment available or needed
- Responder qualifications needed for special assignments
- Provide specific tactical assignments (EMI, 2012a).

The Operations Section Chief (OSC) will also develop and manage tactical operations that will meet incident objectives. They will be responsible for evaluating life safety issues, including deciding the required safety measures, and increasing or decreasing safety perimeters. Safety perimeters can help prevent untrained personnel and non-first responders from entering a potentially hazardous area. Along with monitoring safety measures, they will work with the Planning Section Chief (PSC) to evaluate situations and provide updates on the position, status, and the duty of all resources, the effectiveness of the tactics being utilized, and the type of contingency plans that might be utilized and/or the status of potential contingency plans (EMI, 2012a).

The Operations Section Chief (OSC) is also required to attend various meetings. One of those meetings is the Tactics Meeting. This meeting includes the Command and General Staff members and almost always includes the Operations Section Chief (OSC), the Planning Section Chief (PSC), the Safety Officer (SOFR) and the Incident Commander (IC). The tactics meeting is typically undertaken to review the current and possible future strategies, to discuss current and future tactics, and to help to outline organizational assignments.

The Operations Section Chief (OSC) will also be required to attend Planning Meetings and to hold Section Meetings. The purpose of the Planning Meeting is to review the tactics proposed by the Operations Section Chief (OSC), to determine the strategies to accomplish the strategies that were approved, to assign specific resources that will implement those tactics, and to identify methods for monitoring not only the tactics but also the resources undertaking those tactics. The Section Meetings are called by the Operations Section Chief (OSC) as necessary. These meetings are held to ensure communication and coordination among the entire section, to voice concerns, and to identify contingency plans that are critical to the success of the mission.

When assigned to do so by the Incident Commander (IC), the Operations Section Chief will also be responsible for collaboratively writing the operations portion of Incident Action Plan (IAP) with the Planning Section Chief (PSC). This could include identifying the assignments for each Division or Group, providing input on specific tactical assignments, and identifying what resources may be needed to accomplish both current and potential future assignments. The Operations Section Chief (OSC) will also work with the various individuals in the Planning Section to determine the needs for each operational period and to request additional resources that may be needed (EMI, 2012a).

The Operations Section Chief may also need to manage Investigations and Intelligence Gathering (I/I). Depending on the circumstances, the assignment of the Investigations and Intelligence Gathering (I/I) may be within the Operations Section, based on the needs of the incident. This will most likely be required and employed in incidents that require an abundance of interaction and coordination between the information gathered and the operational tactics that are being pursued. This will be discussed in greater detail in the Investigations and Intelligence Gathering (I/I) Chapter of this book.

The Operational Section Chief (OSC) will work with the Resources Unit Leader (RESL), which is a unit under the control of the Planning Section. The Operations Section Chief (OSC) will notify the Resources Unit Leader (RESL) of Branches, Divisions/Groups, Strike Teams/Task Forces, and single resources that are currently staffed. This would include identifying the location of all resources and the names of each leader. They will also keep the Resources Unit Leader (RESL) up to date on any changes in resource status as they occur (EMI, 2012a).

Among all of the other duties of an Operations Section Chief (OSC), they have the responsibility to coordinate with not only with the Planning Section but also coordinate with other Command and General Staff on other issues such as

- Ensuring that the Operations Section timekeeping, activity logs, and equipment use documents are kept current and up to date ensuring these documents are shared with Planning, Logistics, and Finance/Administration Sections (as is appropriate).
- To make sure that resource ordering and logistical support needs are presented to the Logistics Section in a timely fashion. They also need to guarantee that orders for resources are acted on (in the ordering process).
- To make the Logistics Section aware of any communications problems.
- To make sure that the Planning Section has up-to-date information on resources and situation status.
- To notify the Liaison Officer (LOFR) if any issues arise with cooperating and assisting agency resources.
- To continually maintain communication with the Safety Officer (SOFR) while making tactical decisions.
- To maintain communications with the Incident Commander (IC) to report the status of all operational efforts.
- To coordinate media field visits with the Public Information Officer ([PIO] (EMI, 2012a)

The Operation Section as a whole, and the Operations Section Chief (OSC), is usually the busiest among all of the General Staff positions. While each has their work and duties, the Operations Section is where the rubber meets the road. The work behind the scenes that is performed by other sections is done to maintain the Operations Section and to support their endeavors to ensure that they can bring the incident to a close as quickly, efficiently, and effective as possible.

7.4.2 Logistics Section Chief (LSC)

The Logistics Section will generally manage specialists in medical, supply, ground support, facilities, food, communications, transportation, and security. The leadership component of the Logistics Section is referred to as the Logistics Section Chief (LSC). The responsibilities of the Logistics Section Chief (LSC) are to assure that supplies needed for the incident, especially those in the Operations Section, are on hand when they are needed. Early recognition of the need for a separate Logistics Section can reduce the time and money spent on an incident (EMI, 2012a).

The Logistics Section Chief (LSC) will be an active participant in the creation of the daily Incident Action Plan (IAP). On larger incidents, they may be required to appoint a *Supply Unit Leader (SPUL)* who is responsible for ordering all supplies needed to support an incident, a *Facilities Leader (FACL)* who establishes and manages the facilities needed to support the incident, and a *Ground Support Unit Leader (GSUL)* who ensures that ground transportation needs are met. The Logistics Section Chief (LSC) may also implement a *Communications Unit Leader (COML)* who drafts and manages an incident communication plan, a *Medical Unit Leader (MEDL)* who drafts and implements the Medical Plan for the incident, and a Food Unit Leader who plans for and provides for the nutritional and hydration needs of all personnel assigned to the incident, as well as other units that might be needed (EMI, 2012a).

The Logistics Section Chief (LSC) will be responsible for ensuring that

- An appropriate check-in and check-out system is in place.
- Creating and maintaining the section of the Incident Action Plan (IAP) pertaining to locations of facilities, personnel, transportation, and other support and services that may be needed.
- Maintaining a log of resources on hand.
- Assembling and briefing branch directors on duties, safety, communications, and other field requirements.
- Ensure that command post and field communications are established and remain operational.
- Providing input and reviewing communications plans, medical plans, and traffic plans.
- Coordinating and processing requests for additional resources.
- Meeting with the Information Officer to determine requirements for information center.
- Communicating with chiefs of operations and planning to determine the level of manpower that will be needed and what resources will be needed for next operational period.
- Review contract specifications for supplies that were ordered or that are being ordered.
- Communicate with the Finance (and Administration) Section Chief (FSC) on the preparation of service and equipment contracts.
- Provide input for the Demobilization Plan as required needed by the planning section (EMI, 2012a).

Additionally, the Logistics Section Chief (LSC) will utilize and fill out various ICS forms in the performance of their duties such as ICS Form 205-Incident Communications Plan, ICS Form 206-Incident Medical Plan, ICS Form 213-General Message Form, and ICS Form 214-Unit Log. These forms are an important part of documenting what actions were taken, and they also help protect against litigation.

The job of Logistics Section Chief (LSC) is a critical job in the ICS method of incident management. If the incident does not have readily available supplies, communications, food, manpower, and various other supplies and services, operations will grind to a halt. The Logistics Section Chief (LSC) not only needs to understand the requirement for specific supplies, they must also think far enough ahead to be able to work with the Planning and Intelligence Chief, so they have those supplies delivered and ready for use by the time they may be needed. Should the Logistics Section Chief (LSC) become overwhelmed by the amount of work that they have, they can assign Deputy Logistic Officer's to assist with the administration of the job. Should the job become more than one person can handle, the Deputy Logistics Officer can help alleviate some of the work.

7.4.3 Planning Section Chief (PSC)

The Planning Section is responsible for collecting and evaluating incident situation information, preparing situation status reports, displaying situation information, maintaining status of resources, developing an Incident Action Plan (IAP), and preparing incident-related documentation. This is done under the direction of the Planning Section Chief ([PSC] EMI, 2012b).

The Planning Section is usually responsible for collecting and disseminating intelligence, however, according to FEMA (2017a), intelligence gathering can take place under the Planning Section, or in cases when there will be larger volumes of intelligence gathering and/or investigations are required, it can also be placed in three additional areas. Intelligence functions are usually combined into a single section in most emergency incidents, however, during certain incidents (e.g. terrorist events, major crime scenes, and manmade disasters), the intelligence can be a separate General Staff Section on its own, or even placed in a Command Staff position. These are rare cases, but the flexibility of ICS to adapt to an emergency incident allows for this adaptation (EMI, 2012a). These options will be discussed in the Investigations and Intelligence Gathering (I/I) chapter of this book.

The Planning Section Chief must be aware of the resource status at all times. This individual needs to know what resources are on hand, what their areas of expertise are, and where they are assigned. They must also be aware of current situation status at all times and gather the latest intelligence on the incident if it is placed in the Planning Section. ICS Form 201-Incident Briefing is designed to assist the Planning Officer (and others) in these tasks. They will be responsible for ensuring that the Incident Commander (IC), the Command Staff, and the General Staff receive filled out copies of ICS Form 201 (EMI, 2012a). They are also responsible for ensuring that ICS Form 214 is submitted to the Documentation Unit.

Other forms that are related to the Planning Section Chief's (PSC's) job and their duties include

- ICS Form 202-Incident Objectives
- ICS Form 203-Organization Assignment List
- ICS Form 204-Division Assignment
- ICS Form 207-Organizational Chart
- ICS Form 207 WS-Organizational Chart (wall size)
- ICS Form 209-Incident Status Summary
- ICS Form 211-Check-In List
- ICS Form 213-General Message
- ICS Form 214-Unit Log
- ICS Form 215-Operational Planning Worksheet
- ICS Form 215 WS-Operational Planning Worksheet (wall size [if required])
- ICS Form 215a-Incident Safety Analysis
- ICS Form 215a WS-Incident Safety Analysis (wall size [if required])
- ICS Form 219-1 through 219-19-Resource Status Cards
- ICS Form 220-Air Operations Summary Worksheet (if required)

In some instances, there may be agency-specific forms that are required to be filled out as well. When required and appropriate, the Planning Section Chief (PSC) will ensure that these forms are filled out as well.

The Planning Section Chief (PSC) also has several other responsibilities. They are also responsible for developing alternate strategies and for providing documentation services for a wide variety of Command and General Staff. They are also responsible for creating the demobilization plan in coordination and cooperation of others.

In creating contingency plans or alternate strategies, they will review current and projected resource status, and then propose alternative strategies. They will identify the needed resources and the available resources that can be utilized to implement that contingency plan, and they will provide written alternatives to the Incident Commander (IC) and Operations Section Chief (OSC) so that they these alternative plans can be included in the written IAP (EMI, 2012a).

As part of the intelligence portion of the job position, the Planning Section Chief (PSC) is required to establish and maintain a resource tracking system, advise incident command of any changes in the status of the incident, and either develop or attain incident maps. They will also establish information requirements and reporting schedules for the Incident Command Post (ICP) and field staff, and they will meet with Operations Section Chief (OSC) and/or Incident Commander (IC) prior to the Planning Meetings. This is done to create an agenda that will be used to deliberate the proposed strategies and tactics in an orderly manner. In this meeting, they will also map out the incident organizational structure and resource locations (EMI, 2012a).

The Planning Section Chief (PSC) is also responsible for running the planning meeting. In the planning meeting, there is a set order that should be followed. The Planning Section Chief should follow the order in which the meeting should be conducted according to the NIMS and ICS method to ensure that all issues are addressed in a systematic way. By doing it in this systematic order, it reduces the potential of overlooking any issues. The agenda and the order of these planning meetings should be as follows:

Agenda Item	Presented by
Situation and resource briefing	Planning and Operations Section Chiefs
Discussion of safety issues	Safety Officer
Develop and confirm incident objectives	Incident Commander
Plan location of control lines and division boundaries	Operations Section Chief
Identify tactics for each division/group	Operations Section Chief
Explain/identify resources need for each division/group	Operations and Planning Chief
Identify facilities and reporting locations	Operations, Planning, Logistics Section Chief
Develop resource order	Logistics Section Chief
Communications, medical, and transportation needs discussion	Logistics and Planning Section Chiefs
Provide financial situation update	Finance/Administration Section Chief
Discuss liaison issues	Liaison Officer
Discuss information to be shared and not to be shared	Public Information Officer
Finalize, approve, and implement the plan	All Command and General Staff

Upon completion of the planning meeting, the Planning Section has the responsibility to supervise the preparing and the distribution of the written Incident Action Plan (IAP). The circulation of the written Incident Action Plan (IAP) should include all Command, Command Staff, General Staff, and Operations personnel down to, and including the Division or Group Supervisor level. Depending on the incident and the requirements of the Command Staff (including the Incident Commander [IC]) the written IAP may be distributed to others within the ICS structure (EMI, 2012a).

Based on the written Incident Action Plan (IAP), the Planning Section Chief (PSC) must verify that all support and resource needs are coordinated with Logistics Section. This should be done prior to release of the Incident Action Plan (IAP). The task of verifying resources and support is done to make sure that all resources are readily available to meet the objectives as listed in the Incident Action Plan (IAP). They are also responsible for providing fiscal documentation as requested by the Finance/Administration Section (EMI, 2012a).

Of course, emergency incidents do not always go as planned, and sometimes contingency plans must be implemented. In rare cases, the incident may seem to have a mind of its own and changes in a direction that was not foreseen. In these instances, the Planning Section Chief (PSC) will need to coordinate changes to the Incident Action Plan (IAP) with all Command Staff and General Staff personnel, then distribute written changes as needed, and when appropriate.

Other responsibilities of the Planning Section Chief (PSC) include

- Coordinate the development of Incident Traffic Plans with the Ground Support Unit Leader (GSUL).
- Coordinate the preparation of Safety Messages with Safety Officer (SOFR).
- Coordinate the preparation of the Incident Communications Plan and Medical Plan with Logistics Section Chief (LSC) or their representative.
- Instruct the Planning Section Units about the dissemination of information pertaining to the incident.
- Deliver sporadic predictions that describe potential incident changes.
- When necessary, establish a weather data collection system.
- Identify the need for specialized resources and then discuss those needs with Operations and Command Staff, then enable resource requests with Logistics (EMI, 2012a).

Planning Section Chief's (PSC's) also need to make sure that their section has acceptable coverage and relief as well. As it might be predicted, the tasks of the Planning Section cannot be a one-person show. The work that is done by the Planning Section Chief (PSC) can, and does, affect the whole operation. An overworked or understaffed Planning Section can bring work at an emergency incident to a crawl if not properly staffed. In order to be able to undertake this work, the Planning Section Chief may appoint a Deputy Planning Section Chief to assist with the duties of the Planning Section Chief (PSC).

If the Planning Section Chief (PSC) determines that they can no longer effectively do the job by themselves, then they can expand the organizational structure of the Planning Section to include specific units that are designed assist them with their duties. The four units include the Resources Unit, the Situation Unit, the Documentation Unit, and the Demobilization Unit. Each of these units has a different specialty which will be discussed later in this book.

When the incident and the hierarchy of the Planning Section expands by activating one or more units, then the Planning Section Chief (PSC) will need to be cognizant of the need to communicate and collaborate. As the incident grows, it is imperative that the Planning Section Chief (PSC) facilitates Planning Section meetings with their staff as necessary to ensure communication and coordination among all Planning Section Units (EMI, 2012a).

Perhaps one of the most overlooked jobs of the Planning Section Chief (PSC) is the preparation of a demobilization plan. Demobilization can be defined as the orderly, safe, and efficient return of incident resources to their original location and status. Part of demobilization also includes settling any agreed-upon reimbursement. In smaller incidents, the demobilization plan may be as simple as ensuring that all personnel are accounted for and thanking assisting agencies for their help, then releasing them from the scene; however, larger incidents require a substantial amount of planning to facilitate demobilization. Planning for demobilization in a larger incident should begin at the onset of the incident. This part of the demobilization will begin with tracking and reporting all of the resources involved. This is what is commonly referred to as accountability.

Accountability helps ensure that all first responders are safe and accounted for at all times. Equally important is that accountability is a large part of effectively using incident resources. Accountability is an underlying tenant of the ICS method of incident management that is built into this method. While some may glance over its importance, it is critical for the safety of personnel to understand where their personnel are, what they are doing, and where they are doing it. While accountability is built into the ICS method, this does not mean that it will not take some work as well as an ongoing situational awareness to achieve accountability. Accountability also ensures that as resources are no longer needed, they can be accounted for prior to, and directly after, leaving an incident.

As the objectives of an incident are reached, some or all resources may no longer be needed, and demobilization takes place. Part of the demobilization process is to review a demobilization checklist which is readily available on the Internet. A final incident package must be submitted to the Agency Administrator so that it can be archived or for cases where follow-up after demobilization is required and/or the routing of resources for the return trip may be necessary. This type of follow-up is typically more prevalent on larger incidents (EMI, 2012a).

7.4.4 Finance (and Administration) Section Chief (FSC)

The Finance/Administration Section consists of specialists in accounting, claims administration, and financial management services. In the case of small-scale incidents, this is generally a function of the day-to-day budget process. However, in large-scale or complex incidents, it becomes very important for the Finance/Administration to account for expenditures to guarantee claims recovery from any number of sources. These sources might include cost recovery of disaster monies administered by the US Department of Homeland Security (DHS), which has stringent standards that need to be met in order to recover expenditures. Generally, the function of tracking financial expenditures in larger incidents is accomplished by using four standard units: a Time Unit, a Procurement Unit, a Compensation' Claims Unit, and a Cost Unit. The leader of the Finance/Administration Section is referred to as the Finance (and Administration) Section Chief (FSC).

Personnel assigned as the Finance (and Administration) Section Chief (FSC) should have some form of financial understanding, and it is preferred that they have some type of financial background. This individual should also have historical background and/or experience of working with people from differing organizations. It is extremely important to note that this because of the importance of this key position, and the potential for the response to be funded by other entities or organizations (if properly executed), it critical that the Finance (and Administration) Section Chief (FSC) should be chosen based on experience rather than rank. The number of bars or bugles someone has is not important if they lack the experience that may cost the local agency thousands, or even millions, of dollars. By the same token, if they do not have a comprehensive understanding of the NIMS and ICS method of incident management and have not previously been involved in incident response, this too could have negative effects.

The Finance (and Administration) Section Chief (FSC) is responsible for not only thoroughly understanding the financial side and the incident side (of an incident) but also where those two sides intersect. They must fully understand and have knowledge how one affects the other and how to overcome any hurdles in the finance and administration of an incident.

The Finance (and Administration) Section Chief (FSC) must establish and maintain overall financial-related documentation. This may include the source of funds, what requirements are needed to obtain those funds, the spending limits that have been set by the local authority or other funding sources, the documentation requirements for each portion of the incident, and more. In each operational period, they will also need to identify and document the financial requirements of that period and what sources of funding can be used to cover each aspect of the response and/or recovery.

It is important to note that some sources of funding have requirements that must be met. As an example, if outsourcing work, it may be required that in order to receive financial support from a federal agency, the work might need to be done based on time and materials. On the other hand, a nonprofit or business that will provide funding may require that the work be done utilizing a bidding process and a contract for services. The Finance (and Administration) Section Chief (FSC) will need to know these types of requirements and make sure that all involved know how to proceed. This is just one example of why it is important for the Finance (and Administration) Section Chief (FSC) to maintain effective communications with incident personnel, especially those in key functions.

The Finance (and Administration) Section Chief (FSC) may be required to fill out or assist with filling out multiple ICS Forms. The forms they may be responsible for include

- ICS Form 201-Incident Briefing
- ICS Form 202-Incident Objectives
- ICS Form 203-Organization Assignment List
- ICS Form 204-Assignment List
- ICS Form 205-Communications Plan
- ICS Form 205a-Communications List
- ICS Form 206-Medical Plan
- ICS Form 207-Incident Organizational Chart
- ICS Form 208-Site Safety Plan
- ICS Form 209-Incident Summary Status
- ICS Form 211-Check-In List
- ICS Form 213-General Message
- ICS Form 214-Unit Log
- ICS Form 218-Support Vehicle Inventory
- ICS Form 218-Demobilization Check-Out

Personnel appointed and assuming the Finance (and Administration) Section Chief (FSC) role are generally chosen for their knowledge, skills, and abilities as they pertain to the tasks that need undertaken. The proper selection of the correct person to fill this role is critical to the success in mitigating the incident while being cost effective. Finance (and Administration) Section Chief (FSC) should be familiar with the capabilities and limitations of their or support staff and make every effort to ensure a successful outcome (EMI, 2012a).

Should the Finance (and Administration) Section Chief (FSC) become overwhelmed, or should they foresee that they will not be able to handle their duties on their own, they can activate units within the Finance/Administration Section and/or appoint a Finance/Administration Section Deputy Chief. The units that can be activated include the Procurement Unit, the Compensation/Claims Unit, the Time Unit, and/or the Cost Unit. Each of these units has specific tasks which will be discussed later in this book.

Because it is difficult to find experienced Finance (and Administration) Section Chief (FSC) staff and supervisors, the Finance (and Administration) Section Chief (FSC) should also be constantly looking for ways to improve the capabilities of those under their command (EMI, 2012a). Fostering a desire in subordinates that will help them to aspire to higher levels should be a priority that is not overlooked. It is the role of every public safety leader to foster the development of future staff in their respective fields to ensure that there is never a shortage of qualified personnel.

7.4.5 Investigations and Intelligence Section Chief (ISC) Alternative Placement

In certain instances, Investigations and Intelligence Gathering (I/I) be best utilized as its own section. This option will likely be the best option when there is a large investigation or information gathering to an incident and/or when multiple investigative agencies are part of the investigative process and/or there is a need for classified intelligence. The person who manages the Investigations and Intelligence Gathering Section Chief (ISC).

Personnel assuming the role of Investigations and Intelligence Gathering Section Chief (ISC) are generally chosen for their knowledge, skills, and abilities as they pertain to the tasks of the incident. Their proper selection is vital to the success in investigating and gathering intelligence of the incident. Depending on the size, scope, and complexity of the incident, as well as the jurisdiction of the investigation, the Investigations and Intelligence Section Chief's (ISC) job may be filled by an individual from a local agency, a state agency, or a federal agency. They will be responsible for managing all activities within this section.

While almost any response can have an abundance of chaos, uncertainty, and complexity, the response of Investigations and Intelligence Gathering (I/I) in an incident where they are placed in the General Staff position is often even more complex. In some instances, the incident may be an incident that was not thought of (or could not be prepared for). In these instances, the Investigations and Intelligence Section Chief's (ISC) is responsible for ensuring that the appropriate protocols are in place, which meet the needs of the incident, and ensure that they are created, appropriate, and properly implemented. These protocols can cover a multitude of investigative procedures and identify what information can (and cannot) be released.

The Investigations and Intelligence Gathering Section Chief (ISC) will also give the highest priority to the safety of not only their personnel but all personnel. In some instances, they may receive intelligence that personnel that are working in other functional Sections of the ICS structure may be in danger. In instances like this, they should do all that they can to protect these responders, while still protecting the information that cannot be released. This means that in some instances, they may need to make a decision between releasing classified information or saving the lives of personnel.

The Investigations and Intelligence Section Chief (ISC) is responsible for working with all other Sections of ICS, while at the same time ensuring that classified information is not released. This is especially true if the protocols forbid the release of this information. They will be a participant in the planning meeting and will provide input. The Investigations and Intelligence Gathering Section Chief (ISC) is also responsible for expanding the section (as needed) when the span of control and/or the investigation or gathering of information becomes unmanageable. More in-depth information on Investigations and Intelligence Gathering Section Chief (ISC) will be provided in Chapter 12 of this book.

7.5 Expanding the Hierarchal Structure

As we look at the Command and General Staff, it is obvious that these few individuals cannot command the incident and undertake the physical work by themselves. Their job is not to get in the trenches and to do the work: Their job is to keep everyone safe in the community while responding to and recovering from an incident. The Command Staff and General Staff have the breathtaking and tremendous responsibility of managing and guiding personnel that are in harm's way to protect and save people within a community. Make no mistake these Command and General Staff positions will make choices that could be life or death decisions. One mistake or missing one detail by the Incident Commander, the General Staff, or the Safety Officer (SOFR) could cause death or injury to personnel. This is why span of control is so critical for the safety of response personnel.

As a rule of thumb, the span of control in the NIMS and ICS methods should typically be between three and seven, with five being the optimal number. While the 2017 updates do allow for larger spans of control in some instances, it should be noted that pushing the limits of the span of control can have disastrous results. Span of control is key to safety, but what happens when the span of control reaches limits? What do you do then?

The answer is simple: You expand the hierarchal structure of the incident. There are many working pieces when managing an incident, and expansion of hierarchal structure allows for and creates a manageable span of control. With the ICS method, risks are reduced because of the flexibility for expansion of the ICS methods hierarchal structure. The ICS method of incident management provides a common template for expansion of ICS, while being flexible enough that it can be easily adapted to the needs of the incident.

The expansion of the ICS hierarchical organization to meet the needs of an incident is facilitated through some basic structures in place. These structures allow an expedient expansion to meet the needs of the incident, while increasing the probability of keeping personnel safe. The modular organization of ICS presets roles and responsibilities, plus a flexible organization allows for an expansion with minimal effort.

While we have talked about the higher levels of managing an incident, almost everyone will agree that it takes more than the managers to meet a common objective. The organization must be set up in such a way that it achieves the goals, while protecting the personnel that are responding. In most instances, the expansion of the ICS method comes from the bottom up (the boots on the ground), and it almost always begins with a need to expand operations (FEMA, 2017a).

7.5.1 Modular Organization Supports ICS Expansion

The organizational structures of almost any Incident Management System (IMS) method is developed in a modular fashion and are based on the type of incident and its needs as determined by the Incident Commander (IC). The NIMS method and the companion method of ICS is no exception. As described earlier in this chapter, the common functions of Command, Operations, Planning/Intelligence, Logistics, and Finance/Administration are generally applied on most major incidents. The creating of the Intelligence/Investigations function will depend on the type and scope of an incident as seen by the Incident Commander and/or elected officials. As the needs of the incident grow, the NIMS method and the ICS method will evolve to meet those needs. Within each of the five separate functional sections, branches can be established. Because of the modular organization, the NIMS method and the ICS method can easily and quickly expand or contract.

Many students have been confused by the term "modular organization." Modular organization refers to the division of duties to compartmentalize each specific function separately. By doing so, these tight-knit strategic units can be combined to increase organizational efficiency. When needed, each major function of NIMS and ICS can expand using a module that identifies the organization of that expansion. Each function in the Command and General Staff positions has a specific duty, and with the modular organization of NIMS and ICS, the help they may need to manage the functions is already mapped out. By expanding the organization structure by using modular organization, multiple people with the specific predetermined missions are all reaching for the same goal. These various areas come together for the benefit of the response.

7.5.2 Organizational Flexibility

Every IMS organizational structure (including ICS used in the United States) adheres to a "form follows function" philosophy. In other words, the organization at any given time should employ only the functions that are required to meet planned tactical objectives. The size of the current organization and that of the next operational period are determined by the incident action planning process.

The organizational structure for a given response will be established based upon the management needs of that specific incident, being mindful of the span of control. If one individual acting as the Incident Commander (IC) can manage all major functional areas simultaneously, no further organization is necessary. If one or more of the functional areas require independent management, an individual will (or at least should) be named to be responsible for that specific area. If the functional needs are being met, the number of managers is not relevant.

Within the ICS structure, the first functional management assignments (outside of Safety, Liaison, and Public Information) that will likely be made in a large incident, by the Incident Commander (IC), will normally be (in order): the Operations Section, the Logistics Section, the Planning Section, and finally the Finance/Administration. While the formation of these sections is a general rule during emergency operations, it is not a foregone conclusion. We also need to remember that if Investigations and Intelligence Gathering is needed as a section, it will typically be added when it is first realized it is needed.

In the case of preplanned events or a slowly evolving emergency response, such as a hurricane, the system may deviate considerably from this model. The needs of the incident and the tactical objectives are the only elements that dictate how an incident management structure evolves. It should be noted that every jurisdiction and every incident is different. The arrangement of the command structure should follow some basic rules, but at the same time, the modular organization should be customized to the need and jurisdiction. As long as the core principles are in place and followed, the types of functions assigned, and for the most part how they are named, is not important. In most instances, the jurisdiction should follow the ICS methods modular naming structure, however, if the jurisdiction finds they need to implement a dam shoring branch, they should do so. By the same token, if there were a pandemic, they may want to implement a specific section that deals with providing medicine, enforcing curfews, or in monitoring health. The utility of the ICS method to be flexible is what makes it preferential as a method to organize emergency response in the United States and around the world.

The Incident Commander (IC) will generally establish the General Staff positions as needed, and the person filling those positions should always be called a Section Chief. This Section Chief serves as the Officer-in-Charge (OIC) for a given functional area. Their purpose is to supervise the functional area for which they are responsible in all ways, while keeping the Incident Commander (IC) abreast of the situation and the needs of the incident as they pertain to their specific functional area. Each Section Chief can further delegate management authority within their area as necessary, thereby maintaining a more efficient span of control for that element.

7.6 Conclusion

Not every person in every organization will understand and be competent at all the roles and responsibilities of each operating position in the NIMS/ICS method. It is not realistic to expect that every responder should be prepared to fill every role effectively, but in a marathon operation, it will become necessary at times for people to operate outside their scope of knowledge in a position that needs to be filled until more help arrives.

There are many task books that have been written by FEMA, the US Department of Agriculture, the US Coast Guard, and various other entities. It is important that those who may be called to serve in a Command or General Staff position become familiar with the responsibilities of the job, that they are properly mentored, and that they are qualified and certified. Lives may be at risk, and in the initial phase of an incident (before outside help arrives), qualified and certified Command and General Staff members may be hard to find. Making sure that an agency is being progressive in finding the training, sending the right people to that training, and in getting them qualified and certified in specific task is important. This is an important part in preparedness and planning. If your agency fails to prepare, then they are preparing to fail.

An analogy of how the NIMS/ICS methods work might be compared to the making of a computer. Most computer manufacturers use a modular organization to make these machines that are now commonplace in our society. One manufacturing area may create and assemble the hard drive, while another makes the motherboard, while yet another makes the keyboard, the case, and so on. Separately, they all specialize in a single area, and they make sure quality control and efficiency in their specific area is optimal. The heads of each department ensure that all of the pieces fit together, that all of the components are safe, and that when all of the components are plugged in together, they are compatible. The finance and administration department will make sure that all of the supplies are paid for, that the laborers are paid, and they document all of the serial numbers on each individual piece. The marketing department gets the word out about how wonderful and effective and fast this computer is, and as the computers go out the door, they gather feedback. They learn of the problems that may need addressed, as well as the positives of the new computer based on feedback. All manufacturing areas and nonmanufacturing areas integrate and work toward the same goal of efficient, safe, and effective computer for the end user. Individually, they are all creating a small portion of the overall "big picture" or "end result," and it takes all of those pieces to make a working computer. It is not until all the components are delivered to the assembly area, and put together, that there is a finished and working computer. If one piece of that computer is missing, or if one of the compartmentalized or modular areas did not do their job right, or if there is a problem with compatibility, then that computer will likely fail. By the same token, feedback from the marketing department will probably help determine if the computer is working as expected when it was delivered to its new owner.

Much like the computer company, response to a major incident needs to be modular. Each assignment in the Command and the General Staff is overseeing the making of their specific part. Liaison is making sure that every organization is compatible. Safety is making sure that all the components are safe, while Public Information provides information on the product being produced. Planning is making sure that they provide compatible plans and supplies, and Logistics makes sure that every need in the supply line is met. Finance and Administration are documenting and making sure every vendor is paid, and the Incident Commander (IC) is in the assembly area putting all the components together. The Operations Section is the final product, but at the same time, Operations is also quality control and providing customer feedback.

In both the computer scenario and the ICS scenario, all of the areas of expertise are separated and compartmentalized, each compartment does their own specific work, then all of it is brought together to make a finished product. This is what an ICS and its modular organization does, and how it works. Even while having a modular organization is important, it is equally important in emergency response to have organizational flexibility. This is accomplished by plugging in the module that fits the need of the incident, however, the higher positions that oversee the lower positions in the organizational chart must be in place to prevent freelancing and to increase the use of the Five-Cs.

At this point of the book, you should now have a basic grasp on the NIMS method and the ICS method, especially the highest levels of management. You should also be aware of the key roles and responsibilities that are usually assigned in major incidents, and what these positions do. These are the foundations for understanding the entire system, but there is still much to learn. In the next chapter, we will learn about how to expand the Operations Section when it is needed.

Chapter 7 Quiz

1. **True or False**: A singular Incident Commander can fulfill all Command Staff positions if they choose to do so.

2. _____ _____ allows the command module of the ICS method to make mutually agreed-upon decisions that helps to manage resources better.
   (a) Safety Officers
   (b) Liaison Officers
   (c) Unified Command
   (d) Public Information Officers

3. Unified Command helps to provide a _____ of each agencies capability that would likely not be realized under a singular Incident Commander (IC).

4. All Command Staff leaders except the Incident Commander (IC) will be referred to as:
   (a) A Chief
   (b) A Leader
   (c) An Officer
   (d) None of the above

5. This Command Staff position is responsible for evaluating hazardous or unsafe situations and has the ability and responsibility to stop any unsafe actions.
   (a) Safety Officer
   (b) Public Information Officer
   (c) Liaison Officer

6. This Command Staff is responsible for developing and presenting information about an incident to the media, incident personnel, the general population and other appropriate agencies.
   (a) Safety Officer
   (b) Public Information Officer
   (c) Liaison Officer

7. This Command Staff position is the point of contact for emergency response personnel and has a primary job to coordinate and collaborate with assisting agencies.
   (a) Safety Officer
   (b) Public Information Officer
   (c) Liaison Officer

8. What are the four places that Investigations and Intelligence Gathering (I/I) can be placed?
   (a) Command Staff
   (b) General Staff
   (c) Under the Operations Section
   (d) Under the Planning Section

9. What are the four primary General Staff positions, also identified as sections?

10. All General Staff leaders will be referred to as

(a) A Chief
(b) A Leader
(c) An Officer
(d) None of the above

11 What General Staff Position (or Section) may be added separately if warranted and needed?

12 This General Staff Position (or Section) is responsible for all field elements related to the primary mission and usually encompasses all tactical aspects.

13 **True or False**: A singular Incident Commander can fulfill all Command Staff positions if they choose to do The Operations Section Chief does not need to attend planning meeting.

14 This General Staff Position (or Section) is responsible for collecting information and evaluating incident situation information, preparing situation status reports, displaying situation information, maintain status resources, developing Incident Action Plans, and preparing incident-related documentation.

15 Who is responsible for the preparation of a Demobilization Plan?

16 This General Staff Position (or Section) contains specialists in accounting, claims administration, and financial management.

17 **True or False**: A singular Incident Commander can fulfill all Command Staff positions if they choose to do.
Every person who serves in a Command Staff and/or General Staff position should be able to fill every role in the Incident Command System.

18 **True or False**: A singular Incident Commander can fulfill all Command Staff positions if they choose to do so.
Modular organization refers to the dividing of duties to compartmentalize each specific function separately.

Self-Study

Department of Homeland Security (2013a). NIMS Intelligence/Investigations Function Guidance and Field Operations Guide. Retrieved from https://www.fema.gov/media-library/assets/documents/84807.

Department of Homeland Security [DHS], (2013a). ICS-400: Advanced ICS Command and General Staff-complex incidents. Retrieved from https://www.in.gov/dhs/files/ICS-400_SM.pdf.

FEMA (2017a). National Incident Management System (NIMS). Retrieved from https://www.fema.gov/media-library/assets/documents/148019.

8

Expanding the Operations Section

Do not confuse good tactics with good luck

Unknown

The Operations Section is responsible for the tactical operations of an incident. In essence, this section is responsible for actually mitigating the negative effects of the incident and bringing the incident under control. The Operations Section will be the individuals that have an active, hands-on role in the incident. The Operations Section Chief (OSC) will be directing personnel to positions where they are putting out fires, treating victims, and/or dealing with the hostage situation. Of course, these are not the only jobs or tasks that will be required under the Operations Section. One method of remembering what the Operation Section does would be to be aware that these are the individuals that have direct contact with the incident.

While other sections are often necessary to support the Operations Section, without their support, the incident would in most instances take substantially longer to end. The Operations Section is one of the most dangerous sections to operate under. Expansion of the Operations Section is set up like the rest of Incident Commands System (ICS) in modular format. At the same time, ICS is flexible enough that only those functions that are needed can be expanded upon. There are many ways to expand the Operations Section while still meeting the needs of the incident. In this chapter, we will discuss the common methods of expansion when using ICS.

8.1 Operations Section

Sections are named by the functional area. In the ICS method, the functional area that we will review is the Operations Section. When expansion of the ICS method is needed under the Operations Section, the Operations Section Chief (OSC) will typically create a Branch. Branches are formed and then assigned a Branch Leader who will further delegate individual tasks and expand the branch as needed. With the modular design of the ICS method, predetermined roles and responsibilities guide how to expand a branch and determine what elements will typically be needed.

It is important to note that each activated element must have a person in charge. In some cases, a single supervisor may initially be in charge of more than one element. Elements that have been activated and are clearly no longer needed should be deactivated to decrease organizational size and to free up command-level resources for more incident-pertinent issues.

In the hierarchical organizational chart, the Incident Commander (IC) is ultimately responsible for the response and/or recovery, but as an incident expands, they will usually incorporate

Emergency Incident Management Systems: Fundamentals and Applications, Second Edition.
Mark S. Warnick and Louis N. Molino Sr.
© 2020 John Wiley & Sons, Inc. Published 2020 by John Wiley & Sons, Inc.
Companion Website: www.wiley.com/go/Warnick/EIMS_2e

an Operational Section and Operational Section Chief soon after choosing a Safety Officer (SOFR). The Operations Section will almost always be the first General Staff Position that will be activated. The reason that the Operations Section will typically expand first is because it accomplishes the hands-on work, which is the most labor-intensive work that will be done in an incident.

This Operational Section Chief will guide and maintain span of control over the branches. The Branch Director will manage and oversee the next step down in the hierarchical organizational structure within their span of control, a Division or a Group. A Division or Group Supervisor will supervise the next level below them in the hierarchical organizational structure within their span of control, the Unit. The Unit Leader will oversee strike teams, single resources, and/or task forces within their span of control which also has a Leader for the personnel that work under their guidance. The individuals managing strike teams, single resources, and/or task forces will also be called a Leader. As can be seen from the Table 8.1, most of these positions can also have support staff. The assistants that help them each have a specific identifier that helps separate the level in the hierarchical organization by the use of those specific names.

As the Operations Section Chief sees that the incident is expanding and that they are reaching the limits of their span of control, they should expand the organizational structure to efficiently and effectively manage the influx of additional responders. To do so, they may implement groups, divisions, or branches, depending on how many resources they may need. The expansion of Operations, and how it expands, will be dependent upon the incident. If it is believed that there will be a large influx of responders in a short period of time, they may opt to form a branch. If the incident will remain small to medium, they may decide that they only need to form divisions, groups, or combined divisions and groups.

It should be noted that unless the expectation is that the incident will expand, and expand quickly, then the expansion of the organizational structure to include Branches, Divisions, etc., will only increase bureaucracy and potentially slow the response. A good rule of thumb is to expand the organization structure when the span of control is reaching its limit and to reduce the organization structure when the span of control reaches its lower limits.

The strategic and tactical needs of the incident and the foreseeable operational periods are the predictors of how the organizational structure expands and collapses as that incident progresses. The strategic and tactical needs of the incident alone will dictate the necessity for any expansion or collapse in the organizational structure. It may be necessary for sections to expand beyond the span of control that a Section Chief cannot effectively handle everything so they will appoint a Deputy Chief. If the span of control becomes too large for a singular Deputy Chief to manage, then multiple Deputy Chiefs can be used to expand the number of responders that

Table 8.1 Hierarchical organizational levels and titles.

Organizational level	Title	Support position	Examples
Incident Command	Incident Commander	Deputy	IC
Command Staff	Officer	Assistant	Safety Officer
General Staff (Section)	Chief	Deputy	Operations Section Chief
Branch	Director	Deputy	Branch "I" Director
Division/Group	Supervisor	Not Applicable	Division "A" Supervisor
Unit	Leader	Manager	Explosives Unit
Strike Team/Task Force	Leader	Boss	Ambulance Strike Team

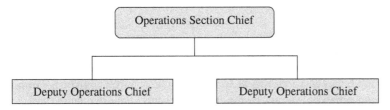

Figure 8.1 Hierarchical placement of Deputy Operations Chief.

could be managed by a section. Before expanding the structure, one should usually first expand the Divisions and Groups before expanding to more Branches. Each Deputy Chief would optimally manage five Branches if assigned to the task. They could manage as few as three or as many as seven Branches, although three to five branches are optimal. In the latest iteration of the ICS method, flexibility is built into the span of control. This means that the individual may be able to have a larger span of control if they have less complicated tasks, and they have sufficient experience to manage more resources.

While Deputy Chiefs can manage those lower in the organizational chart, they are not limited to managing Branches. They can do any job that the Operations Section Chief (OSC) determines would be most useful. When a Deputy Section Chief position is formed, the individual appointed to that position yields the full authority of the Operations Section Chief (OSC), and they report primarily to the Section Chief. The Deputy Operations Chief should only report to others higher up in the hierarchical command structure if directed to do so by the Section Chief. This will reduce confusion and prevent a Deputy Section Chief from potentially undermining the Operations Section Chief (OSC) Figure 8.1).

In most single resource incidents, the organization structure will remain flat and linear in nature, whereas in more complex incidents, a very detailed and expansive hierarchical incident management chart will likely evolve. It is important to note that only the needs of the incident should dictate the structure that is used and the need for expansion and contraction, at every point of each incident. Every incident is unique, and being unique, the needs of the incident should dictate how the organizational structure expands and contracts rather than expanding and contracting based on some other reason. Even so, the Incident Commander (IC) or Unified Command (UC) needs to ensure that the organizational is not top heavy. A top-heavy organizational structure can cause more problems than can be imagined.

These expansions in the organizational structure are possible only because of the modular organization and the organizational flexibility that are purposely built into the ICS method. If this method did not have a modular organizational structure, then there would likely be overlap and/or a duplication of resources. If the structure was not flexible, then there would be a full complement of individuals that responded to nearly every incident, even if they were not needed.

8.1.1 Operations Branches, Divisions/Groups, Strike Teams/Task Forces

While it is important to have General Staff that manage the various sections they cannot accomplish all the tasks on their own. In most instances, they will supervise others who will take actions in their absence, on their behalf. They must continually be conscious of the span of control that is required to meet the objectives of the incident. In this section, we will discuss what the various levels are called and some of the more common duties that are typically assigned under each.

8.1.2 Branches

In the organization structure of the ICS method, a Branch is the organizational level that has functional, geographical, or jurisdictional responsibility for major parts of the incident operations. The Branch level is organizationally between Section level and the Division/Group levels in the Operations Section, and between Section level and the Unit level in the Logistic Section. Branches are identified using Roman Numerals, by function, or by jurisdictional name.

Due to the flexibility of ICS, the use of Branches can also be flexible. In most instances, the optimal span of control is one supervisor is five divisions or groups. While the latest iteration of the ICS method allows the span of control, caution should be used in a chaotic situation to ensure prevent overloading on one person in the span of control. While it is now allowed to expand span of control to more than seven, it is much simpler and easier to remember that the maximum span of control should typically be a maximum of seven with an optimal span of control of five.

In all fairness, span of control can be led by the type of incidents that are being managed. The nature of the incident, the type of tasks being addressed, the geographical location of first responders as well as the hazards and safety issues can play a role in deciding a manageable span of control.

Branches are the supervisory sector of groups or divisions. Branches report to the section chief. In the case of the Operations Section, the Branch Director would report to the Operations Section Chief (OSC). It is important to note that Branches can be a mix of disciplines. As an example, under the Operations Section, the Operations Section Chief (OSC) or the Deputy Operations Chief can create Branches based on their function. If it was a terrorist attack, the Operations Section may create two separate Branches that each separately manages rescue, fire resources, and Emergency Medical Services (EMS) services under Branch I, and Hazardous Material (HazMat) resources, law enforcement resources, and investigation resources would be managed under a different Branch (Figure 8.2).

This type of division would be for medium/moderate-sized incident. This type of Division might be useful in an incident such as 40 car pile-up on an interstate. Branch I would manage the Rescue Division who would be necessary to extricate victims from vehicles, while the Fire Division handles fires that may have broken out and the walking wounded, the EMS Division would meet the needs of medical and medical transport. Branch II would manage the HazMat Division who would take care of any leaking trucks carrying hazardous materials, and the Law Enforcement Division would reroute traffic and secure the scene, while the Investigation Division examines how the pile-up started.

On larger, more complex incidents, Divisions may be split up into geographical or jurisdictional areas. These divisions can be named by the area or jurisdiction that are in (e.g. City of Decatur, City of Macon, South Wheatland Township, and South Macon Township), or they can

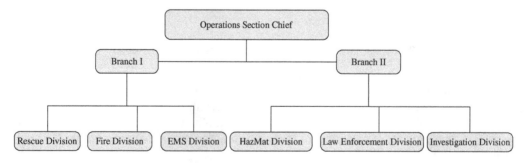

Figure 8.2 Branches separated by function.

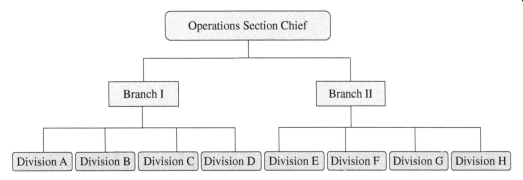

Figure 8.3 Branches identified using roman numerals.

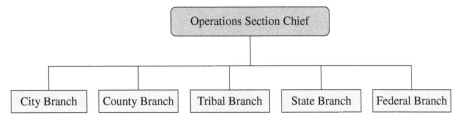

Figure 8.4 Branches identified by type of government.

be divided by the geographical location such as a set number of city blocks, or set areas on a topographical map. When using geographical areas, the Branches would in most instances be arranged by the area that it covers, such as the Northern Branch, the Southern Branch, the Eastern Branch, or the Western Branch (Figure 8.3).

On multijurisdictional responses, the Branches can also be deployed according to the type of assets that are being used. As an example, if there were a response to a major hurricane, there may be a City Branch, a County Branch, a Tribal Branch, a State Branch, and a Federal Branch, based on the resources available (Figure 8.4).

Whenever splitting the Branches in this fashion, you may want to ensure that either every jurisdictional Branch has an Agency Representative in place represented by a person that is knowledgeable in what their assets can and cannot do, as well as the best utilization of those resources.

These are but a few ways that branches can be arranged. As an incident expands, and more resources are brought in, and integrated into the response, one Branch may begin to push the limits of a manageable span of control. When this situation arisis, the flexibility of the ICS method allows for this specific Branch to be split into two separate Branches, thereby maintaining a viable and manageable span of control. By the same token, if resources are dwindling because the incident is coming to a close, or for any other reason, the Branches can be collapsed. In some instances, as Branches are intentionally collapsed, so that they can be blended with other branches or put directly under the control of the Section Chief (or their assigned Deputy Chief).

The flexibility and modular design of the ICS method allows for arranging Branches into a response and recovery in such a way that it fits the need of the incident. It is important to remember that the basic frameworks of the ICS methods should be followed when adjusting the hierarchical organizational chart to meet the needs of the incident. It is important to remember that a basic framework concern, which can also be identified as a common failure in the ICS method, is too large of a span of control.

8.1.3 Additional Branch Considerations

Often, we think about fighting the fire, cleaning up a HazMat spill, or ending the hostage situation, but fail to remember that there are often other underlying issues that may not get addressed. Often, agencies operating during a disaster fail to recognize those that are disabled or elderly in an evacuation; usually these individuals need extra assistance.

If we go back as far as the first World Trade Center bombing in 1993, people with disabilities were told to wait for evacuation; some waited up to nine hours with the average being 3.34 hours (Juillet, 1993). Research on Hurricane Katrina found that 71% of those that died were elderly or disabled, but they represented only 15% of the population (Benson, 2007). Even more recent incidents have identified that evacuating those that are elderly or disabled is rarely a high priority. Similar issues were identified in the 2017 fires in Napa and Sonoma California as well as the 2018 Camp Fire that essentially wiped out Paradise California. In all of these incidents, those with Access and Functional Needs (AFN) suffered a substantially higher death rate than all other any other demographic population.

While evacuating an area of people, it is often easy to forget that there are pets and livestock at risk as well. There have been documented cases where people will not evacuate without their pets, and they choose to die because of it. In fact, depending on which study you read, it has been stated that between 20% and 45% of pet owners will not evacuate without their cat or dog. If they do leave a pet behind, and the incident becomes extended, it has also been seen where pet owners will defy evacuation and exclusion orders, risking their own life to rescue a pet.

Based on these issues and other similar problems, it may be required to enact specific Branches or Groups to deal with these problems and issues. Dedicating first responders to deal with these specialized problems during the initial onset of a disaster may seem counterproductive to mitigating an incident, but they are far from being counterproductive. Life safety should always be the top priority. If this means creating a Branch manned by a combination of first responders and volunteers to evacuate elderly and disabled individuals, then it should be done. Depending on the circumstances, evacuating Access and Functional Needs (AFN) will likely be more important that dealing with the fires after an explosion. If it is considered that 80% of the disabled population can be evacuated in three hours, and firefighting efforts will be ongoing for days and it can be somewhat contained by a smaller crew, then the efforts may be best spent helping those that cannot help themselves. If 20% of people will not evacuate without their pets, the use of personnel to set up a temporary pet shelter may also be a good use of resources. As long as the family knows Fido will be allowed to evacuate and have a safe place to stay, chances are the family will evacuate rather than defy orders.

Perhaps an important part of the response is to provide construction or public services repair. If it follows an Operational objective, then this would be best managed under operations. If restoration of water to 20 000 residents would demand fewer resources to hand out water to those 20 000 residents, then why would we not make them a part of an operational Branch or Group? It is an Operational objective, so we should, at the very least, consider this expansion.

The following section will describe different modules that can be added to the Operations Section to deal with these types of issues. It should be noted that these are but a few examples, and if there is an issue that can undermine the response, the Incident Commander (IC) or their representatives can create Branches or Groups that are focused on a specific task similar to the aforementioned issues.

8.1.3.1 Health and Welfare Branch[1]

In some instances, the ICS method may need to expand to include a Health and Welfare Branch. While this is usually sought after by an Emergency Operations Center (EOC) that is supporting

---

1 This heading is for advanced IMS students.

a major response, it can also be enacted by an Incident Commander (IC) in the field. While the enacting of a Health and Welfare Branch is not common in the US method of ICS, some agencies in the United States and in Canada will activate this Branch in a major response. This Health and Welfare Branch would include sheltering, feeding, and caring for the basic needs of victims (clothes, hygiene, medicines, etc.) as an important part of response. This can include but is not limited to, providing care and shelter for evacuated or displaced personnel, and for ensuring public physical and mental health. The Branch will coordinate the care-giving activities through the unique resources that are available within the community, or by acquiring these services Memorandum of Understanding (MOU) agreements and/or established mutual aid agreements. Specific Units can be activated to handle an area of the response that is focused on a particular problem.

In the United States, the sheltering, feeding, and caring for the basic needs of victims will usually be handled by the American Red Cross; however, even the Red Cross, and anyone assisting them, will need to be included into the ICS structure. Before the American Red Cross arrives on scene, it should be considered who will manage the care and sheltering aspect of the response. Will the agency just tell everyone to evacuate and to fend for themselves, or will they provide prearranged shelter locations where these citizens can find temporary shelter? Even if a temporary shelter is expected to be in operations for only a few hours, someone will need to staff it until the American Red Cross or other agencies get on scene. Moreover, if the evacuation is substantial, or if it encompasses multiple days, the Incident Commander (IC), or possibly even the Operations Section Chief (OSC), will likely need to expand the ICS method to include a Care and Shelter Unit.

By implementing a Care and Shelter Unit, a single Leader with some assistance can (in most instances) manage the entire Unit, thereby relieving stress on operations. This Leader could then request the resources needed, and they could ensure that no resources are wasted, and that there are sufficient personnel to handle the needs of the sheltering operations. By the same token, the Unit Leader may appoint specific individuals to complete specific tasks to ensure that critical attention is given to critical tasks in care and sheltering. An example of this may be a medicine runner. This would be someone who will take trips to local pharmacies to retrieve the medicines needed to maintain the health and quality of life for individuals being sheltered. They may also implement and manage a clergy program to be there for the emotional support of those staying in the shelter. As is common with all of the ICS method, the structure should meet the needs of the incident.

Additional Units that are usually put under the Health and Welfare Branch can include the Public Health Unit and an Animal Care and Welfare Unit. The Public Health Unit will maintain the availability (and viability) of potable water, to assess and maintain sanitation systems, and to assess food for cleanliness, edibility, pest control, and nutritional benefit. The Animal Care and Welfare Unit in the Emergency Operations Center (EOC) will typically focus on maintaining care for animals during the incident, not in rescuing animals from dangerous situations.

8.1.3.2 Construction/Engineering Branch[1]

The Construction and Engineering Branch could be added to the Operations Section to ensure that damage and safety assessments are handled in a quick and orderly fashion as well as making sure that basic services offered by local governments and businesses are addressed as part of the response.

Imagine if you will, a whole section of an urban city without electricity. Tens of thousands of individuals without power. Depending on the situation, opportunists could use this as a reason to commit crimes, no matter if it were in a rural, suburban, or urban setting. Moreover, individuals who have long-term medical needs that cannot be met without power might be struggling to stay alive. In the response phase, while still responding to the original issue, it may be in the best interest of the response (depending on the situation) for the Operations Section to manage these individuals while the Planning and Intelligence section undertakes an ongoing triage of what the priority(s) should be.

The Construction/Engineering Branch will coordinate nearly anything that is infrastructure-related. This can include the coordination of utility companies, public works departments, engineering agencies (and departments), as well as damage and safety inspections (and assessments). This Branch would be responsible for inspecting all utility systems, and restore systems that have been disrupted, including coordinating with utility service providers in the restoration of disrupted services. They would also be responsible for bridge inspections and repairs, and roadways. They are tasked with inspecting all public and private facilities, evaluating the damage caused to these facilities, and being the coordinating Branch that will ensure damage repairs to all critical public facilities.

Depending on the circumstances, it may be required to help mitigate the chaos, this Branch could address the issues based on the

- Utilities Unit
- Damage/Safety Assessment Unit
- Public Works Unit

8.1.3.3 Air Operations Branch (AOB)[1]

It should be noted that on incidents that require a large contingency of air support and aircraft, a separate Air Operations Branch (AOB) can be formed. Much like Investigations and Intelligence Gathering, Air Operations can be placed in multiple levels in the ICS method, but it is important to note that when adding air support to an incident, you should only activate the level that is needed. In essence, you would start at the lowest feasible level and increase the prominence of this operations position as more air resources are needed. Before the span of control becomes unmanageable, air operation should move up the organizational chart from the bottom, staffing only what is needed for the incident. This means that air operation should start at the lowest level as a Single Resource, then as more air resources are needed, and the span of control was reaching its limit, it would become a Division or a Group. When the span of control in a Division or a Group was reaching its limits, then an Air Operations Branch (AOB) would be the next level that would be used. Whether is activated as a Branch, a Group, or a Division, where it will be placed is dependent on the level of air support that is needed. When activated at the Branch level of the ICS method of incident management, there will be a significant amount of aircraft in the operating theater.

The Air Operations Branch (AOB) would, for the most part, follow the same design and requirements as other Branches. This Branch would only be responsible for anything related to air operations, and integrating air resources into the overall incident response. This would be anything that will provide tactical support from the air, including but not limited to fixed wing aircraft, helicopters, and Unmanned Arial Devices (UAD) which are also known as drones.

If activated at the Branch level, this typically means that there is a large contingency of aircraft. In these larger operations, span of control is a major consideration for the safety of pilots, ground crews working on the scene, and the staff that work in air related operations at airfields, bases, and helispots.

When the Air Operations Branch (AOB) is activated, management of the Branch will entail much more than managing the aircraft when they are in the air. The Air Operations Branch Director (AOBD) will need to identify and identify additional aviation related personnel in both management of the Branch and in supporting the mission of the Air Operations Branch. The Branch and all of those in a leadership position will also need to evaluate the requirements to sustain operations, such as food, fuel, relief crews, lodging, transportation, support facilities, and any other aspects that may be needed for support within the Branch. In order to meet those needs, they will need to work with the Operations Section Chief (OSC) and Logistics Section Chief (LSC) to manage and ensure that all needs are met.

Because of the austere conditions that many of these aircraft will operate in, it is important for the Air Operations Branch (AOB) to ensure that aircraft maintenance is a priority and that all support requirements are met. While not always the case, the agency or business that is providing the air asset will provide their own maintenance support for the aircraft. Even when they do provide for their own maintenance support, the Air Operations Branch will need to make certain that the ground support equipment that could not be transported by the agency is made available to them. This ground support equipment might include forklifts, aircraft tow vehicles, stands, generators, man lifts, and more.

The Air Operations Branch (AOB) will also need to provide an exorbitant amount of communication, both from within the incident (to manage aircraft) and outside of the response. Depending on the circumstance, they may need to coordinate with the Federal Aviation Administration (FAA), the United States Northern Command (USNORTHCOM), and/or the Department of Defense (DOD) to help prevent mid-air collisions (or near misses), and to ensure that aircraft operating on the incident do not pass into restricted airspace without permission. The Air Operations Branch will also establish effective Air to Air, and Air to Ground, radio communications and make sure that communications are adequate and have built in redundancies.

Additionally, the Air Operations Branch (AOB) will have to identify safety risks to aircraft. One important factor for safety is the weather. This Branch will need to monitor the weather, both on-scene and approaching weather, and they will need to evaluate what, if any, impacts this may have on the aviation sector. They will also need to identify and manage actual risks (such as power lines, towers, terrain, etc.) and possible risks to flight operations (such as civil unrest and restricted airspace).

Based on risks and communications with other entities, the Air Operations Branch (AOB) may need to enact Temporary Flight Restrictions (TFRs) to protect the aircraft, or to exclude aircraft from entering a specific airspace. In doing so, they may need to release a Notice to Airmen (NOTAM) to advise pilots about a new or temporary restriction or other information such as a busy airfield.

While there are a multitude of other tasks that will need to be undertaken by the Air Operations Branch (AOB), it is plain to see that this Branch, and all of the affiliated work conducted by this Branch, cannot be left to amateurs. According to the US Coast Guard (USCG) Job Book (2017), those assigned to the Air Operations Branch Director (AOBD) should have experience managing aviation and have a background that supports the experience needed. They also state that it is imperative that the Air Operations Branch Director (AOBD) should be based on experience level rather than rank. Other important traits that were identified include good leadership skills, interpersonal and communications skills, and risk-based decision-making. The individual chosen for this position should also have a dependable comprehension of political, social, environmental, and economic issues as well as a good grasp of National Incident Management System (NIMS) and the ICS component of incident management (USCG Job Book, 2017).

8.1.4 Divisions/Groups

Divisions and Groups are established when the amount of resources operating within an incident begins to exceed the Incident Commanders (ICs) or Operations Section Chiefs (OSCs) manageable span of control. Any time you hear the designation of a Division, you can rest assured that this designation has been used to divide an incident geographically or physically. Similarly, you can rest assured that Groups separate functional areas (or job duties) of operations in an incident. The manager of each Division is designated as a Supervisor. The geographical division of a response or recovery is determined by the unique needs of the individual incident. The most common way to identify Divisions is by using alphabetic characters (e.g. A, B, C). Other methods of identifying a Division may be used providing the Division identifiers based on floor levels in a multilevel building.

The use of the two separate terms is necessary to provide clarity of the type of operation that is happening. It does not matter if you are the Incident Commander (IC), or a basic grunt on the scene; when you hear the word "Division," you will always know that it refers to a geographical assignment. Similarly, when you hear the word "Group," your will know that this is a functional assignment. It is important to note that both Divisions and Groups are flexible enough to be used in a simple single agency response or even in a large, complex, multijurisdictional response. The key to success in an incident that uses Divisions and Groups is to maintain proper coordination with those higher (and lower) on the hierarchical organizational chart. It should also be noted that another key to success is to ensure that there is interoperable communications when forming Divisions and Groups. Failing to ensure that communications are interoperable could lead to failure.

8.1.4.1 Divisions (Geographic)

The common method of organizing tactical operations under the Operations Section is for the Incident Commander (IC) to first establish two or more Divisions. As was mentioned in the previous paragraphs, Divisions always refer to geographically defined areas. These geographical areas, or Divisions, might be the area around a building, the inside or floors of a building, an open area, parts of an area, a city, a county, or any incident that can be divided geographically. To help remember that Divisions are a geographical designation, remember that Divisions literally divide up an area, hence the reason for calling it a Division.

If there are four floors in a building, for example, Division A may be chosen the first floor, and Division B the second floor, and so on. If dividing a building from the outside for fire or SWAT Operations, the front side of the building may be identified as Division A, while the remaining three sides may be each given the identifiers of Division B, Division C, and Division D. Initially, divisions may be established to define the geography of an incident and initially may (or may not) include the designation of separate Division Supervisors.

The Incident Commander (IC) or an Operations Section Chief (OSC) has the capability to assign separate supervisors for each Division, or to have one supervisor manage all of the Divisions as long as it does not exceed a manageable span of control. When a span of control is nearing its limits, the Incident Commander (IC) should be informed so that they, or the Operations Section Chief (OSC), can integrate another Division Supervisor into the response.

A division is located within the ICS method organizational chart between the Branch and the Task Force/Strike Team. Should the incident be large enough to implement a Branch, the Division Supervisor will report directly to the Branch Director. If there is no Branch, but there is an Operations Section, then the Division Supervisor will report directly to the Operations Section Chief (OSC). Finally, if there is no Operations Section (and no Branch), then the Division Supervisor will report to the Incident Commander (IC).

8.1.4.2 Groups (Functional: Jobs They Perform)

Another method commonly used is to form Groups. This form of modular organization abandons the geographical area concept and focuses on the job being undertaken. Types of groups that can be formed might include Search and Rescue Groups, HazMat Groups, Explosives

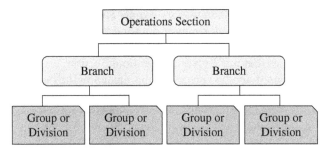

Figure 8.5 Combined divisions or groups.

Groups, Medical Groups, Fire Suppression Groups, and Perimeter Security Groups, just to name a few. Groups are similar to a Divisions because they are managed by supervisors, but they are referred to as Group Supervisors.

Also similar to a Division, the Incident Commander (IC) or an Operations Section Chief (OSC) has the capability to assign separate supervisors for each Group, or to have one supervisor manage all of the Groups, so long as it does not exceed the manageable span of control for the Group Supervisor. When a span of control is nearing its effective limits, the Incident Commander (IC) should be informed so that they or the Operations Section Chief (OSC) can integrate another Group Supervisor into the response.

A Group is located within the ICS method hierarchical organizational chart between the Branch and the Task Force/Strike Team and is identified in the hierarchical organizational chart by the function they perform. Should the incident be large enough to implement a Branch, the Group Supervisor will report directly to the Branch Director. If a Branch has not been activated, but there is an Operations Section, then the Group Supervisor will report directly to the Operations Section Chief (OSC). Finally, if there is no Operations Section (and no Branch), then the Group Supervisor will report to the Incident Commander ([IC]Figure 8.5).

There are times when the use of combined geographic Divisions and functional Groups will enhance a response or recovery. This utilization of resource is commonly used when a specialized functional job is needed to operate across Divisional lines. An example of this might be a specialized Wildlife Rescue Group that is dispatched to a major oil spill. As each division finds wildlife that has been negatively affected, or their life is in danger, they could request that the specialized team to cross geographical response lines, and that this Group should take care of the wildlife found by the Division. Another example might be the use of Federal Emergency Management Agency's (FEMA's) Urban Search and Rescue Team (USAR). The USAR Teams have a specialized equipment that can detect life in collapsed buildings. In incidents similar to Hurricane Katrina or a major earthquake, this specialized Group could cross divisional lines to assist multiple Divisions in finding live victims.

It is important to note that when combined Divisions and Groups are utilized, the Division Supervisors should sustain close communications and coordination. In the hierarchical organizational chart, both Division and Group Supervisors are equal. This means that each Supervisor will have equal authority, so coordination collaboration is a key element.

8.1.5 Single Resources

A Single Resource is defined as an individual, a piece of equipment and its personnel complement, or a crew/team of individuals with an identified work supervisor that can be used on an incident. Single resources are typically used on very small and easily manageable incidents as well as large complex incidents. In the case of small incidents, they typically require either very minimal outside resources or no outside resources at all. In the larger more complex incidents, there may be numerous single resources that undertake the operational and tactical work. The main difference between the two is one reports only to the Incident Commander (IC), while the other is one of many other single resources that make up the Operations Section.

When we speak of single resources, this is an individual piece of equipment and the personnel assigned to that equipment, or a traditional crew or team of individuals with an identified work supervisor that can be utilized on an incident. They will have common communications that they have typically used on a daily basis, so there will usually be no problems with interoperability when it is only a single agency.

Single resources from one agency in a rural fire department would be the entire department, while in an urban area, it may be a station or in some instances, a battalion. When speaking of law enforcement, in a smaller law enforcement agency, it will typically be the entire agency, while in a larger agency, it might be a division. Perhaps the best way to describe a single resource would be a group of individuals and their supervisor that work together on a daily basis. Anything beyond that would in most instances expand the response to a strike team or task force.

An example of a single resource might be three squad cars responding to a domestic dispute. All of the squad cars would come from the patrol division, and they would typically work together on a daily basis. If this reported domestic dispute turned into a larger incident, perhaps a hostage situation that required a state police hostage negotiator, then it would not be a single resource response.

It is important to note that in larger incidents, single resources will likely have problems communicating due to interoperability issues between organizations. This will hold especially true if communications are not preplanned, or at least addressed prior to their arrival. For this reason, single resources will need to ensure that they have a coordination, communication, and cooperation mindset prior to arriving at an incident.

8.1.6 Strike Team

When multiple units are assigned to complete the same tactic together, it is known as a Strike Team. Strike Teams are defined as a set number of resources of the same kind and type operating under a designated leader. These resources communicate and collaborate together representing a known capability, and they are typically highly effective at managing one small aspect of a larger incident. An easy way to remember what a Strike Team does is to realize that they all do the same job and they undertake a joint assault (or strike) against that specific problem.

An example of a Strike Teams would be six different patrol officers, all from differing agencies, working together to storm an active shooter in a mall. What makes them a Strike Team is that they are the same resource (law enforcement officers), with the same or similar capabilities. They collaborate, coordinate, and cooperate, acting as a single resource, to meet the needs of the incident. A similar example might be different fire engines from different jurisdictions who have the same (or similar) capabilities, and they are working together to attack the same fire in the same area.

A strike team may also be comprised of nonpublic safety personnel or a mix of public safety and nonpublic safety personnel. As an example, if a flood was threatening a town, the Incident Commander (IC) could use a mix of street department dump trucks, parks department dump trucks, and private business dump trucks to deliver sand and clay to areas where it can be used to build dykes, or where the community might be filling sandbags to hold back the flood. While the resources being used are provided by different businesses and agencies, they are all the same type of resources doing the same type of job. The term "Strike Team" is used to identify that it is all the exact type of resource.

8.1.7 Task Force

A Task Force is defined as any combination of different resources that are put together to accomplish a specific mission. A Task Force Leader manages those in a given specific task. This allows multiple and/or diverse key resource elements to be managed under one individual's supervision. A way to remember the Task Force concept is they are differing specialties undertaking a task.

An example might be a HazMat incident that involves a fire, and injured individuals that are alive and been trapped in the hot zone (the active area). A Task Force could be put together that utilized a small HazMat Team, firefighters, and/or an extrication team, and EMS personnel that were all protected by HazMat Suits. As a Task Force, each would have their own specialty, be that they would work together (as a team) to save lives and mitigate the incident.

Another way to look at a Task Force is to look at a major sporting event such as a National Association for Stock Car Auto Racing (NASCAR). In larger racetracks, it may be required to have multiple Divisions that cover different geographical areas around the track; however, it may be more advantageous to utilize a mix of individuals with mixed specialties to address any situation that may arise. Rather than dispersing individuals throughout the racetrack, several task forces could be positioned to respond within a given time and handle almost any type of emergency.

8.2 Conclusion

Operations will be the busiest Section in the ICS component of NIMS for almost every circumstance. In reality, the entire structure of ICS is designed to support the Operations Section. Operations Section is where the real work of mitigating the incident is being done, and it does not matter if you are an Incident Commander (IC) or at the hierarchical organizational chart in some other section, the work being done is to support operations.

As we went through this chapter, it was easy to see the modular design that is incorporated into ICS. Whenever a span of control is about to reach a limit, then expansion must occur. Because of the modular design, the user does not have to re-invent the wheel every time they must expand, it is already designed into the ICS method of incident management. They already know what expansions are available, and they place the expansion module that will assist managing the incident into place.

If we compare the modular design of the Operations Section to plug and play computer accessories, a viable analogy can be made. With a computer, you may want to give a presentation, back up your hard drive, add speakers, add a camera, add a microphone, to save a file on a thumb drive, or some other item to adapt your computer and make it more productive. In most instances, you only need to find what you need, and to plug it into a universal serial bus (USB) port. In most instances, it works just fine without doing a bunch of additional work. The same holds true for expanding the Operations Section of the ICS method. If you want to make the Operations Section more productive, you only need to find the resource you need and using the modular design of the ICS method, plug that resource into the area where it will make the response more productive. Just like a computer, the ICS method has protocols in place to make it work.

Real-Life Incident

Around 1990, in a rural Missouri Ozark Mountain area, a rural rescue squad was called out for a man that had a mental impairment lost in the woods around dusk. The man had gone mushroom hunting with his family and wondered away. After an initial search around the house, the new Lieutenant decided to initiate a call for mutual aid assistance from two different counties to help search over 10 000 acres of woods.

Using the newly learned Incident Command System, the new Lieutenant implemented a full Command Staff and a partial General Staff. Before crews even arrived on scene, the Planning Section Chief (PSC), the Logistic Section Chief (LSC), and the Incident Commander (IC) had a game plan in place. A staging area was set up, and the first wave of first responder's included state police, ambulance personnel, over 120 firefighters and rescue personnel, four dog teams, volunteers, and various others. Crews were divided according to their training, and Branches were set up with Divisions and Task Forces. These crews were given maps and flashlights and told which grid pattern to search. The first night the teams did not find the victim. As the media began arriving, the Public Information Officer (PIO) provided interviews and press releases that were approved by the Incident Commander (IC).

After six hours of searching the dense woods, all teams were called back to the Command Post (CP) and allowed to go home. Through the night, the Incident Commander (IC) through the Logistics Section Chief (LSC) lined up air support that would be flying at first light. In the morning, three aircraft searched grid patterns, but they did not find the victim. A second call went out for ground searching assistance from the same two counties later that morning. Once again, ICS was used to manage the crews, and within a few hours of daylight searching, the man was found deceased under a log. He had died of heart attack.

A few days later, a hot-wash was undertaken. A hot-wash evaluates the response to an incident to identifying what went right and what went wrong during an incident. In this hot-wash, the only area where improvement was suggested was a better integration of medical personnel. There was such a large contingency of medically trained personnel, that the various agencies felt that they at least one medically trained person should have been assigned to each team. This young Lieutenant who had only taken an ICS course a few weeks before quickly learned the value of ICS, and how the flexibility and modular organization helped to make a chaotic situation manageable.

Chapter 8 Quiz

1. Who does a Branch Director report to?

2. Fill in the chart below identifying the leader of each level, and what the support position is for each level.

Organizational Level	Title	Support Position
Incident Command		
Command Staff		
General Staff (Section)		
Branch		
Division/Group		
Unit		
Strike Team/Task Force		

Title	Support Position
(a) Director	(a) Deputy
(b) Officer	(b) Assistant
(c) Chief	(c) Deputy
(d) Incident Commander	(d) Not Applicable
(e) Supervisor	(e) Manager
(f) Leader	(f) Boss
(g) Leader	(g) Deputy

3. This is the organizational level that has functional, geographical, or jurisdictional responsibility for major parts of the incident operations. This level is organizationally between Section level and the Division/Group levels in the Operations Section.
   (a) Unit
   (b) Strike Team
   (c) Branch
   (d) Task Force

4. Sections are name by _____
   (a) Roman Numerals
   (b) Numerical values
   (c) Letters
   (d) Functional Names
   (e) Functional Area

5. The section that has direct responsibility for mitigating the incident and who is physically and directly involved in mitigating the incident is the _____ Section:
   (a) Operations
   (b) Logistics
   (c) Planning
   (d) Finance/Administration
   (e) Investigation and Intelligence Gathering

6. When more than one Branch is used, they are identified on the hierarchical organizational chart using _____ or _____.
   (a) Roman Numerals
   (b) Numerical values
   (c) Letters
   (d) Functional Names
   (e) Functional Area

7. When more than one Division is used, they are identified on the hierarchical organizational chart using _____.
   (a) Roman Numeral
   (b) Numerical values
   (c) Letters
   (d) Functional Names
   (e) Functional Area

8 When one or more Groups are used, they are identified on the hierarchical organizational chart using _____.
   (a) Roman Numeral
   (b) Numerical values
   (c) Letters
   (d) Functional Names
   (e) Functional Area

9 All Sections that are implemented in an incident have the primary task of supporting the _____.
   (a) Operations Section
   (b) Logistics Section
   (c) Planning Section
   (d) Finance/Administration Section
   (e) Investigation and Intelligence Gathering Section

10 Define Strike Team:

11 Define Task Force:

12 Define a Single Resource:

Self-Study

FEMA (2007). *All hazards Operations Section Chief task book* (2007). Retrieved from https://training.fema.gov/emiweb/is/icsresource/assets/tb_ops.pdf.

FEMA (2018a). *National Incident Management System (NIMS): Appendix B.* Retrieved from https://training.fema.gov/emiweb/is/icsresource/assets/tb_ops.pdf.

9

Expanding Logistics

The line between disorder and order lies in logistics.

Sun Tzu

The Logistics Section is responsible for providing facilities, services, and material support for the incident. As an incident becomes more complex, it begins to become more critical to have a well-functioning Logistics Section. The Logistics Section Chief (LSC) is the "go to" person for the Incident Commander (IC) and the Operations Section Chief (OSC). The Logistics Section is responsible for supporting the response, especially the operational and tactical response, by providing the resources and services required to make the response and recovery more effective. In doing so, the Logistics Section will regularly need to work with outside agencies.

It could be said that the Logistics Section is the lifeline for the incident. During ongoing operations, the Logistics Section has the duty of providing food, water, and housing to personnel. They are also responsible for ensuring that there are sufficient medical resources, and for procuring and organizing all communication equipment, computers, transportation, and any other item that may be needed to support the incident, for any Section that needs it.

9.1 Logistics Section Expansion

In the Logistics Section, the organizational flow chart will be much smaller than an Operations Section hierarchical flow chart. While it needs to be realized that there is no guarantee that the same resources will be needed for every incident, the typical needs of the incident when it comes to Logistics will include only two Branches with only three Units per Branch. The primary reason behind this is that logistics is not as labor intensive as operations, however, it is equally important toward the success of mitigating and ending the incident (Figure 9.1).

Just like the other General Staff positions, the Incident Commander (IC) may, at their choosing, decide to handle these issues themselves. This may be especially true in smaller incidents. In cases where the response to an incident expands, or even if it is a large incident from the onset, then the Incident Commander (IC) can appoint a Logistics Section Chief (LSC) to ensure logistical support is maintained. Once the Logistics Section Chief (LSC) has been appointed, they will decide if they can maintain the Logistics Section of their own accord, or if the structure of the Logistics Section should expand to include the Branches and/or Units. If the circumstance of the incident dictates that there should be an expansion (to ensure that all duties are given proper attention), they can easily accomplish that task with the modular structure that is built into the Incident Command System (ICS) method.

Emergency Incident Management Systems: Fundamentals and Applications, Second Edition.
Mark S. Warnick and Louis N. Molino Sr.
© 2020 John Wiley & Sons, Inc. Published 2020 by John Wiley & Sons, Inc.
Companion Website: www.wiley.com/go/Warnick/EIMS_2e

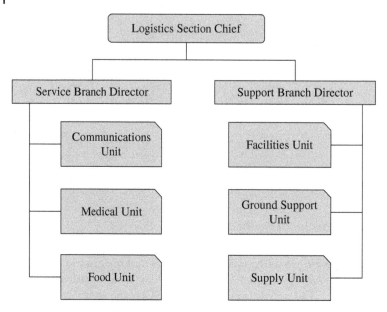

Figure 9.1 Modular construction of Logistics Section.

9.1.1 Logistics Branch Structure

In the typical emergency response after activating the Logistics Section, the ICS method (usually) utilizes only two branches in the Logistic Section when an expansion is needed. With the modular design of the Logistics Section, these Branches will include either a *Service Branch*, a *Support Branch*, or both. By dividing the Logistics Section into these two different Branches, it allows each Branch to focus on providing services or providing support, and it helps to reduce confusion of fluctuating between services and support. It helps those who are managing one of these two branches to stay focused on one mission within their job title. These Branches may or may not create Units that could be needed and that are based on the needs of mitigating the incident. If Units are not needed, then they should not be formed.

The Branch Director will oversee the Units under their command. If they find that there is a need for more individuals, they can have the Unit Leader (the proper name of those managing a Unit) and advise them to expand to meet the span of control. The Branch Director reports directly to the Logistics Section Chief (LSC) Table 9.1).

Table 9.1 Hierarchical command structure for operations.

Organizational element	Leaders is called	Examples
Incident Command	Incident Commander	IC
Section	Section Chief	Logistics Section Chief (LSC)
Branch	Branch Director	Branch III Director
Units	Leader	Medical Unit Leader

9.1.2 Support Branch

The Support Branch will provide provisions for tactical first responders. This Branch is tasked with obtaining the work materials and personnel needed to support the incident. A key responsibility of the Support Unit is to ensure that the needs of the Operations Section are met to fully support mitigation of the incident. If the Incident Action Plan (IAP) calls for individuals or equipment, the Support Branch will do their best to support those needs. This includes, but is not limited to, very specific personnel and very specific equipment. As an example, the Operations Section can request that an additional 20 personnel are needed to look for survivors, or they could ask for a specific Urban Search and Rescue (USAR) team such as California Task Force 7 based in Sacramento, CA, so they will send a formal request on behalf of the Operations Section or the Incident Commander (IC).

As a side note, Federal Emergency Management Agency (FEMA) currently recognizes 28 Urban Search and Rescue (USAR) teams in the United States. Because they follow the National Incident Management System (NIMS) method and the ICS companion method, they are easily integrated into a response with equipment and personnel being delivered.

When an incident needs to expand further and include Units, the hierarchical organization chart and ICS (under the Logistics Section) allow these Units to take on the responsibility they are given. The manager of a Unit is called a Unit Leader, and they report directly to the Branch (if activated), or the Logistics Section Chief (LSC) if Branches are not activated. The names of the Unit describe the task or responsibilities that they are assigned to do.

9.1.2.1 Facilities Unit (Support Branch)

The *Facilities Unit* follows the function of its name. The Facilities Unit is responsible for the layout and activation of incident facilities. The types of facilities that may be used in an incident include

- Incident Command Post (ICP)
- Incident Base
- Incident Camp
- Staging Area
- Incident Helibase
- Helispot
- Drop Points (FEMA, 2017a)

Without the Facilities Unit laying out and creating these facilities, the response to the incident could be greatly delayed and/or create conditions where response personnel are uncomfortable, or even miserable. When utilizing personnel, especially outside personnel, it is important to remember that everyone needs a place to sleep, food, water, and a place to clean up. Facilities provide services that help to provide more effective rest, better nutrition, improved hydration, and a place to relax so that personnel are in optimal condition for the duties they must perform.

An incident also needs a facility to stage resources for deployment. This is a location where resources are placed while awaiting assignment. This allows the supply of personnel and equipment to be highly effective and orderly. From the Staging Area, resources can be quickly deployed, precisely when they are needed. The same holds true for all of activated helicopter facilities. They also need a place to stage all of their supplies and to perform maintenance requirements. They need to be close enough to the incident to be more effective, but far enough away that they are not in danger. The same holds true for Helispots and Drop Points.

In regard to the ICS forms that the Facilities Unit will need to fill out, the primary form that they will be responsible for is the ICS Form 214-Activity Log. While there will be other ICS forms that may be used, especially for communication purposes, the Activity Log describes what the Unit has done and what facilities have been activated (FEMA, 2017a). To better understand how the facilities provided by the Logistics Section and the Facilities Unit help the response, a more in-depth discussion about each type of facility will be described later.

9.1.2.1.1 Incident Command Post (ICP)[1]

An Incident Command Post is the location where the primary multiorganizational command functions are executed. When a facility is needed, it should be the first facility at an incident because it is the primary place where the command functions are performed. While not mandatory, it can be colocated with the Incident Base, provided that some basic concerns are taken into considerations. It is important to note that there should only be one Incident Command Post (ICP) per incident, and unless absolutely necessary, it should not be relocated. Relocating the Incident Command Post (ICP) might add more chaos and confusion to an incident. The exception to that rule is when there is an Incident Command Post (ICP) transition from a temporary facility to a permanent facility.

At the initial onset of an incident, the Incident Command Post (ICP) is usually a temporary facility. It can typically be found at a fire truck, a patrol car, an ambulance, or some other type of vehicle. If the incident requires expansion, then a more permanent Incident Command Post (ICP) will be needed. More often than not, this facility will be a Mobile Command Center (MCC). Other types of Mobile Command Center (MCC) structures that might be included are tents (or shelters) or fixed structures (or buildings) that have been modified; there are some basic requirements that the Facilities Unit will have to take into consideration when planning for the Incident Command Post (ICP).

Because communications are extremely important to managing an incident, there will need to be sufficient communication at the Incident Command Post (ICP) that the command team can easily communicate with everyone. As the incident expands, the planning function is also usually undertaken at this location. For this reason, there must be sufficient room to accommodate these needs. In most instances, when activated, Command Staff (e.g. Safety Officer [SOFR], Liaison Officer [LOFR], and the Public Information Officer [PIO]) will be colocated with the Incident Commander (IC), at least in the beginning of an incident. In the event it is decided to use a Unified Command (UC), then there will also need to be sufficient space to accommodate all of these individuals. Additionally, Agency Representatives (ARs) might be stationed in, or near, the Incident Command Post (ICP). There must also be sufficient room to display Situation Reports (SitRep's) and status displays. The Incident Command Post (ICP) must be large enough to accommodate these tasks plus have essential amenities (e.g. latrines) to facilitate the needs of a potentially large quantity of individuals who might be working there. On larger incidents, and sometimes even on smaller incidents, the Incident Command Post (ICP) will need security to control access to the Incident Commander (IC) and those working with them. Unauthorized personnel such as the media, grieving families, or potentially disruptive individuals should not be able to gain unchallenged access to the Incident Command Post (ICP).

Some basic considerations for placing an Incident Command Post (ICP) include the following:

- Position it away from the noise and confusion of the incident
- Place it out of current and/or expected future hazard areas

---

[1] This heading is for advanced IMS students.

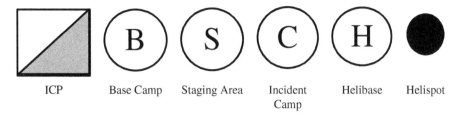

Figure 9.2 Map symbols for incident facilities.

- When possible, it should be in an area where operations can be observed
- It should be located away from tactical and operational personnel

The purpose of locating it away from tactical and operational personnel is to avoid the temptation of personnel jumping (going outside of) the Chain of Command. While jumping the Chain of Command may be innocent in its intent, if personnel are asking questions or getting direction from the Incident Commander (IC) rather than their immediate supervisor, it can add to chaos and confusion, and it can affect accountability. This holds especially true when in the initial stages of the incident.

It is also important to identify the Incident Command Post (ICP). This holds especially true when the Incident Commander (IC) requests outside technical experts, state and federal officials, or various other individuals who can provide knowledge and/or advice. Acceptable methods of identifying the Incident Command Post (ICP) in the United States include using a green and white flag, green flashing lights, or other identifiable markings. When identifying the Incident Command Post (ICP) on an incident map, it should be represented by a square with the bottom right corner triangle of the square filled out in black (see Figure 9.2).

9.1.2.1.2   Incident Base[1]

An Incident Base is the place where primary support activity is carried out, and it is usually a permanent location for the duration of the incident. The Incident Base will be where the Logistics Section operates from, and where it conducts business. The Incident Base must have ample room to facilitate sleeping arrangements, food processing, and a warehouse or temporary warehouse for supplies. In the initial stages of an incident, it can be a preliminary or temporary Staging Area (FEMA, 2017a). It is important to note that if inmate crews are utilized, the Incident Base will require a secure area to house and feed inmates.

There should only be one Incident Base per incident. These Incident Bases can often be identified in the planning stages. An easy way of finding locations in an Incident Base in the planning stages is to preidentify county fairgrounds, local campgrounds, or other large areas or facilities that can be utilized and that will have sufficient room to meet the needs of an incident. When identifying an Incident Base on an incident map, it should be identified by a capital B in a circle.

9.1.2.1.3   Incident Camp[1]

An Incident Camp is temporary, and it can be packed up and moved in a relatively short amount of time. At an Incident Camp, you can find most, if not all, of the services that are found at the Incident Base, The primary differences between the Incident Camp and the Incident Base is that the Incident Camp is usually a smaller version, and they are usually placed at the entrances, exits, and in safe areas substantially closer to the incident than the Incident Base. Incident Camps will provide personnel with critical services that are needed including sleeping, and sanitation facilities. These camps will be managed by the Facilities Unit, who will also oversee and manage the work to maintain the camp through a Base Camp Manager (FEMA, 2017a). The Base Camp Manager reports directly to the Facilities Unit Leader and is responsible for ensuring that there is sufficient security, cleaning services, and general maintenance to keep the camp fully operational. The Base Camp Manager will also ensure that all of the needed personal necessities for personnel are available.

Incident Camps are typically named for their geographical location (e.g. Camp Opossum Ridge, Camp Niagara Falls, Camp Helena [this name is based on the city name] or by numbering them). In some instances, it may be beneficial to place Incident Camp with, or near, Staging Areas. This will hold especially true when there are limited locations where a camp can be set up that may present parking difficulties. Additional benefits of placing an Incident Camp and a Staging Area together is that they can share services (e.g. portable bathrooms, feeding facilities, and shower facilities), which will usually lead to reduced costs and the use of fewer outside resources. Incident Camps are identified on an incident map with a circle and a "C" in the circle.

9.1.2.1.4 Staging Area[1]

Staging Areas are an integral part of the response, and they play a critical role in overcoming the challenges of an incident. Staging Areas are where all available resources will report before they are assigned. It is acceptable to have multiple Staging Areas in different geographic locations, but the majority of the Staging Areas should be within a close proximity to the incident, while still being out of danger.

Locating the Staging Area in close proximity to the incident is done so that if tactical crews need assistance, it is only a short distance away. If need be, Staging Areas can be relocated as the incident expands or contracts. It is important to realize that Staging Area need to be able to respond, and arrive to their assignment, in a relatively short amount of time.

Staging Area's help prevent freelancing of equipment on an incident, and they minimize communication needs because the Staging Manager can direct resources face to face. Additionally, these Staging Areas help prevent the incident being charged for services that were not ordered. This is done by using an ICS Form 260-1 (Resource Order Form) and as needed ICS Form 260-2 (Resource Order Continuation Form). In filling out these forms, the resource is given a specific identification number that they must provide when they arrive at the incident.

The Staging Manager works closely with the Incident Commander (IC) and sometimes the Operations Section Chief (OSC) by advising them how many, and the type of resources that are available for deployment from the Staging Area to tactical positions. In order to reduce blockage, most Staging Areas are set up with different routes for incoming and outgoing resources. In an ideal situation, the Staging Area will have security controls at the entrances and exits, as well as throughout the area to protect the equipment and personnel. Sanitation provisions should also be incorporated into the Staging Area and considerations should be given to either on-site fuel and/or fuel delivery to ensure that equipment is in optimal readiness.

9.1.2.1.5 Helibase[2]

A Helibase is a Landing Zone (LZ) where the helicopters are fueled, serviced, and/or lightly repaired, and where they are parked when not in the air. These Helibase's are also locations where supplies can be retrieved, equipment and personnel can be offloaded, onloaded, or switched out. The Helibase is also a location that fire retardant can be mixed and loaded, and aircraft maintenance and inspection records can be preserved. These Helibases can be located in a field or parking lot (near the incident), or they can be located, or colocated, at an airport. When planning for a Helibase, it should be remembered that these facilities are normally placed in areas away from the flight paths of fixed-wing aircraft. This is done to reduce the chance of an air accident.

---

2 This title signifies highly technical information for advanced students and practitioners.

Helibases can be utilized by government helicopters (e.g. fire department, police department, and FEMA), medevac helicopters, contractual (firefighting) helicopters, or military helicopters. It should be noted that in most instances, military and the National Guard prefer, and will sometimes mandate, that they will only operate from an airport. As an alternative in this type of instance, it is allowable to have multiple Helibases; the military helicopters can be sent to an airport, while other helicopters use remote Helibases.

Individuals who will be in charge of various duties at the Helibase for firefighting incidents can include a Helibase Manager, a Deck Coordinator, a Parking Tender, a Loadmaster, a Helicopter Crew Member, an Aircraft Base Radio Operator, a Take Off and Landing Coordinator, a Mix Master (for mixing fire retardant), and a Helicopter Coordinator. For law enforcement, the Helibase is staffed similarly; however, it is a requirement that firefighting crews and equipment must be present at all Helibases to provide fire protection and crash rescue services for all of these facilities (National Wildfire Coordination Group [NWCG], 2016). The Helibases will be identified on the incident map by a circle with an "H" in the middle.

9.1.2.1.6 Helispot[2]

Helispots are temporary locations that have been deemed safe for helicopters to land and take off. The Helispot Managers are usually required to be on-site to coordinate the landings and take offs, loading and unloading of equipment, and loading and unloading of personnel. These temporary locations are sometimes needed as an incident grows, and they can be located in parking lots, fields, and other large clearings. Helispots are found on the incident map by a black circle with the letters "H" and a designated number outside of the circle. The number is usually in the order that this location was created (e.g. H-1 [first Helispot], H-2 [second Helispot]). It is important to note that once a number for a Helispot is used, it should not be reused at a different location. This is done to avoid confusion. As an example, if H-2 was shut down, only that location should use that identifier if it reopens, but a different location should not use that same number for the duration of the incident.

Personnel that will be used at a Helispot include a Helicopter Crew member, a Helispot Manager, and a Base Radio Operator. Potential crew members could include a Deck Coordinator, a Parking Tender, a Loadmaster, a Helicopter Crew Member, a Take Off and Landing Coordinator, a Mix Master (for mixing fire retardant), and a Helicopter Coordinator (NWCG, 2016).

9.1.2.2 Ground Support Unit (Support Branch)

The Ground Support Unit follows the function of its name, and it provides support to those who are working not only in the Operations Section but also anything that entails ground response. The Ground Response Unit will provide ground support by procuring the materials needed to effectuate a practical response. These items could be anything from the tools and equipment needed by those working under the Operations Section, to providing transportation. This Ground Unit is also responsible for fueling, maintaining, and repairing of vehicles, and they are responsible for the transportation of supplies and the previously mention personnel.

Initially, the Ground Support Unit will work in coordination with the Logistic Section Chief (LSC) and the Service Branch Director to determine the needs of the incident. The Unit will organize and assign personnel in the most effective way to support operations, based on the unique needs of each incident. They will also be required to determine if the resources are sufficient to maintain ground support as the incident expands. A unique duty of the Ground Support Unit is to monitor what supplies are on hand and to identify supplies that are unavailable but still needed. Whenever the issue of unavailable equipment arises, it is the responsibility of the Ground Support Unit to resolve that issue. It is also the responsibility of the Ground Support Unit to monitor work progress and inform the Logistics Section Chief (LSC) of their activities. The Ground Support Unit will need to report their activities using the ICS Form 214-Activity Log.

9.1.2.3 Supply Unit (Support Branch)[1]

The Supply Unit also follows the function of its name. The Supply Unit can be the best friend, or the worst nightmare, of the Incident Commander (IC) and the Operations Section Chief (OSC). This Unit is responsible for ordering personnel, equipment, and supplies. If a Supply Unit does not do their job well, then there will likely be shortages of personnel, equipment, and/or supplies that are needed to mitigate the negative effects of an incident. By the same token, if the Supply Unit is overly aggressive in ordering resources, a much greater cost can be incurred. This is especially true in the case of personnel who have been ordered but are not deployed, and it can have a negative effect on morale. If morale problems arise, it can help to undermine the response.

The Supply Unit is also tasked with receiving and storing all supplies for the incident. They will maintain an inventory of the equipment that will likely be needed. Maintaining the equipment begins when this equipment is delivered, and prior to accepting the supplies from delivery personnel. They will need to ensure that the supplies are received in good working order, and that they are the supplies that were ordered. This is important because sometimes, unscrupulous suppliers may send inferior supplies, or they may send more supplies than were ordered (Illinois Terrorism Task Force, 2011).

The Supply Unit is also responsible for nondisposable supplies. Nondisposable supplies can include fire apparatus, Special Weapons and Tactics (SWAT)-team armored vehicles, heavy equipment, durable medical equipment, mobile command Units, and more. The supply unit is tasked with servicing, and if need be, repairing these supplies and equipment. If they fail to do their job properly, those in operations could be put at risk because of equipment failures. Like the other Units in the Logistics Section, the Support Unit will be required to fill out the ICS Form 214-Activity Log.

9.1.3 Service Branch

The Service Branch will provide key services that tactical first responders may need during the response. To put it simply, those responding to an incident, especially long-term incidents, will need to eat, they may need medical attention, and they will need communications. The Service Branch of the Logistics Section is responsible for providing these services.

An easy way to differentiate the Service Branch and the Supply Branch is to remember that the Service Branch works directly with the operational personnel by personally providing the services they need while the Supply Branch will supply the materials to meet those needs. It is important to realize the differences, especially if you are in any type of supervisory role while operating under the ICS method. By knowing which Unit provides services and which Unit provides supplies, you will be able to save time and prevent frustration by working with the proper Unit the first time, rather than being told you submitted requests to the wrong Unit.

With the modular design of the ICS method of incident management, there are typically three Units under the control of the Service Branch. These three Units will be the Communications Units, the Medical Unit, and the Food Unit. Each of these Units supervise various tasks that are a service to personnel and that ensure that the needs of all personnel are met (Figure 9.3).

9.1.3.1 Food Unit (Service Branch)[1]

As is common with most Units used in the ICS method, the Food Unit follows the function of its name, and when activated, they provide a critical service. The task of planning for supplying food and water is a critical need, and it is an important task that should be undertaken long before an incident. If advanced planning is undertaken, those in a supervisory position working at an incident can evaluate if the incident will last more than two hours, and if so, they can begin activating for those needs several hours before it becomes critical. With preplanning feeding personnel, it provides a roadmap that shows what resources have been preidentified and

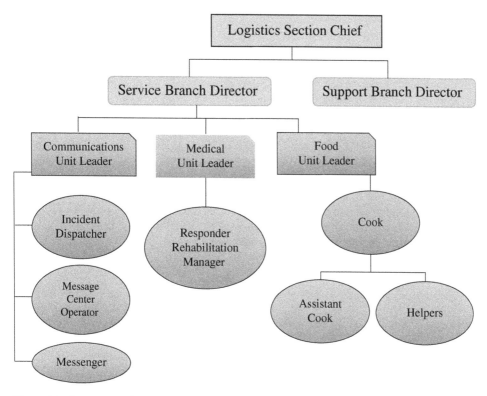

Figure 9.3 Organizational structure of the Service Branch.

contacted, and it should provide a common lead time they need before being able to provide the food.

Hydration is critical when working on an emergency incident, and it will usually need to be supplied in the first hour. The best fluid for hydration is plain and simple water. If it is especially warm, and personnel are excessively perspiring, it might be a good idea to provide sports drinks that will replenish electrolytes. Energy drinks should be avoided if the workers are operating in a stressful incident.

Providing some kind of snack, such as protein bars, protein type snack packs, or snack sandwiches, will typically be needed in the first two to three hours. Because of the energy needs of most personnel, it is best to provide protein or a mix of protein and carbohydrate snacks to provide energy and reduce instances of hypoglycemia (low blood sugar). While snack cakes are not optimal, they will work if no other snacks are available. A meal will be needed in four hours, which should be a somewhat balanced diet. While the local burger shop can provide the first meal, subsequent meals will need to be more balanced to prevent health issues.

On incidents that will last longer than four hours, the Food Unit will need to plan for nutritious meals. They will have to make sure that there is also a variety of options to meet the needs of those with food allergies and religious beliefs. If the incident will require more than eight hours on scene, then there will be a need to supply breakfast, lunch, and dinner meals. High-protein snacks may also be considered for those that may need them (between meals) when working in forward positions. In some instances, depending on the distance of the incident to the facility, the Food Unit may have to pack lunches to go for some of the operational personnel.

The Food Services Unit supplies the food and water for all incident facilities (and personnel). They also supply the necessary equipment and supplies to operate food service facilities at Bases and Camps. The work they do ensures that the nutritional needs of personnel working in austere conditions are met for the work they are doing.

The Food Unit works with the Planning Section (Resources Unit) to anticipate the number of personnel to be fed and to develop plans for supplying food to all incident areas. The Food Unit also interacts with other Logistics Units to locate fixed-feeding sites and when needed, providing catering services to those sites. They are also responsible for anticipating the number of individuals that will be fed at a given meal and ensure the ordering of sufficient food supplies, including for those with special dietary needs. They will also be responsible for transporting food to remote locations.

Whenever feeding areas are set up at fixed locations, the Food Unit is responsible for ensuring that there are sufficient tables, chairs, cooking facilities, menu planning, preparation of food, providing potable water, and providing general maintenance to all official food service areas. When contracting outside caterers to supply food services, the Food Unit will work with the Finance Section to negotiate the contract for these services related to feeding. As with most Units, they must maintain an ICS Form 214.

9.1.3.1.1 Cook[2]

While this, and the following jobs, sound like they are led by a singular person, there can be a multitude of cooks, but all will be identified in the singular. The cook is responsible for cooking food for personnel that are working at the incident. In most instances, cooks will not wander between different areas; they will usually be assigned to one cooking and/or feeding area. The cook may be a contracted caterer, they may be a cook that has been hired for an incident, or they may be someone that is a first responder with experience in cooking and has been given this job.

Cooks are tasked with ensuring that there is sufficient cooked food and snacks that meet the nutritional needs of personnel working on the incident. The cook will also ensure that the cooking facilities meet health standards and that eating areas are cleaned and maintained. They will also ensure that the proper and sufficient utensils (and accessories) are provided. The cook is responsible for supervising assistant cooks and cook helpers as well as maintaining documentation of all feeding activities (ICS Form 214-Activity Log).

9.1.3.1.2 Assistant Cooks[2]

Assistant cooks assist the cook in preparing food for personnel assigned to the incident. In the absence of the cook, the assistant cook will take over the responsibilities of the cook, including but not limited to preparing and serving food and supervising helpers. They will also work to ensure that sanitary conditions are met at all times.

9.1.3.1.3 Cook Helpers[2]

Cook helpers in the food Unit are responsible for assisting the cook and assistant cook in the location for which they are assigned. They will do such tasks as peeling potatoes, chopping vegetables, and the preparation of food ingredients to take some of the stress of the cooks and assistant cooks. They may also responsible for cleaning and supplying eating areas, cooking areas, and the utensils used to cook the food.

9.1.3.2 Medical Unit (Service Branch)[1]

Due to the importance of this position to life safety, the Medical Unit will be covered more in-depth than most other Units. Having the correct people in the Medical Unit can be the difference between life and death. As has been mentioned numerous times in this book, life safety is the number one priority (and should be). While the verbiage that has been written about in this chapter will be helpful, it is strongly encouraged that those who may be planning for larger incidents should seek out additional resources, and they should cultivate individuals who can fulfill these roles.

Much like other Units, the Medical Unit also follows the function of the name. It is primarily responsible for the development of the incidents Medical Plan (ICS Form 206), which specifies and provides well-thought-out information on the best locations to place the incident medical aid stations. ICS Form 206 will also identify what transportation services are available, and it will also provide the location of hospitals, and the medical emergency procedures that address various medical emergencies. This Unit in its entirety will also be responsible for procuring and ensuring that medical aid and transportation is at optimal performance so they can help personnel that have been injured or become ill.

The person that will lead the Medical Unit is called a leader. The Medical Unit Leader (MEDL) will be required to create and develop the Medical Plan and to assign the resources from this Unit to the locations where they are needed. The Medical Plan will become part of the Incident Action Plan (IAP) so that in the event that a medical emergency occurs, all personnel will know how to activate emergency medical protocols and request the needed help. When creating the Medical Plan, the Medical Unit Leader (MEDL) should ensure that detailed information about medical assistance capabilities and locations (at the incident), any potential hazard areas (or conditions), the exact location of off-site medical assistance facilities, and the procedures for handling all complex medical emergencies that may arise.

To be considered to fill Medical Unit Leader (MEDL) position, the individual must meet strict educational and experience requirements. While the type of response (e.g. natural disaster, manmade disaster, and wildfire) typically does not make a difference in the requirements and experience, the individual that is appointed to lead this Unit should be the best candidate for the type of incident. According to the National Wildfire Coordinating Group ([NWCG], 2016), the requirements included for wildland firefighting Medical Unit Leader were the following:

- Introduction to ICS (ICS-100)
- ICS for Single Resources and Initial Action Incidents (ICS-200)
- Intermediate ICS for Expanding Incidents (ICS-300)
- NIMS: An Introduction (IS-700)
- National Response Framework: An Introduction (IS-800)
- Medical Unit Leader (S-359)
- Current Certification as an Emergency Medical Technician (EMT) or Equivalent
- Completion and Certification of NWCG Medical Unit Leader (MEDL) Position Task Book (NWCG, 2016)

While other recommendations are given for training (Basic Wildland Fire Orientation and Fireline Leadership), the requirements listed above are the minimum requirements that should be accepted for the purpose of wildland firefighting. In nonwildland firefighting situations, these guidelines can also be useful in other types of incidents, however, it should be remembered that the NIMS method also provides a list of credentialing standards that should be met to fill a position, especially a task as important as the Medical Unit Leader (MEDL).

Two things are important to note when looking at these requirements. The first is related to the Emergency Medical Technician (EMT) certification. In most instances, this certification must be a professional license or certification by the state that the incident is geographically located in, or they must be a National Registry of Emergency Medical Technician (NREMT) and the state must recognize the National Registry of Emergency Medical Technician (NREMT) certification. It should also be noted that this requirement can be waived by the state in emergency situations or in some states, there is the concept known as reciprocity.

Reciprocity is essentially the concept that the states agree that they will accept each other's credentialing. Personnel credentialed in either state can legally work in both states, provided that they do not operate outside of the scope of duty or exceed the limitations of the license. While this is an oversimplification, there are many different scopes of practice between the states.

The second important issue that should be pointed out is that the training and experience listed is geared toward wildland firefighting. While someone with these qualifications could likely fulfill the requirements in a terrorist attack in an urban setting, we must ask ourselves will they be the best person for the job? It is important for the Authority's Having Jurisdiction (AHJ) to develop multiple people who can fill the Medical Unit Leader (MEDL) position (as well as other positions) using the All Hazards Approach, whereby they plan for the risks that are most likely.

The Medical Unit Leader (MEDL) will assign individuals to Fireline Emergency Medical Technicians (FEMTs) and Responder Rehabilitation Unit. The Medical Unit Leader (MEDL) is also tasked with the preparation of reports and records that are given to the Documentation Unit which is under the supervision of the Planning and Intelligence Chief. Depending on the local laws, the Medical Unit Leader (MEDL) may also be required to submit local and/or state documentation (NWCG, 2016).

The Medical Unit will provide assistance to the Finance/Administration Section as needed and/or required. There will most likely be administrative requirements related to injury compensation for those that are injured while working on the incident. This may include getting written authorizations, bills related to an injury, witness statements, administrative medical documents, and reimbursements as required (NWCG, 2016). The Medical Unit should do everything possible to ensure patient privacy.

9.1.3.2.1 Fireline Emergency Medical Technician[2]

A unique situation arises with the Medical Unit during wildland firefighting situations. When wildland fires occur, Fireline Emergency Medical Technicians (FEMTs) are assigned to the Medical Unit, but they will be assigned temporarily to a tactical supervisor in the Operations Section when they are needed. The Fireline Emergency Medical Technician (FEMT) is placed in forward positions (at an incident) to provide emergency medical care to personnel that are fighting the fire. The reason for this assignment is to provide the fastest possible response to those working in austere conditions (FIRESCOPE, 2014). The number of Fireline Emergency Medical Technicians (FEMTs) that may be assigned to an incident will be dependent on the intricacy, extent, and risks involved with a fire incident.

The Fireline Emergency Medical Technician (FEMT) initially reports to the Medical Unit Leader (MEDL) or the Logistics Section Chief (LSC) if there is no Medical Unit in place at that time. The Fireline Emergency Medical Technicians (FEMTs) will receive an assignment to respond to the location of the tactical team for which they are assigned. The Medical Unit Leader will specify if the Fireline Emergency Medical Technicians (FEMTs) should report to the assignment with, or without, equipment (FIRESCOPE, 2014).

The equipment that may be needed include medical equipment to treat injured personnel and the appropriate strobe light and tape to cordon off a landing zone for an air ambulance (should it be needed). The Fireline Emergency Medical Technician (FEMT) is responsible for bringing their own Personal Protective Equipment (PPE) including a helmet, a Nomex hood, a firefighting protective jacket, firefighting protective pants, and boots that meet National Fire Protection Association (NFPA) standards for wildland firefighters. Depending on the geographic location, and local policy, they may be even be required to have a wildland firefighting smoke filtration mask (FIRESCOPE, 2014).

To qualify for the position of an FEMT, the individual must meet the minimum standard of being a certified/licensed Emergency Medical Technician Intermediate (EMT-I) level in good standing. The Fireline Emergency Medical Technician (FEMT) may also have the advanced certification of an Emergency Medical Technician Paramedic (EMT-P). They must meet all competencies and standards affiliated with their certification, have a current licensure, and they must be physically capable of fulfilling the role of a Fireline Emergency Medical Technician (FEMT). They must also be legally able to function within their scope of practice, regardless of jurisdiction or political boundaries. They must be certified/licensed by the state or by a licensed National Registry of Emergency Medical Technicians (NREMTs) operating in a state that recognizes that licensure.

Upon being activated, the Fireline Emergency Medical Technician (FEMT) will be teamed up with another Fireline Emergency Medical Technician (FEMT), and both will obtain a briefing from the Logistics Section Chief (LSC) or Medical Unit Leader (MEDL). In that they will receive a Situation Report (often called a "SitRep") that includes

- Current situation status
- An analysis of the Medical Plan including the identification of priorities
- The incident communications channels they should use
- An overview of the Fireline Emergency Medical Technician (FEMT) assignment and the potential hazards that the line personnel may face
- The projected incident medical needs that may arise
- What the local medical protocols are including applicable documentation procedures (FIRESCOPE, 2014).

This report is critical in helping these Fireline Emergency Medical Technicians (FEMTs) to understand the unique circumstances they may face at the incident and how to call for assistance. Upon completion of the Situation Report (SitRep), they will receive their assignment and will either be briefed, or they will need to assess the current situation within this assignment, including procuring the needed equipment. At a minimum, they should determine from this briefing or their assessment:

- The number of personnel in assigned area
- The current and expected fire behavior
- The predicted weather conditions
- The terrain where they will be working
- Any natural hazards they and firefighting personnel may encounter
- Any safety alerts
- An evaluation of the most likely medical needs (FIRESCOPE, 2014)

Upon completion of the briefing/assessment, the Fireline Emergency Medical Technician (FEMT) should gather the required medical supplies. While obtaining the medical supplies for this assignment, the Fireline Emergency Medical Technician (FEMT) should use the medical supply list to make sure that no supplies are forgotten.

Rarely will the Fireline Emergency Medical Technician (FEMT) be asked to bring the supplies with them. After gathering the needed medical supplies, the Fireline Emergency Medical Technician (FEMT) should obtain copies of local Emergency Medical Service (EMS) forms and/or any other paperwork that may be available. If no paperwork or forms are available from the local area, then the Fireline Emergency Medical Technician (FEMT) should use their own jurisdictions or agency EMS forms and check their personal inventory to ensure they have the needed supplies to be safe and have all the needed equipment for an extended stay (FIRESCOPE, 2014).

Before leaving for their assignment area, the Fireline Emergency Medical Technician (FEMT) should check out a portable radio from the Accountable Property Section. If the ICS structure has not expanded to include an Accountable Property, they will need to go to obtain these from the Logistics Section. Whomever they receive the radio from will also identify which radio frequencies they should operate on in all types of situations (FIRESCOPE, 2014).

If the incident is a multiday incident, prior to each shift, the Fireline Emergency Medical Technician (FEMT) will obtain a copy of the Incident Action Plan (IAP) and check the Medical Supply Kit as well as their own personal inventory to ensure the maintain a state of readiness. Both in the initial deployment and the ongoing deployment, the Fireline Emergency Medical Technician (FEMT) should be assigned and keep in contact with their assigned tactical supervisor. During that contact, and prior to leaving the check-in location, they will confirm their travel route, their mode and type of transportation, and an estimated time of arrival (FIRESCOPE, 2014).

Fireline Emergency Medical Technician-Paramedic (FEMP) Medical Supplies	Fireline Emergency Medical Technician-Paramedic (FEMP) Personal Inventory
- Airway, Nasopharyngeal Kit (1) - Airway, Oropharyngeal Kit (1) - Bag Valve Mask (1) - Bandage, Sterile 4 × 4 (6) - Bandage, Triangular (2) - Biohazard Bag (2) - Blanket, Space (2) - Cervical Collar, Adjustable (1) - Coban Wraps/Ace Bandage (2 each) - Cold Pack (3) - Dextrose Oral (1) - Dressing, Multi-Trauma (4) - Eye Wash (1 bottle) - Gloves, Exam (box) - Kerlix, Kling, 4.5, Sterile (2) - Mask, Face, Disposable w/eye Shield (1) - Pad, Writing (1) - Pen and Pencil (1 each) - Pen Light (1) - Petroleum Dressing (2) - Scissors, Medic (1) - Sheet, Burn or Equivalent (2) - Sphygmomanometer (1) - Splint, Moldable (1) - Splinter Kit (1) - Stethoscope (1) - Suction, Manual Device (1) - Tape, 1 in., Cloth (2 rolls) - Thermometer, Digital (1) - Tourniquet, Commercially Available (2) - Triage Tags (6) (FIRESCOPE, 2014, p. 8)	- Beacon Strobe, National Fire Equipment System (NFES) 0298 (Incident Cache) - Camp Shoes - Cellular Phone w/DC Adaptor - Chap Stick - Clamshell with Extra Batteries - Cloneable Portable Radio (Required) - Cold Weather Gear - Compass, Silva-Ranger Type - Dispatch Printout (Order, Request Numbers) - Duct Tape, Roll - Ear Plugs - EMS Credentials (Licenses, Certificates) - Fire Starter - Flagging Tape Fluorescent (1 roll) - Food Rations - Glow-Stick (2) - GPS - Hand Tool - Head Lamp - Insect Swabs - Mid-Heavy Weight Hiking Over-Socks - Mini-Binoculars (Optional) - Moleskin - Multi-Tool (Optional) - Nylon Blister Proof Socks - Personal Medications (Tylenol, etc.) - Poison Oak Prophylaxis and Treatment - PPE, Wildland, Web Gear, Full (Required) - Signal Mirror - Sleeping Bag - Sleeping Pad - Sunglasses - Sunscreen - Tent - Toiletries - Topo Maps - Weather Kit, Belt (Optional, Available at Incident Cache) - Whistle (FIRESCOPE, 2014, p. 9)

Upon arrival at their assignment, the Fireline Emergency Medical Technician (FEMT) should meet with their assigned tactical supervisor and obtain a briefing on that day's operations. If they are relieving another Fireline Emergency Medical Technician (FEMT), they should also meet with the technician being relieved to receive a separate briefing. This is done so that the transfer of pertinent information related to the position is comprehensive. When the briefings are complete, they should perform a radio test with their assigned tactical supervisor, the incident Communications Unit, and the Medical Unit if one has been established (FIRESCOPE, 2014).

Throughout their operational period, the Fireline Emergency Medical Technicians (FEMTs) should maintain contact and interact with personnel that are affiliated with their assignment (FIRESCOPE, 2014). During this maintained contact, they should assess the physical condition and medical needs of all of the personnel they are responsible for, and they will evaluate if assistance is needed. If assistance is needed, they will provide the required aid. When a higher level of medical aid is needed, the Fireline Emergency Medical Technician (FEMT) will make requests for transportation through the channels identified in the Medical Plan (ICS Form 206). This is done in this way to effectively and efficiently transport personnel that have fallen ill, or that has become injured. Once the care of the patient has been transferred, the Fireline Emergency Medical Technician (FEMT) will notify the proper individuals within ICS as is documented in the Medical Plan (ICS Form 206) for incident-related illnesses or injury occurrences (FIRESCOPE, 2014).

After each operational period, the Fireline Emergency Medical Technician (FEMT) should instruct the tactical supervisor that they have concluded that operational period before leaving their post. They will also advise if another Fireline Emergency Medical Technician (FEMT) was relieving them before departing. By notifying them that another Fireline Emergency Medical Technician (FEMT) is replacing them, they officially transfer the responsibility of the job to another individual, thereby relieving themselves of any responsibility. Upon leaving, they will immediately report to the Medical Unit Leader (MEDL) for debriefing and to tender all patient care documents (FIRESCOPE, 2014). Other documentation that will need to be filled out may vary from incident to incident and from state to state.

It should be noted that there is also the potential that the name (Fireline Emergency Medical Technician) could be changed to another type of disaster and used in this modular and flexible design. It could be changed to Flood Zone Emergency Medical Technician, or the Collapse Zone Emergency Medical Technician, or some other name. The concept would remain the same, providing Emergency Medical Technicians in close proximity to operations so that there is not a delay in response.

9.1.3.2.2 Responder Rehabilitation Unit[2]

The Responder Rehabilitation Unit is responsible for ensuring the health and well-being of incident personnel who might suffer from the effects of strenuous work or austere and extreme conditions (United States Fire Administration [USFA], 2008). The Responder Rehabilitation Unit reports to the Medical Unit. The Rehabilitation Unit Leader will determine and assign responder rehabilitation sites. They will also assign personnel that fit the needs of the incident in areas that are easily accessible to personnel. These locations will have the exact site broadcast over the radio using the radio name of "Rehab." As the Rehab Team determines the needs of personnel that come to the Unit, they may request medical personnel to assess the medical condition of those being rehabilitated (USFA, 2008).

The Rehab Team can request certain resources such as water, juice, and snacks that are needed for the rehabilitation of personnel (USFA, 2008). They will make those requests to the Food Unit or Logistics Section Chief (LSC) if it is deemed necessary. Once that personnel have recuperated, the Rehab Unit will release those personnel to the Planning Section for reassignment.

Throughout this process, the Rehab Unit will document the actions taken and the responder that was treated. That information will be given to the Unit Leader as required (USFA, 2008).

The documentation that the Responder Rehabilitation Unit will be required to file is the ICS Form 214, the Unit Activity Log (USFA, 2008). It is however important to note that the requirements for additional paperwork in the Responder Rehabilitation Unit may be required by the Authority Having Jurisdiction (AHJ). This is typically because those fulfilling the duties of this position will be evaluating individuals that are working in austere conditions, and the Authority Having Jurisdiction (AHJ) may determine that additional documentation may reduce their liability (USFA, 2008).

9.1.3.3 Communications Unit (Service Branch)[1]

The Communication Unit also follows the function of the name, and they are responsible for facilitating communication. The Communication Unit Leader is responsible supervising the Incident Dispatcher, the Message Center Operator, and Messengers. The Unit as a whole is also responsible for developing plans that maximize the effective use of incident communications equipment, and all of the facilities that support communications (e.g. repeater towers, dispatch centers, and mobile communications vehicles).

The Communications Unit is tasked with the responsibility of installing (and testing) all communications equipment related to the incident, and this Unit is also responsible for managing the Incident Communications Center. The Communication Unit is tasked with the assignment and distribution of communications equipment to incident personnel, as well as the maintenance and repair of this communications equipment. Depending on the requirements of the Authority Having Jurisdiction (AHJ), radio equipment may be handed out through Accountable Property or the Communications Unit (United States Coast Guard [USCG], 2014).

The person in charge of the Communications Unit will be identified as the Communications Unit Leader (COML). They will be responsible for creating and developing plans for the effective use of incident communications equipment and facilities, ensuring the installation and testing of communications equipment is performed, and ensuring that the Incident Communications Center operable and effective. If they are tasked with the responsibility, they will also be responsible for documenting distributed communications equipment to incident personnel is accounted for. They are also responsible for ensuring that communications equipment is quickly repaired and maintained throughout the duration of the incident (USCG, 2014).

One important form that the Communication Unit Leader ([COML] and their staff) will be responsible for is the ICS Form 205-Incident Radio Communications Plan. This plan is important to ensuring that communications are meeting the needs of the incident. Because communications are so important to incident management, the Communications Unit is involved in filling out and maintaining multiple other ICS Forms. Depending on the type of incident, they may be required to assist others in the ICS method with filling out forms that have a communications component. According to the US Coast Guard USCG (2014) Communications Unit Leader ICS Job Aid book, the communications Unit may be required to fill out or assist with filling out the following forms:

- ICS 202-Incident Objectives
- ICS 202A-Command Direction
- ICS 202B-Critical Information Requirements
- ICS 203-Organization Assignment List
- ICS 204-Assignment List
- ICS 204a-Assignment List Attachment
- ICS 205-Incident Radio Communications Plan

- ICS 205a-Communications List
- ICS 206-Medical Plan
- ICS 207-Incident Organization Chart
- ICS 208-Site Safety and Health Plan
- ICS 209-Incident Status Summary
- ICS 210-Status Change Card
- ICS 211-Check-in List
- ICS 213-General Message
- ICS 213RR-Resource Request Message
- ICS 214-Unit Log
- ICS 215-Operational Planning Worksheet
- ICS 215A-Incident Action Plan Safety Analysis
- ICS 217A-Communications Resource Availability Worksheet
- ICS 218-Support Vehicle/Vessel Inventory
- ICS 219-Resource Status Card
- ICS 220-Air Operations Summary
- ICS 221-Demobilization Check-out
- ICS 225-Incident Personnel Performance Rating
- ICS 230-Daily Meeting Schedule
- ICS 232-Resources at Risk Summary
- ICS 232a-Area contingency Plan Site Index
- ICS 233-Incident Open Action Tracker
- ICS 234-Work Analysis Matrix
- ICS 235-Facility Needs Assessment Worksheet
- ICS 237-Incident Mishap Reporting Record
- ICS 261-Incident Accountable Resource Tracking Worksheet (USCG, 2014 pp. 11–12).

It is important to note that some of the forms listed above are optional. Because they are optional, not all ICS Forms are used at every incident. The decision to use optional forms will need to be decided by the individuals that are managing the incident, such as the Incident Commander (IC), General Staff members such as the Logistics Section Chief (LSC), or even the Communications Unit Leader (COML), if they see a need for it.

9.1.3.3.1 Incident Communications Center[2]

Without adequate communication, the sharing of vital information and the notifying personnel of new and emerging safety concerns is critical to the incident. Without the Incident Communication Center, the needed support in emergency incidents could be delayed, or worse yet, individuals could be injured or killed. Incident Communications Centers is the module that helps to facilitate communications with those in tactical positions (and other areas) and they provide a lifeline for them.

The Incident Communication Center Manager is the person responsible for managing this center, and they report directly to the Communications Unit Leader (COML). Depending how this level of ICS is set up, the Incident Communication Center Manager could supervise Radio Operators, Phone Operators, Message Center Clerks, Message Center Runners, and Technical Specialists for communications (USCG, 2014). At the very least, they will always manage Radio Operators and if used, including Auxiliary Emergency Communicators (Ham Radio Operators).

When an Incident Communications Center is opened, the Incident Communications Center should be set up near the Incident Command Post (ICP) in order to facilitate the support of field personnel. The Incident Communications Leader should coordinate where the location with Facilities Unit Leader for optimal placement. It is important to note that the Incident

Communications Center should be placed away from high traffic areas and noise, and whenever possible, away from radio frequency and electronic noise. Also, whenever possible, the Incident Communications Center should not only be located near the Incident Command Post but also located near the Resources Unit and the Operations Section if possible. The reason for this these location preferences is that the individuals working in the Incident Communications Center will be in regular contact with these other modular structures.

The Incident Communications Center can be placed in a building, a portable structure, or it can be placed in mobile Unit that is specifically designed for this purpose. Deciding what type of Incident Communications Center would be best suited for the incident will require an evaluation of the situation. One of the more important considerations when determining the structure type is identifying how many personnel will need to operate the Incident Communications Center. If a large contingency of personnel is needed, then a mobile Incident Communications Center, in most instances, would not be the best option. Perhaps a portable structure (e.g. tent) or a building with sufficient space should be considered. By the same token, if the situation is fluid, and it is possible that the Incident Communications Center may need to move, then a mobile Unit, or even a mobile Unit with a temporary building in close proximity may be the best option. It is important to use common sense when identifying the type and location of the Communications Center.

The Incident Communication Center Manager will create and/or enforce communication protocols for routine requests. Routine requests are nonemergency radio traffic to and from personnel working at the scene on assigned radio channels. Creating these protocols is important to facilitating appropriate, effective, and consistent communications throughout an incident. The Incident Communications Manager will also create protocols for documenting the status, the location, and various accountability information for personnel, and how to relay nonemergent information to incident personnel.

The Incident Communications Manager will also ensure that the Radio Operators who use communication protocols have protocols for emergent situations and that operators handle those call properly. The protocols they create for emergency situations should include the notification procedures for ground ambulance requests, air ambulance requests, aircraft emergencies, evacuation requests (including personnel and public evacuations), requests for Search and Rescue teams, and what should be done when personnel suffer serious injuries. These protocols should also identify what should be done in the event that personnel or members of the public suffer a fatality within the geographical location of the incident.

The Incident Communication Center Manager will, as needed, assign Tactical Dispatchers who are tasked with integrating with the personnel who are undertaking tactical operations and they need close communications with a dispatcher. In doing so, they will be able to manage and dispatch urgent assistance and/or additional resources (or support) for those tactical personnel. This support may include restricting radio traffic for only emergency situations or asking other personnel to use a different frequency during emergency situations.

The Incident Communications Manager will also ensure that all personnel will document the required information and any important or emergency event. This might include ICS Form-309 (Radio Log), and ICS Form-214, the Unit Activity Log.

9.1.3.3.2 Radio Operator[2]

While the title makes this job sound like an individual position, in most incidents, the task of Radio Operator will usually be accomplished by a team of individuals and the subunits that work under their supervision. The supervisor of this team is called the Incident Communications Manager who is responsible for ensuring that there is adequate staffing of Radio Operators. The span of control for a Radio Operator is significantly different from the majority of those operating in the ICS Method. A Radio Operator can usually only monitor and manage three frequencies at an incident. If more frequencies are used, or more frequencies are added, then another Radio Operator should be added to the team.

The Radio Operator team is responsible for receiving and transmitting radio and telephone messages, and they are responsible for documenting all communications that they facilitate. Most of the communications that they handle will be communications between personnel and dispatch, although they will also receive information over the phone and transmit to the appropriate individuals in the field. This may include dispatching additional tactical resources and services to the incident, or potentially even a specific area of an incident.

The Radio Operator will receive and review the Incident Action Plan (IAP) to determine incident organization and the Incident Radio Communications Plan (ICS Form 205). By knowing the organization and the information in the Incident Action Plan (IAP), they can make informed decisions of who to notify or what resources are available should they be requested and needed.

The Radio Operator is responsible for ensuring that the proper equipment is present and set up Incident Communications Center and ensuring that this equipment is functional and fully operational. They have the authority to request service on any inoperable or marginal equipment, if it is needed. The Radio Operators will also maintain General Message files which captures the actions that were taken and when they were taken. They are responsible for keeping and maintaining records on any unusual incidents that occur such as explosions, responder injuries, and other unusual events. While they are not the individuals that maintain records of all the unusual incidents, the information of what was documented by Radio Operators may be helpful in later finding out what happened and why it happened.

At the end of their assigned shift, the Radio Operator is responsible for providing a briefing to the person or persons relieving them. Those briefings will include what current activities are ongoing, the status of equipment, and any unusual communications situations. These unusual situations should include equipment that is not operable, equipment that is not operating to optimal capacity, or new equipment that has been installed. Additionally, the briefing for the oncoming operators should include equipment that is ready to be dispatched from the Staging Area.

Upon the completion of the operational period, the Radio Operator Supervisor will turn in the required documentation to Incident Communications Center Manager or the Communications Unit Leader (COML). One or both of these individuals will review this documentation, and then provide it to the Documentation Unit which is part of the Planning Section.

9.1.3.3.3 Incident Message Center[2]

The Incident Message Center is usually part of the Incident Communications Center; however, it can also be located in an Emergency Operations Center (EOC). Where they are located is dependent on how the incident is set up, and the preference of the local jurisdiction. The Incident Message Center is usually colocated with, or placed near, the Incident Communications Center.

The Incident Message Center receives, records, and routes information. This information may be about resources that are reporting (or has already reported) to the incident, the status of resources (e.g. in staging, facilitating evacuation, direct operations), the situation of resources, tactical information, and some types of administrative information about the incident. The Message Center is managed by an Incident Message Center Manager, who is assisted by Messenger Center Clerks and Message Runners.

Because the message center is colocated within the Incident Communications Center, the staff that make up the Incident Message Center usually work closely with the Radio Operators and Telephone Operators. It is extremely important for there to be a cooperative and collaborative effort among these personnel.

The message center plays a vital role in controlling the flow of information. In most instances, a flow chart should be created and used that shows the routing of messages. This chart should be placed where it can easily be viewed. This will help facilitate better communications throughout the incident and give greater clarity of how information should flow. This holds especially true for personnel that are newly assigned to the Message Center.

When messages are received by Radio or Phone Operators, it is important to guarantee that they are transmitted or distributed as quickly as possible. Whenever possible, messages should be distributed via phone and/or radio (first) because this is the preferred method of communication and usually the quickest form of communication. Messages that cannot be transmitted over the phone and/or radio should be filled out using the ICS Form 213-General Message Form. This form can be filled out by anyone operating at the incident, and it can be forwarded to the Incident Message Center. At the Incident Message Center, the message will be entered into the master message log and a control number will be given to the message. This is done so that the message can be tracked.

If the message cannot be delivered by phone or radio, the written message is given to a runner who will hand-deliver the message to the appropriate individual. The Message Center Manager will maintain files of the filled-out ICS Form 213-General Message Form, to ensure that there is documentation of those messages and to ensure that actions were taken to deliver those messages. These files will be forwarded to the Documentation Unit for final filing at the end of each operation period, or in some instances, in a 24-hour period.

9.1.3.3.4 Auxiliary Emergency Communicators[2]

The Auxiliary Emergency Communicator is a wide-ranging term for any type of volunteer Radio Operator that may be utilized in an incident. In most instances, these will be Licensed Amateur Radio (HAM) Operators, although it could mean anyone that volunteer as a Radio Operator at an incident. Auxiliary Emergency Communicators can also include businesses and family radios and operators that may be able to supplement communications.

There are numerous license-free communication avenues that may help supplement a response. This can include Family Radio Service (FRS), General Mobile Radio Service (GMRS), and various other frequencies that help people communicate. This diverse group of individuals that may be able to facilitate communication will be considered Auxiliary Emergency Communications personnel. Auxiliary Emergency Communications are not only responsible for operating communications equipment but also maintaining that equipment and ensuring they are operating in an optimized capacity.

The physical location of the Auxiliary Emergency Communications Unit will be determined by the Communications Unit Leader (COML), and it will usually be placed under the control of the Communications Center Manager. Depending on the set-up of structure of the ICS method by the Authority Having Jurisdiction (AHJ) the Auxiliary Communications Unit could instead be placed under the Technical Unit in the Communications Unit.

When deciding the best placement for the Auxiliary Emergency Communications Unit, it is important to determine if they will work to supplement the Communications Center, or if they will communicate with other personnel or provide specialized technical assistance. By determining this, it will help with the optimal placement of these volunteers. While this is a huge oversimplification, and only a minimal amount of the potential duties they can fulfill, it gives a basic idea of why this Unit could be placed under different parent Units.

When three or more individuals are in Auxiliary Emergency Communications, or more than one group, then someone must be appointed to manage the group. This person will be identified as Auxiliary Emergency Communications Manager. When an Auxiliary Emergency Communications Manager is appointed, they will be responsible for managing the technical and operational aspects of volunteer communicators. This will likely include creating and sustaining a network of Auxiliary Emergency Communicators, establishing and/or staffing an Auxiliary Emergency Communications Center, and appointing the needed supervising personnel.

9.1.3.3.5 Incident Communications Technician[2]

Almost any time the term "Technician" is used in the ICS method of incident management, it is typically used as a broad phrase to describe jobs that are not specifically addressed and/or identified. The Incident Communications Technician reports to the Communication Unit Leader (COML) and can be responsible for installing, maintaining, and tracking communications equipment. In most instances, these technicians are not the typical first responder that is filling a position. These are typically highly specialized individuals who have advanced knowledge of specific areas of communication. The Incident Communications Technicians are considered support teams, simply because they support the incident and the facilitation of communications. They will support the Communications Unit Leaders (COML) by helping to design communications systems that meet the operational needs of the incident. They can also help the Communications Unit Leader (COML) in determining not only the resource needs of that day or that week but also entire duration of the incident (National Wildfire Coordinating Group [NWCG], 2001).

The Incident Communications Technicians will determine the resource needs for the incident and order the required equipment through the Supply Unit. This can include human resources with specialized capabilities. Upon arrival of equipment, they may be required to set up the equipment and ensure that it is operational, or they may supervise the setting up and maintenance of this equipment (NWCG, 2001). This holds especially true if an outside vendor is used.

The Incident Communications Technicians will typically also assist the Communications Unit Leader (COML) with evaluating and requesting vendor services for communications. Because there is a wide array of communications, this can be challenging. These vendors could include wired telephone, cell phones, satellite phones, satellite Internet, microwave, and a whole host of other forms of communication. Not only will the Incident Communication Technician identify the needs but they will also identify the costs associated, and which vendors have a track record of being reliable.

The Incident Communications Technician will identify where the equipment will be installed, including the location of repeaters, telephone lines, and any satellite equipment needed to receive and transmit signals (NWCG, 2001). This can also include providing communications support for data gathering and the needed imagery to support the incident. When these types of equipment arrive from the Supply Unit, they will either install the communications equipment, or oversee the installation of that equipment much like the previously mentioned installation and maintenance. This will include the testing of all of the equipment to ensure that the incident's communications systems are functional and that they meet the needs of the incident (NWCG, 2001). Common systems that may be used in a large incident can include the following:

- Command repeater
- Logistics repeater
- Remote repeaters or communications
- UHF link
- Aircraft link
- Cloned or programed radios (NWCG, 2001).

The Incident Communications Technician will also establish priorities, sometimes in conjunction with the Communications Unit Leader (COML). When identifying the needs of the incident, the communications needs of tactical personnel is almost always a priority over the communication needs of others working at an incident (NWCG, 2001).

The Incident Communications Technician will also be responsible for documenting and tracking communications equipment that is assigned to them. Considering how expensive communications equipment can be, this can be a critical task in making sure that the response and/or recovery does not go overbudget because of the equipment that "mysteriously" disappears. It is also important that the Incident Communications Technician may need to be assisted by others due to the wide expanse and ever-changing landscape of communications. They may not readily know every little part of communications, and they may need other experts to assist them. Of course, this would be determined by the needs of the incident, and while there is no way to predetermine the assistance that may be needed, it is fair to say that there are vendors as well as state and federal resources that may be able to assist with more reliable, more resilient, and/or the latest technology (NWCG, 2001).

9.1.3.3.6 Communications When a Joint Field Office (JFO) Is Opened

It is important to realize that direct tactical and operational responsibility for incident management rests with the Incident Commander and those operating within the ICS method staff members. Many individuals believe that when the federal government (or FEMA) shows up at an incident, that they take over the situation. This could not be further from the truth. Federal agencies are present only to provide support for the incident, but they will not, and legally cannot, take over except in very rare instances (e.g. never-ending disasters, disaster on federal property). This support is typically facilitated through a Unified Coordination Group. The members of the Unified Coordination Group can change from incident to incident and the members of the group will likely be organized based on the size, scope, and nature of the disaster.

All members of the Unified Coordination Group must have decision-making authority and responsibility from their jurisdiction. The coordination they provide will facilitate more robust operations, planning, public information, and logistics abilities. This group will integrate local, state, tribal, and federal resources into a unified response. This ensures that all levels of government work together to achieve unity of effort. The United Coordinating Group will determine the type staffing need that will be required to provide support for the incident, which may include personnel from state and/or federal agencies, other jurisdictional entities, the private sector, and nongovernmental organizations.

During a Presidential Disaster Declaration (PDD), if an incident is large or extremely complex, the United Coordinating Group will usually create an incident management field structure known as the Joint Field Office (JFO). A Joint Field Office (JFO) is just one part of the much larger Federal Multiagency Coordination System (MAC) that can be provided under the NIMS method.

A Multiagency Coordination System (MAC) allows federal resources to take responsibility and to provide support for the incident, regardless of what type of incident it may be. This support usually provides a combination of facilities, equipment, personnel, procedures, and communications. The NIMS method places the majority of the responsibility for coordinating (and supporting) domestic incident management activities on the federal government, while leaving the operational management and tactical aspects to the local entity. When these Multiagency Coordination System (MAC) resources arrive at the scene of the incident, they are integrated into a common supporting system using the NIMS method to provide support for the incident and ongoing ICS needs. It is important to reiterate that these resources will not take over an incident and they have been taught the mindset that all incidents are the local jurisdictions responsibility.

One primary function of the Multiagency Coordination Systems (MACs) is to support incident management policies and priorities. When Command Staff and General Staff are overwhelmed, they will sometimes make decisions that can lose sight of these policies and priorities that lead to an effective response. The Multiagency Coordination Systems (MACs) helps to lighten the load of the Command Staff and General Staff by communicating with them about resource allocation and resource tracking as well as providing coordination and communication of incident-related information.

One area that the Joint Field Office (JFO) excels at is in providing the architecture to support coordination for incident communications systems integration and information coordination. This assists local agencies and organizations to fully integrate the subsystems of NIMS so that it can support the incident. The Joint Field Office will typically provide a temporary federal facility that enables a central location for the coordination of federal, state, tribal, and local governments. This Joint Field Office (JFO) will be active in ensuring that all stakeholders, including the private sector and nongovernmental organizations, are looped into the communications network. The Joint Field Office (JFO) is planned, staffed, and managed by utilizing federal resources that follow NIMS principles. It shares and supports the ICS method of integrating resources, and they will work in coordination with the Communications Leader. These supporting structures helps to better organize resources which will lead to improved response and recovery methods.

When a Joint Field Office (JFO) is opened, there is almost always a greater need for communications based on the volume of individuals, agencies, and levels of government that will be working on the incident. All of these responding individuals, agencies, and governments will need to integrate, and they will need to communicate.

The Communications Unit Leader (COML) from the Multiagency Coordination System (MACs) will develop a Joint Field Office (JFO) communications plan to make the most effective use of the communications equipment and the facilities that are assigned to the Joint Field Office (JFO) Coordination Group. The Joint Field Office (JFO) Communications Unit will install and test all communications equipment that is needed to support the incident and they will operate and supervise the Joint Field Officer (JFO) communications center. They will also dispense communications equipment as it is needed and recover this equipment when the incident comes to an end. While the Joint Field Office (JFO) is in operation, personnel will be assigned to maintain and repair communications equipment in place, which leads to fewer unexpected breakdowns, and greater resiliency in the communications aspect.

The Joint Field Office (JFO) Communications Unit's major responsibility is effective communications planning for the Joint Field Offices (JFO's) Coordination Group. This is imperative for supporting the Incident Commander's (IC) requirements for radio networks and interagency frequency assignments. These requests are typically made through the Incident Communications Unit Leader (COML) to the Joint Field Office (JFO) Communications Unit, who will work in coordination and collaboration with this leader. This collaborative effort helps to ensure interoperability between all responding agencies and safeguards optimal use of all communications abilities.

Some of the most-used resources that the Joint Field Office (JFO) can provide will be described later. While this is not a comprehensive list of the assistance that the Joint Field Office (JFO) can provide, explaining these NIMS method resources, and how they integrate should provide a better understanding of how these resources integrate and support an incident.

9.1.3.3.7 Network Manager[2]

The Network Manager, much like a network manager in a business, provides automated data processing and computer information technology services, sometimes called IT services. The

Network Manager will ensure that automated data processing and computer Information Technology (IT) services are provided for all facilities that are involved in the response. This includes providing a local area network (LAN) with e-mail access and the standard office suite software used by the Authority Having Jurisdiction ([AHJ] FEMA, 2014).

When a Joint Field Office (JFO) is opened, the Network Manager will meet with the Communications Unit Leader (COML), the Incident Communications Technician, or both. The Network Manager will receive a briefing about the current set-up of communications and what needs for the incident have not been met. From this, the Network Manager will identify and determine the amount of specialized personnel that will be needed to accomplish the goal of providing effective Information Technology (IT) services that meet the needs of the incident. They will also work with the Communications Unit Leader (COML) or the Incident Communications Technician to order any additional personnel that may be needed and any additional equipment and/or services that may be required to meet the needs of the incident. Through coordinating and communicating with each other, they will also determine the best location for the network file servers, where to locate and install the cables and hubs, and they will determine the number of workstations and printers that will be needed to effectively maintain the needs of the response, as well as the Joint Field Office (JFO).

The Network Manager will locate the optimal placement of equipment and identify which individuals should have the rights to submit or change specific information. This is done to protect the integrity of the information being provided. The restriction on data entry could include data that describes roles, positions, and information, and the Network Manager may be required to train these individuals in how to accomplish these data entry tasks.

The Network Manager will also be responsible for setting up workstations, computer hardware, and for installing and configuring software in all facilities that require it. They will test all computers and printers, and they will configure network printer queues for priorities. An effective Network Manager will not just provide what was needed at previous incident(s); they will look at the needs of this particular incident and determine the best solutions. They realize that every incident is unique, therefore, an effective Network Manager will provide what is required by the incident rather than providing software or hardware that may not be effective in the incident they are currently working on (FEMA, 2014).

On a daily basis, the Network Manager (or their assistants) will maintain system operations by performing consistent hardware and network connectivity tests, and if needed, they may upgrade software to better meet the needs of the incident. They will, as needed, monitor and update user accounts, network printers, the server(s), check virus protection, monitor remote access, and various other issues that affect performance, and that need monitored daily. They will also make system backups to ensure that no data is lost, and they will coordinate with the Accountable Property Officer on a daily basis to identify the inventory of Information Technology (IT) equipment that has been used. They will also document all significant tasks undertaken and any significant decisions that were made in the performance of their duties (FEMA, 2014).

As part of the demobilization, the Network Manager will be responsible for dismantling the network and all of the components. As part of that dismantling, they will complete a final system backup and copy any files needed to the regional server. The Network Manager will also inspect all of the equipment that was used, they will document any equipment issues, and they will deactivate the high-speed data transmission line(s). The Network Manager will also work with the Accountable Property Manager in packaging and making shipping arrangements for any remaining IT equipment (FEMA, 2014).

9.1.3.3.8 Network Specialist[2]
The Network Specialist is typically brought in by FEMA upon opening a Joint Field Office (JFO) during a Presidential Disaster Declaration (PDD). It is important to note that this is not always

the case, and in some instances, state resources or vendor services can be utilized to fulfill this need. The Network Specialist is supervised by the Network Manager, and they are responsible for providing automated data processing and computer IT services. Upon being checked in, the Network Specialist will report to the Network Manager and receive a briefing. Their primary job is to support the Network Manager in performing their duties.

These duties could include setting up user accounts for e-mail users, creating and/or providing databases, installing collaboration technologies, and providing various software programs to meet the needs of the incident. It is likely that they will also play an active role in training individuals on the use of the Network Access Form and Network Access Control Systems as well as guiding them through any problems. They will test connectivity of the network and computers to ensure reliability; they will install and configure software, troubleshoot hardware and software problems, and make regular backups of network data (FEMA, 2014).

During the demobilization process, they will assist the Network Manager and Accountable Property in preparing equipment for shipment and/or temporary storage. As part of the demobilization and dismantlement, they will remove information as is appropriate or required. This could include permissions to access files and Digital Access Exchange mail users to name a few. They will not delete information, only the access to them. Along with the Network Manager, during demobilization, the Network Specialist will inspect all equipment and document any issues while shutting down the system (FEMA, 2014).

Telecom Manager[2] The Telecom Manager is typically used when federal resources are needed and used after a Presidential Disaster Declaration (PDD). Like the Network Specialist position, it is important to note that this individual is not always provided by FEMA. In some instances, state resources or vendor services can be utilized to fulfill this need. The Telecom Manager will typically report to the Incident Communications Technician or the Communications Unit Leader (COML), depending on how the Communications Unit is organized. The Telecom Manager is typically activated upon opening a Joint Field Office (JFO). These individuals are responsible for providing telecommunications services for all support facilities in the field. This includes the creation of a supply of cellular and Wi-Fi telephones as well as coordinating frequency management services with the Mount Weather Emergency Assistance Center.

About the Mount Weather Emergency Assistance Center

Mount Weather Emergency Operations Center (MWEOC) is a FEMA facility situated on 564 acres high in the Blue Ridge Mountains. The site originally was opened in the 1800s as a civilian weather site and was acquired by the National Weather Bureau in 1893. Over the years, the government has used this site for various agencies including the US Bureau of Mines, the US Army Corps of Engineers Office of Emergency Preparedness, and the General Services Administration (FEMA, 2015a). From 1958 to 1959, a series of underground tunnels were mined out of the mountain to provide for Continuity of Government (COG) and a five-foot-thick blast door was added (Public Intelligence, 2010). The facility just 64 miles from Washington DC would house high-level government officials in the event of an attack on the United States.

Upon its creation in 1979, the Federal Emergency Management Agency (FEMA) was given space within the facility (FEMA, 2015a). The underground facility is estimated to cover 600 000 square feet, with FEMA using 200 000 square feet to support disaster response (Public Intelligence, 2010). Currently housed within the 435-acre site is

- Office and warehouse space
- Dormitory and training rooms
- A conference room with classified capability up to 125 people

- A café that seats over 280 personnel
- Fully equipped fire department and ambulance service
- A health Unit staffed by medical personnel
- Federal police force providing 24-hour, seven-day-a-week security
- A motor pool: with shuttle service, courier services, and local supply pickups

The MWEOC states a mission of being "ready to support our partners at all times, under all conditions" (FEMA, 2015a, para 4). It provides support for a variety of emergency management undertakings with some of those being classified. As of 2015, the center engaged in a massive five-year infrastructure upgrade to modernize the site (FEMA, 2015a). The MWEOC currently has six major functional government programs related to emergency management that are performed on site:

- National Processing Service Center
- Satellite Teleregistration Center
- Disaster Finance Office
- Disaster Information Systems Clearinghouse
- Disaster Personnel Operations Division
- Agency Logistics Center (Public Intelligence, 2010)

Other emergency management disaster support functions are also housed at this facility as are various functions of the US Government. These are to ensure that the government can still operate in the event of a major incident or attack.

Mount Weather has transformed from a facility that very few people knew about to a facility that is used in almost every Presidential Disaster Declaration (PDD). While the facility is still restricted to the public, it is no longer a hidden secret (Global Security, 2001). It has become an important part of overcoming some of the deadliest and most damaging disasters in the United States.

The main purpose of the Telecom Manager is to support the field operations. Along with providing cell phone and Wi-Fi phone services, they will establish a Phone Bank Exchange (PBX) system to support incident communications. The Telecom Manager will establish a message center that will usually include fax machines. They are also responsible for installing modems when a wide area network (WAN) has not been secured.

As part of their duties, they will provide consultation to the Logistics Section Chief (LSC) and the Communications Unit Leader (COML) regarding the compatibility of telecommunication hardware and software for all proposed and current facilities. They will work with the Logistic Section Chief (LSC) and other involved logistic personnel that are active in creating facility planning, to guarantee that sufficient telecommunications are installed. Upon agreeing on facilities and overcoming any issues, the Telecom Manager will determine the materials needed, which will be based on the availability of proper materials and the unique needs of the incident.

When the appropriate materials are identified and procured, the Telecom Manager will install the wiring and switches that are required, and then install the suitable (to the incident) telephone service at the agreed-upon incident facilities. Upon completion of the installation, the Telecom Manager will be responsible for routine testing and fixing any emergent and nonemergent issues, as well as the maintenance of the equipment and systems. When possible, they will also remove any Plain Old Telephone Systems (POTSs) that are not utilizing high-speed lines from the incidents telecommunication system. The Telecom Manager will be responsible for documenting and justifying all significant decisions related to the job they perform, in writing (FEMA, 2014).

The importance of the Telecom Manager can be seen in how they allow and enable the different facilities to communicate with each other, and the outside world, is significant. Without the ability to easily and effectively communicate with others, the response to the incident could be reduced significantly. An example of this may be that if telecom notifications cannot be given to the Food Unit about an influx of 250 newly expected personnel and the Food Unit cannot in turn notify their suppliers that they need additional food (and supplies), there could be a lot of first responders that might be very hungry. By the same token, if an unusual piece of equipment was needed from an outside resource, without telecom in place, this task could be nearly impossible to accomplish in a timely manner.

9.1.3.3.9 Telecommunications (Telecom) Specialist[2]

The Telecom Specialist is also typically a position that is provided by FEMA after a Presidential Disaster Declaration (PDD) and upon forming a Joint Field Office (JFO). Like other support positions, this is not always the case, and these positions can be provided by state and/or vendor resources. The Telecom Specialist is supervised by the Telecommunications Manager and assists in providing telecommunications services to an incident. Upon arriving on the scene of an incident, the Telecom Specialist should receive a briefing from the Telecommunications Manager.

Under the direction of the Telecommunication Manager, the Telecom Specialist will install, program, and provide all service for telecom equipment for the incident. This includes installing wiring and switches that are needed to support the incident. They will also diagnose telecommunication complications or problems in the hardware and software of the system, and they will repair and maintain telecom equipment. They will also be responsible for providing routine and emergency maintenance of the telecommunication equipment and systems. As with any logistics job, the Telecom Specialist will keep records on any significant actions and/or decisions they were made in the performance of their duties.

As with the Telecommunications Manager, this Telecom Specialist is an extremely important function on large-scale or extended incidents. Telecommunications can be a lifeline between an incident and critical outside suppliers. Without the ability to contact people outside of the incident and within the command system, then the whole operation would slow to a snail's pace, if not grind to a complete halt.

9.1.3.3.10 Internal Communications Manager[2]

The Internal Communications Manager is usually implemented as part of a FEMA response after a Presidential Disaster Declaration (PDD) has been declared. It is important to note that this position is not exclusive to a federal response, although it is most often implemented when a Joint Field Office (JFO) is opened. Much like the previously mentioned positions, this can be contracted or managed by state resources.

The Internal Communications Manager can allow the General Staff and Command Staff to focus on their job of managing the incident while the person appointed as the Internal Communications Director (and any helpers or subordinates they may have) take care of facilitating communications within the incident response. The Internal Communications Manager can be an extremely important resource to have in a major and/or extended incident. Based on their work responsibilities of managing communications, they often prevent silos that bottleneck or stovepipe (or stovepipe) information that comes into the Joint Field Office (JFO).

In most applications, silos of information can be created when there is a hesitancy to share information with other personnel working in other functions of the hierarchical organization. Because personnel within most structures of the ICS method tend to interact with other personnel in their own functional area more than personnel outside their area, they feel more comfortable sharing information that make up their "in-group." Similarly, stovepipes also create a bottleneck that prevents the free sharing of information. This can be caused by trying to prevent certain information from getting into the hands of the media or the public, or it can be caused by an information overload that may lead to the information not reaching all of the individuals that should be privy to it. The Internal Communications Manager and those working for them help to prevent these problems.

The Internal Communications Manager is supervised by the Communication Unit Leader (COML). Their primary task is to facilitate and manage all internal communications. These internal communications, which are critical to the efficiency of the response, might include creating, providing, monitoring, and/or supporting a help desk, a switchboard, a message center, and/or a receptionist desk. They are also responsible for providing copying and printing services to support the Joint Field Office (JFO).

A key element of the job that the Internal Communications Manager is to ensure the creation of a help desk. The Help Desk is created so that all personnel working in the Joint Field Office (JFO) can find the informational support they need. This is significant because when an individual within the Joint Field Office (JFO) needs information, any delay may begin to slow response and recovery. The help desk maintains information of what resource are, where and how to contact that resource, be it human or a physical resource. This minimizes the time it takes to locate the information needed.

About the Robert T. Stafford Disaster Relief and Emergency Assistance Act

The Robert T. Stafford Act was a bill passed by congress and signed by then President Ronald Regan in 1988. The Stafford Act, named after the bill's sponsor Senator Robert Stafford, replaces the Disaster Relief Act of 1974 and puts in place a system whereby a Presidential Disaster Declaration (PDD) activates a physical response from FEMA and financial response from the US government with FEMA administering those funds based on stringent criteria. Based on the terminology of this law, FEMA has the primary responsibility for coordinating governmentwide relief efforts of 28 federal agencies and the American Red Cross; however, the state and local community must verify that the disaster is beyond their combined capabilities to handle.

The Stafford Act provides for the main areas: Individual and Households Program (IHP), Hazard Mitigation (HM), and Public Assistance (PA). IHP program provides direct financial assistance to individuals affected by a disaster for things such as temporary housing, home repairs not covered by insurance, emergency medical needs, funeral costs, moving and storage costs, and other similar urgent needs.

The Hazard Mitigation Assistance grants are to help cover mitigation measure to help strengthen infrastructure to in an effort to reduce the effects of future disasters. These are measures that are above and beyond repairs to strengthen infrastructure to make them more resilient in a future disaster. Infrastructure hazard mitigation can pertain to mitigation measure for drainage, crossings, and bridges, as well as sanitary and storm sewer systems, waste treatment plants, potable water, electric power distribution, above ground storage tanks, and underground pipelines. It also helps with buildings that are part of the infrastructure. These building repairs can include repairing the general effects of flooding, roof repair, shutter repair, anchoring of electrical and mechanical equipment, flexible piping to provide for the expected movement in an earthquake, bracing, and the replacement of glass (FEMA, 2015b).

The Public Assistance Program helps state, tribal, and local governments as well as certain nonprofit organizations to return to predisaster conditions. This assistance is in place to assist with emergency work and the repair or replacement of disaster-damaged infrastructure facilities. This may include but is not limited to debris removal, emergency protective measures, roads and bridges, water control facilities, buildings and equipment, utilities, as well as parks, recreational, and other facilities. When receiving a Public Assistance Grant, the entity is usually responsible for matching funds. While the federal government will pay 75%, it is usually of the local or state entity to bear 25% of the cost for repairs.

What Effects Does the Stafford Act Have on Emergency Incident Management?

The Stafford Act establishes processes and programs for providing federal assistance to state, local, and tribal governments as well as individuals and qualified private nonprofit organizations. Among the federal assistance available, there is technical assistance, providing of goods and services, and financial assistance. FEMA coordinates and issues much of the assistance under the Stafford Act, but other federal agencies also provide assistance, whether from FEMA or on of the other 28 federal agencies.

The Internal Communications Manager should also fill out and maintain a current Phone Directory for each operational period, or to assign that job to appropriate subordinate. The Phone Directory is organized by utilizing ICS Form 205a. The information for this directory should be collected during the check-in process and at the beginning of each operational period. Information that should be collected for the directory at check-in should include their ICS position, the radio frequencies they use, phone numbers, pager numbers, team or vehicle name (e.g. Task Force 3, HazMat 4, Operations Branch I), and any other identifying or contact methods. Upon collecting the information, the Internal Communications Manager or their subordinate will enter the information into ICS Form 205a in alphabetical order. It is important to note that if sensitive contact information is provided, such as cell phone numbers, the header should identify that sensitive information is present. Upon completion of the list, it should be distributed to the appropriate individuals, as instructed by the Communications Leader. In most instances, this will be made available to all Sections and Branches.

Another important responsibility of the Internal Communications Manager is to provide appropriate training (or guidance) and protocols for switchboard operators. They will also establish and maintain emergency contact information for personnel and launch an All-Hands email group to facilitate better communications between personnel. As with any logistics job in the ICS companion method, they must provide written documentation of any significant decisions or facts that were related to their activities as the Internal Communications Leader.

A federal disaster or emergency declarations typically authorizes federal programs to provide monetary and physical assistance in areas such as debris removal, which could help pay for a portion of the response. It should be realized that to receive reimbursement assistance in this manner, there are certain benchmarks that need to be met. To gain more information on reimbursement, research the additional sources section at the end of this chapter.

The Stafford Act also allows FEMA to get ahead of disaster declarations and to prestage federal personnel and resources. This is allowable under the Stafford Act when there is an imminent disaster that threatens human health and safety but has yet to be declared. It should be noted that FEMA cannot provide federal assistance until an emergency or major disaster declaration is made.

The Stafford Act can also be important to incident management because it authorizes FEMA to draw upon temporary personnel to assist in disaster operations. FEMA's staffing model includes four categories of temporary employees; Disaster Assistance Employee (DAE) reservists, local hires, disaster temporary employees, and Cadre of On-Call Response Employees (CORE). DAE reservists are a group of short-term temporary employees who must be activated and told to go to a specific disaster. Disaster temporary employees and local hires typically have less knowledge, skills, abilities, and experience, so they are usually added for a specific disaster response task, such as cleaning up debris, cleaning the interiors of houses that have been flooded, or possibly even as a help desk personnel.

CORE staff are typically assigned to assist with a task(s) on two or four-year appointments and they support disaster response activities.

The use of the incident command structure, including command, control, communication, collaboration and cooperation positively effects federal assistance under the Stafford Act. Utilizing Section 403, Essential Assistance, of the Stafford Act, funding and human resources can pay for work and services to save lives and protect property. This can include funding services that are essential to saving lives and preserving property and public health and safety. While this may seem like a broad and subjective term, some of the activities that can be included are:

- Search and rescue
- Emergency medical care
- Emergency mass care
- Emergency shelter
- Providing the provisions of food and water
- Providing medicine, durable medical equipment, and other essential needs, including movement of supplies or persons
- Providing rescue, care, shelter, and essential needs to individuals with household pets and service animals

These are but a few of the many provisions that can support incident management in a large or long-term disaster. For find more information on how these issues may assist you in incident management, you should use the self-study resources listed at the end of this Chapter.

9.1.3.3.11 Help Desk Operator[2]

The Help Desk Operator is normally implemented when FEMA is brought in to provide disaster assistance under Robert T. Stafford Act funding, and this position is supervised by the Internal Communications Director. While this is typically a position that is provided by FEMA, if the disaster does not meet the criteria for a Presidential Disaster Declaration (PDD), then this can also be a contracted position, or it can be handled by state resources. The Help Desk Operator is not a required position, so it can also not be implemented if not needed.

The Help Desk Operator typically identifies the type of assistance that the individual requires, then they refer them to the appropriate person or function within the ICS method structure. They will also document (log) the contact and track the disposition of their referral actions. The Help Desk Operator is often located in close proximity to the receptionist, the Message Center, and/or the Switchboard Operator. They will often help to set up the Help Desk location.

In the initial onset of activating a Help Desk Operator, if a tracking system is not in place to identify the completion of all needs presented to them, then they will create a system to do so. Tracking these questions or requests to completion is important because it helps to ensure that the needs of the incident, especially operational needs, are not forgotten or mired down in bureaucracy. The Help Desk Operator is the person that is usually the first point of contact when individuals in the Joint Field Office (JFO) needs information.

The Help Desk Operator maintains information of what resource is where, and how to contact that resource. This minimizes the time it takes to find the information that could assist the individual or business that is active in the response, and it reduces the frustration of those individuals who need information and do not know where to find help. The Help Desk Operator is responsible for helping individuals identify the potential support that best fits their needs.

9.1.3.3.12 Message Center/Switchboard Operator/Receptionist[2]

The Message Center, the Switchboard Operator, and the Receptionist are supervised by the Internal Communications Manager. In extremely large disasters, and when a Joint Field Office

(JFO) is opened, these duties can be handled by separate individuals, or in smaller incidents these duties can be combined and handled by one person. As with the other positions that have been identified in the Joint Field Office (JFO), these positions are typically staffed by FEMA personnel, however, they can also be staffed by state resources or by contracting services. This will often be dependent whether the incident meets the threshold of being a Presidential Disaster Declaration (PDD).

The duties of these positions include operating the switchboard, accepting and transferring phone calls, and to transfer messages between Joint Field Office (JFO) personnel. In some circumstances, they may also be required to operate the Reception Desk, or this may be handled by a separate person(s). This will be dependent on how busy the Joint Field Office (JFO) is. Operating the Reception Desk entails greeting and directing visitors to the correct person or area. The Message Center will receive messages, record them, and route that information to the appropriate individuals. More often than not, this information will pertain to resources reporting to the incident, the status of resources, and administrative and tactical messages.

When the Message Center is activated as its own function (rather than a combined function), it should be located near the other previously identified communication functions. The reason for this is that message center staff work closely with the incident dispatchers, Radio Operators, and Telephone Operators. A close working relationship as well as free flowing communications between these personnel is critical. The message center plays a dynamic role in controlling the flow of information.

When messages are received by Operators, those messages are written down and logged. Whenever possible, the Operator will distribute messages to the appropriate person(s) by phone and/or radio to expedite the timely delivery of that message. If there is no way to transmit these messages verbally, then the message is transferred to a written hard copy using the ICS Form-213 (General Message). This form will then be given to the message center, who will then document it in the master message log, and then issue a control number for that specific message. The control number is given so that the message can be tracked. The message is then given to a runner who hand-delivers that message to the appropriate individual and/or Section. Should a return message be needed, then the runner will take appropriate actions, including making sure that the Message Center documents the reply.

The ICS Form-213 (General Message) is a three-part form that is typically filled out using carbon paper. In some instances, this form is premade utilizing three different colors and carbonless paper. When the message is sent, the original message typically stays with the sender, and the copies (parts two and three) are delivered to the intended recipient. This may be different at different incidents depending on how the Authority Having Jurisdiction (AHJ) prefers to be done. In the instances where the two copies are sent to the recipient, they will fill out part two, then return that part to the recipient (usually the Message Center). In instances where the recipient is requesting additional resources, they may also attach ICS Form 213-RR (Resource Request message) and it can be attached to the return message.

The Message Center Manager will maintain files of the messages and retain copies to provide documentation of the messages that were sent/received, and the actions that were taken. These files will be forwarded to the Documentation Unit within the Planning Section for final filing. This is typically done at the end of each operational period; however, these should be given to the Documentation Unit a minimum of once every 24-hour period.

It is important to note that not all messages will be to individuals working within the incident. Some messages may need to be transferred outside of the incident and/or the command system. Additionally, if directed by the Internal Communications Manager, any one of these positions may be responsible for generating, maintaining, and distributing a phone directory. They may also be responsible for requesting additional staff from the Internal Communications

Manager and in maintaining records that document any unusual incidents that occur during the performance of their duties. Upon the completion of their operational period, they will be responsible for briefing the person or persons relieving them, and in submitting the appropriate documents to their immediate supervisors.

9.1.3.3.13 Mobile Emergency Response Support (MERS)

The Mobile Emergency Response Support (MERS) system is a FEMA asset that is only deployed in the United States during a Presidential Disaster Declarations (PDD). FEMA's Mobile Emergency Response Support (MERS) detachments can provide a multitude of resources that are rolling (motorized) Units. These mobile Units can meet many urgent needs that an incident may have. Some of the support that can be provided includes telecommunications, life support, logistics, operational support, and power generation. When needed, these detachments, or specific portions of the detachment, are sent to support the need (FEMA, n.d.a).

FEMA maintains five Mobile Emergency Response Support (MERS) detachments in the United States (FEMA, n.d.a). Each of these detachments is positioned to equally support 2 of the 10 FEMA regions. The geographical location of each detachment includes Bothell, Washington; Denver, Colorado; Maynard, Massachusetts; Denton, Texas; and Thomasville, Georgia. Perhaps the most used function in each of these detachments, particularly during a disaster response, are the communications support Units which are called Mobile Emergency Operations Vehicle ([MEOV] FEMA, n.d.a).

The Mobile Emergency Operations Vehicle (MEOV) is a self-contained mobile communications vehicle that has onboard generators and satellite communications equipment (FEMA, n.d.a). These Units are essentially a mobile office, and they have the capabilities to communicate with other stakeholders in multiple ways. These Units can also provide support in many ways. They can provide something as simple as video teleconferencing for locations that have little (to no) infrastructure left to support the response and recovery, and they can provide almost any type of communication. The mobile Units vary in size from a one-ton vehicle to vehicles that are semi sized with slide-outs on each side. The slide-outs double the size of the semi-trailer when it is parked. These Mobile Emergency Operations Vehicles (MEOVs) provide support and a temporary location for an Incident Command Post (ICP), an Initial Operating Facility (IOF), or a Joint Field Office (JFO). For the purpose of clarification, the Initial Operating Facility is the first facility from which an Incident Management Assistance Team (IMAT) can manage incident-level operations. The Initial Operating Facility (IOF) is a temporary facility until a more suitable facility is secured for the Joint Field Office ([JFO] FEMA, n.d.a).

While not communication-related, other Mobile Emergency Response Support (MERS) Units have the ability to provide logistical support for field operations as well. The logistical support can include providing water, fuel, power generation, Heating Ventilation and Air Conditioning (HVAC), and life support for responding personnel. Mobile Emergency Response Support (MERS) can also provide for the administrative support that may be needed by federal, state, and local responders (FEMA, n.d.a).

The contingency of the Mobile Emergency Response Support (MERS) include approximately 275 mobile Units that stand ready to be deployed (FEMA, n.d.a). These Units can provide local entities that suffer a disaster with a multitude of resources. Imagine having utilities disrupted, and you have no heating or air conditioning at shelters or any of the structures that support incident management. Now imagine the logistical support of that Mobile Emergency Response Support (MERS) Heating, Ventilation and Air Conditioning (HVAC) Unit arriving at the incident, and they generate enough heating or air conditioning for a 16 000 square foot building. Imagine being without power, and the Mobile Emergency Response Support (MERS) truck-mounted generators arrive and can provide power generation and distribution for several large facilities such as a hospital or a large shelter. Shortly thereafter, the detachments transport and distribute fuel with either a 1200, a 2200 or a 3500-gallon tanker to support response and recovery.

These Mobile Emergency Response Support (MERS) detachment can also transport water through 3000-gallon tenders, and they can send one or more reverse osmosis water purification Units to ensure there is clean and fresh drinking water. These reverse osmosis Units can change brackish and saltwater into safe and drinkable water. Looking at these examples, it is easy to see how a Joint Field Office (JFO) and the NIMS methodology supporting structures can support the response and recovery of any incident.

The Mobile Emergency Response System (MERS) can provide services to both the Operations Section and the Logistics Section. In the Logistics Section, they support the Service Branch; however, they will usually reside in the hierarchical command structure as a subsection of the Communications Unit (FEMA, n.d.a).

Real-life Incident

In November of 2016, Gatlinburg TN suffered a massive wildfire. What started as a multi-day, 1-acre fire, turned into 17 000-acre fire that was driven by 60 miles per hour winds. This massive fire killed 14 people, injured over 200 other individuals, and burned over 2500 homes. According to the After-Action Review ([AAR] Guthrie et al. 2017) the response was the largest ever in Tennessee history, and it utilized resources from 50 counties, over 225 agencies, 445 apparatus, and 3535 first responders from across the nation. The influx of personnel for this incident caused an almost immediate radio communications overload on the county radio system. Because of the large amount of radio traffic, and the lack of available radio frequencies, communications were difficult at best. Adding to the problem, the Communications Center received so many calls in a short amount of time, the 911 center was either ringing busy, or callers could not connect at all. Several hours into the incident, the headquarters for the City of Gatlinburg lost electricity, phone lines, and the Internet, which followed shortly thereafter by the cell towers becoming inoperable. This incapacitated many of the emergency notification systems that were in place, and significantly reduced situational awareness about the movement of the fire, and available evacuation routes. Communications between the Emergency Operations Center (EOC) and the Tennessee Emergency Management Agency (TEMA) were sporadic at best, primarily because of the infrastructure damages. This significantly interrupted communications and contributed to mandatory evacuation orders not being sent (Guthrie et al. 2017).

There was also a lack of interoperability among city, county, state, and federal agencies, which caused mutual aid agencies to have problems communicating with each other, and the Emergency Operations Center (EOC) and/or the Communications Center (Guthrie et al. 2017). The original Emergency Operations Center (EOC) as well as other offices of city officials, had to be evacuated to other locations in Gatlinburg. Additionally, one Communication Center had to be relocated because of encroaching fires. Among many recommendations of the After-Action Report (AAR), one reoccurring theme for communications was to implement redundant (interoperable) communications and methods of alerting (Guthrie et al. 2017).

Chapter 9 Quiz

1. Fill in the blanks below

Organizational Element	Leaders is Called
Incident Command	
Section	
Branch	
Units	

2. The _____ will provide provisions for tactical first responders.

3. This Branch is tasked with obtaining the work materials and personnel needed to support the incident.

4. Name the seven types of facilities that may be used in an incident.

5. The _____ will provide support by procuring the materials needed to effectuate a practical response.

6. An easy way to differentiate the Service Branch and the Supply Branch is to remember that the (a) _____ works directly with the operational personnel by personally providing the services they need, while the (b) _____ will supply the materials to meet those needs.

7. There are typically three Units under the control of the Service Branch. These three Units will be

8. **True or False**: Having the correct people in the Medical Unit can be the difference between life and death.

9. The _____ is tasked with the responsibility of installing (and testing) all communications equipment related to the incident.

10. **True or False**: The Unified Coordination Group is responsible for managing the Incident Communications Center.

11. **True or False**: When the federal government (or FEMA) shows up at an incident, they will take over the situation.

12. The members of the Unified Coordination Group can change from incident to incident and the members of the group will likely be organized based on:

13. **True or False**: All members of The Unified Coordination Group must have decision-making authority and responsibility from their jurisdiction.

14 The coordination the _____ provides will facilitate more robust operations, planning, public information, and logistics abilities.

15 **True or False**: The Joint Field Office (JFO) is just one part of the much larger Federal Multiagency Coordination System (MAC) that can be provided under the NIMS method.

16 **True or False**: A Multiagency Coordination System (MAC) allows state and local resources to take responsibility to provide support for the incident.

17 One area that the _____ excels at is in providing the architecture to support coordination for incident communications systems integration, and information coordination.

18 **True or False**: The Joint Field Office (JFO) will be active in ensuring that all stakeholders, including the private sector and nongovernmental organizations, are looped into the communications network.

Self-Study

Federal Emergency Management Agency (2014). *Federal Emergency Management Agency incident management handbook: FEMA B-761*. Government Printing Office, Washington DC.

Federal Emergency Management Agency (2015a). *Fact sheet: Mount Weather emergency operations center*. FEMA Press release. Retrieved from: https://www.fema.gov/media-library-data/1433170060726-5272c667a842f8a56ef9bae118c2ba0a/MWEOCFactsheet.pdf.

Federal Emergency Management Agency (2017a). *National Incident Management System*, (3rd ed.). Retrieved from https://www.fema.gov/media-library-data/1508151197225-ced8c60378c3936adb92c1a3ee6f6564/FINAL_NIMS_2017.pdf.

10

Expanding Planning and Intelligence

I would prefer a sword to fight duel, but a pen to plan a war

Robert Thier

It has been said that failure to plan is planning to fail. In the Incident Commands System (ICS) method of incident management, if we do not plan, we truly are setting ourselves up for failure. This holds especially true when there are multiple working pieces in a fluid incident. Without guidance, we have a haphazard (at best) plan of attack that will most likely have multiple failures or unintended consequences.

In the ICS method of incident management, the Planning and Intelligence Section can be described as the Section that oversees and helps guide the plan of attack. This Section is responsible for the collection, evaluation, and dissemination of information related to the current status of the incident. The Planning and Intelligence Section is also responsible for providing maps (as needed), and the preparation and documenting the Incident Action Plan (IAP), which will be better described in Chapters 14 and 15.

The Planning and Intelligence Section (often referred to as Planning Section) is also responsible for providing intelligence on the anticipated needs of an incident. This can include what the resources needs might be, the status of the resources that are assigned to an incident, and future needs of personnel. It should be noted that the intelligence that is provided under the Planning Section is very different from the Investigations and Intelligence Section described in Chapter 12. Essentially, the Investigations and Intelligence Section is typically law enforcement, public health, or some other discipline that is looking for answers. With a different focus, the Planning and Intelligence Section focuses on incident specific intelligence, such as where a fire is going, what resources are available to end a hostage situation, what progress has been made, the number of people that need evacuated, and similar intelligence important to planning the response.

10.1 Planning and Intelligence Modular Expansion

In a smaller response, Planning and Intelligence can be undertaken by the Incident Commander (IC), if they decide to take on the responsibility. As an incident expands, or in cases where the incident is already large enough that the Incident Commander (IC) does not want to take on this responsibility, then a Planning and Intelligence Section must be formed. In small to midsize incidents, this can be handled by a handful of experienced individuals at an Incident Command Post (ICP). In a large full-scale incident that requires a large contingency of personnel, a much larger structure will be needed. In an incident where only a few people can no longer handle the

Emergency Incident Management Systems: Fundamentals and Applications, Second Edition.
Mark S. Warnick and Louis N. Molino Sr.
© 2020 John Wiley & Sons, Inc. Published 2020 by John Wiley & Sons, Inc.
Companion Website: www.wiley.com/go/Warnick/EIMS_2e

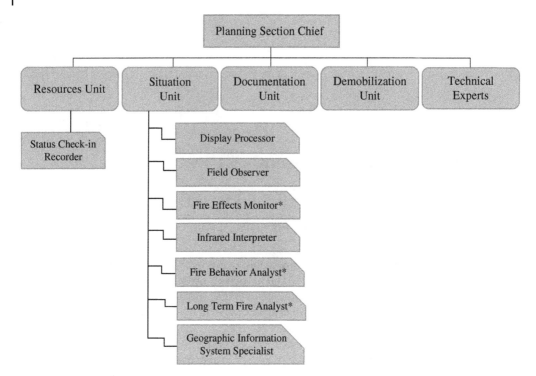

Figure 10.1 Planning Section expansion with the main Units. Note: Positions with an asterisk are mainly used in wildland firefighting.

duties on their own, then four main Units (and three optional Units) can be created (as needed) in the typical incident (Figure 10.1). The four main Units include the following:

- The Situation Unit
- The Resources Unit
- The Documentation Unit
- The Demobilization Unit

In some incidents, Technical Specialists are needed who will advise on technical aspects involved with mitigating the incident. As has been mentioned previously, Technical Specialist is the "catch all" name for individuals who are not specifically named in the National Incident Management System (NIMS) method, or the ICS companion method. When Technical Specialists are brought into the response, they will typically be utilized in specific areas. These will be discussed further in the Technical Specialist portion of this chapter.

While the topic of Planning will be discussed more in-depth in chapters 14 and 15, it is important to have a basic understanding of planning in order to better comprehend what the Planning and Intelligence Section does. Planning is the process of appraising the current situation, defining objectives, choosing a viable strategy, and shaping what would be acceptable outcomes for each unique incident. To accomplish these tasks, the Planning Section must decide which available resources should be used to achieve those goals in the most effective and cost-effective way. At the same time, the Planning and Intelligence Section ensures that the safety of personnel is the highest priority. The Planning and Intelligence Section is however responsible for more than just developing the Incident Action Plan (IAP). It is important to note that they must get approval before disseminating it to the proper individuals. They are also responsible for as many other tasks.

The primary task of the Planning and Intelligence Section is to confirm that there is a plan in place at all times. By doing so, they guarantee that every staff member and responder is on the same page. This includes nonpublic safety personnel that are part of the response, and any spontaneous untrained volunteers.

Just like the other Sections, when the span of control reaches its limits, this Section must expand. The modular design of ICS provides predetermined modules that should be implemented under the command of the Planning (and Intelligence) Section Chief (PSC). Unlike the Operations Section, the expansion of Planning and Intelligence does not typically include a Branch. This Section is so focused on their individual tasks, that it would typically be counterproductive to create anything more than a Unit for the modular predetermined areas. This is not to say that Branches could not be added in a major incident, but this action would need to be approached with caution to prevent repetitive or bureaucratic Branches and Units.

An acronym that many firefighters learn early on in their career is the acronym "KISS." This stand for "Keep It Simple Stupid." This acronym should be applied to the Planning and Intelligence Section, simply because a lot of complication is already present in creating the planning for the incidents, including an exorbitant amount of situational awareness that will need to be gathered. If Units have been formed, and they begin to exceed the span of control, the Unit can be divided into sub-units. In most instances, this will meet the need of the incident and managing the span of control while not adding repetitive or bureaucratic red tape that does not need to be part of the incident.

As we move forward in this chapter, it will be plain to see that the Units used have very specific jobs that will be instrumental in managing the incident. The way these modular Units are designed (when utilizing the ICS method), it provides a divide and conquer framework, with each Unit having very focused jobs. The deliverables from this collaborative work does an amazing job in supporting the incident and the Operations Section especially.

10.1.1 Situation Unit

The Situation Unit reports to the Planning (and Intelligence) Section Chief (PSC) and has the primary responsibility of processing and organizing of all incident information. In many instances, the Situation Unit researches and reports future projections that are directly related to incident growth. They also provide maps and intelligence information that is based on those projections.

The Situation Unit is led by an individual who is identified as the Situation Unit Leader, who will acquire their initial briefing from the Planning Section Chief (PSC). They will be responsible for reviewing the ICS Form-201 (Incident Briefing Form) and other information to obtain the most recent incident status. They will work with subordinates to determine the most effective incident objectives and strategies as well as contingency plans in case the initial objectives and strategies (for the incident) do not work as planned. They also must attend planning meetings to assist in developing the needed plans to guide the incident, especially for the Command Staff and Operations Section.

The Situation Unit Leader will identify all the reporting requirements for the incident, as well as any time schedules that are specific to the incident. This should include external schedules, such as those set by the Planning and Intelligence Section Chief and the Incident Commander, as well as internal schedules that they will meet with the external schedules. They will assign tasks to staff members within the Unit, advise those staff members of the timelines and the format for their work, and they will monitor their progress. Additionally, the Situation Unit Leader will usually assign Field Observers (and potentially others) who will gather intelligence. The Situation Unit Leader is also responsible for requesting Technical Specialists when they are needed to support the Unit. Upon arrival of Technical Specialists or any other new personnel, they will give them a briefing on the current incident status, and they will typically supervise them. The Situation Unit will also work with, and identify, federal and state partners (as necessary) that can help support the unit.

The Situation Unit (as a whole) will be established in an area, building, or tent, where they will collect and analyze situation data. In establishing a base of operation, they will set-up computerized planning systems, create, print, and enter data into computerized forms. They will also obtain or create maps and take pictures. While this may sound like a simple task, the gathering of intelligence in incidents of a large magnitude is anything but easy for the Situation Unit.

As was mentioned previously, the Unit will be required to obtain a copy of the ICS Form-201 (Incident Briefing Form) and then glean what information they can from it. Based on that information, they will have to sort the data into useful categories of information. These could be categories such as the geographical area, the population, the facilities available or designated, the environmental risks, or other pertinent information. They will continue to gather intelligence on these categories during each operational period until the incident comes to a conclusion. During each operational period, they will disseminate the information they retrieved to the appropriate individuals at the Incident Command Post (ICP).

The Situation Unit will also be responsible for regularly reviewing all the data for completeness, accuracy, and relevancy prior to passing it on to those who need it. Throughout the process, they will generate and maintain a log of all the major events that occur, or that have occurred. It does not matter if those events are negative or positive, the purpose is to create accurate information so that informed decisions can be made. This information will be used to create a more complete Incident Action Plan (IAP).

Another responsibility of the Situation Unit will be plotting the borders of the incident, identifying the location of boundaries, identifying the locations of facilities, access routes, and other critical locations, and to create display maps. They will maintain those displays, ensuring that they are kept current and up to date. As required, the unit will also develop other mapping and displays such as weather reports, incident status summaries, geographical divisions, and other locations (or facilities) that they deem important to the incident. Depending on the incident, the Situation Unit may be requested to develop additional or specialized maps that could be needed to better understand the situation and the needs of the incident.

Along with the maps, the Situation Unit provides photographic services. These photographs will usually be taken to document operations and intelligence activities, public information activities, and to document accident investigations. The pictures taken could be from the ground and/or gathering pictures from the air.

The data, maps, and photographs that have been compiled, maintained, and presented to others are typically done to provide status information for the Incident Command Post (ICP) staff. This information will be used to provide an incident appraisal, prediction, and analysis for command and operations. The Situation Unit is also tasked with reviewing current and projected incident and resource status strategies. They will acquire the pre-attack plans and mobilization plans to obtain a more complete picture of what is likely to occur at the incident.

The intelligence that is collected will also aid in preparing information on alternative strategies. This is done for times when the original plans did not provide the expected results. When considering the alternative strategies, the unit must pre-identify the resources that would be required to undertake the contingency plan. Once the contingency plans have been created, the Situation Unit will document those alternatives for presentation to the Incident Commander (IC) and the Operations Section Chief (OSC), so that these contingencies can be included in the written Incident Action Plan (IAP).

The Situation Unit is critical in ensuring that the tactics and the planning are effective. Part of their responsibility is to gather information from Field Observers and various personnel that have just ended their operational period. The purpose of interviewing operational individuals and the Field Observer(s) is to evaluate the effectiveness of the strategies and tactics that have been put into place and to evaluate the work that has been accomplished. They will also be able to gather data on what work is still left to be completed through these interviews. By doing so, they are able to provide estimates related to situational needs. These estimates can be provided to the Command Staff (which promotes a more informed response), the Ground Support Units (so they can provide the assistance needed), the Logistics Section (to project future resource needs), and others that may need this intelligence to provide a more organized response. Portions of the information collected will also be posted on the Situation Unit and Command Post work displays at scheduled intervals.

Forms that the Unit may be required to fill out include the ICS Form 209 (Incident Status Summary) and ICS Form-214 (Unit/Activity Logs). Some states or local governments may require additional specific forms to be filled out by the Situation Unit. The Situation Unit, or their supervisor (the Planning Section Chief [PSC]), should check with the Authority Having Jurisdiction (AHJ) to determine what, if any, additional forms may need to be filled out.

A final consideration of the Unit (as a whole) is the final narrative report. The Situation Unit, in collaboration with others within ICS, will be responsible for ensuring that a final report is written. This written narrative should cover key points of interest from the beginning of the initial attack or actions, until the final demobilization. In order to complete this report, the Situation Unit will need to collect, review, and assemble all the appropriate information to provide a final report. When the report is complete, it should be submitted to the Incident Commander (IC).

The Situation Unit can expand, and usually will on larger incidents. This expansion can include additional staff that will assist the Situation Unit Leader in obtaining the specific information needed to create a more detailed level of situational awareness. As with any function within the ICS method, there will be predesigned modules that can expand to provide for the additional help needed. It is important to note that some of these expansions are for specific types of incidents (e.g. wildland fires, HazMat response), while other expansions will work for almost any type of incident.

10.1.1.1 Field Observer

The Field Observer is the individual (or individuals) that gather intelligence from the field. Their primary responsibility is to collect situation information from personal observations at the incident, and then provide a synopsis of this information to the Situation Unit Leader. This position, and the location for which they are to report and collect information, is assigned by the Situation Unit Leader. The Situation Unit Leader should provide a briefing to the Field Observer that includes the types of information that should be collected to provide greater situational awareness. In most instances, they will also receive a list of all facilities and their locations (Helispots, the Incident Command Post [ICP], Division and Branch boundaries, etc.) and a copy of the Incident Action Plan (IAP) so they can have a better understanding of what should be happening versus what is actually happening on the ground during operations.

In most instances, the equipment and supplies that a Field Observer may need can include a method of communication (radio, cell phone, etc.), personal protective clothing, a camera (digital or disposable), and a way to document what they observed (a tablet, laptop, or pen and paper). Throughout their operational period, they will document and report this information to the Situation Unit Leader in the manner that is required for that unique incident (e.g. radio, phone, text message, written report). Throughout the operational period, the information and intelligence these personnel collect should help to provide more accurate predictions (NWCG, 2018b). If a condition arises that puts any personnel at risk of injury or death, the Field Observer should report this information immediately. Typically, this information would be reported to the Safety Officer (SOFR) or one of the Safety Officers (SOFRs) subordinates. Throughout each operational period, the Field Observer should maintain an ICS Form 214 (Unit/Activity Log).

10.1.1.2 Geographic Information System Specialist (GISS)

The Geographic Information System Specialist (GISS) will collect data from others in the Situation Unit, or those working with the Situation Unit, and transfer those field observations and infrared interpretations into digital geo-referenced format. This is done in a multitude of ways, including downloading Global Positioning System (GPS) data and integrating that data into the incident database. The downloaded data and the field observations are then used to create highly accurate maps (NWCG, 2014b).

Perhaps the most frequent use of this Geographic Information System (GIS) mapping is in the wildland firefighting arena. The mapping provided allows those working on the scene of a wildland fire to have a more comprehensive view of what is occurring and to plan for most scenarios. In the wildfire arena, these maps are used for:

- Incident Action Plan (IAP) mapping
- Situation/Planning Maps
- Operations Maps
- Transportation Maps
- Progression Maps
- Air Operations Maps
- Public Information Maps
- Ownership Maps
- Damage Assessment Maps
- Structural Protection Maps
- Rehabilitation Maps
- Facilities Maps
- Infrared Information Maps
- Fire Perimeter History Maps
- Vegetation Maps
- Fuels Maps
- Contingency Maps (NWCG, 2014b)

In addition to providing these maps, they will likely be required to calculate fire statistics, such as the number of acres that have burned, to what extent the fire has advanced, and the location of the fire line including how long the fire line is, based on the data they collected. To keep an ongoing record, the Geographic Information System Specialist will create daily backups that include all of the incident GIS data. This often is used to document the progression of the fire and offers spatial fire perimeter data for distribution to data and national coordination centers as needed or required. It is important to note that GIS mapping will not only be used for wildland fires but also flooding, earthquakes, and a multitude of other disasters.

In other disasters, the use of a satellite image can be an important factor in understanding the magnitude. Utilizing remote sensors, satellites, infrared imaging and other types of data, the Geographic Information System Specialist can provide maps and data that swiftly estimate the magnitude of an incident. Some of the following examples that are given will be dependent on the resolution of the images received; however, it is important to realize that all are possibilities. GIS mapping can identify the amount of flooding and number of structures affected, the magnitude of a landslide(s), the number of collapsed buildings (in an earthquake or tornado), and they can even identify impassible roads (Department of Homeland Security, 2016). Additionally, the GIS system data can be used to measure the level of ground distortion in the wake of an earthquake, continually expanding sink holes, volcanic eruptions, and the level of erosion that has occurred after a hurricane or tsunami.

The Geographic Information System Specialist can also help to expedite the response after disasters. With GIS information, personnel can use GPS to find the quickest route into specific area within a disaster zone. Depending on the resolution of the imagery, this GIS information can usually identify the hardest hit locations, identify damaged or impassible roads and bridges, as well as help map alternative routes. During the onset of a disaster, those making decisions may need to know the areas that had the highest population, so they can send additional personnel to that area additional help by providing an overlay of the population on a map. GIS can be also be used to glean information from census data to make an assessment of how many people may need temporary shelters, and where the best location might be to place those shelters. GIS mapping can also help to predict wind speed and direction after a hazardous material release or even a radiological/nuclear incident and provide mapping for evacuation zones, sheltering in place, and a multitude of other information (NWCG, 2014a).

When producing GIS maps, the Geographic Information System Specialist will be required to create original copies of the maps for the individuals that might need them. This can include almost anyone involved in the response and those that are coordinating the response from remote locations. Whenever collecting data and creating these maps, the Geographic Information System Specialist will need to ensure that they provide them in a timely manner to meet the demands of that operational period.

While what was written about the Geographic Information System Specialist has focused on response, and a person (or persons) job in the ICS method, it is important to note that GIS mapping can, and usually should be, used in all facets of disaster management. In the planning stages, prior to a disaster, GIS can help gather data that can identify areas of concerns, the amount of the population that might be affected, areas that will be affected by flooding or chemical spills, and a whole host of other information. GIS can also be used to identify those with lower socioeconomic status that are more vulnerable to disasters, creating a focus of planning direction of where more personnel will likely be needed in a response. In the planning phase, GIS information can help guide a risk assessment, even changing the focus of what risks are more pressing. It is possible that the GIS information may lead to more or different mitigation measures that will help build resilience.

In the recovery mode, GIS can help determine what geographical locations should restrict land use, including rebuilding houses and infrastructure in specific areas prone to disaster. GIS can help identify where additional zoning can focus on resiliency and where other mitigation measures may pay off in reducing the effects of the next disaster. GIS can be used to overlay historical disaster data into proposed civil engineering rebuilding projects, showing where historical disasters might affect the completed project in the future. By integrating the GIS information into a rebuild, it creates greater resiliency in the future that will be able to withstand greater natural forces. Introducing GIS mapping into the planning phase of rebuilding levy's and dams can help identify how to plan for a worst-case scenario. GIS mapping can also help document the recovery process and the strides that have been made because of previous mapping efforts.

While these examples are just a few, it is easy to see that GIS data and mapping is an important aspect of planning for, responding to, and recovering from a disaster. In some geographical areas, GIS is a way of life for public safety, while other areas (mostly rural areas) have no idea what it is or what it can do. As public safety becomes more aware of what GIS can do, they will likely become more reliant on it. There should be one word of caution though, and this word of caution should span the spectrum of public safety; while technology has changed the way we work, it is important for us to remember how we did it before technology. We never know when a solar flare, an electromagnetic pulse (EMP), or some other incident could render some or all of our technology useless. For this reason, public safety should utilize new technology but also remember and train on previous technologies as well.

10.1.1.3 Display Processor[1]

The Display Processor is primarily responsible for displaying incident status information. The information that they will display is typically gathered from the Situation Unit and can include information obtained from Field Observers, casualty reports, resource status reports, aerial and drone photos, photos collected from operational staff, Field Observers, social media, news crews, etc., as well as information from other data sources. Display Processors will work with the Situation Unit Leader to identify the equipment and supplies that will needed to complete the task and to set time parameters for developing and presenting the displays needed to support the incidents operation and planning functions (NWCG, 2009).

Upon being appointed the Display Processor, they should receive a briefing from the current Situation Unit Leader (NWCG, 2009). After the briefing, the Display Processor will need to determine how many displays will be needed, the types and locations of the display, and what displays should take priority. The Display Processor will also identify what types of maps will be needed to support the incident and where those maps should be posted. While determining the maps needed for the incident, they should not forget to identify the maps that may also be needed to support the Incident Action Plan (IAP). Display maps for the Incident Action Plan (IAP) can be extremely important to the success of the mission. They will also glean this information from field reports, the Incident Action Plan (IAP), and in working with the Situation Unit Leader to help them to identify and set priorities. Finally, the Display Processor should fill out an ICS Form 214 (Unit/Activity Log) to document the activities they undertook throughout the operational period that they worked (NWCG, 2009).

10.1.1.4 Infrared Interpreter[2]

The Infrared Interpreter is an analyst who is employed when aerial infrared imagery is used. Because of their specialized skills and training, they are able to specialize in interpreting aerial imagery and articulate and explain those findings on aerial photos and/or maps. They will use their analyzation skills to make independent decisions about the data they receive (NWCG, 2018c).

Depending on how the Authority Having Jurisdiction (AHJ) places this position, they will either report to the Situation Unit Leader, or they will be under the supervision of the Field Observer. While being under the direction of the Field Observer is rare, it is practiced in some jurisdictions. Whenever possible, it is best to place them under the direction of the Situation Unit Leader, only because it is the common practice used in the majority of jurisdictions.

As part of their job, they will determine what aircraft and satellites are available to provide infrared data, where that equipment is located, and how to access that equipment. In cases where this can be obtained from available aircraft, they will maintain direct communications with infrared crew liaison and arrange specific missions. In arranging this with the infrared crew liaison, they will outline the objective of each flight, the time(s) of day the mission should be flown, and the number of flights that should be flown each day. They may also be required to set parameters for specific circumstances that may require special or out of the ordinary flights and communicate any specific areas that may need special attention (NWCG, 2018c).

Working in conjunction with the infrared crew liaison, the Infrared Interpreter will also set parameters for where imaging should be delivered. They will also specify how the imagery

---

1 This heading is for advanced IMS students.
2 This title signifies highly technical information for advanced students and practitioners.

should be delivered, and the specific times when it should be delivered in order to provide the incident information on time to the Planning Meeting. This is important because the Infrared Interpreter will work independently to develop the progress of the fire and the efforts made by Operations through GIS mapping resources, and those interpretations to the Planning Section at or before a prearranged time. On complex or extremely large incidents, the Infrared Interpreter may be required to supervise or work in collaboration with one or more data and mapping individuals. Because of the data that is provided, this job usually has a significant effect on the response (NWCG, 2018c). As with almost every other Unit in the ICS method of incident management, they will also be required to maintain the ICS Form-214, Unit/Activity Log.

10.1.1.5 Fire Effect Monitor[2]

The Fire Effects Monitor can be placed under the direction of Field Observer, or they can work equally and collaboratively with a Field Observer as a team. Depending on how the Authority Having Jurisdiction (AHJ) places this position, they will either report to the Situation Unit Leader, or they will be under the supervision of the Field Observer. While being under the direction of the Field Observer is rare, it is practiced in some jurisdictions. Whenever possible, it is best to place them under the direction of the Situation Unit Leader, only because it is the common practice used in the majority of jurisdictions.

The Fire Effect Monitor is responsible for collecting the onsite weather conditions, current fire behavior, and other effects of the fire information that is needed. The purpose of this job is to assess whether the fire is achieving the objectives of the incident response and to report those to the appropriate individuals (Wildland Fire Assessment System, n.d.).

The appropriate person for the position of Fire Effect Monitor will need to be someone that is very detail oriented. Prior to reporting to their work assignment, they will need to review the monitoring plan prior to implementation to ensure that they understand all data that should be recorded for that incident. They will also need to make sure that they monitor the weather, obtain future forecasts, and record all weather data related to an incident. At the same time, they will also monitor and record the fire behavior throughout the entire operation. They will monitor and record smoke information, to identify any issues that may occur with the smoke associated with the fire. Additionally, it may be required of the Fire Data Monitor to collect and record how the fire was affecting environmental issues and record that data so that those in charge of the response can potentially help mitigate any environmental issues (Wildland Fire Assessment System, n.d.)

In cases of prescribed burns, they may be required to undertake reconnaissance before and after the prescribed burn. This is usually done to determine if the prescribed burn had the intended effects. They may also be required to plot the expected burn area and the perimeters on a map prior to, during, and after a prescribed burn. This is done to ensure that the prescribed burn was within the geographical boundaries set prior to the burn and that it remained within those boundaries from the onset (Wildland Fire Assessment System, n.d.).

Whether the Fire Data Monitor is working on an actual fire or prescribed burn, it does not matter. At the end of either type of fire, they will provide a monitoring summary of the fire. This is done so that data from the fire can be analyzed to give a better understanding of fire behaviors that may occur in future fires.

10.1.1.6 Fire Behavior Analyst[2]

A Fire Behavior Analyst is someone that has specialized training and knowledge in predicting wildland fires. They take current and future data and makes predictions about what they believe wildland fire will do. This is especially useful intelligence that can provide the Incident

Commander (IC), Command Staff, and General Staff with intelligence that can allow them to reduce the spread of the fire, or in some instances, keep aircraft and personnel out of dangerous situations (NWCG, 2018d).

Upon reporting to the incident, they will receive a briefing from the Situation Unit Leader or the Planning Section Chief (PSC). Depending on how the Authority Having Jurisdiction (AHJ) places this position, they will either report to the Situation Unit Leader or they will be under the supervision of the Field Observer. While being under the direction of the Field Observer is rare, it is practiced in some jurisdictions.

In order to predict fire behavior, the Fire Behavior Analyst will need certain items that will allow them to collect data, evaluate the situation, and to make accurate predictions. They will need to acquire specific items such as incident maps, a mode of transportation, a two-way radio, and the current weather information. In most instances, these items will be provided to them after receiving their initial briefing. They will then begin the process of collecting the data they need to make the needed predictions (NWCG, 2018d).

As part of their job, they will almost immediately begin to gather information so that they can provide fire behavior predictions. They will monitor the weather, weather predictions, identify the fuels feeding the fire, and how the fire behaving in those conditions. They will also collect information about the origin of the fire, wind characteristics, the moisture in the fuel slope types and angles of slope, fire suppression activities, fire status, and areas of special concern such as private property, structures in the path of the fire, housing developments, etc. From this information, the Fire Behavior Analyst will interpret these factors to make predictions about fire behavior. They will then document and provide summaries of the likely fire behavior. In most instances, they will be asked (or required) to present this information at the planning meeting. These predictions will be important for the planning of the next operational period. These summaries, or updated versions of them will also be presented by them at the operational briefing, prior to sending personnel into the field at the beginning of an operational period (NWCG, 2018d).

In some instances, the Fire Behavior Analyst may need assistance in gathering data. When needed, assistants can be assigned to them. In these instances, they will also be responsible for ensuring that the work area is organized for optimal outputs and that they provide direction and supervision to their subordinates. As with almost every other Unit in the ICS method of incident management, they will also be required to maintain the ICS Form-214, Unit Log.

10.1.1.7 Long-Term Fire Analyst[2]

The Long-Term Fire Analyst is very similar to the Fire Behavior Analyst, and they gather much of the same information and fulfill many of the same duties. They will provide similar reports and fill out most of the same ICS Forms. The primary difference is that they predict long-term expected fire behavior, so they will require additional information. That additional information they collect may be drought predictions, satellite imagery for weather predictions, access to remote sensors from varying government agencies, and more. From this information, they will provide the most probable long-term fire behavior, including direction of spread, and rate of burn, based on local information, topography, historic and current fire spread, and with historic and current fire weather data (NWCG, 2018e).

10.1.1.8 Technical Specialists[1]

Depending on the needs of the incident, the use of a Technical Specialist or multiple Technical Specialists that have and maintain a specialized expertise (or knowledge) may be required. Technical Specialist often fall under the supervisory control of the Situation Unit Leader;

however, they can be assigned wherever their services are needed and best support an incident.

These Technical Specialists are considered Subject Matter Experts (SMEs) in their field, and they can provide valuable input on what may occur if specific actions are (or are not) undertaken. Sometimes, Technical Specialists can be used to enhance an already staffed position with more information, additional resources for a specific position, or even in providing someone with a higher or more current level of experience and/or education. As an example, a Technical Specialist who is considered a Subject Matter Expert (SME) in Geographic Information Systems (GISs), can be used to provide additional support, knowledge, and resources to a Geographical Information System Specialist based on the latest technology.

In most instances, these Technical Specialists will participate in the development of the Incident Action Plan (IAP), especially if their input may change the response in some way. They will often provide input for potential alternative strategies to help mitigate certain issues or to provide information to help bring an unknown situation (within their field of expertise) under control. The Technical Specialist, much like most other individuals in the Incident Command System must fill out an ICS Form 214 (Unit/Activity Log). Some common Technical Specialists utilized include:

- Weather Observer Technical Specialist
- Environmental Impact Technical Specialist
- Training Technical Specialist
- Chaplain Emergency Response Technical Specialist
- Critical Incident Stress Management (CISM) Technical Specialist
- Family Assistance Coordinator Technical Specialist
- Human Resources Technical Specialist
- Salvage and Engineering Technical (SET) Technical Specialist
- Geographic Information System (GIS) Technical Specialist
- Public Health Technical Specialists
- Legal Technical Specialist
- Documentation Technical Specialist

This is not an all-encompassing list. Because there are such a wide variety of specializations, Technical Specialists could come from almost any type of specialization.

As is common with all positions on an expanding incident, the Technical Specialist should initially check in with the Status/Check-In Recorder upon arrival, followed by receiving a briefing from their supervisor. When assigned under the purview of the Situation Unit, they will receive their briefing from the Situation Unit Leader. Whenever assigned to another Section or Unit within ICS, they will receive that briefing from the immediate supervisor. Upon completion of the briefing, the Technical Specialist should be issued the appropriate Personal Protective Equipment (PPE) if they do not have their own. The Technical Specialist's supervisor will need to decide the most viable method of coordination with other Sections, Units, and local agencies, and they will need to establish work area with the needed work materials if required. Once these issues are determined, then the Technical Specialist should work within the given parameters. They will be responsible for filling out forms identified by their supervisor, and they are also demobilized in the same manner as others in the ICS method.

10.1.2 Resources Unit

Maintaining the ongoing status of all resources that are assigned to an incident is an important aspect of managing not only the resources but also the entire response. The Resources Unit is critical to the success of incident response because they supply almost all of the physical needs for resources working on the response. The Resources Unit tracks all resources, maintains records, and keeps a running log of where each resource is at any given time. Knowing where resources are at all times is an extremely important part of ensuring the safety of each individual on the incident.

The Resources Unit is supervised by a Resources Unit Leader (RESL). The Resources Unit Leader (RESL) is in control of ensuring that there is an ongoing maintenance regarding the status of all resources. This includes both primary (tactical) resources and support resources that are assigned to an incident. The Resource Unit Leader (RESL) and their subordinates oversee and maintain the check-in/out of all resources and maintain a status system that maintains a current location and status of all resources. In most instances, this Unit also is assigned with overseeing volunteers. The Resources Unit Leader (RESL) and their staff are responsible for insuring that a master list of all resources is kept and maintained. That list should include key supervisory personnel, primary resources, and support resources, including volunteers.

In smaller incidents, this position may be handled by the Incident Commander (IC) or the Operations Section Chief (OSC); at least in the initial stages. In larger incidents where the hierarchical organizational has expanded, but not enough to warrant a Resource Unit, the Planning Section Chief (PSC) may undertake this job. If the incident has expanded and the span of control is at its limits for the Planning Section Chief's (PSC's), then they will usually create and be the supervisor of the Resources Unit.

Resource Management is an important part of effective incident management. While the Resources Unit will track the resources being used, resource management is more than just identifying the resources that are utilized. According to ICS-300 courses, each Section bears some responsibility for resources management (Illinois Terrorism Task Force, 2011).

- *Command*: Develops incident objectives, approves resource orders, and demobilization.
- *Operations*: Identifies, assigns, and supervises resources needed to accomplish the incident objectives.
- *Planning*: Tracks resources and identifies any resource shortages.
- *Logistics*: Orders and supports the needs of those resources.
- *Finance/Administration*: Pays for resources and any needed items (e.g. fuel, food, housing, maintenance) to support those resources (Illinois Terrorism Task Force, 2011)

While these different functions provide some resource management, the Resources Unit Leader (RESL) and any staff they may have are the individuals who are responsible for tracking resources. Upon an individual being appointed as the Resource Unit Leader (RESL), this individual should check-in with the Logistics Section, then receive a briefing from their direct supervisor. Upon completing the briefing, the Resource Unit Leader (RESL) will begin determining what functions need to be undertaken, if an assistant is needed to assist, as well as how many other staff may be needed. This is determined based on the needs of an incident.

The need for check-in/out areas is a necessary part of the Resources Unit. For this reason, the Resources Unit Leader (RESL) will need to appoint a Status Recorder and determine where check-in/check-out areas should be located. They will often form a Status Recorder Team to document the status of all resources, and if needed, they may initiate a Volunteer Coordination Team. In creating these teams, the Resource Unit Leader (RESL) may find it necessary to appoint Team Leaders in each of the teams, or just some teams, to keep the span of control manageable.

10.1.2.1 Status Recorder

Status (also known as Check-In) Recorders (SCKN) are the individuals who will be placed at each check in location. This is done to ensure that all resources assigned to an incident can be accounted for. Whenever possible and practical, the Demobilization Unit Leader (DMOB) and other Demobilization personnel can be used to fulfill the position of Status/Check-In Recorder (SCKN). This will ensure that all the information that is required to demobilize an incident can be obtained at check-in. In substantially larger incidents, separate personnel will be needed; however, in most instances, the Resources Unit Leader will need to work collaboratively with the Demobilization Unit Leader (DMOB) to ensure all information is obtained.

The Status Recorders (SCKN) are under the direction of the Resources Unit. Status Recorders will provide others working within the ICS method to track the status of each resources. Not all tactical resources that are dispatched to an incident will be available 24 hours per day, seven days per week, during the entire duration of that incident. Resources dispatched to an incident may be designated as temporarily out-of-service for various reasons, including mechanical failures, sickness, rest periods, and more. The Status Recorder will usually classify resource status in one of three ways. Those resource statuses are as follows:

- *Assigned*
  - When a resource is classified as assigned, this resource is working on an assignment, and has a Supervisor.
- *Available*
  - Available resources are typically waiting for an assignment at a staging area. They have been gathered, issued their equipment, and they are ready for deployment.
- *Out-of-Service*
  - These resources are not available to be used at the incident.

Resources may be out-of-service for variety of reasons. In some instances, it may be mandated down time, while in other cases, it may be accidental or preventative. Some of the reasons they may be out of service include:

- Maintenance, repair and/or servicing of vehicles or equipment.
- Downtime needed to ensure the physical and mental health well-being of the personnel working at the incident.
- Weather or other operating conditions such as visibility due to nightfall or fog.
- Not enough qualified personnel available to operate a specific type of equipment or to undertake a specific task.
- Contractual obligations, such as when personnel exceed allowed overtime costs.
- Legal obligations such as labor laws.

Resources that are classified as out-of-service are typically sent to the Incident Base, except in cases when there is a mechanical issue or a staffing reason. The changing of the status for an individual resource is transferred from one status to another status by the supervisor who is, or has been, utilizing this resource. It is important to note that the Resources Unit does not have the authority to change the status of a resource that is assigned to the Operations Section. It is also important to note that in most instances, the resource itself should not be able to change their status. This is done so a resource cannot change its status from being assigned to being out-of-service without their supervisor approving that change or to avoid not knowing that a resource had left while they are counting on that resource to accomplish a task.

Based on the overall incident organization, the size of the incident, and the persons who supervise the resource (either directly or indirectly), the person that authorizes a change in status may vary. In most instances, those who can change the status of resources include

- The person in charge of the single resource.
- A Task Force or Strike Team Leader.

- A Division or Group Supervisor.
- A Branch Director.
- The Operations Section Chief.
- The Incident Commander.

Status changes should be communicated through the appropriate channels in the organization, ensuring that the correct individuals are notified. Exactly who gets notified will vary and will be dependent on the structure of the response and how the organizational hierarchy is setup. The individual who is responsible for making the status change is also responsible for making sure that the change is communicated to those within ICS who need to know this information. As a rule of thumb, this will almost always include the Resources Unit, but others will most likely need to be notified as well. This may be reporting the change in status to those higher in chain of command and/or to the person or Unit responsible for documenting and sustaining the overall resource status list on a given incident.

For common changes in status such as lunch breaks, end of shift, or other expected minor changes of status, the person maintaining the information will only need to report it to their immediate supervisors. If the status change is the result of mechanical issues, a change in conditions, a shortage of equipment or supplies, or something unexpected that might affect the Incident Action Plan (IAP), then the information should be shared more extensively and usually should include notifying the Operations Staff and, in most instances, the Incident Commander (IC). When sharing information about a change of status, this sharing of information should be directed through the Resources Unit so that the change can be documented and kept in the Unit that monitors the ongoing status of every resource. By directing it through the Resources Unit, it reduces the chance that this information will be missed, and it reduces the possibility for confusion because those individuals in Command and/or Operations did not get the information.

With a mandate to track resources, there are varying ways that the Resources Unit can accomplish this task. Usually, the most reliable method is doing so through manual recordkeeping on standard ICS Forms. The most common forms used are as follows:

- ICS Form 201-Incident Briefing Form
- ICS Form 211-Incident Check-In Form
- ICS Form 204-Division/Group Assignment Form Card System
- ICS Form 219-(1–10) Resource Status Card (T-Card)

The ICS Form 219 (Resource Status Card) uses different colors of card stock to make the identification of the type of resource at a glance. This form is T-shaped because the cards are filled out so that the information is recorded, then placed in a rack by the assignment location (see Figure 10.2). The typical colors used represent the following.

- 219-1: Gray Card – This is a header card, used only as way to label the top row of the T-Card racks.
- 219-2: Green Card – Signifies an organized Crew or Team.
- 219-3: Rose Card – Signifies an engine crew.
- 219-4: Blue Card – Signifies a helicopter.
- 219-5: White Card – Signifies personnel.
- 219-6: Orange Card – Orange signifies a fixed-wing aircraft.
- 219-7: Yellow Card – Signifies equipment.
- 219-8: Tan Card – Signifies miscellaneous equipment or a Task Force.
- 219-10: Light Purple Card – Is a generic card for resources not already identified.

ST/Unit:	Name:	Position/Title:

Front

Date/Time Checked In:
Name:
Primary Contact Information:

Manifest: ☐ Yes ☐ No	Total Weight:

Method of Travel to Incident:
☐ AOV ☐ POV ☐ Bus ☐ Air ☐ Other

Home Base:
Departure Point:

ETD:	ETA:

Transportation Needs at Incident:
☐ Vehicle ☐ Bus ☐ Air ☐ Other

Date/Time Ordered:
Remarks:

Prepared by:
Date/Time:

ICS 219-5 PERSONNEL (WHITE CARD)

ST/Unit:	Name:	Position/Title:

Back

Incident Location:	Time:

Status:
☐ Assigned ☐ O/S Rest ☐ O/S Pers
☐ Available ☐ O/S Mech ☐ ETR: ____
Notes:

Incident Location:	Time:

Status:
☐ Assigned ☐ O/S Rest ☐ O/S Pers
☐ Available ☐ O/S Mech ☐ ETR: ____
Notes:

Incident Location:	Time:

Status:
☐ Assigned ☐ O/S Rest ☐ O/S Pers
☐ Available ☐ O/S Mech ☐ ETR: ____
Notes:

Incident Location:	Time:

Status:
☐ Assigned ☐ O/S Rest ☐ O/S Pers
☐ Available ☐ O/S Mech ☐ ETR: ____
Notes:

Prepared by:
Date/Time:

ICS 219-5 PERSONNEL (WHITE CARD)

Figure 10.2 ICS Form 219-5 Personnel Card (FEMA, n.d.b).

The ICS Resource Status Cards are a manual way of tracking resources. Each color of card, based on the type of resource, has different information that requires that specific information be filled out. While some of the information needed between different colors of T-Cards may be similar, they are specifically created to gather differing pertinent information on varying resources at a glance. This is done because the information that may be needed for a fixed wing crew would be different from that of the information needed for an engine company. By the same token, different information is needed for fixed wing aircraft and helicopters.

Besides T-Cards, there are other methods of tracking resources which include the use of computers, including utilizing various hardware and software that can speed up how long it takes to enter information. In some instances, specialized commercial software is used not only to track resources but also in managing the overall incident. These types of incidents can use computer chips, magnetic cards, or other technology to assist in the tracking of resources. When using this type of tracking system, as a resource goes into service for the first time on the incident, they will be asked information and then be handed a plastic card similar to a credit card. The card will have all the pertinent information about the resource, programed into the card. When that resource checks in, checks out, or has a change in status, their card will be swiped or inserted, and it will electronically track that resource from one status to another.

In past instances, some resources working on an incident have been managed by using Microsoft Excel, open source software, and other spreadsheet types of programs. While this type of software can be helpful, it can also be a double-edged sword. Depending on how the program is setup, entering the data for each status change can eat up valuable time that could be used for other tasks. Of course, technology changes quickly, so this could change to being automated in no time.

It should be noted that ICS Form 219 and computer software programs are not the only way to track resources. There are numerous other ways to track resources, including (but not limited to) the use of pen and paper and using a dry erase board, and/or a magnetic board with specialized symbols. Many of these methods were utilized when the ICS method was in its infant stages, and they still work today.

When dealing with resource management, the key thing to remember is that the Status Recorder's (SCKN) job is not just in place to understand resource status for work purposes, but also for life safety reasons as well. Accountability will help to ensure that the whereabouts of personnel are known at all times. In the event that an evacuation is ordered for a specific area of the incident, or if the unforeseen happens (such as an explosion), then it is critical to know where the resources are. It is also important to quickly find personnel should a family emergency happens, and they need to be informed.

Status information should be regularly updated and validated to ensure that when providing resource information for the Incident Action Plan (IAP), that the information is accurate. That information will be documented using ICS Form 209, Status Summary, as well as providing an updated hierarchical command structure using ICS Form-207, Incident Organization Chart.

A key consideration for the Resources Unit and the Status Recorder (SCKN) is to identify the best location(s) to locate check-in/check-out locations and then establish them where there is a higher likelihood that resources will arrive. Common areas that are typically utilized for check-in/check-out locations include:

- The Staging Area
- Boat ramps (typically when working on water)
- Helibases (when helicopters are used)
- Division or Group locations
- Bases and/or Camps

- Incident Command Post (ICP)
- Security Check Points
- Other gathering areas where it is noticed that resources are arriving

At these locations, the Status Recorders (SCKN) will ensure that proper signage is provided to ensure that they are easily identifiable during both day and night to incoming resources. The Status Recorders will also establish a time schedule to communicate a detailed list of the resources they have documented, and the best method to communicate the latest resource information, including the number and type of resources, to the Resources Unit Leader (RESL). The method to communicate should be determined in collaboration with the Communications Unit. The method of communicating this information can include using a fax machine, a telephone, a radio, a computer messaging system, or by email. In some rare instances, the lack of communication or communication infrastructure (at that time) may require the use of a runner.

The Status Recorder (SCKN) position will be open and active until the last external resources have left an incident. While a majority of the check-in/out locations may be closed before all the resources have left, at least one Status Recorder team should remain open. In closing down the locations, it should be considered that the location should not be closed until every resource that checked in, or was assigned, to that location for that operational period has been checked out. Upon closing a location, all data collected from that Status Recorders (SCKN) should also be filed with the Documentation Unit.

10.1.3 The Documentation Unit

The Documentation Unit is responsible for maintaining accurate and complete incident files. They maintain a complete record of the most important steps taken to resolve the incident, provide copying and duplication services for incident personnel, and they file, maintain, organize, and store files related to the incident for legal, logical, analytical, and historical purposes. Typical duties of the Documentation Unit include:

- Proofreading, editing, copying, collating, and assembling the Incident Action Plan (IAP)
- Proofreading the Incident Action Plan (IAP)
- Editing the Incident Action Plan (IAP)
- Copying the Incident Action Plan (IAP)
- Collating the Incident Action Plan (IAP)
- Assembling the Incident Action Plan (IAP)
- Disseminating the Incident Action Plan (including printed, faxed, and electronic dissemination)
- Creating a *history file* and documenting pertinent information to enter into that file
- Make appropriate copies of records
- Scan records into an electronic format
- Organize incident files according to National Archives and Records Administration (NARA) requirements
- Make copies of all written documents
- Make copies of maps needed for the incident and include a copy for the files
- Make copies of electronic records (Compact Disk (CD's), emails, instant messages, etc.)

While this may seem like a lot of responsibilities, this is only part of the Documentation Units duties. As the incident comes to an end, the Documentation Unit is tasked with even more work. Demobilization requires a multitude of work to be completed by the Documentation Unit so the incident can be closed out. This work is critical to making sure that the incident comes to a proper close, and the work done may be extremely important if a lawsuit arises or claims such as negligence are made after the incident is over. For this reason, the Documentation Unit has very specific demobilization duties as well. These duties can include the following:

- Ensuring that all required documentation has been copied and filed.
- Ensuring that all forms have been completely filled out.
- Ensuring that no documentation is missing.
- Transfer all records to the appropriate person, persons, or entity.

Perhaps one of the more important and time sensitive duties of this Unit is to compile and publish the Incident Action Plan (IAP). Because the Incident Action Plan (IAP) is critical to the overall success of mitigating and ending the incident, this is as critical aspect. The Documentation Unit will also be responsible for maintaining these files and records that were created to support the overall Incident Action Plan (IAP), and the planning function.

When looking at the Incident Action Plan (IAP), it is easy to see that the Documentation Unit has an important role in the final appearance and the deliverable information in the plan. After the planning meeting is concluded, the Documentation Unit will be responsible for obtaining the information for the next operational period planning, organizing the plan, proofreading it, organizing the final plan, copying it, and collating it. Although this seems to be a simple task, most personnel do not understand how widespread this Incident Action Plan (IAP) will be disseminated during an incident. Suggestions for the dissemination of the plan and the number of copies for each include:

- *Command Staff*: One copy per Command Staff and one for all of their personnel.
- *Public Information Officer (PIO)*: Delivery of a bundle of multiple copies with the Public Information Officer (PIO) determining the quantity needed.
- *Section Chiefs*: One copy for them plus an additional copy for each assistant and/or trainee they might have.
- *Unit Leaders*: One copy for them plus an additional copy for each assistant and/or trainee they have.
- *Group and Division Supervisors*: One per supervisor.
- *Units*: One per unit leader, and a copy for each assistant.
- *Strike Teams*: One copy per each personnel assigned to the team.
- *Task Force*: One copy per each personnel assigned to the Task Force.
- *Specialized Teams*: One per each personnel assigned to the team (e.g. HazMat, Special Weapons and Tactics (SWAT), Emergency Response Team (ERT's), Evacuation, Explosive Ordinance Disposal [EOD]).
- *Self-propelled Pieces of Operational Equipment (e.g. engine, ambulance, boats, bulldozers, backhoes)*: one copy per piece.
- *Helibase*: Delivery of a bundle that takes into account of one per each active person working the incident.

Additional extra copies will also be needed for incoming personnel, including all ordered or expected resources, and to overcome shortcomings (if any) from the last printing. In looking at this list, it is easy to see that the Documentation Unit could be responsible for assembling hundreds, if not thousands of copies of the Incident Action Plan (IAP) for dissemination in a larger incident. This of course will be dictated by the size and scope of the incident, as well as the number of resources that are active personnel.

One of the more important pieces of equipment that the Documentation Unit requires is a photocopier. Because of the amount of copies that will be needed, a standard small office copy machine will usually be insufficient for the needs of an incident over a long period of time, or that has a large number of personnel. The Documentation Unit will most likely need to create maps that are not standard 8.5 in. by 11 in., but rather 11 in. by 17-in. size. It is usually in the best interest of the Documentation Unit to work collaboratively with the Supply Unit to ensure that a copier is ordered that will meet their needs. This should be done as quickly as possible after the initial formation of the Documentation Unit.

In some instances, the Documentation Unit may need to include the Planning Section Chief (PSC) in the copy machine and supply ordering process to help this Unit identify what the long-term needs of the incident might be. This will help guide decisions about number of copying machines, the capabilities (e.g. pages per minute, dimensions of copies needed, automatic collating) as well as any support or repair services that may be needed from the vendor or supplier. In most instances, the response to the incident will not last more than a few weeks to a month. That is why it is often more cost-effective to rent a copy machine than to buy one.

While waiting on the delivery of the professional copy machine, various other copy machines can usually be found at fire stations, police station, city halls, and sometimes even small businesses that are willing to help in filling the gap in the initial onset. If these resources do not provide sufficient capabilities, a local commercial copying company or print shop could be used in the interim. The Documentation Unit Leader (DOCL) will need to identify alternative potential resources, then include the Planning and Intelligence Section Chief (PSC) to finalize any agreements. The agreements that are made must follow protocols for the incident to ensure that it is done in the proper manner and that it meets the ordering process.

While using temporary or temporarily available machines, it is important to make sure that supplies to support the Documentation Unit are on-hand. This would include sufficient toner, ink, paper, staples, and any other supplies that will be needed. If a town has been evacuated, stores are closed, or the Documentation Unit is working through the night to produce the Incident Action Plan (IAP), then a lack of supplies could impede the response to the incident.

If using a local commercial copy service or a print shop, it is important to ensure that they have the resources, including sufficient staff, to support the Documentation Unit. The issue of how many supplies will be needed for the given period of time should be discussed prior to entering into any agreement with a commercial company. Even if they claim they have sufficient supplies, it may be in the best interest to confirm this by taking a rough inventory of their supplies, or in purchasing additional supplies that can be used when the commercial copiers arrive on scene. This could also be a contingency plan in the event the business cannot meet the required needs.

Depending on the geographical location of the incident, there may be a mobile office service available to support the Documentation Unit. These services usually arrive in a trailer that has an assortment of equipment that is already setup, and in most instances, they are self-sufficient including providing for their own electricity needs. These mobile units are typically outfitted with a document scanner, commercial copiers, a multitude of supplies, and knowledgeable staff capable of operating 24 hours per day, seven days per week, for as long as needed. While typically a more expensive option, this option would require less personnel in the Documentation Unit because the vast majority of the Unit's photocopying tasks can be accomplished by this service. It is important to interface with them on how the documentation products (especially the Incident Action Plan [IAP]) are to be formatted and handled.

When contracting for copying services that will support the Documentation Unit, it is extremely important to have contractual agreements that spell out what happens if the machine or services are not available for a set period of time. While the Documentation Unit Leader (DOCL) will usually arrange the service, the Administrative Contracting Officer (ACO) may need to amend the contract as needed. In most instances, this preapproval will be granted by the Procurement Unit Leader (PROC). Some incidents order complete mobile office services under an Emergency Equipment Rental Agreement (EERA). It is important to make sure that the method of payment and schedule for payment is understood in advance of committing to the service.

Another key consideration for the Documentation Unit is establishing priorities for copying that coincide with the Planning Section timeframes. Producing the Incident Action Plan (IAP) for each operational period should be the highest priority. Providing photocopies for other Sections or Units must take a lower priority at certain times due to the importance of the Incident Action Plan (IAP) and its importance to the operational and tactical response of an incident. If allowing others from outside of the Documentation Unit to use the photocopy machines, someone from the Documentation Unit should provide close supervision, especially if these machines are relied upon for the creation of the Incident Action Plan (IAP). If the machine were to break down, this could jeopardize or potentially set back the response or recovery or the safety involved with the incident. The negative effect could be multiplied if it happened just before producing the Incident Action Plan (IAP). Even more devastating would be if it broke down and the individual did not notify the Documentation Unit of the problem. It is quite possible that the breakdown would not be found out until beginning the production run.

In many larger incidents, and even some smaller incidents, the Documentation Unit may consider initiating a request box for copying services within an incident. In most instances, they will handle those requests as time allows; however, these requests should take a much lower priority than the Incident Action Plan (IAP). It has become somewhat standard procedure when initiating a request box, that the piece to be photocopied is attached to an ICS Form-213 (General Message) using a paperclip. The form is placed in the request box with the following information printed on the ICS Form:

- Number of copies needed
- The timeframe in which they need to be completed
- Contact person and method of contact in case of questions
- Person to notify when completed and the method of contact

When using this method for printing requests, it is important to notify all personnel of this procedure and how the priorities are set. It is also suggested that signage be placed next to the request box so that there are no misunderstandings. If the incident and/or the need for more resources are needed, the Documentation Unit will need to plan for future photocopying needs. This may include, but not be limited to, additional photocopying equipment and personnel that may be needed to support the incident.

Because photocopying services are typically handled by the Documentation Unit, most Sections will come to the Unit to provide request to this service. This additional work can drain the Documentation Unit in especially large incidents. More often than not, the one Section that has historically drained the Documentation Unit of time and resources has been the Finance Section (NWCG, 2008). In the event that the Finance Section is utilizing the photocopier to the point that it becomes a drain on the time and resources of the Documentation Unit, the Documentation Unit Leader (DOCL) should have a conversation with the Planning Section Chief (PSC) about the Finance Section. In that conversation, it could be suggested that the Finance Section might need to order their own photocopier so as not to impede the Incident Action Plan (IAP) or to overtax the personnel in the Documentation Unit. After this conversation, the Planning Section Chief (PSC) may need to discuss the issue with the Finance Section to work out how to overcome this issue.

Additionally, because of the importance of the Incident Action Plan (IAP), the Documentation Unit should create a contingency plan(s) in case there is a problem with the equipment, or an illness that might affect the personnel's ability to produce and disseminate the plan.

Contingency plans to help with copying services may include local fire departments, police departments, schools, local businesses, nonprofits, and similar organizations and companies. Additionally, personnel that are familiar with the equipment, and what is needed to create the Incident Action Plan (IAP), should be preidentified in case they are needed in a hurry. This plan should be documented and shared with the Planning Section Chief (PSC) and the Resource Unit Leader (RESL) so that they also know how to overcome this issue, should it occur. Any suggestions that they may have should be integrated into the written contingency plan for the Documentation Unit so that the Unit can potentially be improved.

While there has been much talk about producing the Incident Action Plan (IAP) and the importance of photocopying, the Documentation Unit has another extremely important mission that they must fulfill. The Documentation Unit is responsible for establishing and documenting the Incident History File. While this may seem like a simple job that almost anyone could do, that supposition would be wrong in most instances, especially on wildfire incidents. On wildfires, incident documentation must be filed and organized according to the guidelines established according to National Archives and Records Administration ([NARA] NWCG, 2009). These are set up to NARA Standards, so the information can be archived should they ever be needed for reference. By using NARA guidelines, the organization and maintenance of all records related to the incident become part of the final documentation package, which can be easily integrated into the National Archives.

In looking at final documentation package for other incidents, Federal Emergency Management Agency (FEMA) has similar yet different requirements for retaining records. It should be noted that these standards may, and sometimes do, change. As of the writing of this book, FEMA uses the National Archives to keep documentation, but they do not always wait for an incident to be closed before sending the documentation to the National Archives. In extended incidents, especially those that last more than a year, the documents are boxed and forwarded to the National Archives on an annual basis. When we consider the length of an incident, this should not only include the response but also the recovery as well. For small-to-midsize incidents, this should not be a huge concern for the Documentation Unit. On larger incidents that last more than a year, the Documentation Unit will need to be cognizant that these records should be forwarded on an annual basis, based on the date of the incident (NWCG, 2009). It should be also noted that in some instances, local jurisdictions may want a full copy of what is being archived, although this is rare.

When documenting electronic documentation to be archived, the primary responsibility falls on the incident Computer Technical Specialist(s) (CTSP). This position is that of a Technical Specialist, and they can be assigned to any section within the ICS method. It is the duty of the Computer Technical Specialist (CTSP) to deliver an incident file on Digital Video Disc (DVD) to the Documentation Unit so that it can become part the official archives. While this will likely include maps used in the incident, a permanent copy of the maps should also be printed by the Documentation Unit that should be folded and filed in the Incident History File.

Electronic Documentation such as DVD's or other electronic documentation data should be filed independently of the official hard copy files. The Documentation Unit needs to ensure that any DVD's are placed in a protective cover, preferably a DVD hard case, and that the case is labeled with the appropriate information. Due to the fast advancement of computer software and electronics, the version of the software used to create the data should also be listed on, or in, the case.

As an incident begins to close out, the Documentation Unit will typically be tasked with assisting in the demobilization. While Demobilization Plans can vary from incident to incident, often the plan identifies the Documentation Unit Leader (DOCL) as someone who "signs off" on the ICS Form 221 (Demobilization Checkout) for those resources that are being demobilized.

The typical duties related to demobilization for the Documentation Unit Leader (DOCL) are to certify that any resources or equipment that were issued to personnel have been turned in (and documented) and that all the documentation needed has been completed to close out that resource.

It is important to realize that the Documentation Unit is an extremely important Unit. Not only does the Unit work to keep the response running by their work on the Incident Action Plan (IAP), but they also are responsible for ensuring that there is historical documentation about the incident and that the documents are organized in such a manner that they can easily be found. It is often voiced that the Documentation Unit in the ICS method is like having an attorney. With the ever-growing litigious society that we live in, the work of the Documentation Unit helps protect those that are managing the response. The documentation that they provide can, and does, protect the initial responding agency, and those who command the incident, by providing a record of what actions were taken to mitigate the emergency at hand. Often, this will provide a record that every action possible was taken. This can be extremely important in a court case, especially in a criminal or civil case that is unfounded. If the incident was the result of a crime, the records kept by the Documentation Unit may also be subpoenaed into court as evidence, and proper documentation could help to secure a conviction, or to set a person free.

10.1.4 The Demobilization Unit

Demobilizing an incident can be extremely complex, tedious, and difficult. The Demobilization Unit is primarily tasked with developing an Incident Demobilization Plan that includes specific instructions for all personnel and resources that will require demobilization, and the Unit is part of the close-out of an incident.

The process of demobilization should begin at the onset of the incident. The initial check-in of resources is as important to demobilization as it is to providing accountability of personnel and resources. This Unit should and usually will initiate the work of demobilizing resources in the initial stages of the incident. As early as possible, they should begin creating rosters of personnel and resources, and locating and identifying any missing information as the check-in process proceeds.

Every resource should be planned for, except those that are local resources. Local resources typically do not require specific demobilization instructions; however, the Incident Commander (IC) may want a Demobilization Plan to include local resources. If it is requested that local resources be included in the Demobilization Plan, then this request should not be ignored, and the Demobilization Unit should demobilize these resources in the same manner.

Demobilization can occur while the incident is still ongoing, or it may occur after the incident is over, or it is coming to a close. On longer incidents, it is not unusual to have an influx of new personnel and equipment to replace (or relieve) personnel and resources that have been involved from the onset, or potentially over a long period of time. This relief can encompass essentially every position and function within the ICS method.

The Demobilization Unit Leader (DMOB) supervises the Demobilization Unit. The method of reporting to the incident and obtaining an initial briefing about the incident is much the same as other Unit Leaders. As the incident builds, the Demobilization Unit Leader (DMOB) will initially review incident reports and records to analyze the resources that will be needed to demobilize the incident. After assessing the needs for the demobilization, they will begin the process of ordering additional personnel, supplies, and the needed workspace to facilitate an orderly and swift demobilization. They will also coordinate demobilization with Agency Representatives (AREP's) from the various agencies that are present at the incident, identify surplus resources, and project potential times of demobilization.

In coordination with the entire Unit, the Demobilization Unit Leader (DMOB) will create a check-out procedure for all resources and personnel, evaluate the transportation and logistics that will be needed to support demobilization, and (as necessary) they will establish communications with facilities and organizations that are not on-site at the incident. The Demobilization Unit Leader will also work with their team (and other functions of the ICS method) to create a detailed Demobilization Plan that describes specific responsibilities as well as the priorities and procedures for the demobilization. They will create maps, instructions, and other required documentation. They will create the orders that effectuate the demobilization and distribute it to the appropriate individuals and agencies, including outside agencies (if needed). As it is required of them, the Unit and/or the Unit leader will provide status reports to the appropriate individuals and ensure that those that responded to the incident understand their specific demobilization roles and responsibilities. In most instances, these are conveyed to those in a supervisory position for those resources being demobilized.

The Demobilization Unit Leader (DMOB) is responsible for supervising execution of the Incident Demobilization Plan. If a problem occurs during demobilization, they will be responsible for creating a solution or a contingency plan. It will be their responsibility to find a solution. At all times, regardless whether the demobilization is going well or it has encountered problems, the Demobilization Unit Leader (DMOB) will regularly brief the Planning Section Chief (PSC) on the progress of the demobilization.

It should be realized that the Demobilization Plan is a wide sweeping document that covers every position mobilized in an incident. The process does not only include those resources and personnel that were involved in tactical work but also every resource that was committed to the response and/or recovery. This can include Command Staff, General Staff, and any others involved in the incident.

While it should be realized that every incident is different, close-out of an incident tends to be a gradual demobilization rather than all personnel and resources being demobilized at the same time. It is important for the Demobilization Unit to know the order in which the incident demobilization should follow. Knowing this information will assist them in writing an efficient and orderly Demobilization Plan. The order for which resources are demobilized can affect the whole incident.

It is typical for the Public Information Officer (PIO) and their staff to have a gradual demobilization. The vast majority of the Public Information Officers (PIO's) staff will be demobilized at the beginning of the demobilization. This is typically possible because as an incident begins to transition from active operations to clean-up, media interest tends to diminish. As the need for information diminishes, so does the need for the staff to provide that information. This lack of less staff can include less need for interagency coordination, fewer press releases, less information being provided, and other similar aspects. In most instances, a skeleton crew can handle the responsibilities of the Public Information Officer (PIO), and if the Joint Information Center (JIC) was activated, they can begin to scale back on the number of individuals needed to provide pertinent information about the incident.

By the same token, as the need for operational resources begins to taper down, so does the need for the staff that support the Safety Officer (SOFR) and the Liaison Officer (LOFR). In most instances, these Officers will need to be available for a postincident review; however, the need for their staff will decrease as the incident transitions from tactical operations to the coming to an end of an incident, which is sometimes called clean-up or mop-up. This decrease will usually occur on a gradual basis. Those that are no longer needed or assigned, and are likely not to be assigned again, should be demobilized.

Similarly, as operations begin to wind down, so should the need for resources and personnel diminish for the Operations Section. As these personnel become no longer needed, the process of demobilization can begin. This demobilization will typically begin with the crews and will include their supervisors and any support staff. This will happen all the way up to through chain of command for those that are under the direct supervision of the Operations Section Chief (OSC). The Operations Section Chief (OSC) will in most instances be the last one in Operations Section to be demobilized because they (and the other section chiefs) will usually be required to remain behind for a postincident review.

The Logistics Section and the Finance and Administration Section will be among the last to completely demobilize. Both of these sections play a major role in the demobilization of resources along with the Planning Section. The Logistics Section has two units that will need to be heavily involved in the demobilization; the Facilities Unit and the Supply Unit. The Supply Unit will be responsible for collecting equipment and other inventory that has been assigned to personnel. The Supply Unit is also responsible for accounting of expendable resources such as fuel, water, food, and other items that have a one-time use. Upon collecting this inventory, they must make arrangements for the nonexpendable resources (radio's, personal protective equipment, etc.) to be put in a state of readiness. This state of readiness can be accomplished through being refurbished, rehabilitated, or replaced.

The replacement of equipment should be a last resort and should only apply to equipment that has been lost, damaged beyond repair, or it can no longer be useful because of some other reason. The nonexpendable resources will need disposed of, and a full accounting will need to be given to the Finance and Administration Section. The disposal of this equipment should be according to Environmental Protection Agency (EPA) guidelines, which are environmentally friendly. In some instances, this equipment must be declassified or decommissioned in such a way that unauthorized personnel or enemies of our country could not have access to information or specific guarded frequencies.

The Facilities Unit is also one of the later Units to be demobilized. The primary reason for this is because they are responsible for demobilizing the various facilities that personnel use. Facilities such as the Helispots, Dropspots, Incident Helibases, and sometimes Staging Areas will usually be demobilized as the need for these facilities diminish. Staging Areas can and will sometimes be used as part of the demobilization of resources to provide a place that is large enough to assist in the orderly demobilization of those resources. This holds especially true when there is an enormous amount of resources. Demobilizing the Staging Areas is usually related to the scope and size of the incident. Camps and other facilities cannot be demobilized until all the individuals utilizing that facility are demobilized or moved to another location.

Additionally, during the incident, personnel may have utilized one or more fixed facilities such as schools, fire departments, city hall, or some other fixed facility. As part of the demobilization process, these fixed facilities must be returned to their original state. The last facilities that usually will be demobilized are the Incident Base and the Incident Command Post (ICP), for obvious reasons. One of the very last Units to be demobilized will be the Facilities Unit. Because of the need for some facilities until the end of the incident, the Facilities Unit will be among the last to be demobilized.

Similarly, the Finance and Administration Section also has duties that need to be completed as the incident begins to end. As part of the Demobilization Plan, the Finance and Administration Section will need to continue to work on financial and administration issues until the very end. The reason for this is that there must be a final accounting, which cannot be completed until the end of the incident.

Demobilization Planning is important for fiscal, safety, and legal reasons. Creating a Demobilization Plan helps to eliminate waste, making it a fiscally responsible task to accomplish. Planning for demobilization also helps to ensure that there is a controlled, safe, efficient, and cost-effective process in place to release resources to their home base. Finally, Demobilization

Planning helps to reduce or remove legal issues that could arise. Legal impacts could be as simple as accusations of waste, or they could be based on injuries or damage to resources, or a whole host of other issues. In the litigious world we live in, mitigating legal issues is always an important factor to consider and to keep in the forefront of your mind.

Demobilization Planning, much like the rest of the ICS method, is a team effort. The key individuals who should be involved in creating the Demobilization Plan include the Incident Commander (IC), a representative of the Operations Section, a representative of the Planning Section (most likely the Demobilization Unit Leader [DMOB]), a representative of the Logistics Section, and a representative of the Finance and Administration Unit.

The Operations Section should begin with identifying excess resources that will no longer be needed for operational response. The Operations Section might identify only a Task Force or Strike Team that are no longer needed, or they may identify the Units, Groups, and/or Divisions that will no longer be needed at the incident. The Incident Commander (IC) will then be required to approve (or deny) the identified resources for demobilization, based on a combination of the Operations Section recommendations and their intimate knowledge of the incident. Once the Incident Commander (IC) signs off on resources that can be demobilized, a shared effort goes into creating the Demobilization Plan. The Logistics Section representative will identify if additional transportation will be needed and suggest how to supply that transportation. The Logistics Section will also develop a transportation inspection program to ensure that all vehicles used in the response are safe to return to their home of origin. The Finance and Administration Section will be responsible for planning the organized and efficient processing of time records, processing reimbursement claims, identifying release priorities based upon financial reasons (union contracts, overtime pay, etc.), and in tallying the entire cost on a resource by resource basis. When each resource has been tallied, then the Finance Section will calculate the cost of the entire incident. The Planning Section will develop and write the plan based on the recommendations of the other Sections.

Once the Incident Demobilization Plan has been developed and approved, the demobilization can begin. It is important to evaluate not only the condition of the vehicles that will be leaving but also the personnel as well. Before being demobilized, they should be well rested and appear mentally stable. On especially disturbing incidents, where personnel may have been exposed to massive death and/or injury, it be necessary to provide a Critical Incident Stress Debriefing (CISD). In some instances, an interview with a mental health professional may also be required if a Critical Incident Stress Debriefing (CISD) did not provide the intended results. Under no circumstances should any individual who has been psychologically affected by an incident, be demobilized until they have received the appropriate help.

As the incident nears the end of demobilization, a transfer of command may be (and usually is) in order, thereby handing the complete incident back to the local jurisdiction. This holds especially true on larger incidents, especially if:

- The response is transitioning from the response mode to recovery mode.
- Command is transitioning from a Unified Command (UC) to a singular Incident Commander (IC).
- An outside Incident Management Team (IMT) is part of the demobilization process.

As with any transfer of command in the ICS method, standard protocols should be followed. In instances where a transfer of command is part of a demobilization, there must be a Close-Out Briefing. A Close-Out Briefing consists of advising the incoming Incident Commander (IC) of the events that have taken place throughout the incident. A Close-Out Briefing consists of five major parts:

- An incident summary.

- Major events that might have lasting implications.
- Documentation related to the incident (including parts of the incident that have not been concluded).
- Concerns that may need to be addressed (for both the outgoing and incoming Incident Commander [IC]).
- A final evaluation of all incident management activities by those in charge of the local agency or government.

Upon completion of the Close-Out Briefing, it is often customary for the Incident Management Team (IMT), and sometimes others involved in the response, to hold a Close-Out Meeting. This Close-Out Meeting is one area where improvement can be realized for future incidents. These meetings are often undertaken to evaluate how the team performed during this incident and to identify how to improve for future incidents. This is done through identifying lessons learned while working the incident. As more incidents are managed, the process of talking about the issues helps to hone the Incident Management Teams (IMTs) skills so that use of the ICS method of incident management can be more effective in future responses.

The final part of demobilization is to undertake an After-Action Review (AAR). This is sometimes called an After-Action Review (AAR) by some agencies. No matter what an agency uses as a name, an AAR consists of asking

- What was the team tasked with?
- What were the actual events that occurred?
- Why did this incident happen?
- What can be done differently in future incidents?
- Was there something in the lessons learned that should be shared to improve future responses?
- What type of follow-up is needed?

In some instances, an After-Action Review (AAR) is done by each Unit, and in other instances, it is undertaken for the whole incident. Sometimes an entire incident review is done by local resources, and other times the After-Action Review is contracted out to an agency or firm. Whenever method is chosen, it is imperative to undertake this process to ensure that we learn from our mistakes and that we build on our successes.

In the end, the Demobilization Unit has a very difficult job in planning for demobilization. If you are interested in becoming more well versed in the demobilization process, you should follow through with the Self-study section of this chapter. There you will find more resources on the Demobilization Unit, as well as sample Demobilization Plans.

10.1.5 Two Optional Units

There are two additional modular Units that can be added to the Planning Section if they are needed. These Units are typically used in wildland firefighting, but if the need arises, they could be implemented in any type of incident. These Units are not well-known, so the description of them will be brief.

10.1.5.1 Strategic Operational Planner (SOPL)[1]

The Strategic Operational Planner (SOPL), when activated, is responsible for developing strategic courses of action on incidents that are long term in duration. This position is mostly used during wildfire events, but the use of a Strategic Operational Planner (SOPL) can be adapted for use in almost any type of event. When the course of action for a wildfire may include

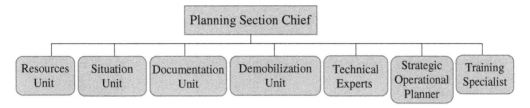

Figure 10.3 All Units that could be activated under the Planning Section expansion.

equally protecting wildlife and natural resources and wildland fuel management objectives, this position is typically activated (NWCG, 2018f) (Figure 10.3).

When activated, the Strategic Operational Planner (SOPL) will typically report to the Planning Section; however, they could also be assigned to the Geographic Area Coordination Center (GACC), or the Incident Management Team (IMT) assigned to the incident (if utilized). Often, the Strategic Operational Planner (SOPL) will work with the Geographical Information System Specialist (GISS). Depending on how the ICS method is activated and set up the Strategic Operational Planner may be required to supervise the Geographical Information System Specialist ([GISS] NWCG, 2018a).

10.1.5.2 Training Specialist (TNSP)[1]

Sometimes, real-life incidents are the best times to teach personnel. In times like these, the Training Specialist (TNSP) can help achieve goals and objectives while overseeing training opportunities at an incident. When utilizing the Training Specialist (TNSP), it is often important to make sure that these training opportunities are not time sensitive or life and death situations. While no training activities involved in public safety are free from risk, every effort should be made that there is sufficient time so as not to create life-safety issues. Another important factor with the Training Specialist is to make sure that all training opportunities are coordinated at all levels, thereby making that opportunity more effective (NWCG, 2018a).

Real-Life Incident

On the morning of 10 June 2014, at Reynolds High School in Troutdale Oregon, a gunman shot and killed a student and injured a teacher and then committed suicide. The response to the incident included local, state, and federal agencies, as well as nongovernmental organizations, who all responded to the shooting incident. The Emergency Operations Center (EOC) was activated during this incident, and it utilized the ICS method for incident management.

A thorough review of the response was undertaken in the After-Action Review (AAR). This report found issues that occurred throughout the incident, including issues in the Planning Section of the Emergency Operations Center (EOC). In the report, it was identified that the Planning Section was understaffed and that there was so much informational flow that future incidents should include the activation of the Situation Unit and the Documentation Unit. They also identified that the Planning Section (and the Logistics Section) were in different locations than the Emergency Operations Center (EOC) decision-makers. This sometimes limited interactions and communications with other functions.

One of the key recommendations for the Planning Section included maintaining a staff roster to fill the staffing needs, including alternate individuals, in the Planning Section. Another recommendation is to provide a space where all activated Sections can freely interact and communicate with each other. There were additional recommendations, but these were chosen as the most important for the Planning Section (Multnomah County, 2015).

Chapter 10 Quiz

1. The _____ Section is responsible for the collection, evaluation, and dissemination of information related to the current status of the incident.
   (a) Logistics
   (b) Planning
   (c) Finance
   (d) Investigations and Intelligence Gathering

2. **True or False**: The Planning (and Intelligence) Section is responsible for providing maps.

3. When expanding the Planning Section, the four main Units used are

4. _____ _____ is the "catch all" name for individuals who are not specifically named in the NIMS method, or the ICS companion method.

5. _____ is the process of appraising the current situation, defining objectives, choosing a viable strategy, and shaping what would be acceptable outcomes.

6. The _____ _____ has the primary responsibility of processing and organizing of all incident information.

7. **True or False**: In many instances, the Situation Unit researches and reports future projections directly related to incident growth.

8. **True or False**: The Situation Unit provides information to the Logistics Section, who will then make maps and provide intelligence outcomes based on their projections.

9. **True or False**: The Situation Unit will be responsible for plotting the borders of the incident, identifying the location of boundaries, identify the locations of facilities, access routes, and other critical locations, and create displays.

10. The _____ _____ is primarily responsible for displaying incident status information.

11. The _____ _____ is the individual (or individuals) that gather intelligence from the field. Their primary responsibility is to collect situation information from personal observations at the incident, and then provide a synopsis of this information to the Situation Unit Leader.

12. The _____ _____ _____ Specialist will collect data from others in the Situation Unit, or those working with the Situation Unit, and transfers those field observations and infrared interpretations into digital geo-referenced format.

13. The _____ _____ tracks all resources, maintains records, and keeps a running log of where each one is at any given time.

14. **True or False**: Knowing where resources are at all times is not an important part of ensuring the safety of each individual operating the incident.

15 _____ _____ are the individuals who will be placed at each check-in location.

16 What is the name used to identify ICS Form-219?

17 The _____ _____ is responsible for maintaining accurate and complete incident files.

18 **True or False**: The Resources Unit provides copying and duplication services for incident personnel.

19 The _____ _____ is primarily tasked with developing a detailed plan that includes specific instructions for all personnel and resources and is part of the close-out of an incident.

20 **True or False**: The process of demobilization should begin at the onset of an incident.

21 **True or False**: The initial check-in of resources is not very important to demobilization process.

22 **True or False**: As the Demobilization process comes to an end, there should not be a transfer of command to the local jurisdiction. The incident should just be closed out.

Self-Study

Federal Emergency Management Agency (n.d.b). ICS 219: Resource Status Card (T-Card). Retrieved from https://www.fema.gov/media-library-data/20130726-1922-25045-2119/ics_forms_219s.pdf.

Federal Emergency Management Agency (n.d.b). ICS 201: Forms Used for the Development of the Incident Action Plan. Retrieved from https://emilms.fema.gov/is201/ics01summary.htm.

Federal Emergency Management Agency (n.d.b). FEMA Director battle book. Retrieved from http://tacsafe.net/resources/Emergency/FEMABattleBook.pdf.

Guthrie et al. (2017). After action review of the November 28, 2016, firestorm. Retrieved from http://gatlinburgtn.gov/pdf/planning/Wildfire/AAR%20of%20the%20Nov%2028%202016%20Firestorm.pdf.

National Wildfire Coordinating Group (2018a). Medical Unit Leader. Retrieved from https://www.nwcg.gov/positions/medl/position-qualification-requirements.

National Wildfire Coordinating Group (2016). Interagency helicopter operations guide. Retrieved from http://ordvac.com/soro/library/Aviation/Operations/2016%20IHOG.pdf.

National Wildfire Coordinating Group (2009). NWCG Task Book for the Position of: Documentation Unit Leader (DOCL). Retrieved on February 24, 2017 from https://www.nwcg.gov/sites/default/files/products/training-products/pms-311-25.pdf.

National Wildfire Coordinating Group (2001). NWCG Task Book for the Position of Incident Communications Technician. Retrieved from https://www.hsdl.org/?view&did=774419.

United States Coast Guard (2017). United States Coast Guard (USCG) job book: Air Operations Branch Director. Retrieved from https://homeport.uscg.mil/Lists/Content/Attachments/2916/AOBD%20Job%20Aid%20July2017.pdf.

United States Coast Guard (2014). Incident Command System: Communications Unit Leader. Retrieved from https://homeport.uscg.mil/Lists/Content/Attachments/2916/COML%20Job%20Aid%20Oct14.pdf.

United States Fire Administration (2008). Emergency incident rehabilitation. Retrieved from https://www.usfa.fema.gov/downloads/pdf/publications/fa_314.pdf.

11

Expanding Finance and Administration

Money is a terrible master but an excellent servant

P.T. Barnum

The Finance and Administration Section is created when the size and scope of an incident increases to the point that the Incident Commander (IC) can no longer effectively handle incident-specific financial matters and/or administrative issues. As an incident become larger in size and/or scope, the Finance and Administrative Section becomes increasingly more important to mitigating the response. The Finance and Administration Section has a scope of responsibility that includes receiving and filing claims (or potential claims), work-related injury information, and addressing work (and rest) guidelines. Their responsibility also revolves around pay issues, financial responsibilities, liabilities, and the procurement costs of the resources and items that are related to the response.

The Finance and Administration Section is led by the Finance and Administration Section Chief (FSC). It should be noted that the acronym FSC is also used by Federal Emergency Management Agency (FEMA) for the Federal Support Center, as well as the Federal Storage Center. In most instances, these acronyms will not overlap; however, if this may be a concern during an incident, there is an alternative way to differentiate between the Finance and Administration Section Chief (FSC) and the similar FEMA acronym. In differentiating these, Typing the incident or the Finance and Administration Section Chief (FSC), is an easy way to differentiate these acronyms. As an example, if the incident and/or the Finance and Administration Section Chief (FSC) are Type-1, then the acronym could be FSC-1 for the Finance and Administration Section Chief (FSC-1). If it were a Type-2 incident and/or Finance and Administration Section Chief (FSC), then it would be FSC-2. Again, this is only an option for incidents where the acronyms may have overlap, or where there may be potential for confusion.

After being appointed the Finance and Administration Section Chief (FSC), the chosen individual will go through much of the same process as the other Section Chiefs. They will report to the Incident Commander (IC) where they will receive a briefing. In that briefing, they will be informed of incident objectives, what agencies are involved in the incident, the complexities of the incident, and how long the incident is anticipated to last. In some incidents, such as wildland fires, they will also be briefed on the topography and weather conditions. The remainder of the briefing will very likely be different from that of the other Section Chiefs. This is because the Finance and Administration Section Chief (FSC) will also be briefed on any political considerations. They will be informed who they will work with from the Authority

Having Jurisdiction (AHJ), and they will be provided the names of any local agency contacts, as well as all agency contacts that responded to the incident. The Incident Commander (IC) will also brief them on the possibility or probability of cost sharing and/or federal financial support.

If the incident has been ongoing, the Finance and Administration Section Chief (FSC) will then attend an Agency Administrator or outgoing Incident Commander (IC) briefing. In that meeting, they will collect all the available incident information, and they will receive the previously set agency and/or organization guidelines and policies. If the incident has not had any guidelines, the Finance and Administration Section Chief (FSC) will work with an Agency Administrator, the Authority Having Jurisdiction (AHJ), and the Incident Commander (if available) to create those guidelines and protocols. This is important part of ensuring that all financial, personnel, and incident policies and/or guidelines are met.

The Finance and Administration Section Chief (FSC) will also receive a copy of Delegation of Authority if one is used at the incident. A Delegation of Authority is a written document that authorizes the Finance and Administration Section Chief (FSC) is a written document that provides a right, and an obligation, to act on behalf of the primary governing body, jurisdiction, or agency. While it does not give them carte blanche authority, it does give them authority to do their job and to make purchases within the scope of their work.

The final part of the Agency Administrator briefing is to share the names, contact numbers, and the positions of all cooperating and assisting agencies involved in the incident. They will usually then receive a briefing from the Incident Commander (IC), and this briefing may a singular face-to-face briefing (one-on-one) or in an Incident Management Team (IMT) meeting, which is most often the case. In this briefing, the Incident Management Team (IMT) and the Finance and Administration Section Chief (FSC) will be informed of the Incident Commander's (IC's) priorities, goals and objectives for the management of the incident. The Incident Commander (IC) will typically provide initial instructions concerning Finance and Administration Section priorities. Before the meeting concludes, the Finance and Administration Section Chief (FSC) will be provided the timeframes that they will need to meet for briefings, planning meetings, and team meetings.

If the incident has been ongoing, the new Finance and Administration Section Chief (FSC) will be responsible for collecting all the necessary information from the Finance and Administration Section Chief (FSC) they are relieving, or other personnel that were responsible for incident prior to their arrival. Additionally, if the incident has been ongoing, they will need to be apprised of the status of the existing Finance and Administration Section, including what each Unit had accomplished, what they are working on, and the hierarchy in each unit, including the Unit Leader. In both an ongoing incident and as an incident is just beginning to use the Finance and Administration Section, the Section Chief will need to take inventory of the necessary personnel and equipment, and if needed, order additional resources to meet the needs of the incident.

One of the first major tasks that the Finance and Administration Section Chief (FSC) will need to undertake is to meet with or have contact with Agency Representatives (AREP's). In most instances, this is already determined by the Incident Commander (IC) and the Operations Section Chief (OSC); however, the Finance and Administration Section Chief (FSC) will need to verify that applicable local, state, and federal laws are followed as well as any work and rest requirements from each individual agency. These specific agency requirements may be contractual agreements between the agencies administration and the workers, and/or they may even be tied to a union contract. If any issues are found, the Finance and Administration Section Chief (FSC) will need to work with the Incident Commander (IC) and the Operations Section Chief (OSC) to create work (and rest) guidelines and ensure that the new guideline is being met.

The Finance and Administration Section Chief (FSC) will also coordinate with the Command, the General Staff, and any agency Human Resources staff to take on the task of determining if there is a need for temporary employees. If there is a need for temporary employees, the Finance and Administration Section Chief (FSC) will usually undertake the task of identifying where to procure these temporary employees. When temporary employees are hired, the Finance and Administration Section Chief (FSC) will ensure that they have skillsets that meet the needs of the job they are to perform. Their job will also include confirming that the needed tax documentation is completed properly, identifying what forms will be used to document time records, and who will keep these time records. They will also make sure that all Sections and the Supply Unit has the proper charge code for the incident. The use of a charge code prevents the incident from being charged for services that were not authorized.

Real-life Incident

Shortly after the threat of Hurricane Katrina had passed, and as the response began, individuals that offered various services began to show up unannounced. In Hancock County Mississippi, individuals began to show up offering potable water, dump trucks, heavy equipment services, and a multitude of other services.

In one instance, a truck that hauled potable water from over 600 miles away presented themselves to an Emergency Support Function (ESF)-4, stating that they had the potable water they needed. As the ESF-4 looked at the paperwork handed to him by this driver, he noticed that there was no charge code on the bill. He immediately presented the driver and the bill to the Finance and Administration Section Chief (FSC), pointing out that there was no charge code. It was quickly determined that this resource was not ordered and that there was already sufficient potable water available. Because the water was not ordered, and was not needed, the driver was told that they would not be paid for their services.

This is but one example of how individuals can try to charge for their services when those services were not ordered. By using a charge code for ordering resources, individuals and companies trying to charge for services that were not ordered can be identified as such and be denied payment. This prevents payment just because they presented themselves at an incident.

The Finance and Administration Section Chief (FSC) is also responsible for attending planning meetings. In these planning meetings, including the Incident Action Plan (IAP) meetings, their duties and responsibilities might include the following:

- Providing financial input
- Providing cost-analysis
- Providing financial summaries on labor
- Providing financial summaries on materials
- Providing financial summaries on services
- Preparing financial forecasts on the cost to complete operations
- Provide cost benefit analysis (as requested)

The Finance and Administration Section Chief (FSC) will work closely with the Planning Section and the Logistics Section. In both instances, the Finance and Administration Section will need to be aware of what is being ordered and what has arrived, as well as being able to reconcile the financial books. This includes reviewing and maintaining vendor contracts and planning for expenses that are related to vendors.

Finance and Administration Section Chief (FSC) will need to monitor the status of the incident as well. Because of the potential costs involved, the Finance and Administration Section

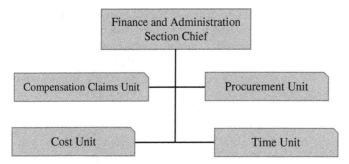

Figure 11.1 Finance and administration hierarchical structure.

Chief (FSC) will need to keep informed about planned operations, any changes in operational objectives, the additional use of personnel, and the potential need for more equipment or aircraft. They also must be ready and available to address any concerns that the Authority Having Jurisdiction (AHJ) or any local appointed officials may have.

The common Incident Command System (ICS) Forms that the Finance and Administration Section will be required to submit include ICS Form-213 (General Message Form), ICS Form-214 (Activity Log), ICS Form-226 (Compensation for Injury Log), ICS Form 227 (Claims Log), and ICS Form 228 (Incident Cost Worksheet). In some instances, the hosting agency may require different forms than those commonly used in ICS, but they may serve the same purpose (e.g. claims logs, injury logs, etc.). In addition, the Finance and Administration Section may use various other forms that are not ICS-based forms. These forms can be premade and available commercially, they may be local agency forms used by the hosting agency, or they may be the forms that have been used on other incidents and are appropriate for the incident at hand. Some of the forms that are not ICS Forms can include insurance specific forms, time sheets or time logs, cost analysis forms, and other similar forms.

In order to support the work that needs to be completed by the Finance and Administration Section, four basic (modular) Units can be activated to support the Finance and Administration Section (Figure 11.1). They are as follows:

- The Compensation Claims Unit
- The Procurement Unit
- The Cost Unit
- The Time Unit

Each of these Units plays a critical role in managing the Finance and Administration Section. Their combined roles allow the Finance and Administration Section Chief (FSC) to have a 360° view of the portion of the incident they are responsible for managing, and it shares the responsibility of ensuring that the most recent and up-to-date information is being utilized.

11.1 Compensation Claim Unit

The Compensation Claims Unit is responsible for addressing financial concerns as they relate to property damage, physical injuries, or fatalities when they are related to the incident. The Compensation Claims Unit is managed by a Compensation and Claims Unit Leader (COMP). Upon being assigned as the Compensation and Claims Unit Leader (COMP), this individual will receive a briefing from the Finance and Administration Section Chief (FSC). In that briefing, the Leader should be given a copy of the Incident Action Plan (IAP) and the names, contact information, as well as the positions and functions of all agencies assisting in the incident. They should also be provided a list of the resources that have been assigned to the incident.

In this briefing, the Compensation and Claims Unit Leader (COMP) should be informed and provided a written copy of the Authority Having Jurisdiction (AHJ) administrative guidelines. These local agency guidelines should include the process and procedures for medical care, any (and all) procedures surrounding compensation, and the methods used to file and process claims.

In reviewing the medical care, based on the Authority Having Jurisdiction (AHJ) administrative guidelines, there may be specific instructions on where nonemergency medical care and aftercare should be performed. The reason this may be important is related to insurance companies that may have medical networks that they work with and that are in their network of local providers. Additionally, the Authority Having Jurisdiction (AHJ) may have agreements with specific medical care facilities that offer discounted, or in some cases free, medical care based on whether the individual was involved in the response or someone not involved in the response. These guidelines may also spell out where rehabilitation or aftercare should be performed.

It is important to note that these medical administrative guidelines should never hinder emergency medical care. If a true, life threatening emergency should occur, the Claims and Compensation Unit should not be involved in where the patient is taken to be treated. In certain incidents, once the patient is stabilized, a transfer may be in order if agreements are already worked out in advance. The Compensation and Claims Unit should only be involved with the decision-making after the injured personnel is deemed stable.

When addressing the tasks of the procedures involved for compensation and the methods to file and process claims, there may be required insurance forms (from the Authority Having Jurisdiction [AHJ] insurer) that will usually need to be completed. In filling out and processing these forms, there may be specific information that will be required for the forms to be processed. Failure to do so may slow the processing by the insurance company. These errors may require that the forms will need to be corrected or a new form may need to be filed again. In the briefing with the Authority Having Jurisdiction (AHJ), these issues should be discussed and clarified.

The Compensation and Claims Unit Leader (COMP) should discuss with the Authority Having Jurisdiction (AHJ) if it might be better if the Compensation and Claims Unit, including all personnel that may be responsible for filing these forms, meet with an insurance representative. If the Authority Having Jurisdiction (AHJ) is in agreement, then an appointment should be made after the briefing which may focus on (in this later meeting with the insurance company) should likely be the exact forms that should be filed, procedures and processes for filing forms, the requirements for filling out these forms, specific information that may be needed, who (if anyone) should review these forms, and where they should be sent or who should they be given to. Although it does not always yield results, at this future meeting with the insurance representative, it should be asked how to expedite the claims process.

Upon completion of the briefing, the Claims and Compensation Unit Leader (COMP) should then order sufficient personnel and brief them to ensure that they perform their required tasks in a timely and efficient manner. Timeframes and deadlines should be set, and a record-keeping method should be established. The Claims and Compensation Unit Leader (COMP) should also procure and review the Incident Medical Plan to familiarize themselves and the Unit of the procedures that will be taken in the event an injury occurs.

In the formation of the Compensation and Claims Unit, the Leader and their staff will develop and maintain a log that will list each claim with basic information that identifies what the claim involves. Part of the claims process, whether personal injury or not, will likely include securing the site where the claim originated, and all the property involved to ensure that there is no tampering with any evidence that might be useful to an investigation. For incidents that do not involve personal injury, the Compensation and Claims Unit will often initiate an investigation. The Unit may also be required to interview witnesses of any claims, and to obtain witness statements to protect the local agency and/or the Authority Having Jurisdiction (AHJ). This might include interacting and coordinating with personnel involved in operations, unit leaders, and in rare instances, section chiefs.

Based on the information that the Unit gathers on each claim, they will be required to develop and create a claims prevention plan that coincides with the ongoing functions of an incident. If requested by the Incident Commander (IC) or the Finance and Administration Section Chief (FSC), the Unit may be required to assist in the implementation of the prevention plan. The assistance in implementing a prevention plan can be similar or very different from incident to incident. For this reason, the Compensation and Claims Unit Leader (COMP) should seek guidance from the Finance and Administration Section Chief (FSC) if requested to assist in the implementation of this plan.

In some instances, especially where personal injury is involved, outside investigators may be required. These outside investigators could include insurance agency investigators, local law enforcement, state law enforcement, railroad police, federal agencies, and/or a multitude of other investigators. When outside investigators are involved, the Compensation and Claims Unit will often be required to coordinate with the investigative team or specific individuals that are investigating the incident.

Those that work in the Compensation and Claims Unit will be required to keep abreast of all investigations, then document and report them to the Compensation and Claims Unit Leader (COMP), including any investigations that have not been completed. They will also be required to document any follow-up actions that might be required and identify any follow-up that the local agency or the Authority Having Jurisdiction (AHJ) may need to perform. The Unit staff will also keep the Compensation and Claims Unit Leader (COMP) informed on the disposition of all existing and potential claims.

In incidents where deaths have occurred, the Compensation and Claims Unit should ensure that an outside independent investigator is utilized. The reason an outside independent investigator should be used is to protect the local agency, the Authority Having Jurisdiction (AHJ), and those involved with the incident from unsubstantiated litigation or from the appearance of wrongdoing when there was none involved. In most instances, the investigators will be from a law enforcement agency that is not local, such as the attorney general's office or the state police, as well as the potential for an insurance investigator.

Occasionally, problems arise that are abnormal or that are beyond the normal protocols of the Claims and Compensation Unit. In these instances, the Unit staff should inform the Compensation and Claims Unit Leader (COMP) of the problem. Should the Unit Leader find that these are beyond their level of authority or that the circumstance is so unusual that their superiors should know about it, then they would report the circumstance to the Finance and Administration Section Chief (FSC), who may or may not present it to the Incident Commander (IC).

Beyond filling out the forms that may be created by the local agency, or that were developed by the Compensation and Claims Unit, the Unit will also be responsible for filling out two ICS Forms. These forms include ICS Form-213 (General Message Form) and ICS Form-214, (Unit Log). Depending on the requirements of the local agency and/or the Incident Commander (IC), the Unit may also be required to fill out ICS Form-221, (Demobilization Check Out).

As the incident comes to a close, just like other Units that are involved in the response, the Compensation and Claims Unit will need to properly demobilize. The Unit Leader will receive instructions on demobilization from the Demobilization Unit (DMOB) or the Finance and Administration Section Chief ([FSC] or both). The Unit Leader will then brief the Unit Staff on the procedures that should be followed and what responsibilities the staff have in the demobilization process. As the demobilization progresses, the Compensation and Claims Unit Leader (COMP) will ensure that all procedures set out in the Demobilization Plan are followed. If required, the Compensation and Claims Unit Leader will fill out ICS Form 221, (Demobilization Check-Out) and turn it in to the appropriate person.

In some instances, the Compensation Claims Unit may need to expand. In these instances, there are two modular positions that are affiliated with this Unit; the Compensation for Injury Specialist (INJR) and a Claims Specialist (CLMS). This expansion essentially divides all duties performed by the Compensation and Claims Unit, which allows the Compensation and Claims Unit Leader (COMP) to have one individual from each team managing approximately half of the work.

11.1.1 Compensation for Injury Specialist (INJR)[1]

The Compensation for Injury Specialist (INJR) may be activated to support the Compensation Claims Unit Leader (COMP). This modular addition will become increasing important in events (and incidents) where there has been a serious injury, or a death, of response personnel. While the title may suggest that their primary duties include suggesting compensation for injuries, their duties go far beyond just compensation.

Compensation for Injury Specialist (INJR) will work under the supervision of the Compensation Claims Unit Leader (COMP), and they should report to them on a regular basis. Whenever possible, the Compensation for Injury Specialist (INJR) should be located near, and work with, the Medical Unit Leader (MEDL) to ensure that there is prompt notification of injuries or death. Additionally, whenever an injury or death occurs, the Compensation for Injury Specialist (INJR) should establish contact with the Safety Officer (SOFR) and Agency Representatives (AREP's) and keep them updated about the status of the injured personnel. In doing so, they will need to be careful not to violate Health Insurance Portability and Accountability (HIPPA) laws. If a death occurs, they will also work closely with the Agency Representatives (AREP) to ensure that every possible need for that agency, and the deceased personnel, is met.

When injured personnel need short-term or long-term treatment, the Compensation for Injury Specialist (INJR) will provide written authority for the approved medical treatment, to those medical professionals that are treating the injured personnel. In doing so, they will be responsible for ensuring that the correct billing forms are used, and they will provide this billing information to the appropriate doctors and/or the hospital(s). This is done to ensure that all liabilities are paid according to the Authority Having Jurisdiction (AHJ) guidelines and/or protocols. The Compensation for Injury Specialist (INJR) will keep informed about the status and condition of any injured personnel and reports on the status while continually being cautious not to violate Health Insurance Portability and Accountability (HIPPA) laws.

Immediately after a death or injury, the Compensation for Injury Specialist (INJR) will usually manage an investigation of the incident (or accident) at the direction of the Compensation Claims Unit Leader (COMP). From the first notification of an injury or accident, the Compensation for Injury Specialist (INJR) will need to ensure that scene of the incident that will be part of the investigation is secured and protected. In cases where serious injury or death occurred, they may need to request outside resources to investigate. Whenever outside resources are called in to investigate, the Compensation for Injury Specialist (INJR) should make every effort to coordinate with the investigation team (as necessary) and to coordinate interviews with on-scene personnel (as needed).

---

1 This heading is for advanced IMS students.

If an outside agency is not brought in to investigate the incident, then the Compensation for Injury Specialist (INJR) will usually need to take the lead in investigating the incident. If the Compensation for Injury (INJR) specialist is tasked with this duty, they should collect witness statements from anyone that was in the vicinity when the person was injured or killed. They will also need to speak with, and possibly obtain witness statements, from all personnel that responded to the injury or death. Upon gathering all the witness statements, they should review them with the Safety Officer (SOFR), the Medical Unit Leader (MEDL), or both. This review is to ensure completeness and to provide an analysis of the injuries.

This report, the analysis, and the outcome should be shared and discussed with the Safety Officer (SOFR) in a coordinated and collaborative manner. In this discussion, mitigation measures should be discussed to help prevent similar incidents from occurring in the future. Whenever needed, the Compensation for Injury Specialist (INJR) will also provide a claims prevention plan with various other Sections and Units, so that future similar incidents can potentially be avoided.

From the report of the first injury or death, the Compensation for Injury Specialist (INJR) should maintain an injury log, documenting all injuries that have occurred on an incident. In instances where the injury is serious, or a death has occurred, they may be responsible for notifying the emergency contact for the personnel that was injured or killed. They will also undertake all administrative paperwork that may be needed to process an injury or death, even though that paperwork will often vary from state to state, and sometimes even from agency to agency.

Another very important task of the Compensation for Injury Specialist (INJR) is to ensure that the Demobilization Plan does not forget those that were injured. In doing so, they will develop a plan of action to ensure that follow-ups for the injured personnel are provided and assigned. They will also coordinate with the appropriate agency(s) to ensure that they will assume responsibility for injured personnel who may remain in the local hospital after the incident is over and demobilization has occurred. During the demobilization process, the Compensation for Injury Specialist (INJR) should also document any incomplete investigations, and any follow-up action that may need to be handled by the local agency. A written document that identifies each task should be given to the local agency to ensure that nothing is missed in the transfer of information.

11.1.2 Claims Specialist (CLMS)[1]

The Claims Specialist (CLMS) typically works on all tasks that the Compensation for Injury Specialist (INJR) does not handle for the Compensation Claims Unit. This usually relates to managing all claims that may be submitted to the Compensation and Claims Unit. These types of claims usually relate to property damage, such as cars that have been damaged by personnel (or equipment) or gates and fences that may have been accidentally or purposely knocked down.

In most accidents where a claim could be filed, the Claims Specialist (CLMS) will usually be required to secure the site where the claim originated, and any of the property involved, to ensure that no evidence is tampered with that might be useful to an investigation. For incidents that do not involve personal injury, the Claims Specialist (CLMS) will often be required to initiate an investigation. Depending on the direction given by the Compensation and Claims Unit Leader (COMP), they may be required to interview witnesses, and to obtain witness statements, in an effort to protect the Authority Having Jurisdiction (AHJ).

In some instances, outside investigators may be required to investigate the circumstances that led to the claim. When outside investigators are utilized, the Claims Specialist (CLMS) will usually coordinate with the investigative team and assist in providing information and/or witnesses to the incident. In this role, they will act as a liaison between those working on the incident and the outside investigators.

11.2   Cost Unit

The Cost Unit has the responsibility of collecting all the data as it relates to the cost of an incident. In doing so, they may be required to perform a cost analysis to determine the return on investment for an incident, providing running cost estimates, as well as recommending cost-saving strategies related to the incident. The Cost Unit is in effect the Unit that makes sure that the response and/or recovery for an incident does not unnecessarily saddle the taxpayer and Authority Having Jurisdiction (AHJ) with costs that are not needed. It is also the Units job to ensure that the incident is not overcharged for the services that are received. The supervisor of the Cost Unit should be identified as the Cost Unit Leader (COST).

Upon being appointed the Cost Unit Leader (COST), the individuals should receive an initial briefing from Finance and Administration Section Chief (FSC). In that briefing, they should be provided with the names, contact information, and position (or function) of the local agency's administrative personnel. They should also receive this same information for every assisting agency.

The Cost Unit Leader (COST) should also be provided with the Authority Having Jurisdictions (AHJ's) administrative guidelines and a list of resources that have been assigned and/or ordered for the incident (including any orders for the Unit). The Cost Unit Leader (COST) should also be provided with operating procedures for the Finance and Administration Section, Cost Unit deadlines, and they should be provided with the expectations for the Unit. The final piece of the initial briefing should be for the Cost Unit Leader (COMP) to be given a copy of Incident Action Plan (IAP) and any other relevant plans that involves spending money on the incident.

If not already in place, the Cost Unit Leader (COST) will need to coordinate with the Finance and Administration Section Chief (FSC) and possibly other units to procure the equipment and resources they will need to undertake their duties. Unlike other Units, the Cost Unit Leader (COST) will need to ensure that there is an overabundance of forms, materials, and supplies to prevent the possibility of a shortage. They will also need to quickly secure the needed radios, telephones, fax machines, computers, and other needed equipment to ensure that the Unit does not cause a delay in operations. Additionally, the Cost Unit Leader (COST) will need to maintain proper Unit staffing, and this will usually entail staff to cover both day and night. Because of the importance of keeping abreast of all expenditures, tallying expenditures, making agreements, researching cost-effective alternatives, and various other duties that related to the expenditure of monies, the need for around the clock staff is usually a necessity, even if the remainder of an incident works only one operational period, or even just one hour per day.

An important responsibility of the Cost Unit is to integrate and to synchronize with the local agency, keeping them apprised of the expenditures that are related to the response. In order to accomplish this task, they will need to coordinate with the Authority Having Jurisdiction (AHJ) regarding when, and how, cost reporting should be provided. This coordination will assist the Cost Unit in setting up, forming, and keeping schedules to meet the requirements of the Authority Having Jurisdiction (AHJ). In order to meet these requirements, they will need to collect all cost data related to the incident, then organize it and document that data, and create cost summaries from the data they collected. This is done to keep the local agency up to date on the entire cost of the incident.

Beyond reporting to the local agency, the Cost Unit must also report to others within the ICS structure to keep them informed of the costs involved with the response and/or recovery. Part of their duties includes preparing cost estimates for the Planning Section. These estimates may include the cost for resources and making cost reduction or saving suggestions to the Finance and Administration Section Chief (FSC). By doing so, they ensure that expenditures are cost-effective and that local, state, and federal funds being expended are the most economical for taxpayers who, in most instances, are paying for the response and/or recovery.

An important job of the Cost Unit is to ensure that all cost-related documents reflect the expenditures in an accurate manner. While some elected or appointed officials may want to alter estimates to meet their agenda the Cost Unit should remain neutral and report the numbers as the data show it to be. It is possible that it be requested that costs be overestimated to gain more funding, or underestimated costs to downplay the expenditures needed. In order to keep the integrity of the incident intact, as well as the integrity of the Unit, Cost Unit staff members should not adjust numbers based on the requests or demands of others. It is imperative that the Cost Unit be ethical and honest. Additionally, the Cost Unit should stay up to date on the overall cost and maintains (and records) a cumulative collective analysis of the incident.

ICS Forms that will be filled out by the Cost Unit may include ICS Form-209 (Status Summary), ICS Form-213 (General Message), and ICS-214 (Unit Log). Much like other Units in the ICS method, it is important to share information with those that need it to effectively do their job. The filled-out ICS Form-209 (Status Summary) allows information to be shared with Command and General Staff, and it allows them to be provided with the latest information. The most current information thereby allows them to make more informed decisions. In some instances, more information beyond what is listed in the ICS Form-209 (Status Summary) may be needed. In cases such as this, the individual needing this information may contact the Cost Unit directly, or they may go through official channels of making contact, such as contacting the Finance and Administration Section Chief (FSC) and making an official request through them.

Demobilizing the Cost Unit involves ensuring that all records are complete and accurate prior to disbanding the Unit. Upon completion of all reports, they should be provided to the Finance and Administration Chief (FSC) for submission to the appropriate individuals and for the purpose of archiving the documents. As with other Units, all property must be returned and signed for by the Demobilization Unit (DMOB) and/or others within the ICS structure that are responsible for ensuring all items are accounted for.

11.3 Procurement Unit

The Procurement Unit is critical to an incident that is long-term or large-scale incident. Primarily, the Procurement Unit holds the key responsibility for all financial matters that deal with vendor contracts and the purchasing process. If a Procurement Unit takes their job seriously and has the best interest of the incident as their primary goal, they can play an important role in making the response and/or recovery substantially more effective and cost efficient at the same time. The actual written mission of the Procurement Unit is "to provide for the administration of all financial services pertaining to purchases and contracts and to maintain contract equipment time records." (Rhode Island Department of Environmental Management, 2016, p. 1).

Upon appointment as the Procurement Unit Leader (PROC), the individual appointed should receive a briefing from the Financial and Administration Section Chief (FSC) to better understand the incident, and to better understand the priorities of the work that the Procurement Unit will perform. At this point, they will make the decision of what the charge code will be (if not already in place), and they will determine who has the authority to promise funds on behalf of the response and/or the Authority Having Jurisdiction (AHJ).

In some instances, the Authority Having Jurisdiction (AHJ) cannot, or will not, provide purchasing authority to the Procurement Unit Leader (PROC) or their representatives. In instances where the authority cannot be granted to the Procurement Unit Leader (PROC), then the Authority Having Jurisdiction (AHJ) will need to assign individuals who have the authority to approve purchases and the commitment of funds for the response and/or recovery of the incident.

The Procurement Unit Leader (PROC) will also need to create an Incident Procurement Plan. Among the many issues addressed in this plan, some of the more important issues include spending limits, the forms needed to officially purchase goods and services, and identifying who has purchasing authority. The Procurement Plan should also include the protocols and processes for gaining approval to exceed spending limits, how to coordinate with Supply Unit, and how to generate emergency purchase orders.

Upon determining purchasing authority and developing a Procurement Plan, the Procurement Unit Leader (PROC) will create a buying team who will assigned to do the ordering and purchasing of all equipment, supplies, and services for the incident. The Procurement Unit Leader (PROC) will be tasked with closely coordinating with this buying team to ensure the most effective use of taxpayer monies.

The Procurement Unit Leader should also create a bid process, to ensure that any items purchased will be on equal footing and at the most reasonable price. It is important to ensure that the bidding process clearly specifies the items or services that are being bid on. Essentially, the Unit needs to make sure that apples are being compared to apples, and oranges are being compared to oranges.

A simple example of how not standardizing the bidding process could provide inferior supplies or services, imagine that bids are being taken for legal note pads. Without specifying the details of the notepad bid, vendor "A" provides a price of $2.25 per legal note pad, while vendor "B" submits a bid of $2.30 per note pad. Although it may appear that vendor "A" has the most reasonable price, further investigation reveals that vendor "A" is providing a price for legal pads that contain only 50 sheets per pad, while vendor "B" provided a price for 200 sheets per pad. When comparing each of these bids (based on the cost), the lower bid of $2.25 would cost four and a half cents per sheet, but the higher bid of $2.30 would cost just a little over a penny per sheet. When comparing the two prices equally, it is obvious that the higher bid is the most economical. This example shows the importance of evaluating services and products based on equal delivery of goods, and the importance of a document that standardizes the bidding process.

Another job of the Procurement Unit and the Procurement Unit Leader (PROC) is to create an approved vendor list. It is important that this list has trusted and established businesses that can provide the items and services needed in a timely manner. Criteria for approved vendors should be established to help decide who should be an approved vendor. While not comprehensive, some basic criteria for approved vendors should include

- How long they have been in business
- The time it takes from order to delivery
- Their ability to continuously provide the service or supply
- The ability to supply all the items needed (a one-stop-shop)

- Flexibility to change (as the needs of the incident changes)
- Availability of experts to work through problems
- Customer service
- Financial stability
- Testimonials and/or references
- Competitive prices
- Payment terms ([cash only, net 30, net 60, etc.], Purchasing & Procurement Center, n.d.)

Other criteria may be added to meet the needs of the incident; however, this list should provide a starting point for evaluating vendors that could be utilized. The Unit or the Unit Leader will also develop vendors that are approved for blanket purchase orders. Blanket purchase orders provide ongoing deliveries of nondurable goods that can be delivered over time using prearranged pricing such as fuel, food, and office supplies.

To keep up to date on the needs of the incident, the Procurement Unit will need to be in regular contact with the Supply Unit. The Supply Unit will be able to provide the Procurement Unit with a list of supplies that will need replenished, and they will also be able to inform the Unit of most of the future needs to support the incident. The Supply Unit will also be knowledgeable on any special requirements to support the incident.

On occasion, an incident may require rental equipment. This could be the rental of equipment that is as basic as a fax or copy machine, or something less usual such as the rental of a bus, crane, or some specialized piece of equipment. The Procurement Unit will be responsible for reviewing all rental agreements to ensure that there are no additional or hidden costs. They will need to ensure that the terms and conditions related to the use of the equipment prior to, or within 24 hours of, the equipment arriving at the incident. They will also be required to forward information such as the hourly or daily rates of this rental equipment to the Cost Unit so that it can be added to the overall cost of the response and/or recovery. The Cost Unit may also need to coordinate with the Compensation and Claims Unit when required.

In some incidents, there may be a requirement or a need to pursue a land-use agreement or a cost sharing agreement. The Procurement Unit will be responsible for preparing and signing contracts that are related to land-use agreements and cost-share agreements when the need arises. In all instances, they should work to negotiate the best price for these services as well as ensuring that the land or cost-sharing will meet the needs of the incident.

From time to time, the Procurement Unit will also be required to draft Memorandums of Understanding (MOU's) or Mutual Aid Agreements (MAA's). These Memorandum of Understanding (MOU) are formal agreements that are often used to establish formal partnerships. They can be used between those utilizing the ICS method to show the intent of other agencies, nonprofits, faith-based organizations, and others to assist with the response and/or recovery of an incident. Whenever a Memorandum of Understanding (MOU) or a Mutual Aid Agreement (MAA) is implemented, the Procurement Unit will need to make sure that it undergoes a legal review and that is approved (and signed) by the Incident Commander (IC) before implementing this agreement.

As needed, the Procurement Unit will also establish and maintain contact with vendors. It is important to realize that they will be the main point of contact with these vendors, and they must develop relationships that are not based on friendships, but rather the needs of the incident. From time to time, some vendors may not be able to provide for the needs of the incident for varying reasons, or a vendor (or vendors) may not be the most economical choice for one or more specific supplies or services. If this does occur, the Procurement Unit will need to determine whether additional vendors may be needed. If additional vendors are needed, or warranted, the Procurement Unit will be responsible for engaging and creating relationships with potential new vendors. The Unit will also be responsible for creating vendor-service agreements.

Additionally, the Procurement Unit will be required to review and interpret contracts and agreements as part of their responsibility. Any issues found should be brought to the attention of the Procurement Unit Leader (PROC) who will either work to resolve those issues or assign someone within the Unit to resolve the issues through delegated authority.

Another important aspect of the Procurement Unit is they are responsible for ensuring that they are not paying for services or supplies that were not performed or provided. For this reason, they should verify all invoices and validate that the services or supplies were satisfactorily provided, and at the level stated. The important task is ensuring that all contractors are accounted for and that their time is documented and aligned with the working hours of the team, unit, or other area to which they were assigned. To verify the status of these contractors, it will often require coordination and information sharing with all Sections. In some incidents, it may be necessary to hire one or more individuals to travel from working location to working location of the incident, so that it can be witnessed and documented that contractors are performing the contracted work and using the equipment stated in the contract.

The Procurement Unit will also administratively monitor and maintain procurement aspects of the incident. As required by agreements, or as specific contractual milestones are met, they will complete the billing and final processing, and then submit the documents to the appropriate person or agency for payment. The Unit will also maintain receiving documents, acquire copies of all vendor invoices, and verify that all contractor and equipment time records are complete. They should maintain a comprehensive audit trail that provides insight into what was procured, as well as where and how it was used. The Unit is also responsible for checking all data entries on vendor invoices to ensure that all documents are complete and that all charges are legitimate. They should compare invoices against procurement documents in an attempt to identify any discrepancies or to identify nonauthorized work from a vendor that took it upon themselves to respond without authorization.

The Procurement Unit Leader (PROC) or their representative will brief the Finance and Administration Section Chief (FSC) about the Units activities in accordance with the Section Chiefs schedule and timeframe. As unusual events occur, the Finance and Administration Section Chief (FSC) should also be informed about these events by the Procurement Unit Leader (PROC) or their representative. All work completed by the Procurement Unit should be documented using ICS Form-214 (Unit Log).

As the incident comes to a close, and the Procurement Unit begins to demobilize, it is important to realize that more work needs to be done. Because of the often-chaotic pace kept by the Procurement Unit during the incident, sometimes documentation may not be ready to close out. Prior to being demobilized, the Unit must review all vendor contracts and complete the terms of those contracts to the best of their ability. If it is not possible to complete them, then they will forward the information and documentation to the person in charge of procurement and/or finance for the Authority Having Jurisdiction (AHJ). Additionally, copies of all contracts will need to be prepared and delivered to agency representatives that will have responsibility for following up on this information.

Another part of demobilization for the Procurement Unit is to ensure that all interagency cost share agreements are complete, up to, and including the last operational period. The Unit will also make sure that all ICS-related documents and forms are completed. Upon completion of these tasks, the documentation for ICS should be delivered to the Documentation Unit. Upon completion of all duties and transferring of information to the appropriate individuals, the Procurement Unit Leader (PROC) should get approval to release all personnel from the Finance and Administration Section Chief (FSC).

11.4 Time Unit

The Time Unit, sometimes referred to as the Time Keeping Unit, is primarily responsible for recording and documenting the time that personnel and hired equipment spent working at an incident. While timekeeping may seem like an easy task, it is complicated by the fact that with multiple resources working with an incident, some resources may only spend an hour working on scene, while other resources may be on-scene for the duration of the incident. The time that each of these resources spends working on the incident must be tracked to ensure an effective and accurate expenditure of funds.

Upon being appointed the Time Unit Leader (TIME), the individual appointed should receive a briefing from the Finance and Administration Section Chief (FSC). In that briefing, the Section Chief and the Unit Leader will determine the requirements for time recording by setting a time reporting schedule, identify the location of where timekeeping activities will be positioned, and in identifying the personnel and rental equipment that will require timekeeping. They will also discuss any special consideration for time keeping from either resources or the Authority Having Jurisdiction (AHJ).

Upon completion of the briefing, the Time Unit Leader (TIME) will begin the process of organizing and staffing the Unit to meet the needs of the incident. They will identify the area of responsibility that each team member will be responsible for, and they will set deadlines for the completion of work that coincides with the Finance and Administration Section Chiefs (FSC) deadline.

The Time Unit will need to contact and coordinate with various Units within the ICS method to ensure that time recording documents are prepared, maintained, in compliance with time unit objectives, and forwarded to the Time Unit in accordance with incident protocols. This can include contacting the Ground Support Unit, Air Support Group (if operational), the Facilities Unit, and other similar units of the requirement to document and report the times of equipment used on the incident. The Time Keeping Unit will also establish and maintain contact with Agency Representatives (AREP's) regarding the time-keeping records and any constraints (usually based on contracts) that each individual agency might have.

While volunteers are typically not paid for their time, it is also important for volunteer and mutual aid resource records to be maintained, even if those hours will not be reimbursed. This allows the Authority Having Jurisdiction (AHJ) to identify and understand the number of hours that were worked on an incident. It may also later be important should a claim be made for an injury.

The Time Unit will also need to establish files and a filing system for time-related records, which should also include security measures to protect these records. Prior to filing records, the Time Unit should review all records for completeness and to ensure that they are filled out according to incident protocols for timekeeping. When filing these records, they should be arranged in such a way that if any issues, disputes, injuries, or investigations, do arise during or after an incident, the Time Unit should be able to find the files in a timely manner. They will also record all overtime hours in a separate log, so that overtime can be reviewed separately if need to be. The Time Unit may also be expanded, especially on larger incidents, to help divide the two time-keeping responsibilities of personnel and equipment.

The Time Unit Leader (TIME) or their representative will be responsible for briefing the Finance and Administration Section Chief (FSC) of any problems that arise during the performance of their duties. They will also make recommendations to the Section Chief and discuss any issues that may arise. Should the incident go on long-term, or if operations are ongoing around the clock, the Time Unit Leader (TIME) will be required to also provide a briefing to the individual who will relieve them about current activities and any unusual events.

In much larger and extended incidents, the Time Keeping Unit may be responsible for establishing Commissary Operations. Commissary Operations is essentially a contracted mobile catering service that provides meals. The responsibility of overseeing Commissary Operations is left to the Time Unit because contractors that supply these services are paid by the number of incident personnel that are fed. The Time Unit will usually provide security and maintain records of the number of crews that utilize this service, then submit these records to the Unit itself based on an operational period. All supplies and stock for the commissary is typically provided by the Supply Unit, with the Time Unit being responsible for keeping an ongoing inventory at all times. This includes the close-out inventory as the commissary shuts down for demobilization.

The Time Unit may be required to fill out time sheets or other types of informational sheets that were created by the local agency or that were developed by the Unit. The Unit is responsible for filling out these forms and they may be responsible for filling out two ICS Forms as well. These additional ICS-Forms include ICS Form-213 (General Message Form) and ICS Form-214 Unit Log. Depending on the requirements of the local agency and/or the Incident Commander, the Unit may also be required to fill out ICS Form-221, (Demobilization Check Out).

The Time Unit will also ensure that all records, for the commissary, all agencies, and all individuals are completed and up-to date prior to the demobilization process. Part of the demobilization process and a responsibility of the Time Unit are to provide time reports specific to each assisting agency prior to demobilization. After completing all the required forms, reports, and other documentation (including providing time records to each agency), the Time Unit Leader (TIME) should receive permission to demobilize and close out that position. Upon approval, all documents should be submitted to the Documentation Unit (unless instructed otherwise) prior to departure.

11.4.1 Personnel Time Recorder (PTRC)[1]

When the duties of the Time Unit are expanded and divided, the Personnel Time Recorder (PTRC) will be activated. Working under the supervision of the Time Unit Leader (TIME), the Personnel Time Recorder (PTRC) is responsible for overseeing the recording of time for all personnel assigned to an incident, auditing those time records, and reporting any issues that may arise that are related to time. They will also identify any special pay considerations. These considerations might include hazard pay, time loss due to injury or illness, required mandatory days off, more than legal-shift lengths, and various other aspects of time- and pay-related issues. They will also identify deduction that should be made, such as commissary debt, time off, and other deductions. The Personnel Time Recorder (PTRC) will also record travel pay (both to and from the incident), and any lost time. They will not only document the staff working at the incident but also document any transfers, promotions, and/or terminations that occur with personnel.

The Personnel Time Recorder (PTRC) may also be required to document volunteer hours. In most incidents, it is not mandatory to document volunteer hours; however, the benefit of knowing the total working hours that volunteers provide might be important in obtaining and/or meeting grant requirements (e.g. in-kind services). Upon completion of their duties, the Personnel Time Recorder (PTRC) will be required to distribute all time documents according to protocols and/or established guidelines.

11.4.2 Equipment Time Recorder (EQTR)[1]

Much like the name implies, the Equipment Time Recorder (EQTR) will record the times of equipment being used. In doing so, they will identify any issues that arise with equipment and notify their supervisor of any concerns. Those issues can revolve around an excessive number of hours that a piece of equipment is being used, equipment that is sitting idle, and any other issues that may relate to use of equipment. While identifying these issues, they may also suggest mitigation measures to remedy these identified problems.

As part of their duties, the Equipment Time Recorder (EQTR) may be required to identify any incomplete or missing equipment documents, missing or incomplete contracts for equipment, and any deduction or claims that may be associated with a given piece of equipment. They will also identify and document the rates that are associated with a given piece of equipment, such as the hours on the equipment's time meter, daily or hourly rates, and any additions or deductions for fuel, parts, or service that was required. They will also ensure that equipment shift tickets are completed, identify the operational periods that lasted longer than the Incident Action Plan (IAP) assignment, and any down time that the equipment may have suffered. They will also document and make comments about special considerations that a piece of equipment may have and obtain approvals and signatures for equipment and or equipment repairs.

In addition to their other responsibilities, the Equipment Time Recorder (EQTR) will also finalize the paperwork for contract resources (i.e. time and pay documents), the close out time, total use of the equipment. They will also document any deductions for a piece of equipment. In most instances, they will be required to attach a completed Equipment Inspections Checklist. Prior to releasing the equipment, they will verify all information with the contractor, obtain and post an estimated travel time, and acquire the required signatures. They will use the Emergency Equipment Rental-Use Envelope as a checklist to help ensure that the payment package is complete, including invoices and the required information regarding pay and hiring documents.

Once the equipment is released, they will attach all the supporting documentation related to that equipment and/or contractor and then complete the Emergency Equipment Use Envelope. The Equipment Time Recorder (EQTR) will then distribute these documents and the envelope according to the established guidelines.

Real-life Incident

After a disaster, it is common for some businesses to price gouge. In the aftermath of Hurricane Harvey, there were reports of were over 5000 complaints about price gouging. According to the Dallas Morning News (Wang, 2017), over 127 gas stations charged more than $3.99 per gallon of gas, with some stations charging as much as $20 per gallon (Garfield, 2017). There were complaints of bottled water that was priced at $99 per case and $8.50 for a single bottle of water. There were also reports of motels that began price gouging, some charging as much as three times the normal rate (Garfield, 2017).

While there were no reports of price gouging, or attempted price gouging, for the Hurricane Harvey response, it is easy to see that Finance and Administration Section of the ICS method is in place to help prevent these types of price gouging. This is why it is so important for those in all of the supporting units of the Finance and Administration Section to review proposals, contracts, and other potential financial expenditures thoroughly. By doing so, they help ensure that they will not be overpaying for the services and supplies that may be needed to support the incident.

Chapter 11 Quiz

1. **True or False**: The initial briefing with the Finance and Administration Section Chief (FSC) will usually include a briefing on any political considerations.

2   The Finance and Administration Section Chief (FSC) will receive a copy of a _____ _____ _____ which is a written document that provides a right, and an obligation, to act on behalf of the primary governing body, jurisdiction, or agency.

3   **True or False**: One of the first major tasks that the Finance and Administration Section Chief (FSC) will need to undertake, is to meet with, or have contact with Agency Representatives (AREP's).

4   **True or False**: The Finance and Administration Section does not need to be aware of what is being ordered and what has arrived.

5   What are the four modular Units that can be activated under the Finance and Administration Section?

6   The _____ _____ _____ is responsible for addressing financial concerns as they relate to property damage, physical injuries, or fatalities when they are related to the incident.

7   The Compensation Claims Unit should check with the _____ _____ _____ has agreements with specific medical care facilities that offer discounted or free medical care.

8   **True or False**: If a true, life threatening emergency should occur, the Claims and Compensation Unit should notify crews where the patient should be taken to be treated.

9   **True or False**: In some instances, especially where personal injury is involved, outside investigators may be required.

10 The _____ _____ has the responsibility of collecting all the data as it relates to the cost of an incident.

11 **True or False**: A primary job of the Cost Unit is to make sure that the response and/or recovery for an incident does not unnecessarily saddle the taxpayer and Authority Having Jurisdiction (AHJ) with costs that are not needed.

12 An important responsibility of the Cost Unit is to integrate and to synchronize with the local agency, keeping them apprised of the expenditures that are related to the response.

13 The _____ _____ holds the key responsibility for all financial matters that deal with vendor contracts and the purchasing process.

14 **True or False**: In some instances, the Authority Having Jurisdiction (AHJ) cannot provide purchasing authority to the Procurement Unit Leader (PROC). In instances where this authority cannot be granted then the Authority Having Jurisdiction (AHJ) will need to assign individuals who have the authority to approve purchases and the commitment of funds.

15 The Procurement Unit Leader should also create a _____ _____ to ensure that any items purchased will be evaluated on equal footing, and at the most reasonable price.

16 The Procurement Unit should create an approved _____ _____ of trusted and established businesses that can provide the items and services needed in a timely manner.

17 The _____ _____ is primarily responsible for recording and documenting the time that personnel and hired equipment spent working at an incident.

18 **True or False**: Because volunteers are typically not paid for their time, the Time Unit does not need to keep volunteer resource records.

Self-Study

United States Coast Guard (2014). Incident management handbook: Incident command system (ICS). Retrieved from https://www.atlanticarea.uscg.mil/Portals/7/Ninth%20District/Documents/USCG_IMH_2014_COMDTPUB_P3120.17B.pdf?ver=2017-06-14-122531-930.

Texas A&M Forest Service & Texas Department of Public Safety-Emergency Management (2014). Incident command system: Finance and admin. Section chief job aid (v2). Retrieved from https://ticc.tamu.edu/Documents/IncidentResponse/AHIMT/JobAid/FSC_JA-07012014.pdf.

NIIMS (n.d.). Compensation/Claims unit leader:1-363 self-paced instruction. Retrieved from http://firescope.caloes.ca.gov/ics-instruction-manuals/I-363.pdf.

U.S. Department of Interior (2013). National incident management system: Incident positions qualification guide (IPQG). Retrieved from https://edit.doi.gov/sites/doi.gov/files/migrated/emergency/upload/DOI-Incident-Positions-Qualification-Guide.pdf.

National Wildfire Coordinating Group (2016). NWCG task book for the positions of: Personnel time recorder (PTRC), equipment time recorder (EQTR) and Commissary manager (CMSY). Retrieved from https://www.nwcg.gov/sites/default/files/products/training-products/pms311-51.pdf.

National Wildfire Coordinating Group (2019). Situation unit leader (SUL). Retrieved from https://www.nwcg.gov/positions/sitl.

National Wildfire Coordinating Group (2019). Resources unit leader (RUL). Retrieved from https://www.nwcg.gov/positions/resl.

Multnomah County (2015). Reynolds High School active shooter response: An analysis of the response to the Reynolds High School Shooting on June 10, 2014. Retrieved from https://multco.us/file/57742/download.

12

ICS Investigations and Intelligence (I/I)

> *War on terror is far less of a military operation and far more of an intelligence-gathering, law-enforcement operation*
>
> John Kerry

It has been mentioned in previous chapters that the US Government evaluates the NIMS/ICS method and provides an ongoing review. As needed, changes are made after opening the proposed changes up to public input and evaluation. This is done for proposed changes for the National Incident Management System (NIMS) method, and the Incident Commands System (ICS) companion method of incident management. These ongoing reviews are undertaken to make the system more useful and effective as an incident management tool, and by incorporating first response personnel in the process, it ensures that it is meeting the needs of incident management where the boots meet the ground.

The review process led to one of the more recent changes to the ICS companion method. The inclusion of Investigations and Intelligence Gathering (I/I), and the various options associated with it has been added and developed over the past 15 or so years. Initially, in the beginning stages of ICS, the primary purpose and function of intelligence gathering was to provide situational awareness. The Planning Section would gather information for situational awareness purposes.

In many instances, the Planning Section still gathers that intelligence and supports the response in this manner. However, as more agencies from a multitude of different disciplines adopted the ICS method, it became apparent that this method was not meeting the needs of all personnel or all disciplines that might respond to an incident. Part of the impetus for these changes was the events that occurred on 11 September 2001, and the response and recovery from Hurricane Katrina in 2005. While these were the impetus, lessons learned from incidents such the Oklahoma City Bombing were reviewed to determine how the ICS method could have been more effective. It became obvious that the ICS method could be improved, so changes were made after gaining the appropriate feedback from first responders.

While the new function of Investigations and Intelligence Gathering (I/I) is useful for unplanned incidents, it can also be utilized in planned events as well. Due to the modular design, it can easily be adapted and incorporated to be a benefit in many different planned events. These planned events included concerts, parades, protests, conventions, sporting events (such as the Super Bowl or the World Series), or national or international events and summits (such as the G8, the UN General Assembly, and the G20 Summits). For planned events, the Investigations and Intelligence Gathering (I/I) operating as a Section can increase collaboration, gather intelligence on threats, and it can even expand or shrink to meet the

---

Emergency Incident Management Systems: Fundamentals and Applications, Second Edition.
Mark S. Warnick and Louis N. Molino Sr.
© 2020 John Wiley & Sons, Inc. Published 2020 by John Wiley & Sons, Inc.
Companion Website: www.wiley.com/go/Warnick/EIMS_2e

needs of the event or emergent circumstances that may arise. The use of the Investigations and Intelligence Gathering (I/I) as a Section in a planned event could prove extremely important in managing resources, especially if there were a breach in security, an escalation, or an attack that occurs during the event. With protocols already in place for an event, the Investigations and Intelligence Gathering (I/I) Section could prove quite helpful should things go awry. These protocols could include credentialing, badging, and additional logistic support which was initiated in the planning phase.

The Investigations and Intelligence Gathering (I/I) function that is not functioning under the Planning Section could be for a multitude of reasons, not just as a law enforcement function. According to the Department of Homeland Security (2013a, p. 7) "The mission of the Investigations and Intelligence Function is to ensure that all intelligence/investigations operations and activities are properly managed, coordinated, and directed." These actions can be taken to address the following:

- Preventing and/or deterring any possible criminal activity, occurrences, and/or attacks.
- Collecting, processing, analyzing, securing, and appropriately disseminating information and intelligence.
- Identifying, documenting, processing, collecting, information and/or evidence.
- Creating a chain of custody for the safeguarding, examination, analysis, and storing of evidence so that it meets or exceed legal standards.
- Conducting thorough and comprehensive investigations that lead to the identification, apprehension, and prosecution of perpetrators.
- Developing a channel to deliver situational awareness about an incident or the outlying factors related to that incident.
- Updating and supporting life safety operations.
- Provide for the safety and security of all response personnel.

Looking at this list, it is easy to see that Investigations and Intelligence Gathering (I/I) is more than just finding a perpetrator. There are a multitude of duties that could be associated with the function of Investigations and Intelligence Gathering (I/I).

The changes that were made were influenced by many other incidents, and when a review of ICS was undertaken, those looking at it utilized critical thinking. Because the investigative process and the acts of gathering intelligence might be needed in so many different ways, and on so many different levels, it was determined that it needed to be more flexible. The flexibility that came about in regard to Investigations and Intelligence Gathering (I/I) evolved over many years. A historical review might help provide a better understanding of the evolution with Investigations and Intelligence Gathering (I/I).

It should be noted that unless otherwise identified, the information for this chapter was gleaned from the National Incident Management System: Intelligence/Investigations Function Guidance and Field Operations Guide (Department of Homeland Security 2013a). This was done to help ensure the highest level of accuracy for this position in the ICS method.

12.1 Historical Overview

The Investigations and Intelligence Gathering Section (I/I) was not originally part of the NIMS method and the companion ICS method. The ICS method was originally developed for firefighters, and there was little, to no, considerations for including investigations and intelligence except for gathering information to enhance situational awareness. Over the years, it slowly became obvious that intelligence gathering and investigations for various incidents was useful for all types of incident management. Moreover, it was realized that these functions were needed in a multitude of situations, by various disciplines (e.g. public health, law enforcement, homeland security), in almost every level of response and recovery.

Incidents such as the Oklahoma City bombing, the 1993 World Trade Center bombing in New York, and the September 11 attacks made it obvious that there were times when Investigation and Intelligence Gathering (I/I) needed to be seamlessly integrated into the response; however, this function was placed under the control and management of the Planning Section Chief (PSC). This sometimes caused time delays and additional problems, especially if this supervisor did not understand the investigative process and investigative strategies. In some instances, these methods seemed counterproductive and were even seen as very disorganized in some cases.

Previously, Investigations and Intelligence Gathering (I/I) was undertaken only as a Planning function. After the Homeland Security Presidential Directive five (HSPD-5) and Homeland Security Presidential Directive eight (HSPD-8) were mandated, critical reviews were undertaken to identify ways of making the ICS method more effective. This proved beneficial for public safety in multiple ways, especially in making it more usable to all first responders.

In reviewing the response and recovery from the September 11 attacks, it became apparent that the integration and the transition of Investigations and Intelligence Gathering (I/I) with other disciplines was sorely needed. It was realized that most law enforcement was not effectively integrated into the response but rather operated independently of other disciplines. This became an issue that was seen to repeat itself in many larger incidents.

After the September 11 attacks, there was a wide array of integration, coordination, cooperation, and collaboration at the Pentagon, while New York City was quite the opposite. Many years after the incident, a report from Harvard University revealed that a Unified Command was used at the Pentagon, and used very effectively, while New York City had multiple command post locations for varying disciplines rather than a single Unified Command (Leonard et al. 2016). This led to delayed assistance during the recovery and demolition of the World Trade Center aftermath. In Hurricane Katrina, because the incident covered such a wide geographical area, many of the same issues were seen, including Unified Command (UC) and response and recovery that did not integrate the multitude of disciplines, especially public health.

After identifying it as a deficiency in the NIMS method and the ICS companion method, the 2008 iteration of these methods was updated, to include that Investigations and Intelligence Gathering (I/I) could be placed as part of the Planning Section (where it was originally placed), or it could be placed as a separate Section that would only be required to report to the Incident Commander (IC). From 2013 through 2017, FEMA guidance on the locations that the Investigations and Intelligence Gathering (I/I) placement in the hierarchal structure was provided in the student manual of the ICS 400 training manual and were taught nationwide.

In the 2017 iteration of the NIMS method and the ICS companion method, additional options were provided. There was the option to leave the Investigations and Intelligence Gathering (I/I) under the Planning Section, or to place it under the Operations Section, or to place it as its own General Staff position (Section), or even in its new placement as part of the Command Staff. The decision of where to place it would be left in the hands of the Incident Commander (IC) or Unified Command (UC). The most common determining factors for this placement included the size, scope, the complexity of the incident, and the number of Investigative and Intelligence Gathering (I/I) personnel that would be utilized (Figure 12.1).

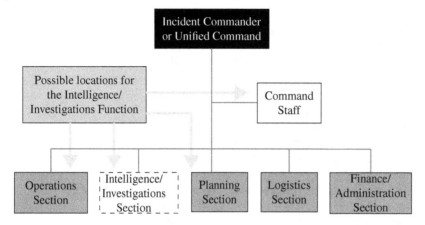

Figure 12.1 Placement of Investigations and Intelligence Gathering (FEMA-Emergency Management Institute 2018).

12.2 More than Law Enforcement

Often when public safety personnel hear the term Investigations and Intelligence Gathering (I/I), they automatically associate it with a law enforcement function. It is important to realize that Investigations and Intelligence Gathering (I/I) can investigate and collect information for a multitude of reasons, and it should not be automatically associated with law enforcement. It could involve a multitude of other disciplines and be equally (if not more) important than law enforcement.

Information collected within this function can also be related to public health emergencies, non-law enforcement cause and determination investigations, and many other types of investigative or intelligence gathering. The Investigations and Intelligence Gathering (I/I) Section can also be responsible for investigating fires, explosions, nuclear (or radiological) incidents, biological incidents, finding intelligence that can prevent crimes, public health outbreaks, public health epidemiological investigation, terrorist events, or a multitude of other incidents. It is important that the student as well as the seasoned professional realize and have a firm grasp on the concept that Investigations and Intelligence Gathering (I/I) is, and can be, much more than a law enforcement function. This is why Incident Commanders (IC), Unified Command (UC), and others should be cautious about limiting this function in the grand scheme of incident management.

Intelligence and investigative information are defined as information that either leads to the detection, prevention, apprehension, and prosecution of criminal activities (or the individuals(s) involved. In looking at this, it could include terrorist incidents or information that leads to determination of the cause of a given incident (regardless of the source). This includes incidents such as public health events or fires with unknown origins.

In some incidents, investigations are part of the equation to mitigate and to bring an incident to an end. In our everchanging world, an incident might have been caused by other than natural circumstances, such as man-made events. There may even be criminal activity during or potentially even after the incident occurred. Most who are old enough to remember the news reports from Hurricane Katrina can remember the looting that went on after the flooding, and some may remember the gang activity after the flood waters rose. There are varied reasons that Investigations and Intelligence Gathering (I/I) may be activated in various locations in the hierarchical organizational chart.

If we look at international incidents that have been perpetrated on American soil, it is easy to see that these types of incidents will usually utilize an investigation. Those in charge of the investigation will want to find answers, while the entire incident response is ongoing. A large part of the reason for this is to understand who committed the crime, why did they do it, were others involved and walking among us, as well as a multitude of other questions. With the new availability to place this function where it is best suited, the Intelligence and Investigations Gathering (I/I) can begin this work while the response is ongoing.

In the case of domestic incidents, there can be various reasons to investigate and a multitude of different agencies that could be involved. In large-scale fires, an investigations section may be needed to determine how that fire started. A few of the agencies that could be involved with a fire investigation might be the local arson (or cause and origin) investigators, local law enforcement, state law enforcement, state arson investigators, the Federal Bureau of Investigations, the Bureau of Alcohol, Tobacco, and Firearms (BATF), as well as many others. Additionally, if fatalities occurred due to that fire, there may be an investigation to determine how these fatalities occurred and may include numerous forensic investigators and coroners. By allowing this function to have separate duties from response, while still being integrated into the response, it makes the ICS method more flexible.

The Investigations and Intelligence Gathering (I/I) function can also be used in transportation incidents. These might include airplane crashes, train derailments, hazardous material transportation incident, tour or cruise boat accidents, and bridge collapses to name a few. The federal agencies that may respond could include the National Transportation Safety Board (NTSB), the Federal Bureau of Investigation (FBI), the Bureau of Alcohol, Tobacco, and Firearms (BATF), and many more federal agencies. Additionally, there will likely be a need to integrate local and state entities involved as well as.

In domestic public health emergencies, the origin of where the incident started and circumstances around the emergency, how it was spread, the likely places it may spread to, as well as questions about if it is intentionally manmade, accidentally manmade or an organic in nature may be part of the Investigations and Intelligence Gathering (I/I) duties. Other duties may include identifying if the problem occurred in other parts of the world or a long list of other items to investigate. A public health emergency may involve local, state, and federal agencies investigating, including the Center for Disease Control (CDC), various branches of the military, forensic scientists, and more. There may be a need for integration of state and local health departments, hospitals, or other health-care providers.

In the case of an epidemic, the Center for Disease Control (CDC), health-care facilities, and public health agencies may work together in an investigative and/or intelligence gathering manner. They may want to investigate where the epidemic started, what was the cause, and what will be the cure. They may want to investigate if it was manmade or if it came about naturally. In some instances, it may be important to locate carriers of the disease, and they may need to find those individuals. They may collect intelligence across the globe to find ongoing evidence of what might prevent the spread of the outbreak, or if a specific medicine is more effective than another. These are only a few of the many pieces of other information that could be gathered to mitigate the effects of outbreak.

12.3 Investigations and Intelligence Gathering (I/I) Information Sharing

An important consideration to realize prior to an incident is that not all information gathered by Investigations and Intelligence Gathering (I/I) can be shared among all personnel. Reasons that this decision can be made may be due to legal constraints, items that relate to national security, or information that only the perpetrator would know, and it is withheld for investigative purposes. Of course, there are many other reasons, but these are the most common reason.

Whenever utilizing Intelligence and Investigations Gathering (I/I), it should always be remembered that information flow may be intentionally restricted. In many instances, this restriction relates to the type and scope of information that can and cannot be shared. This holds especially true when investigating or gathering intelligence on a terrorist event, or in cases where national security may be compromised.

While the withholding of intelligence may be important in some instances, it should never override the primary objective of any incident, life safety. While certain information may be unable to be released, there is usually a way around protecting the public while not releasing the information that should not be released. When intelligence is gathered that might directly affect life safety of personnel and/or the general public, it is a foregone conclusion that this information can and should be at least shared with the Command Staff and General Staff.

Life safety is, and always should be, the primary incident objective. The creation of Investigations and Intelligence Gathering (I/I) as a Command Staff or General Staff position does not, and should not, weaken or change this primary objective under any circumstances. Investigations and Intelligence Gathering (I/I) should always do its best to operate concurrently with life safety operations and provide support for life safety operations while protecting evidence. Evidence collection should in every instance be a secondary objective.

Much like other General Staff functions, life safety operations are and should always be the highest priority in the Intelligence and Investigations Section. The highest priority should be the safety, health, and security of responders first, followed by the safety of the general public. Investigations and Intelligence Gathering (I/I) can be operated in such a manner that it contributes toward a safer, healthier, and more secure operation, placing life safety as the highest priority.

Whenever dispatched to an incident, especially considering the multihazard and threat-laden environment first responders now face, response personnel should be aware of as much going on around them as possible. An increased situational awareness could mean the difference between life and death. Almost any incident might be a scene where something nefarious has been precipitated at an incident.

Prior to an incident, it is in the best interest of law enforcement agencies to train other first responder how to preserve evidence while still doing their job. In doing this, primary responders can take the necessary steps to preserve potential evidence or have the knowledge to at least mentally document what they saw and where they saw it. This can be as simple as training other disciplines in making mental notes or taking a quick picture prior to removing victims or in some other mitigation measure they may be required to start.

The Incident Commander (or the Unified Command) will be responsible for determining if an Investigations and Intelligence Gathering (I/I) Section is needed. This is usually realized when identifying the incident objectives and strategies. This is the same concept and responsibilities that are required with other Sections of ICS. With this guidance, the Investigations and Intelligence Gathering (I/I) group, no matter where the placement is, can be more effective and efficient in their work. While the Incident Commander (IC) may be leading the response, there may be instances where information sharing every detail about the investigative arm may not be possible. Even the Incident Commander (IC) may be left out of the information loop on some issues due to reasons such as national security. The Incident Commander (IC) should realize this from the onset. While not the ideal situation for a response, the Investigations and Intelligence Gathering (I/I) Section should still share as much information as they can.

Priorities that will be set for the Investigations and Intelligence Gathering (I/I) Section should be done through utilizing Management by Objectives (MBO). While some information may be withheld from the Command Staff and the General Staff, Investigations and Intelligence Gathering (I/I) should be part of the planning process and share as much as possible in the Planning Meetings.

During fluid and emergent situations, priorities may need to be determined without the use of the planning function of ICS. While operating under ICS without the benefit of planning should always be avoided, there are times where an incident dictates expedient changes because of a fast paced and fast changing environment. Even in incidents where planning was used, the priorities that were set should have the ability to change as the incident changes. This is important because Investigations and Intelligence Gathering (I/I) may find out information that could change the entire response. Some examples may include toxins found at the scene or that an explosion was actually a dirty bomb.

To efficiently and effectively develop and use the Intelligence and Investigations information in an ICS setting, this function is integrated into the ICS methods structure, when needed. Whenever activated as a Section, they will follow the same framework as all other Sections (Planning, Operations, Logistics, and Finance and Administration). This framework allows scalability to meet the needs of the incident. The ICS method has the flexibility to launch the Intelligence and Investigations Gathering (I/I) function within the organizational structure to meet the needs of incident, thereby fostering better incident management.

12.4 Placement Consideration of Investigations and Intelligence Gathering (I/I)

The decision of where and at what level the Investigations and Intelligence Gathering (I/I) should be based will be determined by the activity level and the importance to the response and recovery of the incident. When Investigations and Intelligence Gathering (I/I) is required, the decision should be based on factors such as the nature of the incident, the level of Investigations and Intelligence Gathering (I/I) activity that will be required, and the relationship of Investigations and Intelligence Gathering (I/I) to other incident activities. The flexibility of the NIMS method and the ICS companion method allows all the flexibility needed so that the Investigations and Intelligence Gathering (I/I) can be placed in the hierarchical organizational structure where they will be most effective. Relatively few other emergency Incident Management Systems (IMS) have addressed this issue as of the writing of this book. Because of the flexible design of the NIMS/ICS method, the integration of Investigations and Intelligence Gathering (I/I) into any level can be implemented almost seamlessly.

There is a multitude of ways that Investigations and Intelligence Gathering (I/I) can be used and a multitude of places within the ICS method where they could be placed. A critical factor in placement and usage of any Investigations and Intelligence Gathering (I/I) within the ICS method is to be forward looking, identifying if (and how) they might be utilized from the onset of the incident. If Investigations and Intelligence Gathering (I/I) is needed in an incident, the earlier that it can be positioned and activated in its final positioning will help with mitigating the response in a quicker manner. For this reason, the initial and ongoing situational awareness is important to the decision-making process.

In the initial stages of an incident, the Incident Commander may determine that the need for Intelligence and Investigations Gathering (I/I) are not warranted. They may choose to create teams under the Planning Section to perform the intelligence function. They may even establish Task Force to support Operations in crime scene processing, data processing (especially for a pandemic), or some other type of Task Force rather than expanding the ICS hierarchal structure. In creating a Task Force under the direction of the Operations Section, the Incident Commander (IC) would essentially incorporate or combine several functions and/or disciplines by creating specialized teams. This would promote faster processing and incident-relevant sharing of information.

While initially the Incident Commander (I/I) may instead determine the need for an Intelligence and Investigations Gathering (I/I) under the direction of the Planning Section, they can also expand this function to be placed in different location by using the ICS methods modular design to move and expand the function.

There are numerous incidents where multiple investigative agencies may collaborate. Incidents such as active shooters, explosions, terrorist attacks, school bus accidents, tour bus accidents, and other incidents causing mass injuries or fatalities are some examples. In these, and other incidents, the function of intelligence and/or investigative resources may be needed to determine the cause and origin of the incident, or they may be even needed to support incident operations by securing a scene over a wide geographical area. This would be a classic case of when to place the Investigations and Intelligence Gathering (I/I) under the direction of the Operations Section. The reason this would be the optimal placement is because the direct contact with the Operations Section reduces the number of people that would need to be involved should there be a need to quickly change strategies.

12.4.1 Investigations and Intelligence Gathering (I/I) as Command Staff

When determining the placement of Investigations and Intelligence Gathering (I/I), the placement within a Command Staff position may be best suited for incidents where there is little tactical input to support the response, and/or the Investigations and Intelligence Gathering (I/I) is handling classified intelligence. The placement of Investigations and Intelligence Gathering (I/I) in the Command Staff position would also be viable when Agency Representatives (AREPs) are providing real-time information to the Command Staff, and the Command Staff is playing an active role in guiding the response.

While the modular design of ICS allows the Investigations and Intelligence Gathering Officer (IIO) can be easily integrated with other functional Command Staff within the ICS hierarchal structure, the scalability and flexibility will allow it to be seamlessly united with other more traditional functions of the ICS method. This framework allows for the incorporation of intelligence and information collection, analysis, and sharing, and when placed in the Command Staff position as an Officer, it can provide real-time information that may be critical to responder safety or to prevent further damage.

A good example of the placement of as an Investigations and Intelligence Gathering Officer (IIO) would be an instance of a cyber-terror attack. If the banking industry, the power grid, the water grid, or some other form of infrastructure were being attacked, then the Investigations and Intelligence Gathering Officer (IIO) can have their staff work on blocking the attack, and the Investigations and Intelligence Gathering Officer (IIO) can share critical information that can help the critical infrastructure under attack from suffering more damage. Real-time information from the Investigations and Intelligence Gathering (I/I) could work to guide how to protect unaffected infrastructure, software that is resistant to the virus or attack, where is the infrastructure and the geographic locations targeted, infrastructure that should be quarantined from the internet, and a multitude of other guidance that the typical person would not understand. Whenever it is beneficial for the Investigations and Intelligence Gathering (I/I) to have direct or immediate access to the Incident Commander (IC) and other Command Staff, then they should be strongly considered for placement as Command Staff. This holds especially true when mission critical intelligence needs to be shared quickly.

Another example may be when an organized terrorist attack has taken place, and more are expected. The Investigations and Intelligence Gathering Officer (IIO) could help guide the response based on intelligence that their staff has gathered, but that cannot be openly shared. Of course, this is theoretical and speculative because there is no evidence that the placement of Investigations and Intelligence Gathering (I/I) in the Command Staff position has happened as of the writing of this book.

The placement of the Investigations and Intelligence Gathering (I/I) as a Command Staff position might also be used if there is a pandemic outbreak. Because the involved public health agencies would need to investigate the spread of the outbreak, and their efforts would guide the overall response and/or put into place mitigation measures, then it may be best to place Investigations and Intelligence Gathering (I/I) as Command Staff function.

When placed in the Command Staff, the person that supervises all Investigations and Intelligence Gathering (I/I) activities will be called the Investigations and Intelligence Gathering Officer (IIO). At the time of writing this book, there were no job aids for the position of the Investigations and Intelligence Gathering Officer (IIO), and it appears that protocols for this position at this time is for local agencies to create protocols for this position that will meet their needs.

12.4.2 Investigations and Intelligence Gathering (I/I) as General Staff

The use of a separate Investigation and Intelligence Gathering (I/I) as a General Staff position would likely be used in instances where there was a suspected crime (e.g. terrorist attack, commercial airline crash, criminal negligence ending in death) and where Investigations and Intelligence would be critical aspect of the response. The Investigation and Intelligence Gathering (I/I) Section might also be used when large areas of evidence need preserved or when they are needed for responder safety.

Similarly, if an explosion occurs at a government building or a manufacturing facility, then it may be best to locate the Investigations and Intelligence Gathering (I/I) into a Section. The reason for this is that while they need to integrate the response with the fire, EMS, and other entities, the information they find will most likely not affect the response. They would (for the most part) be looking to determine whether a terrorist act, a criminal act, or negligence has been committed, and if so, by whom. In this case, the Investigations and Intelligence Gathering (I/I) Section will be an expansion of the existing ICS structure and would likely be placed as its own General Staff Section. Separately, the incident may be based on an investigation and/or the need for intelligence and will utilize the ICS structure to manage those ongoing investigations.

An example of this may be a hostage situation, a noninjury accident involving a government vehicle, or a mass fatality crime. No matter what the incident, or whether it is based in law enforcement, public health, or fire and/or EMS, the Investigations and Intelligence Gathering (I/I) Section Chief's job will, for the most part, be the same.

An example of using Investigation and Intelligence Gathering (I/I) could be an incident that included a terrorist attack, and for responder safety, the Investigation and Intelligence Gathering (I/I) Section is clearing the scene for contaminants of nuclear, radiological, biological, or chemical contaminants, while other responders work in already cleared areas. Of course, in this example, the Investigation and Intelligence Gathering (I/I) Section could also be placed under the Operations Section. Perhaps the biggest factor for deciding where it should be placed (in this scenario) could be determined by the level of interaction needed between Operations and Investigations and Intelligence Gathering (I/I).

12.4.3 Investigations and Intelligence Gathering (I/I) Section Chief

Upon determining that an Investigations and Intelligence Gathering (I/I) Section should be activated, an Investigations and Intelligence Gathering (I/I) Section Chief should be appointed. In most instances, this will be someone from local law enforcement; however, this is not always the case. In some instances, a state or federal agency might be appointed as the Investigations and Intelligence Gathering (I/I) Section Chief. While many factors may be involved, some basic rules to determine whether the Section Chief should be a local, state, or federal representative include the following:

- Is the incident a local incident?
- Does the incident involve state or federal property?
- Does it incident break federal laws?
- Does it involve a federally regulated transportation incident?
- Is it a terrorist event?
- Is it a nationwide event?

If the incident is local, then someone from local law enforcement will likely be the Section Chief of the Investigations and Intelligence Gathering (I/I) Section. In all other cases, there is the possibility that the Section Chief may be a state or federal law enforcement officer or representative that will be appointed. This should not be a foregone conclusion though. In some instances, Incident Command may use local law enforcement, and integrate state and federal resources into the ICS method, or have state or federal law enforcement as Deputy Section Chiefs (immediate subordinates of the Section Chief). There is also the possibility that a local law enforcement agency may initially be appointed, and as the incident goes on, and personnel from a state or federal law enforcement agencies arrive, they may relieve the initial Section Chief and take their place, or the original Section Chief may relinquish command of the Section to another individual.

No matter who is appointed as the Investigations and Intelligence Gathering (I/I) Section Chief, the job requirements are the same. Upon arrival to the incident, they will check-in and then report to the Incident Commander (IC) for a briefing. They will then locate or create their work area and order the needed supplies and resources (if not already in place). They will also organize, order, brief, and assign subordinates. As with any job within the ICS method, it is imperative that the Investigations and Intelligence Gathering (I/I) Section Chief to have accountability for their subordinates. Maintaining accountability will be more challenging in instances when the investigation does not stay within the geographical confines of the incident, or when resources being used should not be known by the public. As an example, if the incident was a bombing, an investigator or investigators may need to go to another geographical area to gather intelligence or to investigate suspects. In some instances, the accountability to manage those off-site investigators may become challenging to say the least. Those that are on the scene of the incident will usually be much easier to account for, manage, and to monitor, thereby helping to ensure their welfare and safety. Off-site personnel will often need specific times to check-in to ensure that their whereabouts and their safety are being monitored. Protocols will also need to be enacted that specifies what to do if personnel do not check-in on time, and there is suspicion they may be harmed.

As the incident expands, or as more ordered personnel arrive on scene, the Investigations and Intelligence Gathering (I/I) Section Chief will need to consider expanding or contracting the organizational structure to meet the needs of the span of control. Under the most recent guidelines and framework of ICS, the Investigations and Intelligence Gathering (I/I) Section is one area of the ICS method that can exceed the recommended span of control of seven;

however, this should be based on the level of supervision that will be needed. Based on the modular design, the Investigations and Intelligence Gathering (I/I) Section can, similar to other Sections, expand or contract to meet the needs of the incident. As they begin to organize the response, the Investigations and Intelligence Gathering (I/I) Section Chief has the capability to appoint Deputy Section Chiefs and to organize this Section into Branches, Groups, and Divisions. Any proposed expansions should be undertaken to meet the needs, size, and complexity of the incident.

Divisions can be created to meet the requirements of span of control when necessary. The expansion of ICS to include the use of Branches in the Investigations and Intelligence Gathering (I/I) Section will usually be required when a large amount of personnel is utilized. Branches will alleviate span of control issues when the number of personnel assigned to the Section expands or grows so large that the Section Chief becomes overwhelmed by the sheer number of Individuals or Groups that are reporting to them.

As the span of control begins to reach the limits, the Section Chief should activate one or more Branches within the Investigations and Intelligence Gathering (I/I) Section and designate a Branch Director for each Branch that has been created. While the creation of an Intelligence and Investigations Branch is a relatively new concept (when looking from a historical perspective), it has already proven to be beneficial in responses where it is needed. The six core Branches that may be activated include the following:

- The Investigative Operations Branch
- The Intelligence Branch
- The Forensic Branch
- The Missing Persons Branch
- Mass Fatality Management Branch
- Investigative Support Branch

It is important to note that it will be a rare instance when all the previously mentioned Branches will be activated at the same time. In most instances, the Branches that will be activated will depend on the type of incident, and if there is a need for that Branch. This statement should not be misconstrued to say that a complete activation of all Branches at the same time is not possible, because it is possible; it is just not probable (Figure 12.2).

Due to the practical nature of the Investigations and Intelligence Gathering (I/I) Section, the Section Chief may also, or in-place of Branches, need to establish Groups based on mission specific areas. If the number of personnel is not sufficient to support Branches, but there is still the need to expand to meet the span of control, the Section Chief may opt to create Groups. The Investigations and Intelligence Gathering (I/I) Section Chief may create one or more Groups within the Section, and the names and functions of each group (in most instances) will mimic the previously mentioned types of Branches. When expanding the Investigations and Intelligence Gathering (I/I) Section to include Groups, the Section Chief must appoint Group Supervisors for each one activated. Once again, this will usually be grounded in the concept of span of control and/or the function of the Group. When the creation of a Group is needed, the Section Chief should notify the Planning Section and, in most instances, the Incident Commander (IC). In reporting this to their supervisor, they should state the quantity of personnel that are assigned to the Section and to each Group. If any of the Groups are not used or have been deactivated, the Section Chief will manage those responsibilities and report these changes as well.

It should be noted that a Group will be used rather than Branch, or Branch/Group. This is done because the Group is the most common activation in the Investigations and Intelligence

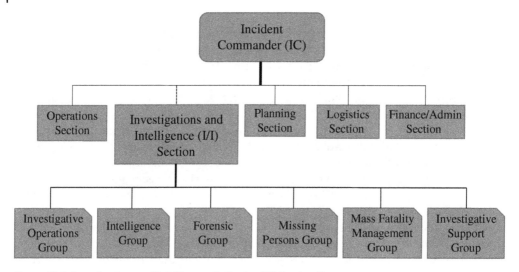

Figure 12.2 Investigations and Intelligence Gathering (I/I) Section Groups.

Gathering (I/I) Section. Additionally, other names may end up being more confusing for the individual reader.

12.4.3.1 Investigative Operations Group

The Investigative Operations Group will be the general investigative team for an incident, when activated. This Group will be the principal investigative arm and will lead the overall effort of investigating the incident. Investigative Operations will make recommendations for specific investigators, or types of investigators, to collect evidence and information. As those additional investigative resources gather the information and/or evidence, Investigative Operations will use that information to accomplish the overall mission of the Investigations and Intelligence Gathering (I/I) Section.

In most instances, when Investigative Operations Group is initiated, the Section Chief will appoint or assign a Primary Case Investigator and a Primary Supervisor. These individuals will be responsible for managing the Group. Investigative Operations is responsible for organizing and guaranteeing many of the distinct aspects of the Investigations and Intelligence Gathering (I/I) function. A primary function is to ensure that an Investigations and Intelligence Gathering (I/I) Plan is developed and implemented. This plan will guide the entire response and provide a systematic approach to ensure that the investigative process is organized and orderly.

Additionally, the Primary Investigator will choose a lead individual for each aspect of Investigative Operations. In doing so, they will ensure that the proper and most qualified personnel is assigned to each assignment and/or priority. The Primary Investigator will also ensure that these individuals are assigned using the appropriate priority and order to make the investigation more efficient.

Often, the Primary Investigator and/or the Primary Supervisor will create and name teams based on the tasks they are assigned (e.g. Interview Team, Interrogation Team, Reconstruction Team). When an individual is assigned as the Lead Investigator of a team, they are responsible for that task, or in some instances, a specific portion of the investigation. As the Lead Investigator, they have the tremendous responsibility of guaranteeing that the task is performed promptly, legally, ethically, and within the confines of the law. Approaching the investigation in this manner reduces the potential of compromising the potential court case (be it criminal or civil). They also have the solemn responsibility of confirming that the investigation was thorough, unbiased, and complete.

In the investigative process, documentation is extremely important. The Primary Investigator, the Primary Supervisor, and the Lead Investigator are all responsible for ensuring that documentation is thorough and that it meets the rigors of evidentiary collection standards. The Lead Investigator will be the primary person to record, or to ensure that the team members record, the tasks they undertook and any findings in the assignment log or database. The Primary Supervisor and/or the Primary Investigator will review these logs to ensure nothing was missed. This redundant review can help to prevent problems should the incident be referred to the prosecutor or in the event it becomes a court case.

The implementation of procedures for documentation and records management is an important part of preserving evidence. Setting priorities, goals, and specific procedures should provide guidance that outlines the actions that should be taken to ensure uniformity and a level of rigor that will strengthen any case that may arise from the investigation. It also ensures that all investigators are on the same page, and they are reporting the information in a manner that is not in conflict with other reports.

In documenting the investigation, the information collected, and the results of each investigative task should be documented. All the associated information and materials should be invoiced, secured, and thoroughly examined. It is important to note that even if the information or evidence collected is contradictory to the current theory of the investigation, it too should be documented. The investigative process should be completely impartial and fair. Failure to document contradictory evidence may compromise the case more than documenting it.

The types of evidence that should be invoiced, secured, and thoroughly examined can be multifaceted. It is important to note that all forensic, digital, multimedia, and investigative evidence is reported to Investigative Operations. This can include documents, interview notes, images, audio recordings, and data, to name a few. Upon completion of collecting and documenting the information, all investigative reports, evidence, and materials related to the investigation should be discussed with the team and the Primary Investigator and/or the Primary Supervisor. It is important to note that these results should not be shared with unauthorized personnel, especially when it may compromise the case. When discussing the reports, materials, and other evidence, an examination and evaluation should also be undertaken to determine whether the assigned investigative task was properly performed and if an exhaustive investigation was performed.

Upon the completion of reviewing the task (and all information associated with it), that specific task should be categorized as closed or open. If categorized as open, then more investigation or other actions are needed. If categorized as closed, it signifies that no further action is needed and that new leads will probably not be generated. Upon closing an investigative task, all related information related to it is recorded in the assignment log or database. Even though the task may be technically closed, the results should be analyzed by the Primary Investigator and/or the Primary Supervisor. If deemed necessary, additional follow-up should be undertaken.

It is important to note that in some instances, especially when a monumental case is being pursued, prosecutors may want to review the evidence as the tasks are closed. In doing so, they too may also require additional follow-up, which may help to more clarify the case. Every effort should be made to work with these prosecutors to ensure that they have every detail possible to have a more solid case. Additionally, maintaining a good relationship with the prosecutor can be a substantial asset. This holds especially true if legal advice is needed, or court proceedings that might be needed such as search, extradition, or arrest warrants. Although their job is technical, a good relationship with a prosecutor may also help expedite the filing of documents, applications, and processes with the courts.

In larger and more complex cases, the use of a display board or Weblog can be especially helpful. The display board or Weblog can be used to provide a chronological record of the significant information, actions, decisions, orders, and outcomes in one place. The display board or Weblog can help to ensure that proper and exhaustive investigative techniques and tactics were completed using the proper priority, order, and sequence, and they can help the investigator to make mental connections that may not have been realized without it.

Communication is an important aspect for the Investigative Operations Group. All groups under the Investigations and Intelligence Gathering (I/I) Section umbrella should utilize an Investigations and Intelligence Gathering (I/I) Operations Center to communicate, coordinate, and collaborate, thereby creating a more thorough investigation. This communication, coordination, and collaboration should be facilitated through either a designated investigative supervisor or an investigator that is assigned to each of the crime scenes. It is also important to include those that were significantly involved in the investigative process. This holds true not only at the crime scene but also at hospitals and other off-incident facilities. On an ongoing basis, the Investigative Operation Group and their supervisors should examine and analyze all facets of the investigation, including all unassigned, assigned, and completed investigative tasks.

There are multiple ways to investigate and/or gather intelligence. The techniques and tactics that are typically used by the Investigative Operations Group may include (but are not limited to):

- Canvasing (both technical and nontechnical)
- Witness interviews
- Interrogations
- Prisoner debriefings
- Identification procedures (including physical and photo lineup)
- Serving search warrants
- Search and seizures
- Database and Record queries (e.g. NCIC, UCR, AIFIS)
- Social Media searches
- Review and analysis of telephone records
- Physical surveillance
- Electronic surveillance
- Procurement and analysis of outside records
- Reviewing evidence from potentially related crimes
- Polygraph examinations
- Undercover officer and confidential informant operations
- Activating and monitoring tip-lines, hotlines, and/or call centers
- Dissemination of alarms such as
  - BOLO's (Be on the Lookout)
  - Amber Alerts
  - Mass text messaging
  - Public alerts, warnings, and notices
- Obtaining and securing of sources of evidentiary data, such as
  - Flight data recorders
  - Cockpit voice recorders
  - Vehicle electronic data recorders
  - GPS data recorders

- Radar data
- 9-1-1 call tapes

In some larger incidents, management of Investigations Operations Group may be complex and may require assigning certain functions to subgroups that manage specific areas of the investigation. This is based upon the scope, complexity, and size of the incident and the Investigations and Intelligence Gathering (I/I) Section and the need to maintain a manageable span of control. Because the ICS organization is flexible and modular, the Incident Commander (IC) or the Investigations and Intelligence Section Chief (with the Incident Commanders approval) can create separate teams to better manage the incident. Some common teams that are initiated include the following:

- Assignment Manager
- Recorder
- Evidence Manager
- Physical Surveillance Coordinator
- Electronic Surveillance Coordinator
- Electronic Communication Records Coordinator
- Tactical Operations Coordinator.

In less complicated incidents, these duties can be combined, or they can be handled by the investigators. The key factors involved in deciding to expand or combine these positions are typically based on the span of control and/or the amount work that has been assigned to one individual. It is important to realize that if one individual has been assigned too much work, they may become overwhelmed and they may begin to miss evidence or information that may be important to the case or the investigation.

In many instances, the Investigative Operations Group acts as a central repository for other Groups that are activated. As forensic information is gathered and analyzed by the Forensic Group, the results will usually be shared or forwarded to the Investigative Operations Group. As more missing persons are reported by the Missing Persons Group, that and other information will usually be shared with Investigative Operations Group. As the Mass Fatality Group identifies deceased individuals, it will also be usually shared with Investigative Operations Group. These examples should help provide a better understanding that the Investigative Operations Group is usually the central hub in the Investigations and Intelligence Gathering (I/I) Section, and they are the one that typically puts all the pieces of the puzzle together to create a more complete picture.

Just like other Branches, Groups, and Units in the ICS method, as the incident comes to a close, the Investigative Operations Group will need to properly demobilize. In most instances, Investigative Operations will be the last Group in the Investigations and Intelligence Gathering (I/I) Section to demobilize. The reason for this is because they are the central hub for all things investigative, so all other Groups will need to demobilize first. The Primary Investigator and/or the Primary Supervisor will usually receive instructions on demobilization from the Demobilization Unit or the Finance and Administration Section Chief ([FSC] or both). They will then brief the Staff on the procedures that should be followed, and what responsibilities the staff have in the demobilization process. Because evidentiary procedures (such as evidentiary chain of custody) must be followed, the demobilization will usually follow the lead agencies protocols for the transfer of evidence and information. As the demobilization of Investigative Operations Group progresses, the Primary Investigator and/or the Primary Supervisor will ensure that all procedures set out in the Demobilization Plan (as well as the lead agencies protocols) are followed. This is done in a systematic order to ensure that any legal actions, whether criminal or civil, are not compromised. If required, the Primary Investigator and/or the Primary Supervisor will fill out ICS Form 221, Demobilization Check-Out, and turned in to the appropriate person.

12.4.3.2 Intelligence Group

The Intelligence Group in most instances reports to Operational Investigations Group; however, in some instances, they will report directly to the Investigations and Intelligence Gathering (I/I) Section Chief. This is often based on the protocols of the lead agency and the circumstances surrounding the incident. In instances of national security, when the intelligence gathered could potentially identify confidential sources (or informants), or when specific information that may separate the real perpetrators from those that are trying to falsely claim responsibility, then compartmentalizing this information may be advised. When these types of circumstances occur, the Intelligence Group and/or the Section Chief may deem information as "Need to Know." Need to know is the basic concept that only those that have a specific role in the case, and their supervisors should know the details of the information collected. Need to know circumstances provide greater protections of individuals and evidence.

The Intelligence (or Intel) Group is a specialized group of individuals that are usually responsible for three major functions;

- Information intake and assessment
- Operations security, operational security, and information security
- Information and intelligence management function

As necessary, these three functions can be subdivided into teams. Each team will be responsible for the specific task they are assigned. Because these duties can be confusing to the common individual, a more thorough explanation was in order to help the reader better understand these duties.

12.4.3.2.1 Information Intake and Assessment Unit[1]

In utilizing the information intake and assessment function, the Unit ensures that incoming information is directly communicated to the Intelligence Group and that the information is documented on an information control form. In most instances, the information is also entered into an information control database. While a matter of preference, most agencies believe that information should never be added to a database alone. The thinking behind this decision is due to the potential of losing the information to an electronic failure or, in rare instances, cyber hackers. The keeping of an additional nonelectronic written documentation kept in a secure location could be important if such an issue occurs.

When evaluating intelligence, it is important to determine the correct information security classification (e.g. classified, sensitive, need to know) and the information security procedures that are associated with the type of classification. After classifying the information, it should be first evaluated and assessed to determine what next steps should be taken. In most instances, this assessment will identify if the intelligence should be forwarded to Investigative Operations Group, or if it should not be kept within the Intelligence Group. When forwarded to Investigative Operations Group, usually one of two tasks will be undertaken by that Group; it will be determined that further investigation is needed, or the Investigative Operations Group will make a final determination and disposition on the information.

When evaluating and categorizing intelligence, the Information Intake and Assessment Unit will almost always assess the information based upon lead agency protocols. While protocols may vary from agency to agency, it is important to evaluate the information to ensure that

---

1 This heading is for advanced IMS students.

it is not repeat information from the same source. This can be done by executing a database or records search and analyzing (through several methods) the information to determine the source. While the same information from the same source would not (usually) be forwarded, the same information from a different source or multiple sources could be highly significant. Information sharing is an important part of helping to put the pieces together and should be a high priority except for instances where the information needs to be compartmentalized,

12.4.3.2.2 The Operations Security, Operational Security, and Information Security Unit

The Operations Security, Operational Security, and Information Security Unit ensure that all actions and activities related to securing received intelligence information are thoroughly implemented. As part of their responsibility, this individual or group of individuals will make sure that classified information is distributed to personnel who have the required clearance, access, and/or have been deemed an individual that is on the "need to know" list. They will ensure that the information is disseminated in compliance with the established protocols, safeguarding against unauthorized dissemination, including leaking of information to unauthorized individuals and organization (e.g. the press).

In addition to protecting information, this Unit ensures that sensitive information is distributed to authorized personnel who are also part of the individuals who are on the "need to know" list. The dissemination of this information should follow strict protocols and, in most instances, strict federal government compliance. This specific Unit guarantees that applicable limits and laws are followed explicitly when it comes to information sharing.

The Operations Security, Operational Security, and Information Security Unit will facilitate and sustain a liaison relationship with the overall intelligence community. By utilizing the appropriate channels, these liaised relationships should (in most circumstances) include federal level agencies (e.g. ATF, FBI, NSA), and when allowable, the intelligence mechanisms of other agencies affected by the incident (e.g. state police, state homeland security, county level law enforcement, local law enforcement), and Fusion Centers. While information sharing is important, the Operations Security, Operational Security, and Information Security Unit should discuss the importance of information security and confidentiality requirements with the Command and General Staff members, or the Investigations and Intelligence Gathering (I/I) Section Chief to prevent the potential of specific information and the origins of that information from being compromised.

12.4.3.2.3 The Information/Intelligence Management Unit[1]

Managing and organizing information and intelligence can sometimes be a daunting task in large, or very fluid situations. One of the primary functions of the Information and Intelligence Management Unit is to provide a critical review of tactical and strategic intelligence/investigations information collection to make sure that these assets are using appropriate, authorized, and lawful methods. They will also ensure that the lead agencies intelligence protocols are followed. This is important for managing and directing collection efforts that are related to intelligence collection. They will perform database and record queries that among other things, looks for the same information from differing sources. They will also undertake language translation and the deciphering and decryption of incoming information.

Another important aspect of Intelligence and Investigations Management Unit is to ensure that information is documented, secured, organized, evaluated, collated, processed, analyzed, and put to beneficial use. By doing so, they ensure that collected information is separated and categorized in such a way that the most important or most credible information is realized and passed on to the appropriate group or individuals for further investigation or action.

The Intelligence and Information Management Unit will also identify gaps in information and intelligence. This will often lead to requests for additional information and intelligence in specific areas. At times, these gaps will reveal specific areas that require a standalone, or an ad hoc investigation, that will not be related to the main investigation or information gathering process. When this type of investigation is needed, it is imperative that the results from this separate ad hoc investigation are documented, analyzed, validated, theories or evidence is formed, and resolved.

The necessity of an ad hoc investigation in a terrorist attack might be an outlier that only one witness observed. As an example, all reports identify that the terrorists took public transportation to the location of the incident. A single witness claims that they observed a man on the roof of a building a block away, watching the incident unfold through binoculars. As law enforcement arrived on scene, this single witness claims that after the event unfolded, they saw the individual exit the roof, then a few minutes later saw them on the street quickly leaving the scene in a white van. This information was only provided by one witness, and it is an outlier from the main investigation of the actual events. The information provided by this witness raises a few questions outside of the main investigation. An ad hoc investigation could be (and probably should be) formed to identify if this was connected to the attack, or if it was just a normal citizen watching for a specific bird, car, or some other thing.

The Information and Intelligence Management Unit is responsible for making requests for information to the appropriate individuals. These individuals can include governmental agencies, nongovernmental organizations, private sector entities/individuals, the media, and the general public. It is important to realize that while there may be easier ways to obtain information, in situations where the Investigations and Intelligence Gathering (I/I) Section has been activated, it will become extremely important to ensure that no stone is left unturned. Besides undertaking the investigation as part of their due diligence, those involved in the investigation will also need to realize that the magnitude or importance of the incident may lead to increased scrutiny. That scrutiny may come from defense attorneys, private investigators, media coverage, conspiracy theorists, and others. It is important that the investigation be thorough, and no part of the investigation has been left to speculation.

Investigations and Intelligence Gathering (I/I) tasks that have been completed, as well as suitable and important raw information, should be documented and produced as needed. Producing this information can include recording it, entering the data into a database, issuing warnings, creating situation reports, briefings, bulletins, and/or assessing the information for others. Information and reports that are below the tear line (such as unclassified or lesser classified reports) are usually produced to release information as necessary or advisable to quell the media's hunger for information and/or to reduce (or contradict) false information and speculation. While it is important to protect some information, it is also important to remember that if some information is not provided to the public and the media, they will often make their own conclusions. In some instances, the media may find someone that claims to be an expert in the field and who will express their own personal beliefs making them sound like facts.

The Information and Intelligence Management Unit should also be responsible for ensuring that classified information and/or access-controlled, sensitive, compartmentalized and/or caveated/restricted information is cleaned-up, sanitized, or redacted as needed. This includes when this information is needed to create any investigate leads, to publish intelligence products, to prepare warrant applications (especially in an open court), and various other tasks that may be needed. When handling this information, it should be considered that not everyone who receives it will be bound by the same protocols that they adhere to.

The Information and Intelligence Management Unit will also disseminate investigative information, documents, investigative requirements, and other information in the proper manner.

They will safeguard the information and ensure that shared material for press releases do not share any need to know information. They will also immediately share any threat information or intelligence with the Investigation and Intelligence Gathering (I/I) Section Chief, the Incident Commander (IC), and/or the Operations Section Chief (OSC). In most instances, they will determine who needs to know the threat information, based on the type of information it is. In rare instances, they may be required to share the information or intelligence with other authorized personnel, such as the Public Information Officer (PIO), so that information can be released to the public. Whenever threat information that may border on need to know information, it is best to have that information vetted and approved by the Investigation and Intelligence Gathering (I/I) Section Chief and the Incident Commander (IC) to ensure that information is not released if it should not be disseminated.

In some instances, Subject Matter Experts (SMEs) may be needed to help better understand the information or intelligence that has been collected. In some instances, Subject Matter Experts (SMEs) may be needed just to determine or apply the mind-set of a group or individual(s), including how the intelligence intersects with their approach or mentality. Before utilizing a Subject Matter Expert (SME), Information and Intelligence will need to make sure that the individual has the appropriate clearance(s) to view the information. They will also need to confirm that they are actually an expert in the areas where consultation is needed.

12.4.3.3 Forensic Group

The Forensic Group is sometimes identified as CSI (Crime Scene Investigations), especially in larger cities. The Forensic Group manages the crime scene or crime scenes if there are multiple scenes. They are responsible for the processing of the forensic evidence, electronic evidence, DNA evidence, and any cadavers. They certify that the proper type of examinations, analyses, tests, comparisons, and enhancements are performed in the proper order, and by the appropriate individuals, businesses, and/or government entities. While the information listed in this chapter about the Forensic Group will be already known by those who will work in this Group, it is important for others to understand what is done and how it integrates into the ICS method of incident management. This will allow a better situational awareness of the constraints and opportunities that the Forensic Group works under.

The Forensic Group will usually share information with Investigative Operations, and they will often coordinate with the Mass Fatality Management Group and the medical examiner (or coroner) when deceased individuals are involved. They will oversee and document the findings of the initial examination, recovery, and transferring of any bodies. This is done so that the location and all pertinent information about all cadavers can be documented and thoroughly support evidentiary rules in a court case.

As part of their duties when responding to pandemic or an outbreak, the Forensic Group may be required to gather and collect specimens. In doing so, they could look for forensic evidence that identified what the pandemic is caused by, where it started, if it is mutating, and much more. They may work within the lab, or they may be sent to the field.

As part of the Forensic Groups duties when responding to a crime, they will be responsible for ensuring that the entire crime scene is preserved, including at all locations if more than one crime scene is involved. If there are fatalities, they are also responsible for quickly identifying the number of fatalities, so the full extent of the incident is swiftly known. They will collect forensic evidence that may identify perpetrators or the method in which they committed the crime.

To preserve evidence, the Forensic Group will ensure that each crime scene and all deceased individuals and/or any forensic evidence is protected and shielded from those that could contaminate the evidence and/or the crime scene. The Forensic Group can personally control access to these crime scenes, or they can request that another agency does so. In controlling a crime scene, it is imperative for the process that access to the scene and/or any deceased individuals is restricted. In some cases, it may be limited to only the Forensic Group or it may be restricted to the Forensic Group and investigators.

It is important that the Forensic Group, or someone that they have appointed, documents anyone that has entered the crime scene, or who has had contact with cadavers. The agency securing the scene should document the rank (or title), the name, the agency, and any other identifier of each individual person who entered the crime scene. This holds especially true if the individual entering the scene touches, examines, disturbs, or moves, anything within the crime scene perimeter. Additionally, anyone that entered the scene prior to limiting access should be documented (such as firefighters or EMTs), and in most instances, the Forensic Group will question them about where they walked, what they moved or disturbed, and various other questions that may help them understand what the area would look like if totally left undisturbed.

The personnel processing crime scenes and/or the deceased individuals must communicate and coordinate with Investigative Operations Group, including the Primary Case Investigator, the Primary Case Supervisor, the medical examiner, and other personnel as necessary. They will also ensure that each of the crime scenes and any deceased individuals are processed in an appropriate and expedited manner. They should also ensure that forensics are accomplished in the proper order, and sequence, and that forensic evidence is given the proper priority.

When working in a forensic capacity, it is imperative that all evidence follows a chain of custody, and all evidence should be processed in a manner that protects the integrity of the evidence. It is also important that the evidence is swiftly processed for multiple reasons. Some of those reasons include the potential degradation of evidence and the potential that the evidence may be called into question if there are large timeline gaps.

If the local agency uses outside sources to perform analytical analysis, then it is vital that the laboratory, analytical service provider, and/or morgue facilities meet all scientific standards that cannot be challenged in court. The Forensic Group should ensure that the contractor or organization (receiving the evidence) will process it using the proper number and types of examinations, analyses, and comparisons. The Forensic Group will also ensure that any contracted work, including tests and analysis, will be performed in the proper sequence.

Communication between the personnel that will be processing crime scene(s), the Primary Case Investigator, the Primary Case Supervisor, and the forensic contractor (or organization), is a key factor in ensuring that the evidence is not compromised. Prior to being transferred to the contractor (or organization), it is compulsory that discussions and agreements are reached regarding the proper length of time the evidence should be stored, secured, retained, and disposed of (or returned to the Forensic Investigators). A proper time frame and a suitable manner must be agreed upon before providing the evidence to them.

In some instances, there may be a need for a bomb squad or explosive ordinance disposal expert assessment of evidence. On occasions when there is a suspicion of explosives, the bomb squad should concentrate on rendering safety to all devices prior to processing any potential explosive or incendiary devices. After an explosion, whether accidental or intentional, it is important that forensic debris documentation and collection procedures are followed and that postblast crime scene forensic measures are implemented according to commonly accepted scientific standards. Crime scene reconstruction techniques and Subject Matter Experts (SMEs) should not be ruled out for any crime scene, especially those involving explosion.

As part of their responsibility, and to ensure that a court case is not compromised, the Forensic Group will document evidence and prepare reports about the forensic evidence. Each piece of forensic evidence should be documented, including the chain of custody, each place it was taken for analysis, and the results of any analysis. When creating reports on forensic evidence that was handled by contractors, it is important that Forensic Group personnel reference and attach a copy of the contractors report to their own report, so that all information is documented every step of the way. It is also important to prove that the forensic evidence is not based on the word of the Forensic Investigator(s) alone.

It is also vital that the Forensic Group and all investigators safeguard all information. From the initial dispatch, until the case is closed and prosecuted, all information should be kept confidential. In some instances, details can be released, while in other circumstances, the release of some information may compromise a case. One example of information that can usually be released would be the name of any deceased individuals; however, it is important to ensure that next of kin is notified prior to releasing the name to the public.

Depending upon the size, scope, complexity, and layout of the Investigation and Intelligence Section, the Forensic Group Supervisor may decide to activate coordinating positions to help manage the span of control. Those coordinating positions may include the following:

- Crime Scene Coordinator
- Bomb Operations Coordinator
- Chemical, Biological, Radiological, Nuclear/Hazardous Materials Evidence Coordinator
- Forensic Evidence Analysis Manager (including digital and multimedia evidence).

When initiating these coordinating positions, it is extremely important to choose the most qualified person. In some instances, a superior officer may believe they should oversee or manage an area, but they may not have the required expertise, or perhaps they don't even have the basic experience needed. Appointing someone into a coordinating position that does not have experience or expertise could compromise a case. For this reason, it is strongly suggested that the appointment of a coordinating position be based on the most qualified, rather than the highest rank.

It is also important to note that the Forensic Group may be required to work in close collaboration with the Missing Persons Group. This is especially true when there has been a significant loss of life due to a disaster such as a major hurricane, an earthquake, or a terrorist attack. In large-scale events, there can be a multitude of missing individuals, and forensic evidence collection may be necessary to identify those deceased after such an event. Whenever this occurs, the Forensic Group will likely be asked to cooperate and collaborate with the Missing Persons Group to identify those that are could already be in the morgue and are awaiting identification.

12.4.3.4 Missing Persons Group

When a manmade or natural disaster occurs, there is always the possibility that there will be people that will be classified as missing. In major disasters, locating and/or identifying missing individual can be a monumental task. An individual only needs to think about the September 11 attacks on the World Trade Center, the Pentagon, and the crash site in Shanksville Pennsylvania to get an idea of how difficult this can be. They only need to remember the widespread devastation caused by Hurricane Katrina or Superstorm Sandy to realize that finding and/or identifying missing persons can be an enormous task.

The formation of a Missing Persons Group under the direction of the Investigations and Intelligence Gathering (I/I) Section is the ideal location within ICS method to undertake this task. The Missing Persons Group oversees, and leads, all operations and activities associated with a missing person. Their primary job is not to perform actual information and intelligence gathering in the field, but their job is to take reports and gather needed items from friends and family members (of the missing person) that may help with identification process. They will work closely with the Mass Fatality Group who will undertake the actual work in the field.

The Missing Persons Group will coordinate and collaborate with the Mass Fatality Group and the medical examiner (or coroner) to provide base line information about the missing individual. This base line information will come from working with the friends and families of those missing. In most instances, the Missing Persons Group will create a Family Assistance Center, which is a place where families can report the missing person and where the friends and family can provide and/or receive information about their loved ones.

The Missing Persons Group should also maintain documentation of the overall number of individuals reported missing, and all data associated, including but not limited to potential missing persons, actual missing persons, and when a missing person is located. The Group will accumulate and process information such as data on the missing person, any records associated with the individual (e.g. dental, X-rays, criminal, and other records), pictures of the individual, and when possible, DNA reference samples. They will also collect and compile investigative evidence, forensic evidence, digital and multimedia evidence, and nonevidence property. In most instances, they will collect these items at one or more Family Assistance Centers or facilities that have been opened for this purpose.

Real-Life Incident

On 11 September 2001, terrorists crashed planes into the World Trade Center (WTC) in New York City. Fifty-six minutes after the incident began, the first WTC tower crumbled and collapsed. Many people were still in the building when they collapsed. While a few survivors were found, there were 2952 individuals who were thought to be missing (Thomas & Bode, 2011). The daunting task of finding their remains would take eight months, but the forensic identification of body parts and matching them with a person would take nearly three years to complete.

Because the WTC collapse left such a large debris field, the site was partitioned into four separate Divisions. A Division Chief (for each shift worked) was put in charge of each Division (Barry, 2001). As the debris was removed, and loaded on trucks, strike teams and task forces (sometimes with a cadaver dog) scanned the rubble for body parts. This was done on the ground where the rubble lay, and where trucks were being loaded. The debris was then transported to the Fresh Kills Landfill in Staten Island NY where the FBI and various other agencies also looked for remains (Hirschkorn, 2005). All remains that were located were sent to New York Medical Examiner's Office for identification although multiple agencies, contractors, and organizations collaborated in this location and identification endeavor.

Some of the methods used to identify body remains included dental records, fingerprints, and X-rays; however due to the violent nature of the collapse, most remains were identified through DNA analysis (Thomas & Bode, 2011). As remains were identified, the Missing Persons Group would notify families.

No single agency did it all. Thousands of individuals, from the site of the collapse, to the land fill, to the medical examiner's office, to the forensic contractors, to the Missing Persons Task Force collaborated and cooperated to find remains, identify them, and to notify the families of their findings. It took local, state, and federal government as well as public and private partnerships to mitigate the effects of this disaster and to identify those missing and presumed dead. As of the writing of this book, 18 years after the incident, there are still small pieces of bones being found at the landfill, and DNA is being tested. As of 2016, no remains have been found for approximately 40% of those that were reported missing (Abramson, 2017)

It was surmised that the vast majority of them died in the collapse. As the rubble was removed, it was checked numerous times for body parts, including at the scene of the incident, as it was loaded onto barges, and at the landfill, various body parts were found. Forensic analysis positively identified 1638 individuals from DNA analysis. After a positive identification was made, the information was

12.4 Placement Consideration of Investigations and Intelligence Gathering (I/I) | 335

then shared with those that were maintaining the missing persons' roster. This was prior to the initiation of the Investigations and Intelligence Section in ICS, and part of the impetus to include this module in ICS.

Similarly, in the 2005 aftermath of Hurricane Katrina, there were reports of almost 6600 people that went missing.

The Missing Persons Group will also report nonspecific information to the proper individuals or Groups within ICS structure, as needed, as part of situational awareness. As additional information or intelligence is collected, they will assess the information, consolidate it, and categorize that information. They will also be responsible for the tracking, storage, and dissemination of nonspecific information. Nonspecific information may provide how many individuals are missing, where they were at the time of the incident, and various other broad information; however, they will not give information that violates the privacy of the victim, or their friends and family.

As conclusive information is found about the disposition of a missing person (e.g. deceased, in the hospital, found alive), the Missing Persons Group will notify the family of that disposition. It is imperative that this information be given to the family prior to release of information to any other agency or organization. This is done so the family does not receive the information second-hand. As each person that has been reported as missing is located, a person from the Missing Persons Group (possibly in conjunction with clergy or a mental health professional) should make a face-to-face notification. These notifications should be made in a respectful and timely manner, and they should be made to the appropriate next of kin. Upon completion of the notification to the next of kin, the Missing Person Group member that made the notification should document the date, time, place, who was notified, and any other individuals that were present when the notifications were made.

The Missing Person Group should also communicate and coordinate with the Public Information Officer (PIO). In doing so, they will provide approved information and instructions on the proper actions to report a missing person. The Public Information Officer (PIO) will then release this information to the media, the public, governmental agencies, nongovernmental organizations, and private entities/individuals as they deem necessary. The Group should also provide regular updates to keep the public informed of any developments that have occurred once those items have been approved for release.

Depending upon the size, complexity, and scope of the incident, the Missing Persons Group Supervisor may determine that the span of control is reaching unmanageable limits. In cases such as these, the Missing Persons Group Supervisor may subdivide the Group to help better manage their duties and responsibilities. When this division of resources is needed, it is common to split the workforce into two teams, the Missing Persons Team (managed by the Missing Persons Coordinator) and the Family Assistance Center Team (managed by the Family Assistance Coordinator).

The Missing Persons Group Supervisor is responsible for ensuring that coordination and information sharing are established. Those individuals that they will share information with include the Forensic Group, the Mass Fatality Management Group, the medical examiner, the Mass Fatality Branch and the Public Information Officer (PIO). They will also notify the Investigations and Intelligence Operations Section Chief that these teams have been activated.

12.4.3.5 Mass Fatality Management Group

The Mass Fatality Management Group is responsible for investigating and gathering intelligence when an incident has multiple deaths. While the Missing Persons Group deals mostly with collecting information from friends and families, they also collect intelligence and information from the Mass Fatality Group, and these two groups work collaboratively. The Mass Fatality Group is the operational arm that is responsible for field operations when a mass fatality management incident has occurred.

The Mass Fatality Management Group is responsible for ensuring that mass fatality management operations, protocols, and activities are implemented. These protocols should fall in line with local procedures and standard evidentiary guidelines. In some instances, especially if there may be a federal criminal case, this Group may need to follow federal protocols.

About Family Assistance Centers

Immediately following a Mass Casualty Incident (MCI), the families and friends of potential victims will desperately seek answers about their loved ones. In most instances, they will assemble at places where they believe they will find their loved one, or where they may be able to find out information about them. This usually means that they will assemble at the site of the incident and/or local hospitals. Due to high emotions and anxiety, these areas where they assemble can see outward manifestations of anger, fear, and anxiety, especially if they feel their concerns are not being addressed. The formation of a Family Assistance Center can help provide compassion, reduce these feelings of anger, as well as support the family.

In most instances, a dedicated structure for the duration of the incident should be opened. This can include an office building, a section of a police department, fire department, or some other governmental building, a house of worship, or other suitable structure. Services that should be included when creating a Family Assistance Center should include the following:

- A call center where individuals can call for information
- Reception and information desk that can direct friends and family to the proper resource
- Spiritual care (inclusive of all religions)
- Mental health services
- First aid and/or temporary medical services (to address physical problems that may arise from the stress)
- Interpretation services and interpretation cards to help identify the language
- Child care
- Food services
- If necessary, housing assistance and/or sheltering
- A lead Investigation Center to follow up on leads
- Internet-based status update site
- A list of resources that may be needed by family members

The Family Assistance Center will typically be led by one organization, the Missing Persons Group; however, it will also work collaboratively with other organizations such as the American Red Cross, the Salvation Army, the medical examiner's office, the Forensic Group, and other public, private, and nongovernmental resources to meet the needs of the families (Forrester et al. 2008). In the initial stages of an incident, the Family Assistance Center is an afterthought rather than a priority.

The Mass Fatality Management Group will oversee all aspects of deceased individuals in a mass fatality situation. They will ensure that all documentation is undertaken (according to protocols), and they will report this information to the appropriate individuals. They will assess, categorize, and consolidate all information and evidence, including where forensic evidence was found and where evidence was taken for testing. Even though the Forensic Group has the responsibility for managing that evidence, outsourcing of testing evidence is allowable, providing the contractor or agency meets required standards for evidence processing. The Group will also track where the deceased was taken and where they are/were stored (e.g. taken to mobile morgue, taken to medical examiner, released to family).

In larger incidents with numerous deceased individuals, the Disaster Mortuary Operational Response Teams ([DMORT] or a similar resource) may be needed to help handle and process the deceased. It is the responsibility of the Mass Fatality Management Group to identify when these resources are needed, and to use the appropriate channels to request them. Upon arrival at the incident, the Mass Fatality Management Group will work closely with the Disaster Mortuary Operational Response Teams (DMORT) or other resources used to undertake these tasks.

Depending on the circumstances, there may be a need to implement debris sifting operations to find any bodies or body parts. When debris sifting is necessary, the Mass Fatality Management Group should be responsible for setting protocols, documenting where the deceased person (or body part) was located, and for overseeing the recovery of the body from the debris in a respectful manner. They will also ensure that the body or body parts are safeguarded for evidentiary purposes, usually in conjunction with the Forensic Group, and that a clear and concise chain of custody is maintained and documented.

When managing large debris sifting operations, the Mass Fatality Group should be included in every briefing for every operational period that debris sifting is ongoing. They should instruct those working in and around the debris about the proper protocols to follow when a body or body parts are found. The purpose of this briefing is to ensure that clear and concise directions and documentation is undertaken and to preserve any forensic evidence that the body or body parts may provide.

The most obvious cases of debris sifting operations that usually come to mind are terrorist attacks, such as the September 11 attacks on the Pentagon and the World Trade Center, and the Oklahoma City Bombing. It should be realized that debris sifting operations for fatalities is needed in a multitude of incidents, even those that do not involve a criminal case. Other types of incidents that may need debris sifting operations might include earthquakes, industrial accidents, hurricanes, mudslides, and other instances where people may be buried in debris. While many of these instances may not be related to criminal activities, in larger incidents, the oversight of cadavers in debris sifting operations could still be assigned to the Mass Fatality Management Group under the Investigations and Intelligence Gathering Section of the ICS method. This is dependent however in how the Incident Commander (IC) arranges the response to the incident, and the structure of the hierarchal organizational chart.

The Mass Fatality Management Group is also responsible for ensuring that mass fatality-related public health hazards are mitigated. This should include creating protocols that mandate the use of Personal Protective Equipment (PPE) to ensure that body substances do not come in direct contact with first responders or anyone else who may be working on scene. When creating these protocols, it would be well advised to incorporate the local public health agency for guidance. Depending on the circumstance surrounding the incident, the public health agency may suggest medical gloves, eye protection, respiratory protection, or a multitude of other PPE measures.

The Mass Fatality Management Group is also responsible for ensuring that all fatalities are identified using proper scientific methods and that identification is completed in a timely manner. While this Group is not typically responsible for notifying families that the individual has been identified and is deceased, they are usually responsible for notifying the proper individuals within their chain of command. The Mass Fatality Group will usually be required to contact the Missing Persons Group who will (in most instances) then notify the family. It is important to understand that these notifications should be done in a timely manner and notification should be done according to protocols. If the Mass Fatality Group does make a notification, it should be documented who was informed, the method utilized to inform them, and anyone that was present during that notification.

The Mass Fatality Management Group will also work closely with the medical examiner. In working with the medical examiner, they will ensure that the cause and manner of death is identified in a swift manner. In some instances, the Mass Fatality Management Group will request that the cause of death for a specific cadaver and/or forensic evidence collected from a specific body or body part be expedited. This expedited request can be for a variety of reasons, but it often related to information that will help solve the case, or potential information that will direct the investigators to new leads. The Group will also ensure that a final disposition of each of the deceased and that a death certificate is issued in a swift and acceptable manner.

The Mass Fatality Management Group will work with the Missing Persons Group to ensure that the needed information, data, records, images, DNA reference samples, and various other materials that may be needed to identify an individual, are collected. The Mass Fatality Group will then keep, or forward, these items to the appropriate individuals. These items will usually be collected at a Family Assistance Centers and/or other identified facilities, to ensure that the friends and families of the individuals do not have direct contact with deceased individuals until absolutely necessary. This Group will document where the items provided by friends or family are stored, so that these items can be quickly retrieved when needed.

Depending upon the size, complexity, and scope of the Investigations and Intelligence Gathering (I/I) Section, as well as the size, scope, and complexity of the incident, the Mass Fatality Management Group Supervisor may need to initiate a separate team or teams to maintain a proper span of control. The supervisor of each team will be identified as coordinators, and in most instances, they will coordinate with other individuals and teams not in the Mass Fatality Management Group. Common coordinators that are initiated include the following:

- Mass Fatality Management Coordinator
- Field Site/Recovery Coordinator
- Morgue/Postmortem Examinations Coordinator
- Victim Identification Coordinator
- Family Assistance Center Coordinator
- Quality Assurance Coordinator.

Because ICS has a modular design, the organization can be flexible, and these Coordinators can be added as modules. The Mass Fatality Supervisor or the Investigations and Intelligence Gathering (I/I) Section Chief may, with Incident Command (IC) approval, choose to combine these functions as a singular group or they can create individual teams to undertake these duties. As was previously mentioned, the decision to expand or contract is usually based on span of control. Under some circumstances, the decision will be made based on integration of Mass Fatality Management Group with other Sections or Groups within the ICS structure.

It is imperative that evidence collected by the Mass Fatality Management Group is shared with the Forensic Group in a timely manner. This is due to the fact that some evidence degrades with time. It is the responsibility of the Mass Fatality Management Group Supervisor to guarantee that coordination is ongoing and that collected evidence is quickly forwarded to the Forensic Group for analysis. It is also a widely accepted practice for the Mass Fatality Group Supervisor to establish coordination and information sharing between the Missing Persons Group, the Forensic Group, and the Mass Fatality Management Group.

12.4.3.6 Investigative Support Group

Depending on the size, scope, and complexity of an incident, the Investigation and Intelligence Section Chief and/or Group Supervisors may become overwhelmed with the multitude of tasks that they must complete. Those operating under the Investigations and Intelligence Gathering (I/I) Section may require specialized operational and support resources that will help to ease that burden. In these instances, the initiation of the Investigative Support Group may be warranted.

Among other tasks, the Investigative Support Group is responsible for working closely with the Command and General Staffs. In most instances, this Group will work the closest with the Logistics Section and Planning Section, to ensure that the resources, services, and support needed are secured for the Investigations and Intelligence Gathering (I/I) Section.

One of the many duties of the Investigative Support Group would be the responsibility of guaranteeing that Investigation and Intelligence Gathering Section staging areas are initiated, and each staging area is appropriately located, identified, and staffed. Depending on the situation, the Staging Area can be incorporated into already established Staging Areas, or the Investigations and Intelligence Gathering (I/I) Section Staging Areas may need to be separated from those areas already established. Instances that require separate Staging Areas may include resources that are not widely known to the general public (e.g. undercover officers, specialized teams), the integration of military resources, or instances where confidential or secret individuals or equipment may be used (e.g. counterintelligence officers, military equipment).

Upon realizing that a resource will be needed, the Investigative Support Group (if activated) will order that resource in an expedient manner. This may include technical and nontechnical services and support, as well as confidential and secret resources that are not known by the general public. Upon ordering the resource, they will obtain an estimated time of arrival and provide Staging Area details to that resource. If separate from already established Staging Areas, the resource should be notified of the address of the Investigations and Intelligence Gathering Staging Area that they should report to. If this Staging Area needs to be concealed, they should inform them if they are not to share the location.

When a Staging Area is needed, one Staging Area Manager from the Investigative Support Group will be appointed to manage each activated Investigations and Intelligence Gathering (I/I) staging area, or to provide a presence in already established Staging Area. All incoming operational and support resources that have been ordered will be staged, to ensure the availability and quick response of these resources as they are needed. Types of operational and support resources that may be required include personnel, equipment, vehicles, aircraft, watercraft, supplies, facilities, infrastructure, secure networks, as well as other resources.

From the moment the resource arrives until the resource is demobilized, accountability procedures must be implemented. These will commonly be managed by the Planning Section; however, if certain resources should not be known or published by the media (or others), it is acceptable for the Investigative Support Group, or even the direct Supervisor of the Group they are assigned to, to handle their accountability.

If needed, the Investigative Support Group may be required to provide food and beverages to personnel. This is especially true if these resources are in the field and not operating with other Section personnel. An example of when this Group may need to provide food and beverages might include a bomb squad that is sweeping a large area for an extended amount of time, or operations that are water based.

The Investigative Support Group is also responsible for ensuring that resources are maintained, repaired, or replaced when necessary. They are also charged with safeguarding, tracking, and documenting, all resources. They will ensure that each resource is properly utilized and retrieved or demobilized when the incident is complete. As the incident comes to a close, or the Investigations and Intelligence Group begins to decrease in size, the Investigative Support Group will ensure that all human resources have physically and mentally recovered prior to demobilization. Resources should not be demobilized until they are no longer needed, and they have had down time to ensure they are well rested. They will also maintain the appropriate records and reports that document the activities undertaken by the Investigative Support Group and any related activities that were performed.

Depending upon the size, complexity, and scope of the Investigations and Intelligence Gathering Section, the Investigative Support Group Supervisor (or the Section Chief) with Incident Command approval may choose to activate one or more subgroups to help better support the mission of this Section. This is usually based on the supposition that additional support would be more expedient and organized. If they expand, the Investigative Support Group will be managed by the Investigative Support Supervisor, and the teams that are typically created include the Investigative Support Group Staging Area Manager (or Managers for multiple Staging Areas), the Intelligence and Investigation Work Area Manager, the Resource Coordinator, the Communications Coordinator, and the Physical Security Coordinator.

12.4.3.6.1 *Staging Area Manager*[1]

When the position of the Investigations and Intelligence Gathering (I/I) Staging Area Manager is implemented, one Staging Manager should be appointed for each Staging Area that is opened and active. In the case of utilizing existing Staging Areas that are implemented for all other resources, an Investigations and Intelligence Staging Manager should be positioned at those Staging Areas to ensure that when resources are needed, the proper credentialed resource is provided.

Operating as the Investigations and Intelligence Gathering (I/I) Staging Manager, the individual has the responsibility of knowing and properly documenting information about requested resources. As those resources arrive at the Staging Area, the Manager will categorize and separate these resources based on one (or more) of the following conditions:

- The agency's jurisdiction and legal authority (are they commissioned in that state, or have provisions been made to give them authority)
- The technical skills of personnel
- The nontechnical skills of personnel
- Security clearance and access clearance of personnel
- Proficiency of the personnel
- Types of equipment and resources that each resource brought and can provide

The Investigations and Intelligence Section Staging Manager is responsible for verifying that staged personnel have the correct credentials and ensure that they have proper and acceptable identification. They may also be required to evaluate for access and entry control for the resource and ensure that badging measures are implemented. The Staging Manager should also confirm that all personnel are equipped and wearing the required Personal Protective Equipment for the task they will be assigned.

The Investigations and Intelligence Gathering (I/I) Staging Manager will be required to organize all personnel and ensure that they receive a briefing about the incident. In that briefing, the Staging Manager should give a brief overview of the entire incident, and a more specific summary of the intelligence gathering and investigations portion of the incident. In that briefing, the Staging Manager will provide initial instructions, give directions, provide information and data, as well as identify any hazards and/or precautions.

At the completion of the briefing, the Investigations and Intelligence Gathering (I/I) Staging Manager will assign the resource. The Staging Area Manager will also specify the work requirements that should be performed and who that resource will report to. It is important to note that while some personnel might be assigned or be deployed to a specific assignment, other personnel may be assigned to remain at the Staging Area as reserves. All resources, whether deployed or held in reserve should be tracked to ensure proper accountability is in place.

12.4.3.6.2 Work Area Manager[1]

The Investigations and Intelligence Gathering (I/I) Section Work Area Manager makes sure that all work areas where personnel may be assigned is maintained, and safe, for personnel. They will coordinate with the Logistics Section to affirm that all utilities, wired and wireless communication services, sanitation, accommodations, infrastructure, and other essential services are satisfactory and safe. If the services required for a work area are not satisfactory, or are they unsafe, the Work Area Manager will collaborate with the Logistics Section and/or the Communications Coordinator (if activated) to ensure that the situation is rectified.

12.4.3.6.3 Resource Coordinator[1]

The Investigations and Intelligence Gathering (I/I) Resource Coordinator is responsible for ensuring that all needed resources are ordered. They are typically activated when a considerable number of Investigations and Intelligence Gathering (I/I) resources are required, or when an incident that is already complex, may require resources for an extended period. To successfully undertake this job, the Resource Coordinator will work directly with their counterparts in the Logistics Section. In collaboration and cooperation with each other, and their respective jobs, they will work to order the needed resources. Once resources arrive, these resources will usually be tracked by the Planning Section to ensure accountability for all resources. In rare exceptions, when the actual resource is assigned on a need to know protocol, the Resource Coordinator or the Group Supervisor to which that resource is assigned will have the responsibility of accountability. The Resource Coordinator will also ensure that accountability procedures are implemented and followed for all operational and support resources, regardless of who is tracking the resource.

The Investigations and Intelligence Gathering (I/I) Resource Coordinator will work collaboratively and cooperatively with Group Supervisors and act as the intermediary between those supervisors and the Logistics and/or Planning Sections of the ICS method to ensure that Group Supervisors have the additional resources they need and that they are provided in a timely manner. This means that they will work diligently to provide the technical and nontechnical services and support needed for these resources, and they will be tracked from the time they are ordered until demobilization. The Resource Coordinator will also ensure that physical resources are maintained and repaired or replaced when necessary. They will also pay close attention so they can ensure that these resources are secured, accounted for, documented, used for the proper purpose, and decommissioned or demobilized as required.

12.4.3.6.4 Communications Coordinator

The Investigations and Intelligence Gathering (I/I) Communications Coordinator will work directly with the Logistics Section. As part of their responsibility, they will ensure that all audio, data, image, and text communications procedures and activities are implemented to support the efforts of the Investigations and Intelligence Gathering (I/I) Section. They will continuously evaluate communications to ensure that an adequate amount of communication devices have been issued to personnel, including secure communication devices. If there are an insufficient number of communications devices, it will be their responsibility to secure more. They will also maintain, repair or replace those devices, as required, to support the Investigations and Intelligence Gathering (I/I) Section.

It will also be the responsibility of the Investigations and Intelligence Communications Coordinator to make sure that measures are taken to safeguard, properly distribute, to track, and to document all communications equipment used by the Investigations and Intelligence Gathering (I/I) Section. They will be responsible for ensuring that all communication equipment are properly used and returned when demobilization occurs.

As part of their responsibilities, the Investigations and Intelligence Gathering (I/I) Section Communications Coordinator will also establish the designated radio channels for communications, and point-to-point radio channels that will be used for the incident. They will create and regularly update the Investigations and Intelligence Gathering (I/I) Section Communications Plan and submit it to the Logistics Section for approval. As part of their duties, they will also be required to monitor all radio channels that are being utilized by the Investigations and Intelligence Gathering (I/I) Section.

12.4.3.6.5 *Physical Security Coordinator*[1]

The Investigations and Intelligence Gathering (I/I) Physical Security Coordinator ensures that adequate physical security measures are in place; however, it should be realized that they do not have the power or authority to implement site security actions on their own. They will identify, investigate, and provide possible resolutions to actual and perceived threats. They will then present these resolutions and/or new physical security measures to the Logistics Section, the Operations Section, and the Safety Officer (SOFR) within ICS for approval. To be able to implement the proposed measures, they must collaborate and cooperate with their counterparts in those Sections about potential plans, proposed procedures, and any activities they believe should be taken to protect personnel assigned to the Investigations and Intelligence Section.

As part of their duty, they will work to secure force protection, security measures, health, safety measures, and to minimize or eradicate any environmental hazards. As risks in the Investigations and Intelligence Gathering (I/I) Section areas are identified, the Physical Security Coordinator will work to ensure that the issues are addressed and fixed.

If an emergent threat arises, the Physical Security Coordinator and/or those who report to them and/or those who are assigned to physical security, will identify threats, isolate those threats, control the situation, and safely mitigate that threat. The types of threats that may be faced can include dangerous people, unauthorized or contraband weapons, dangerous devices (e.g. explosive, chemical, gas), dangerous animals, or any other condition that may cause a risk to personnel.

12.4.4 Investigations and Intelligence Gathering (I/I) in the Operations Section

In some circumstances, it may be best to place the Investigations and Intelligence Gathering (I/I) under the control of the Operations Section when using the ICS method. This placement usually coincides with incidents that require an advanced level of interaction, linkage, communication, and/or coordination. This holds especially true when the investigative information and the operational tactics that are being employed are synonymous or on the same path. This option is also used when protection is needed for other non-law enforcement personnel or when areas need to be secured to protect personnel and citizens.

Because of the flexibility of the ICS method, different disciplines can work alongside of each other, doing different jobs with the same end goal in mind. In placing the Investigations and Intelligence Gathering (I/I) under the direction of the Operations Section, these resources can be utilized as a Task Force, or a Strike Team. A key element that should help inspire placing them under the Operations Section's control is that a high level (usually directly) of interaction will be needed. As an example, if looting and gunfire is taking place after a disaster, the Investigations and Intelligence Gathering (I/I) strike team can respond with local fire and EMS crews to ensure their safety as they operate, while gathering the intelligence they need about the situation. In rioting situations, a local strike team or task force can provide safety, and if needed cover fire, for medical crews trying to rescue those that are injured while gathering intelligence or in getting "grab and go" crime scene photographs.

In a law enforcement context, Investigations and Intelligence Gathering (I/I) the Incident Command (IC) or Unified Command (UC) will need to ask themselves some questions before placing them under the direction of the Operations Section. While these questions should not be the only questions asked, it gives them at least a starting point:

1. Will placing Investigations and Intelligence Gathering (I/I) under the direction of the Operations Section prevent or possibly deter additional activity?
2. Will placing Investigations and Intelligence Gathering (I/I) under the direction of the Operations Section help to prevent another incident?
3. Will placing Investigations and Intelligence Gathering (I/I) under the direction of the Operations Section prevent or deter attacks on personnel?
4. Will placing Investigations and Intelligence Gathering (I/I) under the direction of the Operations Section help to prevent another attack?

If the answer to any of these questions is in "Yes," then careful consideration should be given to placing Investigations and Intelligence Gathering (I/I) under the management and direction of the Operations Section. However, this is not the only option of integrating a law enforcement Investigations and Intelligence Gathering (I/I) into Operations.

If beneficial to the mitigation of the incident, the Investigations and Intelligence Gathering (I/I) can operate as Section and loan-specific personnel with specific jobs to the Operations Section. A good example of this might be loaning experts in Improvised Explosive Devices (IEDs) to the Operations Section after a bombing attack. As Operations personnel search for survivors and victims, the specialists from the Investigations and Intelligence Gathering (I/I) Section can identify or ensure that there are no secondary devices.

Looking from a public health perspective, it may be advantageous to place specialists in public health with EMS and fire personnel during an outbreak or pandemic. In these type of instances, the placement could be under the direction of the Operations Section or as part of loaning specialists from the Investigations and Intelligence Gathering (I/I) Section. Imbedding these personnel can be done to ensure that all precautions are taken to identify quarantine areas, to gather intelligence on who the patient has been in contact with, or a whole host of other intelligence gathering, or investigations that may be related to the incident. While this might be unusual, it is possible if an outbreak was spreading quickly, or there is a quick mortality rate that makes gathering information difficult.

In looking at a cyber incident, if critical infrastructure, such as the power grid, was attacked by cyber terrorists, this might be another instance when placing personnel under Operations, or loaning them to Operations, might be equally advantageous. Specialists from the Investigations and Intelligence Gathering (I/I) Section might be a benefit if imbedded with the Operations Section that is working to restore portions of the power grid by using these specialists to undergo forensic cyber security measures on site.

While there are no set rules of what should constitute placing Investigations and Intelligence Gathering (I/I) under Operations. On a case by case basis, it will need to be determined based on the level of interaction that is needed, as well as the role that Investigations and Intelligence Gathering (I/I) will have in the entire response.

12.4.5 Investigations and Intelligence Gathering (I/I) in the Planning Section

In the early stages of the ICS method, operational information and situational intelligence were management functions that were located in the Planning Section. In many instances, Investigations and Intelligence Gathering (I/I) is still placed in the hierarchal structure as reporting to the Planning Section. In this placement, there is a focus on three incident intelligence areas:

situation status, resource status, and anticipated incident status or escalation (e.g. weather forecasts, location of supplies, etc.). This intelligence is utilized to assist with incident management decision-making process

When utilizing the Investigations and Intelligence Gathering (I/I) under the direction of the Planning Section, the Incident Commander should establish a system for the collection, analysis, and sharing. The information that is received will help those making decisions to have a higher level of situational awareness. It is also noteworthy that information developed during these efforts will support the process of planning for the next operational period.

Because of the redundancy and the overlap of duties that is inherently built into the ICS method of incident management, this was already discussed in Chapter 10. If you need more information, or you need to refresh your memory, you should review the information written about the Situation Unit. This will explain how Investigations and Intelligence Gathering (I/I) is utilized under the direction of the Planning Section.

12.5 Conclusion

It is important to remember that in some instances, the Incident Commander (IC) may determine that the Investigations and Intelligence Section may not be required. As part of the decision-making process, they should identify whether the Investigations and Intelligence Gathering (I/I) portion will consist of only a few personnel, or if there is a multitude of personnel. If a larger contingency is expected, implementing Investigations and Intelligence Gathering (I/I) as a Section could manage hundreds of personnel. They will also need to determine whether Investigations and Intelligence Gathering (I/I) will need to be right beside the decision-makers (Investigations and Intelligence Gathering [I/I] Officer), if they should have a multitude of tasks (Investigations and Intelligence Gathering [I/I] Section), if they will need to work closely with the Operations Section (a Task Force or a Strike Team), or if they will just increase situational awareness (Situation Unit of the Planning Section).

Another important thing to remember is that Investigations and Intelligence Gathering (I/I) can start as a lower function, and as an incident becomes more complex, or it is suspected that a criminal or intelligence aspect of the incident is present, then the Incident Commander (IC) or the Unified Command (UC) can modify their placement in the hierarchal organization chart. By the same token, the Incident Commander (IC) may decide that they may want to reduce Investigations and Intelligence Gathering (I/I) from a Section to incorporating intelligence under the Planning Section of the ICS method.

The size, scope, and complexity of an incident can also help decide at what level Investigations and Intelligence Gathering (I/I) should be initiated. Once the Investigations and Intelligence Section is initiated, the size, scope, and complexity of an incident will also determine if more resources are needed. As more resources are needed, the organizational structure of this Investigations and Intelligence Gathering (I/I) Section should expand as well, keeping in mind the importance of span of control.

Unless it is recognized that there will be a large influx of personnel in a brief period of time, the expansion of the ICS method to include Investigations and Intelligence Gathering (I/I) at a higher level should be gradual. It should also be remembered that span of control also has a lower limit, and this too may help decide when to collapse the Section. The rule of thumb is that no less than three individuals should be managed by one supervisor. The configuration of the ICS organization is flexible, so at some point during certain incidents, the Incident Command (IC) may choose to combine some functions under one title. This holds true especially if there are insufficient personnel.

The mission of Investigations and Intelligence Gathering (I/I) is to ensure that all investigative and intelligence operations, functions, and activities within the incident response are properly managed, coordinated, and directed. The multitude of activities they are responsible for include

- Preventing or discouraging additional activity.
- Preventing or discouraging additional incidents.
- Preventing or discouraging additional attacks.
- Collecting, processing, analyzing, and appropriately disseminate intelligence information.
- Conduct a thorough and comprehensive investigation.
- Identifying, processing, collecting, and creating a chain of custody for all evidence or potential evidence.
- Safeguard, evaluate, analyze, and store all related intelligence and/or evidence.
- Gathering DNA from family and friend to help identify victims.
- Share information about the disposition of victims with families.

While there may be many other duties, those identified are the most common activities of Investigations and Intelligence Gathering (I/I) in the ICS method of incident management. Investigations and Intelligence Gathering (I/I) as a function has responsibilities that have critical importance to all disciplines and agencies, and the work they do could save lives.

In closing, there should be a reminder about classified information. Information that requires a security clearance, extremely sensitive information, or specific investigative tactics should be at the forefront of the Investigations and Intelligence Gathering (I/I) personnel's mind. Any information that could compromise the investigation, compromise the identity of informants, or compromise a case, should only be shared with those who have the appropriate security clearance and/or need to know. Even when restricting an information flow, life safety should be the highest priority.

Real-Life Incident

At 9:30 in the morning on 19 April 1995, the Murrah Federal Building in Oklahoma City, Oklahoma, was the target of a domestic terrorist attack. Initially, emergency personnel and volunteers rushed in to assist the walking wounded and to find those trapped by the explosion.

At 10:15 a.m., it was thought that there was a secondary device. Reports are conflicting whether somebody thought they saw a secondary device or if it was a realization that there may be a secondary device. No matter what the circumstance, once a threat was realized, all personnel were immediately evacuated. After law enforcement and bomb units gave the "all clear", personnel were allowed to re-enter that area and resume work.

This real-life scenario shows how the Incident Commander, or Unified Command, should react when there is a threat to the well-being of personnel and/or the general public. It also provides insight of how Investigations and Intelligence Gathering (I/I) can be beneficial by integrating with the Operations Section.

Chapter 12 Quiz

1. Prior to Homeland Security Presidential Directive five (HSPD-5) and Homeland Security Presidential Directive eight (HSPD-8), Investigations and Intelligence Gathering (I/I) were undertaken only as a _____ function.

2. What four locations can Investigations and Intelligence Gathering (I/I) be placed in the ICS method?

3   **True or False:** Investigations and Intelligence Gathering (I/I) staff should never be put under the control of the Operations Section.

4   **True or False:** Since the changes in 2013, the function of Investigations and Intelligence Gathering (I/I) should never be placed in the Planning Section.

5   **True or False:** Need to know or classified information should not be compromised for life safety.

6   **True or False:** As expansion of ICS is needed in the Investigations and Intelligence Gathering (I/I) Section, the creation of a Group should be avoided.

7   What are the six core Groups that are typically activated as the structure of ICS in the Investigations and Intelligence Gathering (I/I) Section expands?

8   **True or False:** The Investigative Operations Group (or Branch) is usually used as a central repository for all investigative and intelligence information, and most (if not all) teams submit findings to this Group.

9   When needed, what are three of the seven common teams that can be initiated to assist under the Investigative Operations Group/Branch?

10 What are the three major functions of the Intelligence Group?

11 The Forensic Group manages the crime scene (or crime scenes) and is responsible for _____.

12 **True or False:** The Missing Person's Group is responsible for field work and looking for missing individuals and/or identifying the remains of a missing person.

13 When this division of resources is needed in the Missing Persons Group, it is common to split the workforce into two teams. What are the common names used to identify those teams and the common names for the manager/supervisor of each team?

14 A task of the Missing Person's Group is to collect data from family members. Name three types of data that could be helpful.

15 **True or False:** The Mass Fatality Group/Branch should work closely with the Forensics Group and the Medical Examiner.

16 **True or False:** Whenever the Investigations and Intelligence Gathering (I/I) Section is activated, then it is a requirement that the Investigative Support Group be activated.

17 **True or False:** The Investigative Support Group is responsible for (among other things) for providing a Staging Manager at established Staging Areas and/or establishing and managing new Staging Areas specifically for Investigations and Intelligence Section incoming resources.

18 **True or False:** The Investigations and Intelligence Gathering (I/I) can and when possible should, loan personnel to the Operations Section if they request that assistance.

Self-Study

Barry, (September 24, 2001). A nation challenged: The site; at the Scene of Random Devastation, a Most Orderly Mission. The New York Times. Retrieved 2019-04-26 from https://www.nytimes.com/2001/09/24/nyregion/nation-challenged-site-scene-random-devastation-most-orderly-mission.html.

Department of Homeland Security (2013a). National incident management system: Intelligence/Investigations function guidance and field operations guide. Retrieved from https://www.fema.gov/media-library-data/1382093786350-411d33add2602da9c867a4fbcc7ff20e/NIMS_Intel_Invest_Function_Guidance_FINAL.pdf.

Forrester et al. (2008). Managing Mass Fatalities: A Toolkit for Planning. Santa Clara County Public Health Department. Retrieved from https://www.sccgov.org/sites/phd-p/programs/BePrepared/Pages/managing-mass-fatalities.aspx.

Hirschkorn, (February 23, 2005). Identification of 9/11 remains comes to end. CNN News. Retrieved from http://www.cnn.com/2005/US/02/22/wtc.identifications/.

Abramson, (2017, September 11). These 9/11 families still don't have their relatives' remains 16 years later. Time Magazine [online]. Retrieved from http://time.com/4932331/911-anniversary-remains-families/.

Leonard et al. (2016). Command under attack: What we've learned since 9/11 about managing crises. *The Conversation*. Retrieved from http://theconversation.com/command-under-attack-what-weve-learned-since-9-11-about-managing-crises-64517.

13

The Agency Administrator, Common Agency Representatives, and a Basic Overview of the Planning Process

> *In a very real way, ownership is the essence of leadership. When you are 'ridiculously in charge,' then you own whatever happens in a company, school, et cetera.*
>
> Dr. Henry Cloud

13.1 The Agency Administrator

When outside agencies respond to an incident, an Agency Administrator will be needed to provide direction to all resources via the Incident Command staff. Agency Administrators are the managing officer of an agency, division, or jurisdiction having statutory responsibility for incident mitigation and management. In some instances, this will be an elected official such as a mayor, a fire chief, a police chief, a park superintendent, a state forest officer, a tribal chairperson, or essentially any individual who has the initial responsibility for the response area. The Agency Administrator (sometimes called the Executive Official or the Senior Official) is the person responsible for administering policy for an agency or jurisdiction.

The Agency Administrator is the person that (by default) is responsible for the protection of the citizens. Along with the responsibility of protecting citizens, this individual (or individuals) will have the authority to make decisions for the community, or a geographic area, as a whole. They will be able to commit resources and obligate funds. They will have the legal authority needed to guide the resources within the policies and priorities of protecting the population, stopping the spread of damage, and protecting the environment (FEMA, 2019).

The responsibility of an Agency Administrator, especially in planning and mitigating natural and man-made incidents, should begin on the first day of accepting the position and measures should (hopefully) be in place prior to a complex incident. In preparing for disasters or incidents, it is the responsibility of the Agency Administrator to ensure that their jurisdiction has established resource management systems and whole community preparedness plans. In these plans, they should have a method in place for identifying, inventorying, requesting, and tracking of resources, as well as initiating and dispatching those resources (FEMA, 2013b).

Additional planning should be undertaken including creating and/or pre-established local volunteer resources. This can include Community Emergency Response Teams (CERT), neighborhood watch or patrols, civil air patrols, and more. In addition to planning for mobilizing these resources, the preparedness plans should also address demobilizing or recalling resources, and the process that they will need to undertake to make them financially whole again after a deployment is complete. Preplanning and training on what is needed for financial tracking of expenses, reimbursement process for volunteer resources, and reporting/justification of expenses incurred. This also applies to mutual aid resources as well.

Emergency Incident Management Systems: Fundamentals and Applications, Second Edition.
Mark S. Warnick and Louis N. Molino Sr.
© 2020 John Wiley & Sons, Inc. Published 2020 by John Wiley & Sons, Inc.
Companion Website: www.wiley.com/go/Warnick/EIMS_2e

As was described multiple times in this book, mutual aid is the voluntary sharing of resources between agencies or organizations. The mutual aid partners commit to assisting each other when they have an incident where their own resources are not suitable or are not sufficient for the incident at hand. Mutual aid allows jurisdictions to share and pool resources between their mutual aid partners, thereby saving expense and providing more value for their constituency. Mutual also enhances protection at little to no cost to the community.

It is important that the Agency Administrator realizes the benefit of Mutual Aid Agreements (MAAs) and how critical it is to operate under the National Incident Management System (NIMS) method and the Incident Command System (ICS) companion method of incident management. This holds especially true if the Agency Administrator is an elected official, a city manager, or someone that has not been actively involved in a complex or large incident. These individuals should know, understand, and they should have practiced NIMS/ICS principles long before a large or complex incident strikes their jurisdiction.

While the Agency Administrator may have the authority and responsibility for citizens, it does not mean that they should assume a command role over the incident. Instead, they should provide policy guidance on the priorities, objectives, and directives. This guidance should be based on situational awareness and the Emergency Operations Plan (EOP) rather than being based on emotions or feelings they have at that time.

Many believe that Agency Administrator should not be involved in on-scene commanding at all, nor should they be placed in the Incident Command Center of an incident. This is based on the supposition that they may make decisions based on emotion rather than situational awareness and facts. Instead, the best placement of the Agency Administrator during a large or complex incident is in the Emergency Operations Center (EOC), and they should be tasked with resource coordination and the support to the on-scene command. While the Agency Administrator should not be at the incident, they must still have the ability to communicate and meet with the Incident Commander (IC) as necessary.

The responsibility for the final outcome of an incident remains with the chief elected official, chief executive officer, or Agency Administrator. It is therefore imperative that the Agency Administrator should undertake a role of an active participant, supporter, supervisor, and evaluator of the Incident Commander (IC). By the same token, the Agency Administrators should also be aware that a visit to the scene by them could become easily disruptive. Their presence could draw more media and bystanders into the hazardous area, thereby putting those individuals at risk. The existence of a chief elected official could also cause confusion with personnel and the media about who is in charge of tactical operations. For these reasons and more, visits to any personnel working on the incident should be coordinated with the Incident Commander (IC) in an effort to not jeopardize the response or to sidetrack workers.

Additionally, visiting dignitaries can also cause additional chaos, and siphon resources away from tactical operations to attend to the needs of a visiting dignitary. While these visits from both Agency Administrators and other dignitaries may be a morale booster, it is also important to avoid pulling resources away from the response. By pre-scheduling their visits, outside resources are not needed to operate on the scene of the incident to undertake protection duties, or to relieve those undertaking tactical operations so that they can meet with the dignitary.

The Agency Administrator undertakes some important responsibilities. Among them is providing a Delegation of Authority. Simply put, a Delegation of Authority is a right, or an obligation, to take actions for a department, agency, or jurisdiction. In most jurisdictions, the obligation for protecting citizens is the responsibility of the chief elected official who will often be an Agency Administrator. These elected officials have been given the authority to make decisions, commit resources, obligate funds, and command the resources (usually through being voted in) to protect the population. This may include making every effort possible to stopping the damage caused to life, property, and the environment through incident stabilization.

The Delegation of Authority is a statement provided to the Incident Commander (IC) or Unified Command (UC) by the Chief Elected Official. This statement gives authority to the incoming Incident Commander (IC) and their staff. The Delegation of Authority is a formal document that assigns the task of managing an incident to the incoming Command and General Staff. This will be discussed in greater detail in Chapter 15.

Going back to the analogy of the ICS method being similar to a car, the Agency Administrator and the Delegation of Authority can be compared to the ignition key. In order to give permission to the car to start and for most of the other parts of the car to operate, you must first have the permission of the proper key. The proper key is the Agency Administrator and handing the key to someone so they can drive is the Delegation of Authority. Without the key, there is no authority to operate the vehicle. If someone wants to drive the car without the key, they will pull the ignition and hotwire it. In relationship, this would be the equivalent of an Incident Management Team (IMT) managing an incident without a Delegation of Authority. They are making the car work, using the gas that someone else paid for, and they put wear and tear on an automobile for which they do not have the authority to operate.

13.2 Agency Administrator Representatives

When talking about the Delegation of Authority, we need to understand that the Agency Administrator is only one person, and they may need help to undertake all of the tasks affiliated with the incident. They too need to be mindful of the span of control, and they may want, or even mandate, that specific advisors should be kept in the loop, or even advise and be active in specific areas of managing an incident. These individuals are identified as Agency Administrator Representatives. If the outside Incident Management Team (IMT) is required to work with these individuals, it is because the Agency Administrator trusts these individuals in their area of expertise.

If you are, or potentially will be, an Agency Administrator, it is important to realize that these representatives should be trained in the NIMS method and the ICS companion method so that they can integrate into the command structure and immediately go to work rather than fumbling around and trying to figure out what to do. They should also know about the concept of span of control. Because it is rare occasion that these individuals are mandated, a brief synopsis of each job is listed in the following.

13.2.1 Resource Advisor[1]

The Resource Advisor when appointed can play a key role during the management of an incident. This holds especially true in wildfire and flooding incidents, where mitigation efforts can help protect those resources. The Resource Advisor will typically provide input for natural and cultural resources by utilizing their professional knowledge and expertise. Their knowledge base is usually derived from attending higher educational institutions for specific classes or degrees, specialized training courses, and/or on-the-job training. Essentially, they are trained to advocate for specific natural or cultural resources. As an advocate for the natural and cultural resources, they must integrate with the Incident Management Team (IMT) as an advisor. A Resource Advisor provides the Incident Management Team (IMT) with data about areas

---

1 This heading is for advanced IMS students.

that may be impacted by the incident and advocates for protection of the resources they have identified. This allows the Incident Management Team (IMT) to plan the response to, and around, these sensitive areas as well as plan or undertake mitigation measures.

It should be noted that the Resource Advisor has no authority over the Incident Management Team (IMT); they are only in an advisory position which is based on a very specialized field, not incident management. An effective and well-trained Resource Advisor can support the Incident Management Team (IMT) by not only reducing the negative effects on natural and cultural resources but also (in some instances) in helping to reduce the general response costs.

In complex incidents, the Resource Advisor may have Assistants as well. In these cases, the Resource Advisor will become the Resource Coordinator. As the Resource Coordinator, they will provide leadership and supervision as they manage a team of Resource Advisors and/or Technical Specialists. Functioning as the lead Resource Advisor, they will become the single point of contact for the Incident Management Team (IMT) on issues that they are experts on. As the Resource Coordinator, they will work with the Planning Section Chief (PSC), Operations Section Chief (OSC), and Safety Officer (SOFR), to provide guidance on protecting resources and alternative methods (or plans) to help ensure the preservation of resources.

The Resource Coordinator can make recommendations to the Incident Management Team (IMT), especially if they feel that there is a need for more Resource Advisors or Technical Specialists. When doing so, they should provide specifics regarding the type of expertise that these advisors or specialists should have, and how their input will support incident objectives in the Incident Action Plan (IAP).

The Resource Coordinator should stay informed about threats to natural and cultural resources and current conditions near these resources. They will need to be organized, accessible, and involved in the incident response as it relates to natural and/or cultural resources. As the Resource Coordinator, they may be required to manage Fireline Resource Advisors, All-Hazards Resource Advisors, and/or Technical Specialists. The Resource Coordinator and all Resource Advisors should be trained in the ICS method, and they should remember, and follow, the guidelines for span of control. The Resource Coordinator is also responsible for ensuring the safety and accountability of those that are working under their directions (NWCG, 2017).

13.2.2 Historic Preservation Advisor[2]

While rarely used at the local level, Historic Preservation Advisors can be mandated by the Agency Administrator. When this position is mandated by the Agency Administrator, it is typically someone that has been working on local historical identification and preservation. The local Historic Preservation Specialist should not be confused with the National Incident Management System or the Federal Emergency Management Agency (FEMA) Environmental and Historical Preservation Specialist that is required for a Presidentially Declared Disaster (PDD).

According to Preserve America (2008), at the local level, the Historic Preservation Advisor is typically the most knowledgeable about public and private archeological sites, historical sites, historic buildings, museums, antiquities collections, and other important historical resources. They usually have an intimate knowledge about the area's history and have specifically worked on site-specific plans and preservation (Preserve America, 2008).

Essentially, these individuals advocate on behalf of historical places and/or historical items (Preserve America, 2008). As an advocate, they must integrate with the Incident Management

---

2 This title signifies highly technical information for advanced students and practitioners.

Team (IMT) as an advisor. The Historical Advisor will provide the Incident Management Team (IMT) with data about areas that have historical significance that are in the path of impact by the incident. They advocate for protection of historical assets within the response area. This allows the Incident Management Team (IMT) to plan the response to, and around, these historical areas and to do their best to take mitigation measures. They will work with the Planning Section Chief (PSC), Operations Section Chief (OSC), and Safety Officer (SOFR), to provide guidance on protecting historical assets and potentially provide plans to help ensure the preservation of those assets into the Incident Action Plan ([IAP], Preserve America, 2008).

The Historical Coordinator heads up all efforts related to historical assets (Preserve America, 2008). As a Historical Coordinator, some of the work they may be tasked with prior to the arrival of FEMA's Environmental and Historic Preservation Specialist could be as follows:

- Advising the Command and General Staff
- Setting up site-staging areas for volunteers and equipment to remove or protect sites
- Providing plans for removing and safeguarding collections
- Providing damage assessments
- Planning for salvage of items already damaged
- Prioritizing postdisaster recovery activities for historical assets and sites
- Managing and training of volunteers when removal of collections is required (Preserve America, 2008)

While not mandatory, it is strongly suggested that individuals that might be chosen to be appointed to this position should take, at a minimum, some basic emergency management courses. More advanced courses are preferred. They should also work in collaboration and cooperation with the local emergency management agency and create networks of volunteers and organizations that may be able to help in a disaster.

13.2.3 Incident Business Advisor[2]

The Incident Business Advisor is both a liaison and advisor to the Agency Administrator. They work directly for the Agency Administrator and acts as a bridge between the Agency Administrator and the Incident Management Team (IMT) and other incident support functions. The Incident Business Advisor has responsibility for administrative/fiscal oversight and work under the direct supervision of the Agency Administrator while still coordinating with the Incident Management Team (IMT).

The job of being the Incident Business Advisor is a very specialized job, and one that requires a substantial amount of training in most instances. This holds especially true if the individual who may be required to fill this job wants to be good at it.

The principal duties of the Incident Business Advisor will be to provide the Agency Administrator with an overview of incident management business practices, then make suggestions that will cut costs or make improvements to the processes of conducting business. In doing so, they will make recommendations for improvements to the Agency Administrator, and then communicate the Agency Administrators direction and/or wishes to the Incident Management Team (IMT) and/or other resources that are supporting the incident (NWCG, 2018f).

The Incident Business Advisor will work with the Finance and Administration Section Chief (FSC). In most instances, they will also work those individuals responsible for buying, ordering, hiring, or anything that commits the agency to ordering resources. This can include the Procurement Unit Leader, the Buying Team Leader, the Buying Team Member, and the Compensation/Claims Unit Leader or any other individuals or teams that deal with finances and/or purchases (NWCG, 2017).

The Incident Management Team is responsible for ensuring that the appropriate information and an adequate amount of documentation is undertaken in business management of the incident. This may include documenting business management orders, resolving financial problems and administrative issues that arise, and providing the incident agency with an Incident Financial Plan. An Incident Financial Plan will spell out the procedures that will facilitate payments, processing of claims, and resolving outstanding issues (NWCG, 2017).

The Incident Financial Plan is totally separate from, and a unique assortment of documentation, that is gleaned from the incident records maintained by the Planning Section (NWCG, 2017). The Incident Business Advisor will in most instances meet with these team members to discuss agency-specific policies and procedures. This is done to ensure that those in positions that make financial decisions are good stewards of the money that has been provided for the incident. In essence, they guide others in accounting for every dime spent, and they look for ways to prevent wasting money. This is done by providing written policies and procedures, not only for those handling finances but also the Incident Management Team (IMT). As the incident progresses, they may need to inform others about the availability (or increase) of financial support, or the need to find more financial support (running low on funds).

The Incident Business Advisor will also provide written guidelines regarding the Incident Finance Package requirements and the performance standards that may need to be enacted. It is important to note that the Incident Management Team (IMT) should be given the Incident Finance Package at the initial briefing. The reason for providing the Incident Finance Package from the onset is so that the documentation process can be established from the beginning of the incident, rather than playing catch-up later in the incident. There are numerous agencies that provide Incident Finance Package guidelines and incident guideline examples which can be customized for the local government. The National Wildfire Coordinating Group (2018g) provides a task book with resources for the local agency. These guidelines should be adjusted to meet the specific needs of the incident agency.

The Incident Business Advisor may also be tasked with finding ways to recuperate money. As is required with many federal responses, the Agency Business Advisor may want to consider a recycling program to recoup some of the money spent on the incident. Of course, this would be dependent on the size and scope of the incident, and whether it would be a financially sound decision for the effort to collect recyclables. In federal responses, Federal Executive Order No. 12873 delivers a mandate for federal agencies to encourage practical waste reduction by recycling reusable materials in a government response. As a side note, Public Law 103-329, (H.R. 4539) (1996), Sec. 608 permits all federal agencies to receive and use funds resulting from the sale of materials recovered through recycling or waste prevention programs (NWCG, 2018g).

This type of recycling generally takes place at the Incident Base, and it usually requires coordination between the incident agency and the Logistics Section. In most instances, the Incident Business Advisor will provide information on recycling procedures and requirements to the Incident Management Team (IMT) during the Agency Administrator briefing or later in the response if the agency sees that there are more goods to recycle than previously anticipated. The Logistics Section will usually be responsible for managing incident recycling.

The incident agency will also provide information to the Incident Management Team (IMT) about a multitude of other issues as well. This may include items such as the following:

- Areas where there may be potential claims
- Standing cooperative agreements
- Standing mutual aid agreements
- Contact information for those already involved in the incident

- Cost sharing criteria.
- Payment criteria and procedures
- National Guard Troops requested (or on scene)
- Military operations that were requested (or who are on scene)
- Procedures for requesting services and supplies
- Property Management Guidelines
- Geographic Area Supplement (prevailing equipment rates for that specific area)
- Incident Management Team Release and Demobilization

The Agency Administrator should consider what business management requirements are needed when considering the release of the Incident Management Team (IMT). The Agency Administrator or the Incident Business Advisor will organize a closeout session with the Finance/Administration Section at the end of an incident. During this closeout session, the Finance and Administration Section will review the Incident Financial Plan. This close coordination for the closeout session should also be undertaken when and Incident Management Team (IMT) is being replaced. Whenever possible, the outgoing and incoming team should meet with the Incident Business Advisor at the same time. This allows a debriefing and briefing session that helps to confirm a complete transfer of records.

In some instances, the Incident Management Team (IMT) or the responsible agency may desire, or even require, a close-out report to be provided to the incident agency by functional areas. This type of report typically provides summarized information by function, and it will typically be used by the responsible agency after the Incident Management Team (IMT) is released and the incident is completed.

Submission of the Incident Financial Plan and expenditures based on that plan (in accordance with established guidelines) is required prior to release of the Incident Management Team (IMT). The Incident Management Team (IMT) will also provide the Agency Administrator, the Incident Business Advisor, or another Administrative Representative with a list of Finance/Administration and Logistics Section members' home unit addresses and telephone numbers. This is done so if there are any questions or concerns that arise later, these individuals can be contacted for clarification (NWCG, 2018g).

13.2.4 Buying Teams[2]

At the discretion of the agency responsible for the incident, a Buying Team can be created and implemented. Buying Teams can be created (especially if a smaller incident) or they can be ordered (especially in larger incidents). This is undertaken by the agency responsible for the incident, and these teams report directly to the Agency Administrator or other designated personnel as directed by the Agency Administrator. Buying Teams will work with the local clerical staff to support incident-related purchasing and contracts.

In most instances, buying teams are ordered. The location of the incident usually determines where Buying Teams will come from. The United States is divided into geographical regions, and each region trains and has organized at least one Buying Team that can be ready for national dispatch.

The Buying Team Leader is tasked with coordinating between the incident agency personnel and the Incident Management Team (IMT). They are utilized in a capacity that confirms that goods and services are purchased in a manner that follow the incident agency's policy, and they ensure that proper documentation is maintained. The Buying Team will be responsible for reporting the acquisition of property that should be accounted for to the incident agency.

It is important to note that Buying Teams should not be used as payment teams by default, especially if the incident is a large or complex incident. The incident agency may want to order an Administrative Payment Team if the incident agency does not have the personnel to process payments in a timely manner, or if it is expected to take some time to bring the incident under control.

Buying Teams will ensure that the purchase of items will meet time deadlines for arrival, and they will ensure that it will meet the requirements of the Incident Management Team (IMT) and the Logistics Sections request. As part of the process, they will also ensure that a fair price is negotiated and that there is no price gouging on that item. This is done through a multitude of emails, phone calls, and fax transmissions. Once they are satisfied with the price, they will order the requested items.

13.2.5 Administrative Payment Team[2]

Administrative Payment Team is ordered by the incident agency and report to the Agency Administrator or other designated personnel. Administrative Payment Teams work with the local administrative staff to accelerate payments that are associated with an incident. An Administrative Payment Team should only be ordered when the length of the incident will be long, and the incident agency does not have the needed individuals who can process payments in a timely manner.

The Administrative Payment Team is authorized to make payment for supplies, materials, services, equipment rental, and casual hires that have been used on an incident. The Administrative Payment Team instructs others on the payment package, the audit process, and processing requirements. Those that they may instruct, or provide guidance to, might include the Finance and Administration Section Chief (FSC), the Buying Team, and incident agency administrative staff. Administrative Payment Teams will almost always employ a Contracting Officer for interpreting contracts and agreements while processing incident payments. If the configuration of the Administrative Payment Team does not include a Contracting Officer, the incident agency must provide one.

It is also important to note that if this is a Presidentially Declared Disaster (PDD), under the NIMS framework, FEMA will usually send a Procurement Disaster Assistance Team. A Procurement Disaster Assistance Team will ensure and guide agencies on federal procurement standards applicable to FEMA's Public Assistance disaster grants. This is done to enable compliance with these standards and to reduce the likelihood of being denied these funds for failure to adhere to FEMA requirements. The Procurement Disaster Assistance Team are lawyers, and they are sent not only to ensure that nonprofit, local, state, tribal, regional, and national emergency management personnel are familiar with the requirements but also that FEMA employees understand the guidelines.

13.2.6 Other Teams[2]

Based on the type and magnitude of an incident, the incident agency may determine that they need to utilize other special teams to ensure that, e.g. Burned Area Emergency Rehabilitation (BAER), Prevention and Education, Cost Review, Fire and Aviation Safety (FAST), Investigation, or other such teams. The Agency Administrator or their designee coordinates with Incident Management Team (IMT) and other personnel to assist in meeting the objectives of these special teams.

13.3 An Overview of the ICS Planning Process

The planning process, not to be confused with preplanning (Preparedness), is an important part of mitigating any type of incident. Imagine if you will, an Incident Commander (IC) is leading an emergency response, and they make all the decisions without the input of others. Everything related to that response, and the decisions that will be made, will be based on one person's ideas and from that singular perspective. While having one person making decision may be acceptable on small (daily) emergencies, it is not wise on complex, or large incidents, or an incident that entails a multiagency response. A decision made in these situations, without the benefit of input from other, can be dangerous and possibly even deadly.

By having a planning process that utilizes multiple people, there is a greater chance of having a 360° view of the incident. Not only will alternative ideas be introduced and investigated but also the design and methodology of the planning process used in the ICS method allows others to help build on any and all strategies presented in the planning meeting. With proper situational awareness while creating an Incident Action Plan (IAP), effective strategies and tactics can be implemented to bring order out of chaos in an orderly and efficient manner.

In the preliminary stages of creating and developing the ICS method, it was recognized in field studies that planning was not applied, or it was haphazardly applied, while performing incident operations. This led to ineffective and more costly strategies, increased issues regarding safety, a less efficient use of resources, and reduced effectiveness. The developers of the ICS method quickly came to the realization that these issues need to be addressed. Their solution was to develop a process that was not complicated, yet in-depth at the same time; something that could be used on both large-scale and small incidents. They ensured that this process could be used with a single resource or a multitude of resources.

The solution created was the process of creating an Incident Action Plan (IAP). An Incident Action Plan (IAP) can be used for planned events or emergent incidents. It follows a process that ensures that planning is a priority no matter what the size, scope, or complexity the incident may be. This new process identified established steps that should be taken and the staffing requirement to develop an effective written Incident Action Plan (IAP).

13.3.1 The Incident Action Plan (IAP)

The planning process in the ICS method was designed in such a way that the overall incident objectives were/are broken down into much smaller tactical assignments for each operational period. This process leads to the creation of an Incident Action Plan (IAP) which guides the response and recovery. While hands-on tactics are being undertaken during an operational period, in a separate location, the next operational period tactics and operational guidance are being planned and developed behind the scenes. In most instances, the work involved with the planning process is undertaken in a safe area, away from the incident itself.

In the Incident Action Planning (IAP) process, it is important to note that the overall approach to establishing incident objectives should be designed to cover the course of an operational period, rather than the entire time it takes to end the incident. Incident objectives are very different from the goals of an incident. Incident objectives are basic incident strategy, the command emphasis and/or priorities. This process also will identify safety concerns and safety mitigation measures that should be used in the next operational period. In essence, those creating the Incident Action Plan (IAP) will be looking at short-term strategies that will be used for each operational period, by identifying the resources and the tactics needed for the next operational period. During the creation of the Incident Action Plan (IAP), those involved in the process must ensure that the incident objectives must obey the legal obligations and management objectives of all affected agencies.

Except for planned events, the Incident Commander (IC) will make multiple decisions. These decisions will be derived from their training and experience, and sometimes they are even based on "gut feelings". This holds especially true during the initial stages of an incident, while it is a single resource response. Often, the utilization of agency protocols and preplanning for an incident will be a foundation for making these quick decisions during the beginning of an incident.

If and when it is realized that the incident will require additional resources, or that the incident may be a long-term event, then the Incident Commander (IC) will request the additional resources. In most instances, this is the time period in which the Incident Commander (IC) begins to appoint personnel to fill Command Staff and General Staff positions. As these positions are filled, the Command Staff and General Staff will provide input to assist the Incident Commander (IC) with developing a more robust Incident Action Plan (IAP). In some instances, as more resources arrive, the leadership of the response may transition from a singular Incident Commander (IC) to a Unified Command (UC) structure, thereby adding more individuals who can provide more alternatives and ideas to overcome the unique challenges of that incident.

13.3.2 General and Command Staff Planning Responsibilities-the Basics

In the initial stages of a complex incident, the Incident Commander (IC) will be wise to initiate the Safety Officer (SOFR) at the onset of the incident. Depending on how the incident unfolds, it may be wise to also appoint an Operations Section Chief (OSC) to manage the tactical considerations. In cases such as these, the Incident Commander (IC), the Safety Officer (SOFR), and Operations Section Chief (OSC), as well as any other Command Staff or General Staff appointed will cooperate and collaborate to create the plan of attack. The first (written) Incident Action Plan (IAP) will usually require revisions from the initial unwritten plans that the Incident Commander (IC) used, but this is not always the case.

As Command Staff and General Staff positions are assigned, the process of creating an Incident Action Plan (IAP) should be initiated. To initialize the planning process, the Incident Commander (IC) will fill out ICS Form 201 (Incident Briefing) and submit copies of it to all Command and General Staff prior to, or at, the initial planning meeting. The planning team will then begin to create a written formal operational plan. This will include the determination of what constitutes an operational period.

The determination of an operational period will be dictated by the type of incident and safety concerns. In incidents such as a wildland fire, an acceptable operational period could be 8 to 12 hours long depending on the circumstances and be only during daylight hours. By the same token, a HazMat situation, or a radiological or nuclear incident, may require a two- or four-hour operational period. The length of an operational period should be (primarily) based on the safety of personnel. Responder safety should always be the highest priority when determining operational periods. It is important to note that an operational period should never be longer than 12 hours. This too is related to the safety of personnel.

Early on in the process, planners should develop a simple operational plan that can be communicated in a succinct verbal briefing. Normally, this type of plan must be developed in a short amount of time and with only basic situational awareness information. As the hierarchal structure of the ICS and the requested resources and/or incident management personnel expand, then there will be additional time and resources available for a more comprehensive and effective plan. As the availability of staff is expanded, so should information systems and various technologies that will help to enable a more detailed planning process. This should include the ability to document and catalogue the events and actions taken. The documentation from the Incident Action Plan (IAP) will also assist in completing an After-Action Review (AAR) that can take "lessons learned" from the incident and apply them to future events.

Planning for the response to an incident is not a haphazard process. There are specific issues that must be reviewed and discussed to accomplish the planning process. To effectively plan for an incident, the planners must do the following:

- Evaluate the situation using a variety of information and intelligence
- Develop incident objectives (using SMART Objectives discussed in Chapter 14)
- Select a strategy or strategies that will efficiently and effectively mitigate the incident
- Determine which resources would be best utilized to achieve the objectives in the safest, most efficient, and cost-effective manner
- Create the safest and most cost-effective strategies

In creating the Incident Action Plan (IAP), all Command and General Staff will play a critical role in ensuring that the plan is comprehensive and that it meets the needs of the incident. The General Staff will be the individuals that will provide most of the input into the planning process and create the vast majority of the strategies and tactics that will be utilized.

The Planning Section Chief (PSC) and their subordinates will be responsible for providing the most up-to-date intelligence on the incident, including the current incident status, for the planning process. The Planning Section Chief (PSC) should also be the person that manages the overall planning process. They will work with various individuals throughout the ICS structure to provide a complete view (from all angles) of the response. Prior to the Planning Meeting, they will attend the Tactics Meeting with Operations Section Chief (OSC), the Safety Officer (SOFR), and the Incident Commander (IC) to review strategy, discuss tactics, and summarize organizational assignments. The Planning Section Chief ([PSC] and their subordinates) are responsible for filling out and/or collecting a multitude of ICS Forms that help develop the Incident Action Plan (IAP). Some of those ICS Forms include the following:

- ICS Form 201-Incident Briefing
- ICS Form 202-Incident Objectives
- ICS Form 203-Organization Assignment List
- ICS Form 204-Assignment List
- ICS Form 205-Incident Radio Communications Plan
- ICS Form 206-Medical Plan
- ICS Form 208-Safety Message/Plan
- ICS Form 209-Incident Status Summary (depending on the structure of the incident this may be filled out by others)
- ICS Form 213-General Message
- ICS Form 215A-Incident Action Plan Safety Analysis
- ICS Form 221-Demobilization Plan (FEMA n.d.b,e)

When an approved plan is completed, the Planning Section Chief (PSC) will physically produce the approved plan, publish it, and disseminate it to the appropriate individuals. This published plan will be the plan of action that guides the tactics to overcome the incident.

The Logistics Section Chief (LSC) and their subordinates will order resources and develop the Transportation Plan or Traffic Plan, the Incident Communications Plan (ICS Form 205), and the Medical Plan (ICS Form 206). In the planning meeting, the Logistics Section Chief (LSC) will identify or validate what resources are available for the next operational period. They will also voice support or opposition to the plan based on their area of responsibilities, and they may be required to offer alternatives to overcome any logistical issues in the Incident Action Plan (IAP).

In the planning process, the Operations Section Chief (OSC) and their subordinates will assist with developing the strategy and tactics for the next operational period. They will attend the Tactics Meeting with Planning Section Chief (PSC), the Safety Officer (SOFR), and the Incident Commander (IC). The Tactics Meeting is before the Planning Meeting and is used to review strategies, discuss tactics, and outline organizational assignments. These individuals are responsible for briefing those in the Planning Meeting on the situation status and the resource status as well as plotting control lines and boundaries of Division.

The Operations Section Chief (OSC) will also identify the needed resources to undertake the proposed tactics and (along with Planning and Logistics) identify the check-in locations and the various facilities that are being utilized. The most common ICS Forms used in the Planning Meeting will be filled out by the Operations Section Chief (OSC), or in cooperation with other Sections Chief's, include ICS Form 204 (Assignment List), ICS Form 215 (Operational Planning Sheet), and possibly ICS Form 220 (Air Operations Summary Worksheet) if there are Air Operations being utilized on the incident.

The Finance and Administration Section Chief (FSC) and their subordinates will develop a cost analyses for the Incident Action Plan (IAP), and they will ensure that it is in line with the financial constraints that were established by the Incident Commander (IC) and/or the Authority Having Jurisdiction (AHJ). They will also provide a financial update of the cost of the incident and identify any concerns regarding contracts and/or payment for services and/or resources. In the Planning Meeting, the Finance and Administration Section Chief (FSC) will not be required to provide data or information on any ICS Forms except for possibly ICS Form 214 (Activity Log).

The Investigation and Intelligence Gathering (I/I) Section Chief when activated will participate in Planning Section meetings; however, it should be noted that some of their responsibilities can be considerably different from other Sections. They may also change based on where they are placed in the hierarchal structure of the ICS method. Similar to other Sections in the Planning Meeting, they will usually assist in the review of incident priorities, help establish incident objectives, and the formulation and preparation of the Incident Action Plan (IAP). As necessary and/or required, they will be responsible for providing the Investigations and Intelligence Gathering (I/I) organizational chart, the Section's supporting plans, and materials that may be relevant (e.g. maps, data, images, situation reports). In the Planning Meeting, they will usually ensure that the needs of the Investigations and Intelligence Gathering (I/I) Section are considered when incident objectives and strategies are formed for the Incident Action Plan (IAP).

Perhaps the most critical aspect of the Investigations and Intelligence Gathering (I/I) Section Chief is to ensure that planning activities do not violate operations security, operational security, or information security procedures, measures, or activities (DHS, 2013a). There usually are no required ICS Forms that the Investigations and Intelligence Gathering (I/I) Section is required to fill out (in most instances) for the Planning Meeting. It should be noted that, depending on the incident and the duties of the Investigation and Intelligence Gathering (I/I) Section (e.g. site security [only], terrorist investigation, criminal investigation, combination of multiple duties), the Investigations and Intelligence Gathering (I/I) Section Chief may not need to attend planning meetings, but rather report the information to the Planning Section Chief (PSC) who will present it on their behalf in the Planning Meeting (DHS, 2013a). It should however be mentioned that this should be discouraged.

In the Planning Meeting, the Safety Officer (SOFR) will identify any safety concerns and provide a briefing on any safety issues that have arisen. Prior to the Planning Meeting, they will review the Medical Plan and voice any concerns and make suggestions for changing the plan if needed. As the overall Incident Action Plan (IAP) develops, they will also voice any concerns they have with the plan. Should they have an issue with the safety of personnel and/or the public, they will make recommendations to reduce the risk. The Safety Officer (SOFR) will be required to fill out ICS Form 215A (Incident Action Plan Safety Analysis), and if required ICS Form 208 (Safety Message/Plan).

The Liaison Officer (LOFR) will brief those in the Planning Meeting about any issues regarding interagency coordination and cooperation. If it is found that some agencies are having issues with the response or how the incident is going, then the Liaison Officer (LOFR) may make suggestions about how to overcome those issues. This may include specific assignments, identifying mitigation measures to overcome the issues, and in rare cases, recommending that these resources be demobilized. The Liaison Officer (LOFR) is not required to fill out any ICS Forms for the Planning Meeting; however, they may consider having a copy of ICS Form 214 (Activity Log), especially if there have been issues in liaising with an agency.

The Public Information Officer (PIO) will provide an overview of the actions that have been undertaken regarding public information and make suggestions for future public information releases. In most instances, this is not a severely critical aspect of the Incident Action Plan (IAP) except during mass evacuations or the release of special instructions to citizens. It should be recognized that it is *extremely* important when the Investigations and Intelligence Gathering (I/I) Section has been activated, approval should be gained from the Section Chief before providing press releases. The release of unapproved (or need to know) information could potentially undermine a criminal case and/or make the Investigations and Intelligence Gathering (I/I) process substantially more difficult.

The Incident Commander (IC) or Unified Command ([UC] if present) will develop the overall incident objectives and strategies based on the recommendations of the Command and General Staff. They will approve resource orders that support the mission, and they will approve the Incident Action Plan (IAP) and then physically sign it. This signature is provided so that there is documentation that the plan was approved. As the incident begins to close out, the Incident Commander (IC) will also approve the demobilization plan by signature. When handing the command off to a Unified Command (UC) or an Incident Management Team (IMT), the Incident Commander (IC) will be required to fill out the ICS Form 201 (Incident Briefing). Throughout the incident, they will also maintain the ICS Form 214 (Activity Log) to document what activities they undertook.

13.3.3 The Benefits of the Planning Process

The Planning Process is a critical factor in ending and/or mitigating an incident. Thorough and sensible planning creates a foundation for effective incident management. This planning is applicable to both emergency incidents and to planned events. Utilizing the NIMS method and the ICS companion method for the planning process will provide a template that will guide operational, tactical, and strategic planning while allowing a degree of freedom when making decisions. This process will direct Command Staff, General Staff, and the Incident Commander (IC) with the steps they should take to develop an effective Incident Action Plan (IAP). The formalized steps taken while creating the plan will give guidance to subordinate personnel on how their actions should be performed in order to end or mitigate the incident.

To write an effective plan, some key information must be obtained. First and foremost, there must be current, accurate, and relevant information that describes the situation. Without accurate and current information related to the incident, those managing the incident will have to guess what the situation is and base the plan on incomplete information. When forced to base a plan on incomplete information, it can be and normally is a recipe for disaster.

Additionally, those in charge of the incident will need to reasonably forecast how the incident will progress. By being able to predict what will likely happen, mitigation measures can be taken to reduce the effects. Along with creating strategies to reduce the damage and destruction of the incident, they will need to provide alternative plans that would address what to do if the initial tactics did not work.

13.3.4 The Planning Process-an Overview

It is important to note that this chapter is only a brief overview to help novice users better understand the overall process. There will be a more in-depth analysis in Chapter 15. The Planning Process can be fluid and have a lot of moving parts. For this reason, it was decided that a basic overview should be provided to gain an uncomplicated description in this chapter, and more comprehensive review in Chapter 15.

During the creation and development stages of ICS, the planning process of an (expanding) incident was found to be extremely challenging. The developers wanted to find a way to make the planning process consistent, regardless whether the incident was small, midsize, or an extremely large response. To make the planning process uniform (in the application to any incident), the developers of the ICS method designed a way to ensure that no steps were overlooked or missed. This process became known as the *Planning P*.

The Planning P (See Figure 13.1) is used as a guide. If the Command and General Staff adhere to the Planning P, they will have a roadmap for greater success. If they followed the steps of the Planning P procedure that is mapped out, then they will follow a proven method of planning for all types of incidents, especially large and/or complex incidents. While the initial stages of the Planning P are used only one time for each event or incident, the circular part of the Planning P is a constant and ongoing process that continues until the end of the incident. The Planning P can even be used to plan the demobilization of all resources.

Each step, beginning at the base of the P, is mapped out and provides basic instructions for the Incident Commander (IC) and those involved in the response. This guide ensures that no steps are missed and that each aspect has the appropriate focus and attention in the planning process. While there are multiple steps to be taken in the Planning P, it can be divided into five distinct phases with four of those Phases being reoccurring. They are as follows:

1) Initial Understanding of the Situation
2) Establish Incident Objectives and Strategies
3) Develop a Plan
4) Prepare and Disseminate the Plan
5) Execute, Evaluate, and Revise the Plan

These processes ensure that the Incident Action Plan (IAP) is effective. This is accomplished through ongoing situational awareness, and the ability of the Incident Action Plan (IAP) to be flexible enough to meet the unique needs of the incident.

13.3.4.1 Initial Understanding of the Situation

A key component of making decisions to end or mitigate an incident is to have complete situational awareness. This is typically done through Situation Reports. In some jurisdictions, this is known as a SitRep. While not all jurisdictions use this terminology, the concept is still the same. The purpose of the Situation Report is to provide current and relevant information so that those in charge know exactly what is happening. Ongoing SitReps help the Command and General

Staff understand what progress has or has not been made, and it helps them to determine what tactics are effective and what tactics need refinement or not be used.

The act of providing a Situation Report is designed in such a way as to provide specific information about the incident. To create a SitRep, the Planning Section Chief ([PSC] or their representative) will collect all the relevant information from every available operational unit in the field. A good situation report will provide an amazing amount of benefits to the response, especially in the ongoing. In the ongoing awareness of the situation, SitReps will help to predict what efforts are working, what efforts are not working, and what is changing in areas that do not yet have response personnel.

In the initial phase of an incident, the SitRep will provide information on as much of the incident as is possible. This may include gathering the information, recording the information, analyzing the situation, identifying resources, and presenting a complete picture of the situation to the Command and General Staff. The SitRep will also usually provide information about the potential impact of the incident to ensure that all future hazards are identified. All these factors will help those in command make more informed decision.

Although SitReps are an important part of the planning process, spot reports can be just as important to the success of the response. Gathering information for spot reports can be obtained and produced by any individual that is physically at the incident. Spot reports provide information on any unusual or even expected occurrences, or any conditions that may have a direct and important effect on the operations or tactics that are currently being employed. Spot reports often provide information that is important to the incident objectives and tactics. These might include a change in weather, wind direction, and/or worsening conditions. Spot reports often support incident action planning by providing timely information on critical characteristics of the incident.

While maintaining situational awareness is important throughout the life cycle of the incident, the steps in Phase One of the Planning P are performed only one time. Once these steps are completed, incident management transitions into a cycle of ongoing situational awareness that continues and is repeated throughout each operational period. While Phase One is depicted in the leg of the "P," it automatically transitions into the Operations "O," which is represented inside the Planning P (See Figure 13.1).

13.3.4.2 Establishing Incident Objectives and Strategies

The second phase of the planning process involves creating incident objectives and strategies. Incident objectives drive the incident organization as it conducts response, recovery, and mitigation activities, while strategies are the approach that will be taken to meet the objectives. Establishing objectives and strategies deals with forming and prioritizing measurable outcomes. If priorities are not set, then the most critical issues that should have been addressed may worsen. Similarly, if there is no way to measure strategies, then there is no way to determine if the efforts that are being taken are effective.

The incident priorities that were initially established by the Incident Commander (IC) in the initial response should be reviewed. In some instances, the priorities initially chosen may need to be changed. The determination of whether these priorities should be changed must be reviewed during each operational period. As an emergency incident changes, so might the priorities. This type of review in each operational period is done to help to clarify the order of importance regarding incident objectives. While priorities are used to help determine incident objectives, they should be listed on the Incident Action Plan – Incident Objectives (ICS Form 202). Incident objectives are based on obligations, requirements, and strategies. These incident objectives will help guide the distribution of resources to meet those objectives (FEMA, 2012).

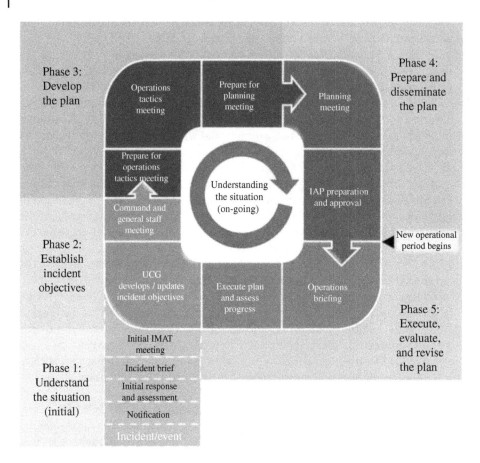

Figure 13.1 The *Planning P* for the ICS method (FEMA, 2012).

It is important to note that incident objectives and strategies must follow the legal guidance of the Authority Having Jurisdiction (AHJ). Subsequently, policies such as the Standard Operating Procedures (SOPs) or Standard Operating Guidelines (SOGs) will need to be followed as well. Additionally, because the Authority Having Jurisdiction (AHJ) is responsible for the incident, a change of objective(s) should in most instances be consulted before changing an objective. It should also be noted that while not typical, the primary objective may be to restore or protect critical infrastructure as the highest or a higher priority over other areas of the incident. While it is rare for this to be the case, it can happen.

In establishing the incident objectives and strategies, all options should be identified, analyzed, and evaluated to decide on the most effective strategy for the incident. Additionally, alternative plans should be made in the event that the initial strategies do not work out as planned. Contingency plans ensure that whenever a strategy does not work out, the operational and tactical factors will not have to wait for new guidance from the Incident Commander (IC) because an alternative strategy has already been vetted.

In the evaluation process, the criteria that should be used include public health and safety, the environmental impact, and the cost of the efforts proposed. While evaluating those aspects, those creating the Incident Action Plan should also be cognizant of all legal and political aspects. These legal and political aspects should first give priority to the local laws and political environment, followed by the state and federal laws. This is important because all incidents are considered local in nature.

Management by Objectives (MBO, better described in Chapter 14) is a key characteristic of the NIMS method and the ICS companion method. Management by Objectives (MBO) is the recommended way of creating objectives. Incident objectives are a statement of guidance and the direction that is necessary for the selection of the appropriate strategies and tactical decisions. Incident objectives should be based on realistic outlooks of what can be accomplished when all available resources have been effectively utilized.

When creating incident objectives, it is important to remember that those objectives must be flexible enough to allow for strategic and tactical alternatives. Incident objectives provide guidance and strategic direction but should not address the tactics that may be involved. Incident objectives should help drive response and recovery activities while being specific to what the desired outcome should be.

13.3.4.3 Develop a Plan

The third phase of the Planning Process involves deciding which tactical direction the response should take. In doing so, an evaluation of the available resources, reserve resources, and the specific resources that will be needed should be undertaken. By doing so, the Incident Commander (IC), the Command Staff, and the General Staff ensure that the tactical objectives can be met by the resources that are available for that operational period. Each member of the Command and General Staff has set responsibilities (described earlier) that they must undertake before the formal Planning Meeting. They should make sure that the areas they are responsible for is accomplished for each operation period prior to the Planning Meeting.

Operational planning is a continuous process; however, the process should begin immediately following the Command and General Staff meeting. For the planning meeting, the Operations Section Chief (OSC) will develop strategies to meet incident objectives. This is done by recommending tactics that they believe will be beneficial and effective in accomplishing the set strategies. The Operations Section will match the required work to be accomplished to resources that are proficient in that type of work, and this Section will create work assignments. Those assignments will be documented on the ICS Form-215 (the Operational Planning Worksheet). As these work assignments are established or presented in a planning meeting, the Command and General Staff should review the work assignments and provide input as necessary.

13.3.4.4 Prepare and Disseminate the Plan

Upon the completion of the planning meeting, Phase Four of the planning process begins. This is typically identified by the Operations Section Chief (OSC) providing the approved ICS Form 215 (Operational Planning Worksheet[s]) to the Planning Section Chief (PSC). The Planning Section is responsible for completing the appropriate Incident Action Plan (IAP) forms and assembling the completed plan. While the Planning Section is responsible for copying and collating this information, it is important to note that other Sections within the hierarchal structure of the ICS are also responsible for contributing certain pieces of information and supporting documents.

A team effort is needed to ensure that the Planning Meeting and the ensuing Incident Action Plan (IAP) is a success. Furthermore, the preparation and the lead up to the Planning Meeting are equally as important as the Planning Meeting itself to ensure that the meeting is a success. As an example, the Planning Section is tasked with collecting the various parts of the Incident Action Plan (IAP) and compiling them during the time period between the Operations Tactics Meeting and the actual planning meeting. Without the proposed operational tactics in hand for the Planning Meeting, valuable time would be lost discussing potential tactics. As part of the team effort that goes into planning, the Planning Section Chief (PSC) is tasked with constructing the draft Incident Action Plan (IAP) after the meeting. The Planning Section Chief (PSC) will be tasked with the responsibility of tracking any changes made to the draft during the final review and approval process. In completing this task, the Planning Section Chief (PSC) ensures that the Incident Action Plan (IAP) is a thorough and complete plan.

It should be noted that the Resources Unit also plays an important role in preparing the Incident Action Plan (IAP). Besides filling out the ICS form 215 (Operational Planning Worksheets) prior to the Planning Meeting, this Unit is responsible for preparing and documenting a number of important forms after meeting is finished. These forms are critical to the overall operation so that everyone in the incident knows their jobs, responsibilities, and where and when they should report. Some of the forms related to the Incident Action Plan (IAP) that the Resources Unit is responsible for include the following:

- Organizational Assignment List (ICS Form 203)
- Division/Group Assignment Lists (ICS Form 204)
- The incident Organizational Chart (ICS Form 207)

It is plain to see that there must be a cooperative and collaborative effort in order to have a highly efficient and effective incident response and Incident Action Plan (IAP). The planning that was discussed in this chapter is only the basic precepts of the planning process in the ICS method of incident management. These are just the basics. To fully understand the intricacies of planning for an incident, a thorough review of the Planning P will be presented in Chapter 15.

Real-Life Incident

On 5 August 2014, an Indiana (volunteer) assistant fire chief died after being caught in a roof collapse while fighting a fire in a commercial septic tank cleaning company. The fire was in a pole barn-style building with metal siding and a roof with wood-truss supports and a pan ceiling (a metal ceiling that blocks the view of trusses). Heavy fire was showing through the roof on two sides of the structure when the first fire company arrived.

The National Institute for Occupational Safety and Health (NIOSH, 2018) investigated the incident that contributed to this firefighters' death. One of the issues identified was that there was not an effective incident management system implemented at this fire, and there was not a proper Incident Command Post (ICP) setup. It was stated in the report (2018) that the noise created by an Incident Command Post (ICP) that was set up in the front yard, likely led to missing the mayday call of the firefighter down. Additionally, the report stated that the agency was not utilizing the planning process including undertaking a Risk Assessment (RA) and developing an Incident Action Plan (IAP). The planning process was probably needed since the NIOSH report (2018) noted that water shuttling to the scene with water tenders, there were reports of running out of water. This likely which contributed to this death. Had planning measures been implemented, even rudimentary planning measures, this death possibly could have been avoided.

The NIOSH report (2018) also stated that the tactics being used did not match the actions the firefighters undertook. While the Incident Commander (IC) ordered a defensive attack, three firefighters continued to enter the structure, including the victim. Other findings of the 2018 report identified that there were no mutual aid Standard Operating Procedures (SOPs) for fireground operations, which included the key factors of incident management, communications, and joint operations. Furthermore, these agencies never jointly trained on those key factors.

Perhaps the most disturbing finding of the NIOSH report (2018) was that no Safety Officer was appointed for this incident. Considering there were multiple agencies and probably more than 30 firefighters on scene, there should have been at least four Safety Officers (SO) and probably more. In this instance, it was suggested that there should be a Safety Officer (SO) posted at every entrance to the structure, and because the response was supposed to be in the defensive mode, a collapse zone should have been established to protect all personnel (NIOSH, 2018).

While we should make every effort not to second guess the response or the tactics from incidents such as these, we still need to learn from them. By doing so, we can help to prevent similar future issues in the incidents we respond to.

Chapter 13 Quiz

1. An _____ _____ is the managing officer of an agency, division, or jurisdiction having statutory responsibility for incident mitigation and management.

2. **True or False:** The Agency Administrator is the person that (by default) is responsible for the protection of the citizens.

3. The _____ _____ has the responsibility for protecting citizens and the authority to make decisions for the community (as a whole). They will also be able to commit resources and obligate funds.

4. **True or False:** The Agency Administrator does not have the legal authority to guide the resources to protect the population, stop the spread of damage, and to protect the environment.

5. **True or False:** The Agency Administrator is responsible for planning and mitigating natural and man-made incidents.

6. **True or False:** The Agency Administrator Representatives have the authority to ensure that a jurisdiction has established resource management systems and whole community preparedness plans.

7. The Agency Administrator should have a method in place for identifying, inventorying, requesting, and _____ of resources, as well as initiating and dispatching those resources.

8. **True or False:** The Agency Administrator should assume a command role over the incident.

9. A _____ ___ _____ is a right, or an obligation, to take actions for a department, agency, or jurisdiction.

10. The Agency Administrator also needs to be mindful of the span of control. They may mandate, that specific advisors should be kept in the loop, or even advise and be active in specific areas of managing an incident. These individuals are identified as:

11. By having a planning process that utilizes multiple people, then there is a greater chance of having a _____ _____ view of an incident.

12. **True or False:** The planning process used in the ICS method of incident management allows other individuals to help build on the strategies presented in the planning meeting.

13. With proper situational awareness while creating an _____ _____ _____, effective strategies and tactics can be implemented to bring order out of chaos in an orderly and efficient manner.

14. The planning process in the ICS method was designed in such a way that the overall _____ _____ were/are broken down into much smaller tactical assignments for each operational period.

15. Those creating the Incident Action Plan (IAP) will be looking at short-term strategies that will be used for each operational period, by _____ the resources and the tactics needed for the next operational period.

16. **True or False:** As Command Staff and General Staff positions are assigned, there is no need to create an Incident Action Plan (IAP)

17. **True or False:** Thorough and sensible planning creates a foundation for effective incident management.

18. What are the Five Phases of the Planning P?

19. **True or False:** Situational awareness is an important part of the planning process.

20. **True or False:** The preparation and the lead up to the Planning Meeting are equally as important as the Planning Meeting.

Self-Study

Federal Emergency Management Agency (2019). ICS 402: Overview for executives/senior officials G0402 [Presentation]. Retrieved from https://training.fema.gov/gstate/xcr3wnlp/g0402%20-%20ics-402%20-%20incident%20command%20system%20(ics)%20overview%20for%20executives%20and%20senior%20officials/ics-402_ig.pdf.

Federal Emergency Management Agency (2012). FEMA Incident Action Planning Guide. Retrieved from https://www.fema.gov/media-library-data/20130726-1822-25045-1815/incident_action_planning_guide_1_26_2012.pdf.

National Institute for Occupational Safety and Health (2018). Volunteer Assistant Chief Killed and One Fire Fighter Injured by Roof Collapse in a Commercial Storage Building-Indiana: Report # F2014-18. Center for Disease Control, NIOSH Fire Fighter Fatality Investigations and Prevention Program. Retrieved from https://www.cdc.gov/niosh/fire/pdfs/face201418.pdf.

National Wildfire Coordinating Group (2018g). NWCG standards for interagency incident business management: PMS 902. Retrieved from https://www.nwcg.gov/sites/default/files/publications/pms902.pdf.

National Wildfire Coordinating Group (2017). Resource Advisor Guide: PMS 313. Retrieved from https://www.nwcg.gov/sites/default/files/publications/pms313.pdf.

Preserve America (2008). Preparing to preserve: An action plan to integrate historic preservation into tribal, state, and local emergency management plans. Retrieved from https://www.doi.gov/sites/doi.gov/files/migrated/pmb/oepc/rppr/upload/12-18-08-Preparing-To-Preserve.pdf.

14

Management by Objectives – SMART Goals

If you are not making the progress that you would like to make and are capable of making, it is simply because your goals are not clearly defined.

Paul J. Meyer

If you have ever attended a class on the Incident Commands System (ICS) method of incident management, one area that seems to always be lacking is the description of Management by Objectives (MBO) and SMART Goals. When teaching about these highly important factors, instructors will in most instances just give a brief description of these concepts and then move on. Considering these objectives are the underpinning of the planning process, as well as the basis for the tactics that will be employed for an operational period, it became obvious that this book should cover this topic thoroughly. For that reason, this chapter will really get into the nuts and bolts of Management by Objectives and SMART Goals. Trying to understand this concept can be difficult for some people. An old parable might be one of the best ways to describe what Management by Objectives accomplishes.

A US citizen goes into the back country of Africa on a safari, and he sees native man that had recently killed a wild elephant. The American citizen questions the African native why he killed the elephant. He responds that his family was hungry, and that they needed food to survive. The American looked at the massive elephant in front of him and became overwhelmed by its size. He then asked the African man, "How in the world can you eat that big elephant?" to which the African man replies, "We eat it one piece at a time."

While this parable is slightly humorous, there is a parallel meaning that coincides with Management by Objectives. When an incident arises, we can easily become overwhelmed by the magnitude of that incident. If we look at the overall magnitude of the incident, it is easy to be confused by the chaos that surrounds it. In order to successfully manage an incident, no matter how large or small, we must proceed like we are eating an elephant: We need to take it one piece at a time. While we need to be mindful of the totality of the incident and the overarching intent, we need to break it down into smaller objectives. When doing this, we take one bite (or a manageable objective) of the overall incident for each operational period.

During an operational period, the Incident Commander (IC) or the Unified Command (UC) as well as the General Staff and Command Staff must determine the objectives for the next operational period. During the first operational planning period, the Command Staff and the General Staff (including the Incident Commander [IC] or Unified Commanders [UC's]) will identify the overall goal. Usually these goals are simple. They may include containing the wildland fire, ending the hostage situation without a loss of life, treating and transporting all patients, or something similar or along these lines. When these goals are identified, then it will be broken down into smaller objectives that will help attain the overall goals for the incident. Management by Objectives utilizing SMART Objectives is the way this is accomplished.

Subsequent operational period planning meetings will build on the previous operational periods while still keeping the overall goals in mind. The initial part of this process is to evaluate the most recent objectives and determine if they are having the intended effect. If they are not having the intended effect, then revisions should be made to make them more effective for the next operational period. In order to determine if the strategy previously decided is effective, some basic questions should be asked (and answered) to evaluate the objectives. These questions include the following:

- Is the incident stable?
- Is the incident increasing in size and/or complexity or has it reduced in size and/or complexity?
- What are the current incident objectives, strategy, and tactics?
- Have any safety issues been identified?
- Are the current objectives relevant to the big picture and effective?
- Would a change of course improve tactics or create more chaos?
- Was the time frame for completion of these objectives reached?
- What is the current status of the resources available?
- Are the current resources in good condition or should they be replaced?
- Are there sufficient available resources?

In asking these questions, the Incident Commander (IC), or the Unified Command (UC), and the staff can discuss the response and make informed decisions about the current objectives being used and ongoing situational awareness. This re-evaluation of the objectives should take place in every operational period. Should the Incident Commander (IC) determine that there must be a change in objectives, those concerns should be conveyed to the Command Staff and General Staff in a meeting as soon as possible.

It is important to note that developing objectives is more than stating "I want to do this" or "I want to do that." In the ICS method of incident management, the development of objectives is an organized process. This organized process needs to have attention to detail, with situational awareness being the guiding force behind setting the objectives and identifying the tactics that should be used. Thorough intelligence regarding what the current incident status is and where it is anticipated to be at the end of the operational period is imperative to overcoming any challenges.

A key factor in incident management is that first responders should not let the incident manage them, but rather they should manage the incident. If the incident manages those responding to the incident, then there will be more chaos, confusion, and uncertainty. On the other hand, if the incident is managed by first responders, then there will be less chaos, confusion, and uncertainty.

An important part of actively controlling and managing the incident is to develop practical objectives that are comprised of strategies that will guide your operational response planning (FEMA, 2010). Practical strategies plus an effective operational response plan creates tactics that mitigate the incident, providing you make those goals *SMART*.

Control of an incident that reduces chaos, complexity, and confusion is gained by creating incident goals and objectives. This can take considerable planning, especially when the response is a large-scale or complex incident. The proper use of Management by Objectives (MBO) allows the individual commanding an incident to break a large crisis into much smaller, manageable pieces and create effective response tactics. Management by Objective is a tool that assists in creating systematic goals and objectives that are actionable and manageable in a short

Practical Strategies
+ Operational Response Plan
= Response Tactics

Figure 14.1 How to develop response tactics.

timeframe. If those creating the Incident Action Plan (IAP) will create practical strategies and then add a compartmentalized and thorough operational response plans, then these two factors will provide the response tactics that will be needed to overcome the challenges of the incident (Figure 14.1).

There are multiple examples that could be provided that can help to better understand the concept. An example of how Management by Objectives in action can be helpful might include the response to a house fire. Because life safety is the highest priority, the first objective might be to complete a search of the entire house for survivors within a set amount of time: perhaps that goal for this search would be a total of seven minutes. The second objective might be to stabilize the incident by putting out the fire within 30 minutes. Once the fire is out, the third objective might be to salvage the occupant's personal belongings from the structure, so we can reduce the loss for the homeowner. In using Management by Objectives, each phase of a complex situation is broken down into more manageable tasks.

In law enforcement, an example of Management by Objectives might be an accident reconstruction scene. The first priority might be to secure the area where the accident occurred and to reroute traffic. The second objective might be to take photos of the entire scene. The third objective might be to map out the accident area and the location of each potential piece of evidence. The fourth objective might be to take pictures and collect forensic evidence. The fifth and final objective might be to clean-up the accident and to reopen the road. Each major task is part of an objective, or a bite of the elephant, that gets you that much closer to the overall goal of the incident. That is how Management by Objectives works; it takes the larger incident and breaks it down into smaller systematic steps that build on each other to end the overall incident. Essentially, you will take one bite at a time.

Just like more practice of driving a car makes you a better driver, the utilization of Management by Objectives over a period of times, and at multiple incidents, helps the Incident Commander (IC) and the Command and General Staff to become better at managing an incident. As they become more versed in commanding an incident, and in utilizing Management by Objectives, they will begin to undertake these tasks more naturally.

In most instances, Incident Commanders (ICs) will not have to identify each individual objective in smaller incidents. The reason for this is that over time, they have conditioned themselves in the order things should be done. In the initial phase of a smaller or less complex incident, the experienced Incident Commander (IC) will just know what should be done, and they will direct it to be done. This holds especially true if it is a routine or normal situation. As an incident occurs that is outside of the normal day-to-day response, the Incident Commander (IC) might have to think through what order these objectives should be accomplished to mitigate the situation. This holds especially true on especially large or complex incidents. One thing that is important to remember is that there is no shame in creating the objectives and mapping out the response using a pen and paper or a computer. Where trouble begins in an incident response is when the Incident Commander (IC) flies by the seat of their pants, or they do not have objectives to meet their goals.

14.1 Underlying Factors for Determining Incident Objectives and Strategies

The determination of incident objectives and the strategies that are needed should be based on the continuous situational awareness and size-up process. An Incident Commander (IC) should use these practices in every emergency response situation and for each operational period. This is why it is so important to set these incident objectives as part of the planning process for each operational period. While every effort should be made to create and follow the designated incident objectives, it should also be realized that not all incidents go as planned.

A change in the conditions based on situational awareness of the current conditions might lead to objectives being altered and/or updated as the incident progresses. This change or update to tactical objectives is often referred to as a *corrective action*. Corrective actions cannot be made if the Incident Commander (IC) is suffering from what is commonly called tunnel vision. Tunnel vision can be classified as the primary enemy of situational awareness. Tunnel vision happens when those in charge of an incident will focus on one goal, or a set of goals, and they fail to look at the big picture or the overall situation. Tunnel vision prevents the Incident Commander (IC), and potentially others, from being able to make neutral appraisals of the situation, which in turn inhibits the ability to engage in the corrective actions that might be necessary to successfully mitigate the incident.

Tunnel vision is not exclusive to one discipline, and it can happen at any time. It does not matter if the incident is large and complex, small and simple, or if it is an uncommon incident or something mundane. Tunnel vision can be deadly to those that work in public safety. An officer making a normal traffic stop could have tunnel vision focusing on the driver that is acting odd, while the passenger shoots the officer with a pistol. A firefighter rescuing people from a collapsed building could become impaled by a piece of debris because they had tunnel vision focused on those that needed saved rather than identifying the hazards before proceeding. Let us not forget about the Emergency Medical Services (EMS) personnel or emergency management personnel that have tunnel vision, and they are so focused on rescuing people that they did not notice the chemical plume that developed.

Although tunnel vision can be dangerous or deadly to the individual who works in public safety, the effects that tunnel vision has on a leader, such as an Incident Commander (IC), can be multiplied. In most of the previously mentioned scenarios of tunnel vision, only those that had the tunnel vision were in danger. If an Incident Commander (IC) has tunnel vision, the effect of that tunnel vision will likely be felt by many that are under their command. That is why it is so important to ensure that objectives and strategies do not have any form of tunnel vision associated with them.

In any type of incident, it is important to understand that your goals and strategies must meet the agency's policy and direction. Understanding these concepts prior to an incident will help frame and/or guide your response. These policies that the Incident Commander (IC) will operate under can be either verbal or written, although written policies are more encouraged. Written policies such as Standard Operating Guidelines (SOGs) or Standard Operating Procedures (SOPs) can come from a variety places such as an agency administrator, a chief elected official, a board of directors, a city council, a commissioners' court, or some other governmental body. Where the policy comes from can be diverse, and it will usually be based on the way that the local government is organized.

In rare instances, the Authority Having Jurisdiction (AHJ) may not be a government agency. They might be an organization, a nonprofit, or even a business. In fact, many civilian contractors that support the United States government utilize the same ICS method that public safety uses. Knowing that Standard Operating Procedures (SOPs) and Standard Operating Guidelines (SOGs) can come from multiple places, we need to realize that the key takeaway is that this policy and direction is how you and your team are expected to respond. This also allows the Incident Commander to establish incident objectives based on the way that the Authority Having Jurisdiction (AHJ) prefers to respond.

It should be noted that policy and directives may not always agree with what may be best for the incident. In some instances, those in command might have to make a judgment call and accept the ramifications of a decision that is outside of the policy. Conversely, in keeping within policy requirements, they could go against their intuition and forego a tactic that would have been the right action. Ultimately, the decision is the Incident Commanders (IC's), and they must base their decision on the unique circumstances that have been presented.

14.2 Establishing Immediate Incident Objective Priorities

When first arriving on scene, the initial Incident Commander (IC) will establish immediate priorities which will be used to create the objectives. Those objectives will assist them in mounting a tactical response. The tactical response, if done properly, should be based on the premise of resource allocation and situational awareness. If it is determined that additional resources are needed, they have already established incident priorities, objectives, and tactics, and these can be used as a reference point to determine what additional resources they may need throughout the entire incident. It is important to note that sometimes resources are not available. It is also important to note that objectives may change.

In instances such as these, corrective actions must be taken to change the tactical response, and then approach the incident in a different manner. The same strategies hold true when working on simple and less complex incidents or those incidents that are especially complex or that will have extended operations. The only difference is that on complex or extended incidents, the objectives are usually written, while on smaller incidents, they may only be in the Incident Commander's (IC's) head. While technically the goals and objectives should always be written down, anyone who has ever served in public safety know that this is not always possible or practical.

Routine incidents such as dumpster fires, simple house fires, basic law enforcement investigations, and other routine calls that are typically single agency responses are handled by mentally acknowledging goals and priorities. While handling incidents by mentally making goals and objectives is widely accepted on routine incidents, this does not make it necessarily the acceptable way to handle an incident. Prioritizing objectives is critical to handling a situation in an efficient and orderly fashion. It is also a way to make sure that tasks are undertaken in order of importance or priority.

Prioritization of what tactics should be used is extremely important. With the wrong priorities, incident can deteriorate quickly, and moral can go from good-to-bad in an instant. When we discuss priorities, it is important to realize that the acronym LIPE can be an important prioritization tool. In order, LIPE stands for and puts the priority on

1. Life safety
2. Incident stabilization
3. Property conservation
4. Environmental/economic protection

These are the four main priority considerations for any incident response situation. It does not matter what the origin of the incident is, or the background of the personnel responding to it. By controlling these four elements in the order they are presented, the Incident Commander (IC) and those involved in the response will have the ability to regulate the important factors that determine how an incident will progress.

In some jurisdictions, the acronym used has only three priorities and uses the acronym LIP and uses only the first three priorities in order. While LIP is a viable way of managing an incident, this acronym fails to recognize that an incident usually does have a substantial (negative) economic and/or environmental impacts on a community (or an individual), leading to increased recovery costs and time. As will be discussed later in this chapter, environmental and economic impacts are a consideration in everything from a simple house fire (or burglary) to large incidents such as wildland fires, floods, and hurricanes.

14.2.1 Life Safety

A primary concern relating to control of a chaotic situation is life safety. The Incident Commander (IC) and first responders *must* make life safety the highest priority. If there is a life in danger, then above all else the first objective should be to initiate a life safety measure such as a rescue. This holds true on most instances except in instances where there is a high probability that the first responder will lose their life in the process.

First responder safety should always be more important than a victim's life. The Incident Commander (IC) and all personnel should have the mind-set that they did not put the victims into that situation and that they should not trade a First Responders life for a victims' life. Whenever possible, we should provide reasonable attempts to save lives; however, trading a life for a life is unacceptable. Supervisors and officers need to ensure that everyone that responded to help someone in need must be able to get home safely after the incident is over.

Most who have never served in public safety may not realize that there are protocols in place to protect the lives of first responders. In order for an Emergency Medical Technician (EMT) (be they a basic, advanced, or paramedic) to become licensed, they must pass written and practical tests. In the practical portion, when given their scenario, they must first ask the question "Is the scene safe?" and then advise the evaluator that they have Body Substance Isolation (BSI) in place before doing anything else. If they do not do this, they cannot proceed with testing, and in most instances, they fail the exam. The testing is done this way so that it is drilled into their head that if the incident is not safe for them, they should not be there. First responder safety should be the highest priority.

When fire departments are dispatched, if the scene might be dangerous (e.g. a domestic dispute, shots fired, potential explosives), then the dispatcher will advise the fire department that the scene is not safe, and they should stage a safe distance away from the scene. The fire department will stage away from the scene until they are notified that the scene was made secure by law enforcement. The point being made is that when we go to our public safety job, there are protocols in place to ensure that we go home when our shift is over. While there are specialized EMS and fire personnel that have been trained as combat medics, the vast majority should avoid putting their life in danger.

If the situation does not involve immediate life safety (or rescue), but life safety might be compromised without mitigation measures, first responders should work to eliminate the undesirable factors that could be dangerous to citizens and personnel alike. These incidents may include potential threats seen with fires, leaks, perpetrators, downed power lines, and/or other hazards that cause injury to civilians and responders alike.

14.2 Establishing Immediate Incident Objective Priorities

An example of this type of strategy might include a chemical release. If an anhydrous ammonia leak occurs near a populated area, it may be more prudent to stop the leak than to evacuate the public. Having the affected area shelter in place and stopping this leak may be a more efficient way of providing life safety than evacuating the area and putting those evacuated at risk. This would likely be a judgment call by the Incident Commander (IC) based on the unique intricacies of that incident.

It is also important to note that life safety does not include retrieving deceased individuals. In the event a victim or victims are deceased, and the incident becomes a body recovery scenario, then actions to recover the body in most instances should not be taken. First responders should never risk injury or death to recover a body, at least not until the incident has been stabilized and there is no risk posed to the responder. While there may be exceptions, they are rare. Even if an incident is stabilized, a body recovery may need to wait until investigators give permission to move the body. If the victim is already deceased, the Incident Commander (IC) should weigh the risk to personnel in recovering the body. If there is no reason to risk life for a body recovery, then it should not be done.

Perhaps one of the most scenarios an Incident Commander (IC) can face is a fire with children that are deceased. While any loss of life is hard to accept, the loss of a child seems to be among the most difficult to deal with. Whenever arriving to an incident with children involved, the Incident Commander (IC) must evaluate the probability of survival. If the Incident Commander (IC) deems that there could be no chance of survival, then they will in most instances focus their attention on stopping whatever was the main cause of the incident. As an Incident Commander (IC), this may be a very unpopular decision, especially if a child is involved. In instances such as these, the Incident Commander (IC) may need to ensure that no personnel freelance and attempt a rescue on their own. While a difficult situation, it may be prudent to post officers or other individual at entrance and egress areas to prevent first responders from entering the area. As has been said before, life safety of personnel should be the first and highest priority.

Keeping in mind that life safety is the immediate priority, those trying to bring an incident under control should consider what is most beneficial for the most people. In some incidents, the Incident Commander (IC) and other response personnel may be faced with a situation in which not every person can be saved. While somewhat unpopular in the public safety realm, the Incident Commander (IC) and response personnel should take a utilitarian view: doing the most good for the most people. Unfortunately, in public safety things do not always go according to the textbook. Sometimes those controlling a situation will need to make a tough decision, one that chooses the life safety of many over the life safety of just a few.

For example, in a hostage situation involving the local SWAT Team, the perpetrator becomes violent, fatally injuring innocent victims. The team decides to make entry to free the hostages because it is believed the perpetrator will kill or injure all involved. In planning the tactics, the SWAT Team evaluation shows that a small number of hostages may be injured or killed when making entry; however, if nothing is done, or entry is delayed, there is a high probability that all the hostages will be killed by the perpetrator. The final decision must evaluate and choose the option that provides the greatest amount of life safety for the most individuals.

The same type of life safety decision is needed in a Mass Casualty Incident (MCI) such as a bombing, a stadium grandstand collapse, etc. In Mass Casualty Incidents (MCIs), there are usually limited resources during the onset of the incident, and in some cases throughout the entire incident. Those controlling the incident will likely implement a triage system to sort victims. A triage system allows patients to be sorted into four groups: injured individuals that are beyond help, injured individuals that need immediate help, injured individuals that can wait for help, and the walking wounded (those who appear to have minor injuries). Those patients that have only a slight chance of surviving or that would require a large amount of personnel to save their life would be labeled as being beyond help. They would be placed in an area where they would be helped once viable patients with a higher likelihood of surviving or that would use less resources were first taken care of. This allows the limited available resources to focus on those that have a higher probability of living. While it may sound cruel, the time and resources that would be used to save one patient labeled as being beyond help could potentially save three or four other individuals in the same timeframe. Incident management is not an easy or sometimes even a popular task. It often involves making hard calls that can determine the fate of many.

To make sure there is no mistake in attitude, it should be reiterated that while life safety is the highest priority, there also must be common sense applied. Rushing into a situation without evaluating the risks may do more to harm, or potentially even put the life of emergency personnel at risk for death. It is important to remember that life safety can have a point of diminishing returns, and it is incumbent upon both the Incident Commander (IC) and emergency response personnel to identify when those instances occur. It is critical that common sense should be used when determining priorities rather than emotion. If emotion is the driving force, then on-scene personnel may become victims.

14.2.2 Incident Stabilization

Much like making utilitarian decisions that can relate to life safety, the same can hold true for incident stabilization. Incident stabilization will slow or stop the chaos, confusion, and uncertainty and replace those negative emotions with the positive emotions of calm, understanding, and certainty. In some incidents, incident stabilization may only require shutting off a gas meter, having the electricity shut off, or in surrounding a perpetrators residence. In other emergencies, incident stabilization may be more complicated, and detail oriented such as surrounding a Hazardous Material (HazMat) spill with sand (or some other absorbent), evacuating individuals and pets from a large geographical area, shutting down an interstate, containing a wildland fire, or in starting negotiations with a hostage taker. The correct way to proceed will depend on the unique circumstances that are faced.

Incident stabilization is critical in ensuring that the incident does not deteriorate. Efforts to stabilize an incident may be direct or indirect. In direct stabilization, emergency personnel will be solely responsible for stabilizing the incident. Some examples of this may be in containing a wildland fire, sandbagging berms, removing potential hostages, and other similar strategies. If the incident is stabilized by first responders, then this is a direct incident stabilization.

Indirect stabilization occurs when those that are not first responders are needed to assist in stabilizing the incident. As an example, if a gas line ruptures and explodes, and it is feeding a massive fire, first responders will need the gas feeding the incident shut off before they can stabilize the incident. No matter how hard they try to put out the flames, the burning gas would continue to ignite the nearby structures. It is important to note that for safety reasons, they should not put out the burning gas because this would lead to more safety concerns, including another explosion that could possibly be larger and more damaging.

Another example of indirect incident stabilization might be a vehicle fleeing from the police that hits a power pole, and live electrical lines are on the vehicle or the surrounding ground. The incident will need to be stabilized by the power company prior to removing the person from the car. This same concept holds true in varying incidents, and these are only a few examples to help the reader understand how indirect help may be needed to facilitate incident stabilization.

When looking at what incident stabilization should accomplish, we need to realize that it contains the incident to ensure that it does not expand into a larger or more serious incident. Incident stabilization is a critical part of managing an incident because it will ultimately lead to an incident being controlled and managed rather than the incident being out of control and unmanaged, thereby causing more damage.

14.2.3 Property Conservation

Much like making utilitarian decisions can relate to life safety and incident stabilization, the same can hold true for property conservation. Sometimes the best method to conserve property may not be the most popular. In fact, there are instances where more damage must be done in order to prevent widespread damage.

In most wildland fire situations, a well-known practice is to take crews well beyond the fire-line and to remove a large strip of vegetation with a bulldozer in the direction that the fire is moving. Once the bulldozer has created a path of dirt that will intercept the ongoing wildfire, firefighters will typically light backfires. Sometimes, roads and tributaries are used as firebreaks, and backburning will begin at these points. Many of those outside of the fire service call this "Fighting fire with fire." While this practice may cause acreage and wildlife to be lost in the backburn area, these tactics can save tens of thousands of acres by taking away the wildland fires fuel, thereby potentially saving thousands of acres and the wildlife that are supported by those acres. Once again, decisions may need to be based on doing the most good in the long term.

Prioritizing property conservation should look at the effect that it will have on individuals and businesses. While some would say this is an economic effect, this does not always hold true. In many instances, economic factors are tied to the loss of property, but there is also sentimental value that cannot be considered an economic factor. When looking at property conservation, we need to be mindful that some things cannot be replaced by insurance. The pipe that grandpa used to smoke, wedding pictures of various family members, pictures of great-grandparents, and almost anything else you can think of. There are also families that are fourth or fifth generation farmers or ranchers who are living in the same homestead that their ancestors lived in. Rather than having a dollar value, they may have a high sentimental value of which cannot be replaced.

When looking at property conservation, we need to look beyond dollar values. In a wildland fire situation, crews are often dispatched to houses in the path of the fire. They are sent to remove foliage, to wet the house down, or to even fight the fire as it approaches the structure they are protecting. This is also part of property conservation. Other types of property conservation may be to assist a bus service in moving a large quantity of buses out of the path of a flood, or to assist a car lot in moving vehicles out of a flood or fire zone. While property conservation is the third on the list of priorities, it is nonetheless an important factor, especially for those whose property you preserve.

Real-Life Incident
Sometime during the 1990s, a fire broke out in a commercial business building in rural Missouri. The buildings affected by these fires were constructed in the 1920s. Over the years, renovations were made to the adjoining businesses, and those renovations compromised the fire integrity of shared walls. Wires, ventilation shafts, and other items left holes in the fire-break wall, leaving holes where a fire could extend. Besides the holes in the walls that were meant to block fire, there were also multiple layered ceilings, which added to the problem of trying to get to the fire. In some buildings, the ceilings had three to four layers of ceiling directly on top of each other.

When a fire erupted in one storefront, the flames quickly extended to the next store by means of the compromised (shared) walls and became difficult to access due to the multiple ceilings. Building after building caught fire due to the modifications over the years. Firefighters were frustrated, and they saw no conceivable way to get ahead of the fire. Finally, in order to prevent an entire block of the business district from being decimated, the fire chief called for a local contractor with heavy equipment to demolish the next building in this row of building.

The chief told the business owner of his intentions to demolish their business and offered firefighters and other volunteers who could assist in remove the inventory from the store as quickly as possible. By the time the heavy equipment arrived on scene, the business was already on fire, so the chief had the building cleared and told the heavy equipment operator to demolish the building. While important documents, equipment, and inventory were saved prior to demolition, the building owner did lose their business. Ultimately though, the decision that the fire chief made led to saving the entire block, around 12 buildings, from destruction. In doing this, he made his decision based on what would do the most good for the most people.

14.2.4 Environmental and Economic Protection

Many people fail to recognize that incidents, both major and small, can have an adverse effect on a geographical location or a community. Depending on the magnitude of the incident, it can also have an effect on the state, or even a nationwide effect. One only needs to look to the historical events of Hurricane Katrina, Superstorm Sandy, severe flooding along the Mississippi River, and several major wildfires to understand how an emergency can affect the environment and local, state, and federal governments.

If we take a historical look at the 1993 flooding along the Mississippi River, it is plain to see that this disaster had a huge impact on the ecological system and the economic system. It severely taxed the resources of local, state, and federal resources. According to Kolva (2002), the 1993 Midwest floods caused more than $20 billion in economic damages. This same flood damaged or destroyed more than 50 000 homes and killed 38 people. If you look beyond the cost and the death toll, there was severe ecological damage as well.

The U.S. Fish and Wildlife Service stated that this prolonged flooding caused many bottom-land hardwood forests and backwater wetlands stands to tip over. The flooding also caused a large loss of ground cover, eroded islands, destroyed vegetation beds, damaged or impacted dikes, and destroyed moist soil plants. Wildlife was also severely affected by these floods. Many birds suffered a decrease in hatchlings because their foraging areas were flooded. Freshwater mussel populations were found buried in 1–2 feet of sand and mammals that were displaced had higher than normal death rates (Kolva, 2002).

When we deal with emergency incidents, no matter how large or how small, we need to realize that these incidents usually have both an economic impact and an environmental impact. When we look at a simple house fire, we can see that it has an economic impact on the family that lives in the house, and the fire releases contaminants into the atmosphere. An area that suffers from high crime rates will typically have an economic impact as well. When individuals hear of crime in a specific area, a certain amount of the general public will avoid this area for fear of becoming a victim. The environmental impact that typically follows can in some cases be seen in graffiti on the walls, unkept streets, and homes that become dilapidated and a breeding ground for rodents.

Going beyond typical thinking, a recent study found that burglaries in England and Wales led to 4 million tons of carbon dioxide emissions when the items that were stolen had to be replaced (Skudder et al. 2016). It would be safe to assume that there would be a similar environmental impact in the United States, or any other country. It would be difficult to create objectives to mitigate environmental impacts based on the future impact of replacing these items; however, we need to be cognizant that every incident has an environmental and economic impact.

As an incident grows in size, the environmental and economic impact it has on the community usually grows as well. In some cases, these impacts affect the state or the nation. The loss of one home in a city reduces the amount of money that can be collected in taxes. If 100 homes are affected in a county, it will affect the county, and in most instances, the city. This also reduces the taxes that can be collected at the state level. Then there are the major disasters that not only affect the local and state governments but also the federal government as well. It was estimated that Superstorm Sandy cost the federal government $169 billion (FEMA, 2015c), and it was estimated that the total loss of Hurricane Katrina was $125 billion (Pender, 2005). The environmental impact of each was also devastating.

When identifying the economic and environmental impacts of an incident, the local jurisdiction will need to make the determination which one should take priority: Economic impact, or environmental impact. In some instances, the priority may change based on the type of incident being addressed. No matter what direction the local government decides, the key thing to remember is economic and environmental concerns are not as important as life safety, incident stabilization, and property conservation.

14.3 Management by Objectives

Now that we have discussed the LIPE prioritization method, we should address how we will manage the incident using these priorities as an underlying guide. If these guiding principles are not integrated into the response, then there is the possibility that they be overlooked, or certain aspects of the response may be prioritized wrong.

Under the National Incident Management System (NIMS) method and the ICS companion method (as well as other IMS Systems), first responders should utilize the Management by Objectives framework as the method to identify what goals should be accomplished, and in identifying the highest priority(s). Once the objectives are determined, personnel should use those objectives for managing the incident, unless of course corrective actions are needed. While many entities teach that Management by Objectives should be used, they unfortunately do not properly explain how Management by Objectives works nor do they explain how to properly write these goals and objectives.

Management by Objective is not a new concept. The Management by Objective framework was first introduced in the book *The Practice of Management* (Drucker, 1954, 2010). In this book, Drucker (1954) identified common attributes that allow the Management by Objective method to be effective. Over the years, the Management by Objective framework has been adapted to many diverse types of businesses and organizations. The widespread adoption of Management by Objective across a wide array of businesses is due primarily to the flexibility. This framework helps business to create an outline of what needs should be done and in what order. The main overarching concept of Management by Objective is to divide a larger task into multiple smaller tasks, all of which will eventually lead to the completion of the overall larger project. Key benefits of Management by Objective include the following:

- *Motivation*: Superiors and subordinates plan the goals which create empowerment, job satisfaction, and commitment. This creates motivation.
- *Increased Communication and Coordination*: Regular reviews and exchanges between superiors and subordinates help maintain a better relationship. It also increases problem solving, communication, and coordination of subordinates.
- *Clarity of Goals*: Everyone is on the same page and working to complete the written objectives. This provides clear and concise goals.

- *Common Goals*: With everyone working on common goals, it increases unity.
- *Commitment*: Subordinates are more committed to success because they helped create the goals.
- *In Line with Organizational Objectives*: Superiors ensure that localized or smaller objectives are designed to support and enhance the overall goals related to the "big picture" (Drucker, 1954, 2010).

Those that have responded to an emergency incident can testify that most incidents are complex, uncertain, and chaotic. On larger incidents, the chaos, complexity, and uncertainty can be magnified by the many tasks that may need addressed. The need to complete so many tasks can easily become overwhelming.

The Management by Objective framework has become a necessity to make larger incidents more manageable by eating the proverbial elephant one piece at a time. By taking one bite of a large or complex incident at a time, you prioritize each portion of the response and undertake the most important tasks first. By using the acronym of LIPE to identify priorities in conjunction with Management by Objective framework, the Incident Commander (IC) and their staff can systematically take on the most critical tasks first. By doing so, it can and does reduce chaos, confusion, and complexity and replaces those feelings with calm, order, and simplicity.

While the LIPE acronym states that life safety is the highest priority, it is important to realize that in some instances incident stabilization, property conservation, or economic or environmental impact can be integrated into the primary goal. As an example, if a person is alive but trapped behind a wall of flames, response crews may need to knock down the fire in order to reach them, thereby having the dual goal of saving lives and preserving property at the same time. Similarly, if there are sufficient personnel on scene, there could be two different goals of saving lives and incident stabilization at the same time. While this would require two separate written objectives, it is still a viable option.

A prime example of this would be a mentally stable person in a house, firing shots, both inside and outside of the house. Upon arrival to the scene, one law enforcement agency could set up a perimeter around the house and begin negotiations, while mutual aid agency begins evacuating homes and transporting individuals to a safe zone.

Management by Objectives that is used in a normal business setting has substantial benefits because all players are brought together to create the goals and objectives they wish to reach. In looking at the benefits of Management by Objective in an ICS setting, the end of an incident is everyone's goal, although other objectives (such as preventing floods from spilling over a bank, or containing a fire or perpetrator) may help crews conquer short term goals. After the initial response, the objectives of how to reach specific goals can be supplemented by the lower levels of the organizational chart sending ideas up the chain of command. Should those in the mid to upper levels determine that the idea has merit, then they too can take it up the chain of command. After going up the chain of command, if it is thought to still have merit, it might be discussed in multiple meetings, including the Objectives Meeting, the Tactics Meeting, and the Planning Meeting. If the Incident Commander (IC), the Command Staff, and the General Staff are in agreement, then they can (at their discretion) initiate these new ideas and/or objectives.

It is important to note that these suggestions are not normally solicited, but any suggestions from the field should be brought to the attention of the Incident Commander (IC) or the General Staff, or Command Staffs, attention. As has been mentioned before, no single person can come up with the best idea all the time. The sharing of these ideas also creates a better communication and coordination system because each level of the hierarchical organizational command chart has someone they report to. Additionally, because the lower levels of the organizational chart can suggest different tactics and objectives, it provides a method that may better meet the needs of the incident. This allows a top down, bottom up method of managing an incident.

Because the goals and objective are written and are shared, all levels of the organizational command chart have clarity in what these goals and objectives are. They share common goals, have definitive objectives, and everyone involved in the response is committed to the same results: Ending the incident, or in mitigating the effects of the incident. Because of the bottom up and top down incident management style, the incident response is also in line with the organizational objectives set by the local government.

It should be noted that it is extremely important to recognize that in a major or extremely chaotic incident response, incident goals and objectives should be written down and disseminated to prevent confusion and/or a misunderstanding of what should be done. Failure to write these goals and objectives, disseminate them to all personnel, and to brief them on what those written orders entail, could lead to even more chaos and confusion.

14.4 Writing Goals and Objectives for the Incident Action Plan

As is the case with most incident management tasks, writing the goals and objectives for an incident when utilizing Management by Objectives becomes less difficult as you get more practice. To clearly write your objectives, you first have to write the goals for the incident. In writing a goal, it should be realized that a goal in the ICS context is a general statement of purpose that specifies what you want to accomplish. This general statement specifies a location, a population, or portion of the incident, and it provides direction of what you want to accomplish.

Some guiding factors that can assist you in writing a goal are "What do we want to happen?" and "What are our priorities?" In the process of writing goals, it is important to use action verbs such as stop, eliminate, contain, ensure, and other similar action verbs. The incident goals should be written prior to writing the objective itself. As the goals are written, there should be an emphasis on the results of finishing that specific goal, and how it fits into the larger picture of stabilizing the incident. Some examples of goals might be as follows:

- Contain the wildland fire on the leading edge.
- Contain rioters.
- Evacuate the courthouse.
- Contain the HazMat spill.
- Establish Communication with the media.
- Stop civilians from entering the perimeter.
- Search Main Street for survivors and fatalities.

After establishing and writing your goal (or goals), you should then write your objective (or objectives). Often objectives will be similar or the same as your goals, but this does not always hold true. Resources can also affect how objectives are written. Depending on the number of personnel involved, the goal may be split up into multiple objectives that is being assumed by multiple different crews.

As an example, if an objective is to set up roadblocks in an effort to catch an escaped prisoner, there may be multiple roadblocks set up at various locations while tracking dogs and strike teams of officers search the contained area. By the same token, searching for trapped victims after a tornado may have the same goals and objectives, but be identified in different geographical locations.

When writing objectives, they are best written and utilized using the SMART objective method. The acronym SMART refers to goals that are

- **S**pecific,
- **M**easurable,

- Achievable,
- Relevant, and
- Time-based/Time-framed

In filling out the *Specific* portion of the SMART Objectives, the person writing them should provide the goals for the objective. These goals should be detailed and include the desired achievement that should be reached. In writing the specific section, the writer will want to write goals that can be given to subordinates, and they will have no doubts about what they are supposed to do. Examples of writing a goal might include the following:

- Find victims in rubble.
- Contain the wildland fire.
- Contain rioters.
- Evacuate the courthouse.
- Contain the HazMat spill.
- Establish Communication.
- Search for fatalities.
- Transport the injured.

In the ***Measurable*** portion of SMART Objectives, the individual or group will describe how to identify that the specific goal has been met. While rare, in some large-scale or ongoing incidents, there may be no way to complete a task. In those instances, measuring the success may need to measure success by not allowing the incident to worsen. Often, measuring the success of an operation or a tactic can be accomplished by identifying that a specific task was completed. The previous examples in the specific section did not provide a location for completion. Some examples of measuring the task in the objective writing stage might include the following:

- The wildland fire will be contained on the leading edge.
- The rioters will be contained rioters to the six-block area.
- The courthouse will be evacuated.
- The HazMat spill will be contained.
- The Public Information Officer (PIO) will establish communication with the media.
- Main Street will be checked for survivors and fatalities.

If you will look at the difference between the specific list and the measurable list, you can see that each specific action now has a way of measuring success. By combining the specific and measurable portions, we see a plan of action that can lead to success. Whether that success is keeping a fire in a specific zone, or rioters in a specific area, or any other task, it is easy to see that we can measure the success (or failure) based on whether the task was completed.

In the ***Actionable*** portion of SMART objectives, the objective is evaluated to ensure that there are sufficient resources in place to achieve the goal. To set up an objective without available resources to achieve that objective is futile exercise. In essence, it would be like having a wish list to buy a new car or truck today, knowing full well that you could not even come up with the down payment for that vehicle for several years. Some examples of questions to ask when writing actionable objectives might include the following:

- Is there enough manpower and equipment to contain the wildland fire to the leading edge in the specified operational period?
- Are there enough personnel to contain the rioters to the six-block area during this operational period?
- Are there enough resources and personnel to evacuate the courthouse in the allotted time?

- Are there enough personnel and HazMat equipment on hand to contain the spill during this operational period?
- Is there a qualified Public Information Officer (PIO) who can establish communication with the media in the time given?
- Is there enough equipment and personnel on scene to check Main Street for survivors and fatalities during the operational period?

If the goal is not achievable, then it would be advisable to divide the goal into several smaller, more actionable, parts. By breaking the goal into smaller, more actionable, parts, it will in most instances create a more achievable goal that can be reached with the resources that are available at that time. The Table 14.1 will provide some examples of splitting goals and potential objectives to make them more achievable.

As can be seen by the chart, there are many ways to change a goal to make a more manageable objective. Making a smaller goal is only limited by the imagination of those writing them. If splitting a goal is a necessity based on a lack of resources, it is extremely important to remember the previously mentioned acronym of LIPE. In doing so, you will focus on what will do the most good for the most people.

In the **Relevant** portion of SMART Objectives, the objective is evaluated to ensure that it falls in line with the overall goal of ending (or mitigating) the incident. If the objective you wrote does not help with the overall goal, then it is a bad objective that does not meet the needs of the incident, and it should be reconsidered. Some examples of asking if it is a relevant objective include the following:

- Does containing the leading edge of the wildland fire contribute to the overall goal of extinguishing the larger fire?
- By containing the rioters to a six-block area, does it contribute to the overall goal of preventing citywide rioting and destruction of property?
- By evacuating the courthouse, does it contribute to the overall goal of locating and disarming a potential explosive device?
- By containing the HazMat spill, does it contribute to the overall goal of stopping the spill from causing more environmental damage before the cleanup team arrives on scene?
- By establishing communication with the media, will it contribute to the overall goal of evacuating the city before the hurricane arrives?
- By checking Main Street, will it contribute to the overall goal of finding all survivors of the tornado?

Table 14.1 Dividing objectives so they can be achieved with available resources.

Unachievable	Alternative Goal No. 1	Alternative Goal No. 2
Do we have enough (and the right) people to write a press release in the allotted time?	Have an official provide a verbal press release. **(Smaller goal)**	Remove the goal of a written press release until resources are available. **(Delayed goal)**
Are there sufficient personnel on scene to evacuate all four floors of the courthouse in the allotted time?	Evacuate the first two floors of the courthouse. **(Equally split goal)**	Evacuate the top two floors of the courthouse. **(Equally split goal)**
Do we have enough people to search Main Street in an operational period?	Search the 2800–3400 block of Main Street **(Split goals by block numbers)**	Search Moss Road to Hibbard Road on Main Street **(Split goals by geography)**

If the objective does not contribute to the overall goal, then that objective should be rewritten so that it does help contribute to the overarching goal. If there is no way to make an objective relevant, then you should consider removing that objective from the response altogether.

In the ***Time-Based*** or ***Timeframe*** portion of SMART Objectives, the writer will want to provide a time when the task should be accomplished. It is extremely important to make sure that the timeframe provided is achievable. Failing to put a timeline on completing a goal is one of the most common mistakes in the planning process. When holding planning meetings, the Planning Section Chief (PSC) or the Incident Commander (IC) should solicit opinions as to how long the task should take to ensure that the timeframe is an achievable benchmark.

If this incident is a major incident, the agency will likely want to plan for an operational period of not more than 12 hours. If it is a public relations incident, the writer may want to set the timeframe on 24-hour timeframes. In some instances, such as establishing contact with the media, it may be best to split an operational period into two halves and have the first objective of establishing contact with the media in the first six hours of the operational period and then separately have an objective of providing a detailed press release during the next six hours of an operational period. An operational period can be divided in any manner, providing it is beneficial to the overall incident and does not last more than 12 hours.

14.4.1 SMART Objective Worksheets

When creating SMART Objectives for Management by Objectives, it might be helpful to locate a worksheet online. There are a multitude of various kinds of SMART Objectives Worksheets that can help to ensure that your goals and objectives meet the needs of the incident. There is not a specific type that is more effective than another, providing they cover the SMART Objectives by the letter of the acronym.

Providing a SMART Objective worksheet in this book was unfortunately not possible. When filling out a SMART Objectives Worksheet, you should not use the term "immediately" or "as soon as possible." These are unacceptable timeframes that do not set goals that everyone should work toward. The timeframe box of a worksheet should be based on minutes or hours. It is very important that there should be a set time that creates a definitive end time. Timeframes should never be open ended and left to the interpretation of the boots on the ground.

14.5 Management by Objective for Never-Ending Incidents[1]

This book would be remiss if it did not mention that some incidents will never end or that will last more than five years. While these incidents are rare, and they are not specifically addressed in the NIMS method or ICS companion method of incident management, they do occasionally occur. Emergency response agencies and individuals who may respond to these types of incidents need to be made aware of actions that should be undertaken to control such incidents to the best of their ability.

It is important to note that never-ending incidents can sometimes make an agency feel defeated and deflated. If a never-ending incident should occur, the agency should do their best to utilize Management by Objectives to mitigate the effects of the incident until they are relieved or released by another agency. This may include evacuations, refusing admittance to the area, managing the media, or any other actions may be required to ensure that life safety is preserved and that the incident is stabilized to the best of the agency's ability is performed.

---

1 This heading is for advanced IMS students.

It is extremely important not only to use written SMART Objectives, and to respond under the NIMS method and the ICS companion method, but to also document every facet of the emergency response. In some instances, those who are the responsible party for the never-ending incident (e.g. chemical companies, mining companies) may want to place the blame, or a portion of the blame, on response agencies. Thorough documentation of all actions taken can go a long way in protecting the agencies good name. It can also help save the agency's insurance agency from spending a large sum of money.

In other instances, there will be victims of the never-ending incident (or even a minor incident) who believe litigation is a way to provide relief for their suffering and/or loss. Often when litigation is initiated, the attorneys will name nearly every person and agency involved with the incident. They use this strategy in an effort to improve their clients potential financial gains. By creating thorough documentation throughout the entirety of the incident, the agency can help exonerate their agency, other agencies, and the local government, or in some instances even reduce their liability. By the same token, a lack of documentation can lead to an insufficient defense and potentially being found negligent. Unfortunately, we live in a litigious world, and even though public safety personnel risk their lives to save others, there are some people who will still file lawsuits, even if the agency or the personnel did everything correctly.

In nearly every instance of a never-ending incident, a state or federal agency will at some point take total control of the incident. The agency that will take total control will usually be based on the regulatory agency who is responsible for enforcing laws in that industry. When transferring command over to a state or federal agency, it would be in the Authority Having Jurisdictions (AHJs) and the Incident Commanders (ICs) best interest to obtain documentation in writing that the agency is assuming command.

The documentation transferring control of the incident should be done to legally protect the Authority Having Jurisdiction (AHJ) and, in some cases, the Incident Commander (IC). This can be done using a written transfer of command that identifies the incident and the current conditions. Make multiple copies of the written transfer of command and make sure that the agency and the Incident Commander (IC) obtain a copy with an original of the signature rather than a photocopy of the transfer. This once again is a measure to protect the agency and the Incident Commander (IC) from litigation or in extremely rare instances, the potential for being charged with abandonment of an incident.

While never-ending incidents are rare, they have occurred. Some of the more notable incident affected Straitsville Ohio, Times Beach Missouri, Picher Oklahoma, Niagara Falls New York, Centralia Pennsylvania, and other cities and towns in the United States. Outside of the United States, perhaps the most notorious instances of ongoing incidents would be the Chernobyl Nuclear Power Plant incident (in the Ukraine), and the Fukushima Nuclear Power Plant (in Japan) (Table 14.2).

14.6 The Importance of SMART Objectives in the Planning Process

The creation of SMART objectives is undertaken as part of the planning process. If we look at the Planning P (Figure 14.2), we can see that identifying the objectives (in the lower left-hand corner of the continuous circle) is the foundation for which everything else in planning will be based on. When the objectives are created, they typically address the next Operational Period; however, this is not always the case. Depending on the incident, the same objectives can be used for multiple Operational Periods.

The Incident Commander (IC) or Unified Command (UC) establishes the incident objectives. In larger and complex incidents, the objectives may be so large or complex, that it may

Table 14.2 Never ending incidents in the United States.

Location	Circumstances	Year it occurred
Straitsville, OH	Mine fire underground with no ability to extinguish it	1884
Times Beach, MO	Widespread dioxin contamination while oiling the roads	1972[a]
Picher, OK	Unrestricted lead and zinc mining caused water contamination as ongoing collapsing of the ground	1983
Love Canal Subdivision: Niagara Falls, NY	A toxic dump site was discovered underneath a subdivision. The developer of the subdivision was aware of the site but backfilled it and built houses on top of the toxic site	
Centralia, PA	A coal seam fire that continues to burn. Smoke is seeping from the ground and polluting the environment (as well as collapsing ground) to this day	1962

a) While the incident occurred in 1972, the identification of the toxin, making it a Superfund site, and the subsequent evacuation did not occur until 1982. This site is now a wildlife refuge.

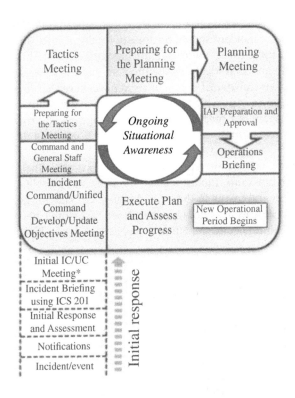

Figure 14.2 The Planning P.

take multiple operation periods just to complete one objective. When it takes more than one operational period to accomplish one of the incident objectives, there should be set goals provided to help hold crews accountable.

Imagine if you will, a mine collapse with miners trapped 400 feet underground. Initial intelligence and ongoing situational awareness have determined that the mine entrance was unstable and is impassable. After weighing all the options, the Incident Commander (IC) determines that the best plan of action is to drill a rescue hole with specialized large drilling equipment so that rescue pods can be sent down to evacuate the miners. Because of how large the hole will need to be, the specialized drilling machinery can only drill 100 feet per operational period. The SMART objective would be the same over a period of four days, including the requirement of reaching 100 feet per day. This is not to say that other objectives could not be written and tasked to the appropriate personnel at the same time. While these instances are not common, we need to realize that they can occur, and we need to at least have a cursory understanding of how to proceed.

The recurring planning process is intended to take the overall incident objectives and break them down into smaller tactical assignments for an operational period. While objectives are reviewed at each operational period, the initial establishment of overall incident objectives will help establish the course of the incident, which in turn will direct each operational period, based on the accomplishment of the previous operational period objectives.

The setting of the Incident Objectives is done through the Objectives Meeting. The Objectives Meeting is typically led by the Incident Commander (IC), or in the case of a Unified Command (UC), the person that was chosen as the primary leader. It is not uncommon for the Operations Section Chief (OSC) and the Planning Section Chief (PSC) to attend this meeting as well. This meeting is held to evaluate the current situation and to determine the following:

- A cooperative perception of the incident's issues.
- The key necessities to facilitate response and recovery.
- A general consensus of incident priorities and objectives (for the next Operational Period.)
- Identifying constraints and limitations that can affect success of incident objectives.
- Providing operational guidance that achieves incident objectives and that identifies and mitigates any problems.

From this meeting, a documented list of incident objectives should be formed. It is important to note that these objectives should be approached in a systematic way that addresses the full scope of the incident. Failure to evaluate the entire scope of the incident could lead to failure, or it could require changing the priorities (or possibly even objectives) in the middle of an operational period. Should this happen, it will likely increase the chaos, confusion, and uncertainty, if this mistake is made.

14.6.1 Completion of SMART Objectives

Upon completion of creating the SMART Objectives by the Incident Commander (IC) or the Unified Command (UC), a combined Command and General Staff (C&GS) Meeting will be held. This is typically done on a set time schedule. On major incidents where state and federal resources are brought in to assist, this Command and General Staff (C&GS) meeting can be extremely important. In this meeting, the Incident Management Team (IMT) or the Authority Having Jurisdiction (AHJ) can be informed about the available state and federal resources. This can include resources already in place and resources that can be ordered.

As has been identified numerous times in this chapter, the creation of SMART Objectives is the foundation of the overall response. Once it is determined what needs to be accomplished that success can be measured, that the resources needed are available, that the objectives will play a role in ending the incident, and there is a timeframe to complete the objective, then the entire response will be guided by these objectives.

While looking at it one piece at a time by using individual objectives, those planning and leading the response and/or recovery have a less likelihood of being distracted or in becoming mesmerized by tunnel vision. By focusing on the pipeline feeding the fire rather than the flames, they can systematically and metaphorically take another bite the elephant that they need to eat. By using the LIPE acronym and evacuating the perimeter before focusing on disarming the Improvised Explosive Device (IED), the law enforcement agency managing the incident is taking their figurative bite from the elephant. While there will likely be many more bites to take, they ensure that they take the most important bite first.

In the next chapter, we will look at what is built from SMART objectives, and how the building of the planning process is done. While it may seem like a difficult process, or even potentially a confusing or complicated process, it is not. In fact, once you understand the concept, it seems to be common sense. Even so, just like everything else in the NIMS method and the ICS companion method of incident management, the more you do it, the more proficient you will become at it.

Chapter 14 Quiz

1. **True or False**: Developing objectives is essentially stating "I want to do this" or "I want to do that."

2. In the ICS method of incident management, the development of objectives is an _____ _____.

3. **True or False**: A key factor in incident management is that first responders shouldn't let the incident manage them, but rather they should manage the incident.

4. The acronym LIPE stands for:

5. **True or False**: Economic impacts from an incident rarely occurs.

6. _____ ___ _____ allows the individual commanding an incident to break a large crisis into much smaller, manageable pieces and create effective response tactics.

7. **True or False**: An Incident Commander (IC) should use the size-up process and situational awareness practices only in large-scale or complex emergency response situations.

8. A change in the conditions and altered or changed tactical objectives is often referred to as a _____ _____.

9. A method to take a complex incident and to break it down into smaller, more manageable tasks is called:
    (a) Manageable Decision Making (MDK)
    (b) Breakdown Incident Analysis (BIA)
    (c) Management by Objectives (MBO)
    (d) Eating an Elephant (EAE)

10 _____ _____ occurs when those in charge of an incident will focus on one goal, or a set of goals, and they fail to look at the big picture or the overall situation.

11 **True or False**: Handling incidents by mentally making goals and objectives is widely an acceptable way to handle an incident.

12 The Incident Commander (IC) and first responders must make _____ _____ the highest priority.

13 **True or False**: If there is a life in danger, the first objective should be to initiate life safety measures, even in instances where there is a high probability that the first responder will lose their life in the process.

14 **True or False**: When writing objectives, they are best written and utilized using the SMART objective method.

15 When writing objectives, they are best written and utilized using the SMART objective method. The acronym SMART refers to goals that are:

16 Using the _____ portion of the SMART Objectives, the person writing them should provide the goals for the objective. These goals should be detailed and include the desired achievement that should be reached.

17 In the _____ portion of SMART Objectives, the individual or group will describe how to identify that the specific goal has been met.

18 In the portion of SMART objectives, the objective is evaluated to ensure that there are sufficient resources in place to achieve the goal.

19 In the _____ portion of SMART Objectives, the objective is evaluated to ensure that it falls in line with the overall goal of ending (or mitigating) the incident.

20 In the _____ portion of SMART Objectives, the writer will want to provide a time when the task should be accomplished. Failing to put a timeline on completing a goal is one of the most common mistakes in the planning process.

Self-Study

Miller, L. (2012). All hazard incidents: Setting objectives and strategies. *Fire Engineering*. Retrieved from https://www.fireengineering.com/articles/print/volume-165/issue-2/features/all-hazard-incidents-setting-objectives-and-strategies.html.

National Academies (n.d.). SMART objectives. Retrieved from http://www.nationalacademies.org/hmd/About-IOM/Making-a-Difference/Community-Outreach/Smart-Bites-Toolkit/~/media/17F1CD0E451449538025EBFE5B1441D3.pdf.

United States Environmental Protection Agency (n.d.). Superfund. Retrieved from https://www.epa.gov/superfund.

15

The Planning P-In Depth

Productivity is never an accident. It is always the result of a commitment to excellence, intelligent planning, and focused effort.

Paul J. Meyer

The Planning P refers to a diagram of the Planning Process that is shaped in the form of the alphabetical letter (Figure 14.2). The Planning P diagram should be the guide for the planning process when using the Incident Command System (ICS) method of incident management. It can be very helpful to keep a copy of the Planning P diagram (Figure 14.2) in the Incident Command Center (ICC) and meeting rooms for a reference point so that none of the steps involved in planning the response to an incident are missed. Make no mistake, when the response is complete it does not mean that the command structure is automatically shut down.

In some instances, command will be turned over to the Authority Having Jurisdiction (AHJ), so they can continue working on completely closing the incident, but this is not always the case. In other instances, especially on extremely damaging incidents, the focus will go from response to recovery. This will usually entail a different type of Incident Management Team (IMT), one that either has experience or that is specially trained for recovery efforts, and they too will use the Planning P as a guide for recovery efforts.

In looking at the Planning P, the initial response begins at the bottom leg (or base) of the P. The actions taken during this time period will only be one-time actions during an incident. The initial response covers

- The beginning of the incident
- Notifications
- Initial Response and Assessment
- The Incident Briefing (utilizing ICS Form 201)
- The initial Incident Command or Unified Command (UC) Meeting (See Figure 15.1).

15.1 The Beginning of the Incident and Notifications

After an incident has happened, the notification that an incident has occurred should begin the planning process. As the agency is dispatched and is responding to the incident, the initial Incident Commander (IC) should begin forming strategies and identifying what factors should be considered as part of a size-up. The initial size-up will be based on the situational awareness that was provided in the initial dispatch. In most instances, the initial stages of planning that occur while enroute will reduce the amount of time it will take to make decisions when on

Figure 15.1 The beginning of the Planning P.

*This should occur any time there is a transfer of command

scene. This holds especially true when it comes to ordering additional resources and the types of tactics that may be used in the initial response (based on the available resources).

15.2 Initial Response and Assessment

As the initial Incident Commander arrives on scene, they should start with a thorough size-up. This size-up should be continual throughout the entire incident, regardless of whether the incident involves a single resource or multiple agencies. This size-up provides the information needed to make informed incident management decisions.

The initial response will typically be a single agency response; however, in some instances, the incident may be dispatched as a multiagency response from the onset of notification. Of course, this will depend on the information provided to the dispatcher. This request of additional resources will usually be dependent on the size and scope of the incident when first reported, as well as the availability of personnel to address the perceived threat of the incident. In a single-agency response, the highest-ranking officer will usually take command of the situation and direct crews when they arrive on-scene. If more resources are needed, they will request them.

In a multiagency response, the first arriving unit will assume command until a more qualified individual arrives. It is important to note that a multiagency response can turn into a fiasco if Mutual Aid Agreements (MAAs) are not already in place prior to an incident. This Mutual Aid Agreement (MAA) should be based on common courtesy and common sense, including identifying who is in charge of an incident. In the Mutual Aid Agreement (MAA) it should spell out how the command structure will be arranged, and in some instances, it may even identify who should fill key positions. Predetermining the command structure will include who should be in command. Some examples include the first agency arriving on scene, the agency of the primary response area, a Unified Command (UC) from the onset, or some other criteria. No matter how the command structure is set up, the first arriving units and their Incident Commander (IC) will initially need to take command until other more qualified personnel arrive on scene and size-up (or assess) the incident.

In the initial size-up (assessment) of the incident, the initial Incident Commander (IC) may determine that additional resources or an expansion of ICS is needed from the onset. This is typically done by undergoing a complexity analysis, which will help determine the type of response that is needed.

15.2.1 Complexity Analysis-Incident Typing

An incident may have many factors that create chaos, confusion, and uncertainty. Anyone that has been an Incident Commander (IC), or the acting Incident Commander (IC), can tell you that there are times when they arrive on scene, or conditions change, and they have that "Oh crap" moment. In the firefighting community, this is also referred to as the "pucker factor" and is based on how much you pucker your butt (so you do not fill your pants). While these jokes are regularly made among fire crews, these types of situations are serious business, and they can seriously derail an effective response if caution is not taken.

In order to overcome emotions and fears, the process of a Complexity Analysis was created. A Complexity Analysis is a tool to remove emotions and fears from the decision-making process, and it allows the Incident Commander (IC) to make decisions based on the facts. It is undertaken to identify if the incident can be handled by the resources on scene, or if additional resources are needed. It also allows the initial Incident Commander (IC) to identify if there is a need to enact certain functions within the ICS method of incident management. It also allows the Incident Command to identify the level of response needed, also known as "typing" an incident.

15.2.2 Incident Complexity

Analyzing an incident's complexity can help to identify the resources needed to appropriately handle the incident. A Complexity Analysis can also help to determine if the existing incident management structure is suitable for the incident. Key factors to consider in a complexity analysis are as follows:

- Life safety of the community and personnel.
- The impact of the incident to life, property, the economy, and the environment.
- The potential of a hazardous materials release or an already identified release.
- The weather and other environmental factors.
- The potential for cascading events.
- The potential that the incident may be a crime scene.
- The potential that the incident may be a terrorist event.
- The incident has (or may have) potential political sensitivity.
- The incident may have external influences (e.g. ATF or FBI investigation into a local incident.)
- A special interest taken by the media or a need for increased media relations.
- The incident crosses jurisdictional boundaries.
- The availability of resources to meet the needs of the incident.

Identifying the complexity of the incident is based on a numerical incident typing (see Incident Management Teams in the following). This numerical typing system utilizes a Type 5 incident as the lowest level of incident based on the resources needed, and a Type 1 incident as the highest level of resources needed. Simply put, a Complexity Analysis undertaken by the initial Incident Commander (IC) will assist in identifying how many resources will be needed, and if needed, the type of Incident Management Team (IMT) will be most effective.

Guidelines for completing a complexity analysis and typing of an incident are as follows:

- Type 5 Incident – Typically requires only a local level incident response
  - The incident can be handled with one or two single resources and will utilize no more than six personnel.
  - Only the Incident Commander (IC) is activated and there is no need for Command and General Staff positions
  - An Incident Action Plan (IAP) is not required.
  - The incident can be limited to the first operational period, and frequently within a few hours after the initial resources arrive on scene.

- Type 4 Incident – Typically requires a city-wide or county-wide response
  - Command Staff and General Staff functions are started only if needed.
  - Several additional resources beyond local resources are required to mitigate the incident. This might also include a task force or strike team.
  - The incident is usually limited to one operational period in the response phase, or in bringing the incident under control.
  - The Agency Administrator may provide briefings and confirm the Complexity Analysis and a Delegation of Authority (DOA) is updated.
  - No written Incident Action Plan (IAP) is needed, but a documented operational briefing will be needed for all incoming resources.
  - The role of the agency administrator includes operational plans with objectives and priorities
  - The response has the potential to become a Type 3 incident, requiring a Type 3 Incident Management Team
- Type 3 Incident – Typically requires a large metropolitan area or state assisted response
  - When the needs of the incident exceed abilities of the Incident Commander (IC) to handle it on their own, the appropriate ICS positions should be activated to meet the complexity of the incident.
  - Some, or all, of the Command and General Staff positions have been activated, as well as a Division or Group supervisor and/or a Unit Leader position(s).
  - The incident will take an extended amount of time for the response phase or until it is contained.
  - The incident was (or needs to be) managed with a substantial amount of resources, or the incident is still expanding.
  - The incident will, in most likelihood, extend into multiple operational periods.
  - A written Incident Action Plan (IAP) is required for each operational period.
- Type 2 Incident – Typically involves a state and national level response
  - The incident extends beyond the local responder's capabilities and is expected to go into multiple operational periods.
  - The incident will likely require the response of resources from outside of the area, which may include regional, state, and/or national resources, to effectively manage the Command, Operations, and General staffing.
  - Most, if not all, Command and General staff positions are filled.
  - A written Incident Action Plan (IAP) is required for each operational period.
  - Most of the functional units are needed and staffed.
  - As a guideline, not more than 200 operational personnel are utilized during an operational period, and the total amount of incident personnel does not exceed 500.
  - The Agency Administrator is responsible for the incident Complexity Analysis, Agency Administration Briefings, and the written Delegation of Authority.
- Type 1 Incident – A large national response
  - This type of incident is the most complex, requiring national resources for safe and effective management and operations.
  - All Command and General Staff positions are filled.
  - Operations personnel often exceed 500 per operational period, and total personnel will usually exceed 1000.
  - Branches need to be established.
  - A written Incident Action Plan (IAP) is required for each operational period.
  - The Agency Administrator will have briefings and ensure that the Complexity Analysis and Delegation of Authority are updated.

- Use of resource advisors at the Incident Base is recommended.
- There is a high impact on the local jurisdiction, requiring additional staff for office administrative and support functions.

While these guidelines help to identify the need for expanding an incident, the ultimate decision is made by the initial Incident Commander (IC) or the person who replaced them in this role. Many who have filled the Incident Commanders (IC's) role in numerous incidents will tell you that it is far better to err on the side of caution and call for additional resources (early) than it is to try to catch up with the incident because you failed to make the call for assistance early. If the Incident Commander (IC) delays this decision, then the incident can have the upper hand for a long period of time. It is far better to have too many resources on hand and not utilize all of them than to not have the resources which will cause the situation to worsen.

15.2.3 Typing the Incident – Incident Management Team (IMT) Typing

A somewhat recent trend (era 2000) in managing complex incidents has been to bring in highly trained Incident Management Teams (IMTs). An Incident Management Team (IMT) does not take over the incident or usurp the local levels authority. Instead, they allow the Authority Having Jurisdiction (AHJ) to make decisions while the Incident Management Team (IMT) strengthens the response by providing organizational assistance to the local agency. These teams are often brought in during Type-3, Type-2, and Type-1 incidents to ensure that the incident is managed in highly efficient way. These Incident Management Teams (IMTs) are often called All Hazards Incident Management Teams (AHIMT) because of their ability to manage any type of hazard (e.g. hurricanes, earthquakes, fires).

Incident Management Teams (IMTs) are a unique group of individuals that have advanced training in managing emergency incidents. In most instances, Type-4 and Type-5 incidents can be handled by local-level responders. Type-3 Incident Management Teams (IMTs) are typically used to support local incidents utilizing state-level resources. Type-1 and Type-2 teams are typically state- or national-level teams, and they are typically utilized in the most complex types of incident response.

Incident Management Teams (IMTs) perform a vital role in responding to, and the managing of local, regional, and national incidents. These teams are a comprehensive resource that can enhance ongoing operations by providing the ICS method with strong infrastructure support. They can also transition to an incident management function that includes all components of the Command and General Staff position. They are trained and certified as Command and General Staff members so that they can be the needed support personnel for an incident. These teams are given statutory authority by the Agency Administrator or the Authority Having Jurisdiction (AHJ) through a Delegation of Authority, and they operate under formal written response requirements and responsibilities. Each member of the team has prearranged roles and responsibilities for which they will perform.

It would be easy to surmise that these Incident Management Teams (IMT) would come with five or six individuals but that would be a wrong assumption. A Type-3 (the lowest level) brings a cadre of individuals to support the response. A team of 10–20 trained personnel are deployed as a Type-3 team. The team will usually represent multiple disciplines and will manage complex incidents that require a sizeable number of local, state, and/or tribal resources. Type-3 teams are usually called to manage incidents that extend into multiple operational periods and that require a written Incident Action Plan (IAP). In some instances, such as expanding incidents (e.g. wildland fires, ongoing floods, aftershocks from earthquakes), they may initially manage an incident which later morphs into a more complex incident, or it transitions into a national-level incident, requiring a more advanced Incident Management Team (IMT).

Type-2 Incident Management Teams are self-contained, all-hazard (or wildland specific) teams that are recognized at both the national and the state level. A Type-2 Incident Management Team (IMT) is deployed with 20–35 team members to manage incidents that are even more complex. This type of incident usually requires more than 200 personnel during an operational period and does not exceed a total of 500 personnel working on the entire incident. These Type-2 Incident Management Teams (IMTs) are requested by the affected state and coordinated by the Geographic Area Coordination Center (GACC), or the National Interagency Fire Center (NIFC).

Type-1 Incident Management Teams (IMTs) are also self-contained. A Type-1 Incident Management Team (IMT) is deployed as a team of 35–50 to manage incidents that are more complex incidents. A Type-1 team is required when 500 or more personnel are involved in the response during an operational period or 1000 or more personnel are working on the entire incident. This will usually involve incidents that require a large number of local, regional, state, and national resources. These Type-1 Incident Management Teams (IMTs) are typically requested by the affected state and coordinated by the Geographic Area Coordination Center (GACC) or the National Interagency Fire Center (NIFC) the same as a Type-2 incident. Type 1 Incident Management Team members can effectively manage a Type 1, Type 2, or a Type 3 incident.

FEMA also provides similar teams for Type-1 and Type-2; however, instead of being called All Hazards Incident Management Teams (AHIMT), they are called Incident Management Assistance Teams (IMATs). These teams will have the unique ability beyond being able to manage an incident (FEMA, 2017b); they can also work with already deployed Incident Management Teams (IMTs) and advise them of federal resources available (FEMA, 2017a, 2017b).

National standards have been implemented for All Hazards Incident Management Teams (AHIMT). Just to be on a Type-3 Incident Management Team (IMT), the individual must complete a multitude of ICS method training, and they must successfully pass the position-specific training that they will serve as. They must also be mentored on an incident by another team member who has experience in being in the specific functional position that they wish to fill. Type-1 and Type-2 have even more specialized training, and they have worked hands-on for a specified amount of time at the next lower level. In essence, they must graduate up through the different types of incidents in order to fill that position.

15.3 Incident Briefing – Preparing for a Transfer of Command

After completing a complexity analysis and typing the incident, the Incident Commander (IC) may come to the realization that they need help to manage the incident. As an incident transitions from being a relatively simple response into a more complex response, including covering multiple operational periods, then a transfer of command is usually needed. A transfer of command is the process of moving the responsibility for incident command from one Incident Commander (IC) to a different Incident Commander (IC). Examples of when a transfer of command is needed include the transition from a single Incident Commander (IC) into a Unified Command (UC) or if an Incident Management Team (IMT) has been brought in. A transfer of command might also be necessary if the current Incident Commander (IC) needs time off or becomes ill to name a few reasons. The actions that need to be taken are important to facilitate a smooth transition in a transfer of command. As an incident expands, there will likely be at least one transfer of command, and potentially more.

If the structure of the incident needs to expand, the initial Incident Commander (IC) will usually fill out the ICS Form 201 (Initial Briefing). This is done so they can provide the incoming

Incident Commander (IC) and the Command Staff with the information they need to begin to manage the incident. The information provided will include the incident situation, safety concerns, and the resources in place at that time. The use of ICS Form-201 creates a permanent record of the initial response to the incident as well as describes the most recent conditions of the incident. The ICS Form-201 can also be used to transfer command from being led by one incident management team to the next (higher or lower) team. This is because if the form is properly filled out, it will provide the information that the incoming Incident Commander (IC), or a Unified Command (UC), will need to begin managing the incident as it relates to the expansion or contraction of the Incident Command structure.

15.3.1 Filling Out the ICS 201[1]

While it will not be described how to fill out every ICS Form, the ICS Form-201 (Incident Briefing) will be well covered. Whenever filling out ICS Forms, it is important to note that these forms have instructions on how to fill out that specific form. Remember, these forms are legal documents. They will be part of a permanent record of the incident, and should litigation or an investigation arise, the forms can be used as evidence. They can be used to protect the Incident Commander (IC), or anyone in the Command and/or General Staff.

Filling out the ICS Form 201 (FEMA, n.d.c) (Incident Briefing) is simple, but a sufficient amount of detail must be included to ensure that the incoming Incident Commander (IC) and their staff can hit the ground running. Box number one of the ICS Form-201 should be filled out with the given name of the incident. This is often given as the location of the incident such as *Main Street Explosion, City of San Fernando earthquake, Gatlinburg Wildfire* or something similar. This is not to say that this format must be followed, it is just a common practice that most Incident Commanders (ICs) will typically follow.

On a side note, it is important to realize that in every ICS Form, no matter what number it is, the information required in box number one will be the same for the duration of the incident. This information will also be provided on any additional pages of the ICS Form 201 (FEMA, n.d.c). This is done so that the incident can be quickly and easily identified.

Box number two of ICS Form-201 will be filled in with the incident number. The incident number is a unique number that the local agency designates through a numbering system. While this ICS Form and others ask for the incident number, not all ICS Forms require the incident number, and they may require different information in box number two. This information will also be provided on all subsequent pages of ICS Form 201 (FEMA, n.d.c).

Box number three is the date and time of the initial incident. It is important to note that this box requires the initial dispatch of the incident, not the current time. By providing the time of the initial call, it allows the incoming Incident Commander (IC) to see the progression made on scene from the time of dispatch, or it can help them understand how slowly or quickly the incident is deteriorating or improving. The time entered should be based on the 24-hour clock, meaning that one in the afternoon would be 1300 hours, and six o'clock in the evening would be 1800 hours, and so on. All the information provided to this point is basic information and requires the initial Incident Commander (IC) to provide historical information about the incident. The time initiated will also be provided on additional pages of the ICS Form 201 (FEMA, n.d.c).

In box number four or the ICS Form 201 (FEMA, n.d.c), the Incident Commander ([IC] or the person they appoint) should draw a map or sketch of the incident. In this map or sketch, they should include the current status (e.g. the boundaries of the incident), resource assignments,

---

1 This heading is for advanced IMS students.

incident facilities, and other special information that may be helpful to the incoming Incident Commander (IC). In creating this map or sketch, the person should utilize commonly accepted ICS map symbology identified in Figure 9.3. Additionally, when creating the map, north should always be placed the top of the page unless specifically noted. If specific geospatial location reference points are needed regarding the incident's location, or perhaps an area outside the ICS organization at the incident, this additional information should be submitted on the ICS Form 209 (Incident Status Summary).

In box number five of ICS Form-201, a situation summary including any health and safety issues should be addressed. This box is critical for the transfer of command and ensuring that all information is transferred. In the initial part of filling out this box, the person providing the information for this form should provide all major details of the incident up to that point and time. The information entered should not talk about every action taken, but rather should describe the incident itself and any significant changes or concerns about the incident.

An example may be that a local oil field tank farm was struck by lightning, thereby causing an explosion in one tank. While fighting the tank fire, a nearby tank began to expand and looked as if it may BLEVE (explode). This information is part of the incident, and dually, it is part of recognizing health and safety issues. Not only should this box recognize all potential incident health and safety hazards but it should also provide the measures taken, or ongoing measures being taken to mitigate those hazards. Examples of mitigation measures can include (but are not limited to) the following:

- Removing the hazard
- Providing personal protective equipment
- Warning people of the hazard
- Securing the area
- Allowing limited access to the area through specific entry points
- Evacuation (ongoing or completed)
- Perimeter are or were set up

Box number six of the ICS Form-201, the name of the person preparing the form should be printed in the first part of the box, followed by their position within the ICS organization, then their signature and the time that it was completed. This is an important part of making this a legal document. Failure to provide this information could undermine the legality of the document and the transfer of command.

In box number seven of the ICS Form-201, the current objectives and any planned objectives should be identified. It is important to note that these objectives should be in line with overall incident stabilization and ending of incident. If the objective does not align with incident stabilization or ending of the incident, it should not be included. Additionally, any proposed objectives should be identified as such so that the incoming Incident Commander (IC) knows that the objective has not been implemented.

In box number eight of the ICS Form-201, the Incident Commander (IC) should enter the current tactics, strategies, and actions being utilized (including the time at which they were implemented) and any future planned actions. Listing these actions will require identifying what resources are in place and working at that time. It is important to list all actions taken so that the incoming Incident Commander (IC) does not repeat any actions unnecessarily. If an objective was achieved, a description of that achieved objective, and the time that it was achieved, should also be listed. In many instances, especially on larger incidents, additional pages will be needed to cover all the tactics, strategies, and actions. If additional pages are needed, blank sheets of paper can be used, or additional copies of page two may be inserted and used to document these actions. It is important to note that the numbering of pages should be adhered to, so if additional pages are added, the numbering of pages (in the lower left corner) should be changed to reflect the changes throughout the remainder of the document.

In box number nine, the person filling out ICS Form 201 (FEMA, n.d.c) should provide the structure of the current organization. It is not necessary to list every individual on scene, just those in key positions. Those positions include

- Incident Commander ([IC] or the Unified Command [UC])
- Liaison Officer (LOFR)
- Safety Officer (SOFR)
- Public Information Officer (PIO)
- Planning Section Chief (PSC)
- Operations Section Chief (OSC)
- Finance/Administration Section Chief (FSC)
- Logistics Section Chief (LSC)

You should enter the names of individuals assigned to each position on the organizational chart. In the case of a Unified Command (UC), not only should the name of the person be entered but also the agency that they represent. There is a space to the left of the Incident Commander (IC) box where this can be accomplished, and a line should be drawn to connect those names with the Incident Commander (IC) box of the organizational chart. It is also important to make a document all Command Staff, General Staff, their Assistants, and Agency Representatives (AREP's). The ICS Form-201 has sufficient room not only to modify and list those names but also the ability to chart them necessary by adding lines, boxes, and spaces needed to include them under the appropriate position. If a position on the organizational chart is not filled, it might be helpful to identify that it has not been filled by designating so in the associated box or boxes.

In box number 10, the person filling out the form will identify resources that have been assigned to the incident. The reason for this information revolves around the fact that sound planning to determine your resources, and your resource needs, is an essential issue at all stages of the incident. Resource management is usually most critical during the initial stages of an incident. Mistakes made early in the incident can compound and complicate the response, sometimes even causing cascading events. When filling out box number 10, you should know that much like page two of this form, additional pages may be added by utilizing one or more pages of blank paper, or additional pages that contain box number 10 (page four). It is important that pages are appropriately renumbered in order. By doing so, there will be less confusion regarding the numerical order of pages.

When filling out box number 10, you should first list the quantity and appropriate category, kind, or type of resource that has been ordered in the "Resource" column. As you move horizontally across box number 10, you should also provide a unique resource identifier for the resource in the "Resource Identifier" column. In the next horizontal box (the Date/Time Column), you will enter the date, and time (using a 24-hour clock, and date formatted as xx/xx/xxxx) that the resource, or resources, were ordered. Moving horizontally to the next box, you will enter the estimated time of arrival (in the ETA column) to the incident (still utilizing a 24-hour clock). If the resource has already arrived, you will put an "X" or a checkmark in the next (horizontal) box which is the "Arrived" column. In the last box, the "Notes" column, you will list any notes as they relate to that resource. Notes can include where this resource is intended to be used, that the resource arrived and where it was deployed (between the writing of the ICS Form 201 (FEMA, n.d.c) and the arrival of the incoming Incident Commander), and other critical information.

15.3.2 Potential Other Forms for Transfer of Command

In some instances, additional forms may need to be filled out. The determination of what other forms may need to be filled out should be evaluated on a case-by-case basis. Some forms that may help the incoming Incident Commander (IC) include ICS Form-203 (Organization Assignment List), ICS Form-205 (Incident Radio Communications Plan), the ICS Form-205A (Communications List), ICS Form-206 (Incident Medical Plan), ICS Form-208 (Safety Message/Plan), ICS Form-215A (Incident Safety Plan Analysis), and ICS Form-220 (Air Operations Summary). While these are the main forms that may be helpful, essentially any form that may provide more situational awareness to the incoming Incident Commander (IC) might also be helpful.

The use of these additional forms will usually be based on how complex the incident has become, and if there are issues, then that should be addressed more in-depth beyond the use of ICS Form-201. While the use of these additional forms is a judgement call, there may be instances where it makes sense to fill out these forms so that the incoming Incident Commander (IC) has more clarity of the situation.

15.4 Delegation of Authority (DOA)

Another key issue to cover prior to the transfer of command is the Delegation of Authority (DOA). While Delegation of Authority (DOA) has been briefly discussed in previous chapters, it is so important (to an effective response) that it will be discussed in-depth in this chapter. A Delegation of Authority (DOA) gives permission and creates an obligation to take actions on behalf of the Authority Having Jurisdiction (AHJ).

In most jurisdictions, the obligation for protecting citizens is the responsibility of the chief elected official. These elected officials have been given the authority to make decisions, commit resources, obligate funds, and command the resources (usually through being voted in) to protect the population. This may include making every effort possible to stopping the damage caused to life, property, the economic conditions, and the environment through creating incident stabilization.

The Delegation of Authority is a written statement that is provided to the Incident Commander (IC) or Unified Command (UC) by the Chief Elected Official, Agency Administrator, or the Authority Having Jurisdiction. In this written document, it gives authority to the incoming Incident Commander (IC) and their staff to manage the incident as if it were their own incident, within certain constraints (e.g. within financial constraints, Standard Operating Procedures [SOPs], Standard Operating Guidelines [SOGs]). This Delegation of Authority (DOA) assigns the task of managing the incident to the incoming Command and General Staff. This is important for several reasons.

The first, and probably most important reason is all incidents are local (unless on federal property). A Delegation of Authority (DOA) allows the incident to be managed by professionals who have managed larger incidents, while still allowing the local government to still be involved in the decision-making process and to manage those that are managing the incident.

Secondly, a Delegation of Authority allows the incoming Incident Management Team (IMT) to have the legal authority to act in the best interests of the local government. The general public often mistakenly thinks that FEMA or that a state agency comes into a major incident and will take over, but they are wrong. A Delegation of Authority (DOA) is essentially a statement that the local government is still in charge, but they hired (or allowed them to volunteer) the Incident Management Team (IMT) for their expertise. A Delegation of Authority (DOA) does not transfer authority to a different level of government; it allows the local government to use subject matter experts (SMEs) that will help bring the incident under control.

Imagine if you would, that a local government has a major flood, such as we saw when Hurricane (and later Tropical Depression) Harvey struck the Houston TX area. Now imagine that Houston was doing their level best to respond to this disaster when the federal government stepped in and said, "Get out of the way, we are taking over%" This would cause a huge uproar from the local and state level governments, mainly because the authority at the local level (and possibly the state level) was seized. A Delegation of Authority (DOA) is the mechanism in place for the local level to say, "I will allow you to work for us, because you're an expert at this. Even though you work for us, you will do so for the best interests of our local citizens and government, and you will ask questions and answer to the Chief Elected Official." This is the reason that a Delegation of Authority (DOA) is so important%

In the Delegation of Authority (DOA) document, the Chief Elected Official, Agency Administrator, or Authority Having Jurisdiction (AHJ) can touch on many different aspects of the response, and how they want it managed. They can specify detailed objectives and priorities, and they can document their expectations and any constraints. If they have special considerations (such as protecting priority areas), then they can also address those issues in the Delegation of Authority (DOA). They can also provide guidelines of where that authority stops, or what actions need approval prior to acting.

15.4.1 Delegation of Authority Briefing

While not a requirement, it is strongly urged that there should be a Delegation of Authority (DOA) Briefing between the incoming Incident Management Team (IMT) and the Agency Administrator. This should not occur until after all the Incident Management Team (IMT) members have arrived, but it should take place prior to the Incident Management Team (IMT) taking command of the incident.

This briefing should be led by the Chief Elected Official or their representative (such as a city or county manager). This person should be identified as the Agency Administrator. It is important to note that the Agency Administrator can also be identified as the Chief Elected Official; however, the Agency Administrator name is typically preferred when utilizing the ICS method of incident management or in utilizing the term of the Authority having Jurisdiction (AHJ). This provides the Incident Management Team (IMT) a standardized term for the person that has administrative responsibilities that ensure the safety of the local community.

The Delegation of Authority (DOA) Briefing will be attended by the Agency Administrator and by all the members of the Command and General Staff for the incoming Incident Management Team (IMT). In this briefing, the Agency Administrator and the incoming Command and General Staff will discuss and make agreements on a multitude of issues. They will work to come to a common understanding regarding the environmental, social, political, economic, and other issues that are applicable to this specific incident and the geographical locations that the Delegation of Authority (DOA) encompasses.

This briefing is also a chance for the Agency Administrator to inform the Incident Management Team (IMT) of the history, the current status, and any major actions that have been taken on the incident from the Agency Administrators perspective. This is important because the Agency Administrator may have a different view of the response than the current Incident Commander (IC). In providing this information, it will enhance situational awareness, and it may provide information that will not be presented in the transfer of command.

The Agency Administrator should provide documentation from the incident. This documentation may include intelligence reports, maps, photos, and GIS documentation. If known, they should also share both long- and short-term weather forecasts, contact lists (including phone and radio contacts), any agreements that are already in place, and any plans or previous plans (including failed plans) that have been undertaken. Another important document that should probably be prepared for this briefing by the Agency Administrator should be the ICS Form-209 (Incident Summary Status). The ICS Form-209 provides instructions on how to fill the form out, and it is for the most part self-explanatory. This form can also be used as a guide to help the Agency Administrator remember all the aspects that need to be covered in this briefing.

With all the information and documentation in hand, the Agency Administrator and the Command and General Staff of the incoming Incident Management Team (IMT) should discuss and develop the appropriate Delegation of Authority (DOA) letter for this specific incident. This will be a letter from the Agency Administrator to the incoming Incident Commander (IC) or Unified Command (UC). In this letter, the Agency Administrator will recognize the key agency personnel (if any) who will also be imbedded in the Incident Management Team (IMT) to provide guidance in the absence of the Agency Administrator. The letter will help to provide guidance on what authority they actually have.

In the briefing portion of the meeting, it is extremely important for the Agency Administrator to convey what they need in a short amount of time. The Agency Administrator may be concerned about saving lives, or the possibility of litigation, or they may be worried about financial, or even historical aspects of managing the incident. By quickly conveying their priorities, the Agency Administrator provides guidance to the Incident Management Team (IMT), and it gives them their marching orders. The list of priorities provided can vary, and there could be a long list, or it could be quite small.

When creating this letter, it is important to realize that the incoming Incident Commander (IC) or the Unified Command (UC) needs to have enough authority to do their job. If they are continually asking for additional authority to manage the incident, then the response will be slowed. Additionally, it will likely become frustrating for those in charge of managing the incident. Both the Agency Administrator and the Incident Management Team (IMT) must communicate their needs, and then come to a mutual agreement for the proper authority to meet the needs of the incident.

In most instances, the Incident Commander (IC) and the Agency Administrator cannot cover every aspect that may come up during an incident response, so they will need to also communicate throughout the incident. Each should provide the other with emergency contact information, as well as prescheduled specific times for discussing the ongoing status of the incident. There may be times when the Agency Administrator may even need to approve something that is outside of the Delegation of Authority (DOA). These approvals should be done in writing so that there is no question that approval was given.

As an example, let us say that the Agency Administrator allowed the Incident Management Team (IMT) to purchase equipment, providing it did not cost over $5000, and they had to stay in a budget of $100 000 for purchases within the whole response. The Incident Management Team (IMT) decides they desperately need a specialized piece of equipment; however, the only piece of equipment that is readily available will cost $7000. The Incident Commander (from the Incident Management Team [IMT]) will most likely need to visit the Agency Administrator and make their case for purchasing this equipment. If approved, the Incident Commander (IC) will ask the Agency Administrator to put the approval and the price stated in a letter giving them authority to purchase that single item. This protects both the Incident Commander (IC) and the Agency Administrator.

Another important issue that the Agency Administrator and the Incident Commander (IC) should discuss is how, and what, information should be released to local VIPs, and they should identify who those VIPs are. Does the Agency Administrator want the Incident Commander (IC) to share (or not share) information with the City or County Council? What about political contacts from the state or federal governments? Should information be shared with them? Are there certain issues that should not be discussed with certain VIPs? They will also need to discuss what information should be shared with the media, and other forms or public information. By discussing this prior to the Delegation of Authority (DOA), most aspects can be worked out in advance, which will likely cause fewer problems later on in the incident.

An important discussion that should take place in the Delegation of Authority (DOA) Briefing should include how resources should be ordered and any monetary considerations associated with purchases. This should include discussion about what is the maximum that should be spent on an item (without prior approval), buying at the best value, or even if ordering durable equipment, should the best or the cheapest equipment be ordered. Perhaps the Agency Administrator will want office supplies such as paper, pens, mapping services, etc. purchased on their account at the local office store, or they may even want every price negotiated. Of course, these are basic considerations, and there could be a long list of limitations and constraints that the Agency Administrator may bring up. Remember, this is their jurisdiction and their incident; the Incident Management Teams (IMTs) are just a visitor giving them a hand.

In the briefing, they should also discuss and create the standards required to hand the incident back over to the Agency Administrator or the Authority Having Jurisdiction (AHJ). This may include what benchmarks that should be reached, what resources should be left and/or returned, demobilization benchmarks, and other issues that can be identified by both sides. At some point, unless it is an ongoing incident, the Incident Management Team (IMT) will return the management of the incident to the local agency, and both parties will need to discuss how that will transpire as well.

The Agency Administrator and the Incident Management Team (IMT) will also discuss any special safety awareness, any concerns, and they will also work on expectations management. In some instances, the Agency Administrator will want their specialists involved with the Incident Management Team (IMT) so that they can have input on specific issues within the response. If this occurs, the Agency Administrator should list any individuals that may need to have input and speak on behalf of the agency. This may include the Agency Administrator (or their representative[s]), and local specialists in a specific field such as a resource advisor, a historic preservation advisor, an incident business advisor, or other technical specialists (as discussed in Chapter 13).

At the conclusion of the briefing between the Agency Administrator and the incoming Incident Management Team ([IMT] including all Command and General Staff), a Delegation of Authority (DOA) letter should be prepared by the local jurisdiction or by the Planning Section of the incoming Incident Management Team (IMT). Once drawn up to everyone's satisfaction, the Delegation of Authority (DOA) letter should be signed by both parties. It is essential that this occurs prior to the Incident Management Team (IMT) briefing with the current Incident Commander (IC) so that there is a clear understanding of the roles and responsibility of the Incident Management Team (IMT) when assuming command of the incident. This Delegation of Authority (DOA) letter should, whenever possible, be printed on the local agency or jurisdictions letterhead.

The following sample Delegation of Authority (DOA) letter should provide a basic idea of what is needed. Your agency may desire something different than the following example letter; however, this should provide an idea of what will be needed.

DELEGATION OF AUTHORITY LETTER

(Insert Name) is assigned as Incident Commander on the (insert incident name).

This Delegation of Authority Letter allows the above-named Incident Commander to have full authority and responsibility to manage the incident activities within the framework of agency policy and direction. Your primary responsibility is to organize and direct your assigned and ordered resources for efficient and effective control of the incident.

You are accountable and will report to (Insert Agency Administrators name) of (List the name of the jurisdiction they represent) or his/her designated representatives. If (Agency Administrators Name) is not available, the listed representatives should be contacted in the order they are listed:

1. (Insert name, title, and phone number)
2. (Insert name, title, and phone number)
3. (Insert name, title, and phone number) (Note: More or less names can be entered)

The financial limitations will be compatible with the most responsible actions for the assets at risk. Specific direction for this incident covering management and other concerns are:

1. The safety of the public, first responders, and other personnel.
2. Establishing accountability of resources, including but not limited to check-in areas, staging areas, and other accountability methods.
3. Maintaining all planning for operations, including Incident Action Plans and operational period briefings.
4. The responsibility of ordering resources for the incident; managing resources in an effective and efficient manner while ensuring fiduciary responsibility.
5. Managing the incident within National Incident Management Systems guidelines
6. Providing support to the Emergency Operations Center and the local elected officials as needed.
7. Provide management support to non-profits, stakeholders and others (as requested) in providing direction for meals, clothing, and other Points of Distribution.

This authority becomes effective at: (Insert time) (Insert Date)

This authority shall remain in effect until a Returned Delegation of Authority is filled out and signed by both parties listed below.

_____   _____
Agency Administrator Signature   Printed Name

_____   _____
Incident Commander   Printed Name

_____
Date and Time Signed

Although identifying the particulars of a Delegation of Authority (DOA) is important, it is equally important that the authority given to the Incident Management Team (IMT) is returned to the Agency Administrator when the incident is over. This can be done through a simple letter, and many variations of these letters are available on the internet. The important factors that should be in this letter include the name of the Incident Management Team (IMT) returning the authority, the Agency Administrator accepting the authority, the type of disaster, that both parties agree that the Incident Management Team (IMT) has completed their objectives, the names of both parties, and the time and date that the authority was returned. It is also important that both individuals sign the return of authority, and it may be helpful to have those names printed below the signature, should there ever be any reason to review who signed this return of authority. The following is a sample Return of Delegated Authority.

RETURN OF DELEGATED AUTHORITY

By all parties signing this document, it officially returns the authority and responsibility for the management of **(insert incident name)** and incident number **(insert incident number if available)** to the local jurisdictions having responsibility for the previously mentioned geographical location in which the **(type of disaster)** required the assistance of **(insert name of Incident Management Team)**. It is mutually agreed upon by **(Insert Name)**, Agency Administrator for **(Insert Jurisdiction Name)** and **(Insert Incident Commanders Name or Unified Commander Representatives Name)**, the Incident Command (or Unified Command Spokesperson), that the objectives and management direction have been met and the **(Incident Management Team name)** is hereby released on **(insert date)**, **(insert time)**.

Signed:

Agency Administrator: _____

Printed Name: _____

Incident Commander: _____

Printed Name: _____

15.4.2 Transfer of Command

The next step in having an Incident Management Team (IMT) assist is the procedure of moving the responsibility for incident command from one Incident Commander (IC) to another and is referred to as a "transfer of command." Transferring command should never be considered a failure on the part of the initial (or previous) Incident Commander (IC). It should be recognized that as the need for resources are expanding, so is the need for individuals that have more experience in managing large incidents. The transfer of command on a large or an increasing incident should be expected. This is because the incoming Incident Commander (IC) has been there and done that, and they know many of the nuances of managing larger incidents. They are considered an expert in the field of managing incidents.

Transfer of command can be from one Incident Commander (IC) to another, or the initial Incident Commander (IC) to a Unified Command (UC), or it can be from a single Incident Commander (IC) or Unified Command (UC) to an Incident Management Team (IMT). In some instances, especially on highly complex incidents that require federal state (and potentially even a local or tribal response), specialized Incident Management Teams (IMTs) that are familiar with integrating a federal response will be required. The important factor to remember when transferring command is to ensure that they have the experience and training needed to handle the incident.

When transferring command, there are typically five important steps that can help ensure that the individual assuming command can do so effectively. The first step for the incoming Incident Commander (IC) is to personally perform an assessment of the incident with the existing Incident Commander (IC). If this cannot be accomplished with both the incoming and the outgoing Incident Commander (IC), at the very least, the incoming Incident Commander (IC) should do the assessment on their own. The reason for this assessment is to give the incoming Incident Commander (IC) a first-hand visualization of the actual event, rather than getting it second-hand from someone else. This will help them to better visualize the needs, the resources, and the actions that are currently being taken, and it is an important part of situational awareness. The purpose of having the outgoing Incident Commander (IC) accompany them (if possible) is to point out any issues that were mistakenly missed in the ICS Form-201 and/or to provide a more comprehensive view of the overall response and any idiosyncrasies.

The second step of the transfer of command is that the incoming Incident Commander (IC) must be properly briefed. This proper briefing should be presented by the current Incident Commander (IC), and if at all possible, it should be a face-to-face briefing. Items that must be covered in this face-to-face briefing should include the following:

- The incident history (what has happened)
- The priorities and objectives
- The current plan
- All the resource assignments
- The incident organization/organizational chart
- Resources that were needed and/or ordered
- What facilities have been established
- The status and an overview of the current communications plan
- Any media releases or news reports
- Any identified constraints or limitations
- The potential of the incident (what would happen if the incident was not addressed)
- Delegation of Authority (DOA)
- Any special requests from agency representatives
- An introduction of the Command Staff and General Staff

When providing this briefing, the outgoing Incident Command staff should utilize the ICS Form-201 to provide this briefing. This form was specifically designed to assist with the incident briefings and should be used whenever possible. Besides providing the primary information for the transfer of command, it also provides a written record of the incident as of the time it was prepared, and it is considered by most as a legal document.

After completing the incident briefing, the transfer of command will go to step three. In step three, the incoming Incident Commander (IC) should determine the proper time for the transfer of command. The incoming Incident Commander (IC) should try to schedule the transfer of command as quickly as possible while still doing so in a timeframe that will not be disruptive to the response. Upon making the decision of when transfer should occur, the information should be disseminated to Command and General Staff personnel so they can do any work necessary to support the transfer of command by the allotted time.

Step four in the transfer of command is the actual transfer of command, including notifications to the following:

- Agency headquarters (by utilizing dispatch)
- All appointed General Staff members
- All appointed Command Staff members
- All incident personnel.

After assuming command, a broadcast should be sent out over every radio frequency used on the incident, announcing a formal transfer of command. This is done to ensure that everyone in the chain of command is aware of the transfer of command to another individual. It may also help them to understand that there potentially may be a change in tactics, resources, or actions due to new leadership.

The fifth step in the transfer of command is only a recommendation and should not be viewed as a requirement. The incoming Incident Commander (IC) should consider giving the previous Incident Commander (IC) an assignment on the incident, preferably as part of the Command Staff or General Staff. Depending on their level of experience, they could fill a function, or they could be support staff on a function, including an advisor or a specialist.

Some may wonder why you would include the outgoing Incident Commander (IC) in the continued response, and it should be noted that there are multiple advantages of doing so. The initial Incident Commander (IC) has first-hand knowledge of everything that has happened at the incident. They know which tactics have been tried, which ones worked, and which ones did not work. They also (usually) have a better knowledge of the area because they are usually a member of that community. They may know ways to get from "point A" to "point B" without passing through heavy traffic or hazard (such as a chemical spill or a wildland fire) that may be approaching traveled roads. If they are a person that lives in the community, they may know where to find nontraditional resources, such as where tools can be obtained, who could haul additional water, who in the area may have the needed equipment for law enforcement, and a multitude of other resources. This approach of keeping the outgoing Incident Commander (IC) involved also allows them to witness the progress of the incident, and it allows them to gain experience in managing future incidents.

15.4.3 Initial Incident Command/Unified Command Meeting

At the completion of the transfer of command, an initial planning meeting should be undertaken. If you are looking at the Planning "P", this is at the top of the leg, and it will be the beginning of the first operational planning period cycle. While a multitude of vital tasks will need to be completed during the initial planning period, this meeting needs to take place so that the Command Staff and the General Staff can gain an understanding of the situation, and they can gain understanding of what issues might be incident priorities. The Incident Commander (IC) will use the ICS Form-201 to assist with the briefing, but they will also use their own personal knowledge of what they saw when reviewing the incident with the initial Incident Commander (IC).

Acquiring a thorough understanding of the incident includes gathering, recording, analyzing, and displaying information that spells out the magnitude, complexity, and potential impact of the incident. In this instance, the new Incident Commander (IC) may want to have the initial Incident Commander (IC) to make a presentation to add important facts to the meeting, or to correct any information that may be wrong. Understanding the entirety of an incident is crucial to the process of developing and implementing an effective Incident Action Plan (IAP).

When utilizing federal resources and opening a Joint Field Office (JFO), this is also the point in which the initial Incident Management Assistance Team (IMAT) meeting is conducted. This meeting will be used to confirm that incident management personnel understand the Unified Coordination Group (UCG) expectations for the incident and initial strategies. It also set in motion the plans for integrating other Unified Coordination Group (UCG) members into the Incident Management Assistance Team (IMAT).

15.4.3.1 Establish Core Planning Meeting Principles for the Incident

From the transfer of command from a single agency response to a multiagency response will require much planning in order for the response to be effective. Most of the planning decisions made will be done in meetings. As an important element to proper supervision and ongoing incident management, these meetings are held to determine and disseminate vital information that is required outside of the Command Staff and General Staff. In most instances, these meetings are intended to be clear and concise, direct and to the point. Unless absolutely necessary, they should not include lengthy discussions, unless there is a need for complex decision-making. Their intended purpose is to allow the General and Command Staff to communicate specific information and outlooks for the upcoming operational period, to provide tactical assignments at the operational level, and to identify issues (safety issues, media concerns, security issues, etc.) and answer questions or concerns that may be presented.

Due to the potential for meetings to be derailed, certain key elements must be part of each meeting. These added key elements will help to ensure that the meeting addresses the issues pertaining to the meeting, rather than being sidetracked. While the potential for a trained Incident Management Team (IMT) to become sidetracked is rare, there have been actual instances where meetings have been sidetracked by discussing things such as hunting, fishing, and yes, even politics. In an effort to reduce these sidetracking issues, in the ICS method planning process, certain items have been mandated for meetings. Among those are the use of a facilitator, a written agenda, ground rules for every meeting, and a requirement of documentation at every meeting.

15.4.4 Facilitating (Ongoing) Meetings

Each meeting should have a facilitator. It is important to note that a facilitator is not the same as a leader or the President of a Board of Directors. A facilitator might be better described as an assistant who helps those attending a meeting to accomplish a common task. That common task is to complete the agenda in the time allotted while still making the necessary decisions and plans that need to be accomplished.

The facilitator does not make any decisions, but they do suggest ways that will help the group to achieve their goals in the length of time given for each meeting. They perform their duties in such a way that those involved in the meeting are aware that the facilitator is in charge, but each person seated at the table knows they have a role to play in the outcome of the plan. The facilitator's core responsibility is to the group, rather than themselves or to the individuals within the group. If the meeting becomes sidetracked, it is the responsibility of the facilitator to bring it back to the subject.

As the facilitator, the Planning Section Chief (PSC) should make it clear to all participants prior to the meeting that they should come prepared for the meeting. While there may be the occasional issue or information that someone may not be able to provide, it should be rare instance that this problem occurs. If it does appear as a reoccurring problem, either the

Incident Commander (IC) or the Planning Section Chief (PSC) should discuss the issue with any individuals who continually do not come prepared.

To reduce the time needed for a meeting, the Planning Section Chief (PSC) can also prepare a seating arrangement so that people that will be presenting at the meeting will be toward the front of the room. If a conference table is used, then the presenters will sit at the table, while non-presenters may sit or stand nearby. Along with the seating arrangement, the Planning Section Chief (PSC) should identify and briefly discuss the role of the presenter before the meeting. This is done to reduce confusion and to keep the flow of the meeting continual and to reduce dead silence when someone could be presenting. Directly prior to the meeting, the Planning Section Chief (PSC) may also want to discuss with first time participants how that particular meeting process works in an effort to reduce confusion and/or interruptions.

A key part to facilitating any meeting during the planning process is to stay focused and keep meetings as brief as possible while still covering everything that needs to be addressed. To prevent meetings from getting sidetracked, they should always start on time, and ground rules should always be given at every meeting. Additionally, the Incident Commander (IC) should be present at most meetings to provide an air of authority, and every meeting should focus on its intended purpose.

Occasionally, there will be side discussions or questions asked that pertain to the incident but that do not pertain to the meeting. The Planning Section Chief (PSC) and others can keep the meeting on track by suggesting after-meetings for individuals that need to discuss something that is important, yet off topic or irrelevant for the current meeting. This prevents multiple individuals from listening to a discussion between two people while the clock keeps ticking.

As part of the facilitating meetings, schedules should be set to ensure that work is not slowed or delayed. For that reason, the Planning Section Chief (PSC) will predetermine the meeting schedule and get it approved by the Incident Commander (IC). They will also usually identify an estimated amount of time the meeting should take and provide that information to those that need to be present. An example of this can be seen in Table 15.1.

Because the Planning Section Chief (PSC) is responsible for facilitating meetings, they have the authority to implement different policies and to enact various requirements. Their goals for these meetings should be to keep all meetings direct and to the point. By doing so, they will facilitate the most important aspect of these meetings; making them effective and complete. The implementation of these policies may seem a bit rigid to individuals who have not previously been through this process, especially landowners, business owners, and various others that might sometimes be present. It should be noted that these rules are put in place to help ensure that emergency response planning gets accomplished in a manageable timeframe, which will also prevent work delays in the response and recovery.

Table 15.1 Meeting and briefing schedule example.

Meeting	Length	Time
Command and General Staff meeting	30 minutes	At 0900 hours
Tactics meeting	30 minutes	At 1100 hours
Planning meeting	30 minutes	At 1300 hours
Final IAP parts submitted to Planning Section Chief (PSC)		By 1500 hours
Operational Period Briefing	30–45 minutes	0700 hours
Command and General Staff meeting	30 minutes	0900 hours

15.4.4.1 Ground Rules

While each incident and each Planning Section Chief (PSC) is different, they will usually want you to adhere to similar ground rules. Ground rules are usually quite similar at almost every incident, and they are important for keeping a meeting on track. Ground rules allow the meeting to be orderly and efficient, thereby taking less time. These rules also allow the facilitator of the meeting to keep everyone on track by ensuring that the participants discuss only what is relevant to the meeting, thereby overcoming distractions.

Ground rules allow the facilitator to advise those involved in the meeting what is expected of them, and what is not acceptable. The Planning Section Chief (PSC) has the freedom to create and implement new ground rules. This freedom to do so can be extremely useful if previous meetings have been bogged down with issues that were not previously a part of the ground rules. As a rule of thumb, ground rules should reduce or eliminate distractions that could occur during a meeting.

Common ground rules usually include that all cell phones and tablets should be silenced. Unless an emergency, phone calls should not be taken during the meeting, and if it is an emergency, the recipient should step outside to take the call. Ground rules also usually cover text messages as well. While some Planning Section Chief's (PSCs) enact a rule that answering text messages is not acceptable during a meeting, others require that answering text messages should be restricted to only responding to incident-related communications during a meeting. Of course, these decisions should be based on how it affects the incident.

The ground rule that radios and pagers should be shut off (so that they do not interrupt the meeting) is also usually one of the ground rules. Additional ground rules will usually include that there should be no side conversations and that taking notes at the meeting is a requirement. While the requirement to take notes is optional, mandating so reduces the chances of mistakes or misunderstandings outside of the meeting, which might lead to delays. Another helpful ground rule that is sometimes made is to make it the responsibility of the participants to assist in keeping the meeting organized and on track. While not technically a ground rule, it is also a good practice for the Planning Section Chief (PSC), as the facilitator, to inform participants to ask questions if they do not understand something as part of the ground rules.

15.4.4.2 Agenda

In preparing for meetings, the Planning Section Chief (PSC) can reduce the time that the meeting will take, simply by undertaking some basic tasks. Among them is preparing a written agenda that should be passed out to every participant of the meeting. Some Planning Section Chiefs (PSCs) even put estimated times next to the presenters name to identify the time allotted for them; however, caution should be used when using this method. In some instances, providing the amount of time that an individual can speak might have a detrimental effect, especially if the presenter feels rushed for time. Timeframes such as these can cause a presenter to forget something or they may feel that they do not have time to discuss an area that they wanted to cover. If this occurs, important situational awareness information could not be presented.

The agenda is an outline of the meeting. The agenda provides the order and sequence of items that will be discussed in the meeting, and it is typically sent out prior to the meeting. By sending out the agenda in advance, it allows all participants to gather the information they may need to present at the meeting. The agenda also allows those who will be involved in the meeting to be familiar with the topics that will be discussed. It also allows the facilitator to convey what outcomes are expected from that meeting.

There are several key parts that are crucial factors in the agenda. Each agenda should share the purpose of the meeting. If the meeting is to discuss Objectives for the operational period, then it should be plainly stated in the purpose of the agenda. As an example, the listed purpose might be listed as such:

Purpose: Determine new objectives for the next operational period and the remainder of the incident.

By clearly identifying the purpose of the meeting, it discourages rogue discussions that can disrupt the meeting and shift the focus to other issues. It also allows participants to gather specific information that is relevant to the type of meeting prior to the meeting being initiated. The designation of the purpose also allows others involved in the planning process to prepare information related to that purpose prior to the meeting

Another key part of the agenda is the attendance list. The attendance list allows those that should be involved in the meeting to be preidentified and to notify them that their presence is required. As a rule of thumb, there should be no observers listed in the attendance list. Observers can add confusion and potentially lead to questions being asked that would already be known by the collective group. For a truly effective meeting, your attendance list should be only those that are required to be at each specific meeting as dictated by the Planning P, with the exception of technical experts who should only be involved if their expertise is needed.

The agenda should also name the facilitator of the meeting. This allows all attendees to have a point of contact if they have questions or concerns. This also allows the attendees who are not capable of making the meeting to contact the facilitator and inform them of who will be taking their place.

The agenda should mention the *ground rules* for the meeting. The presentation of ground rules is undertaken by the facilitator, which in most instances is the Planning Section Chief (PSC). Ground rules are used to keep meetings on track, so it is important to set and verbalize the ground rules at the beginning of each meeting.

The next part of the agenda should include a briefing on the current objectives. This can be presented by the Incident Commander (IC) or the Planning Section Chief (PSC). The individual chosen will present new incident objectives or they will confirm existing incident objectives.

Upon completion of hearing from everyone listed, and discussion on all matters related to the meeting, the Incident Commander (IC) will reiterate all operational concerns and direct which resources will be deployed. This is done so that if there are misunderstandings, they are addressed in the meeting rather than in the field where personnel might be injured, or that time is wasted by setting up for one operation then having to transition to another different type of operation.

At the end of the meeting, the facilitator announces the next meeting for the purpose that was addressed. At the conclusion of the meeting, the facilitator will ask if there are any questions or comments. If there are questions or comments, they will be heard. The facilitator will then set the time for the next Operational Period and then adjourns the meeting.

15.4.4.3 Vetting Visitors

It is critical for a speedy and effective meeting that only those outside individuals who are needed are in the meeting. Sometimes landowners and businesses may be needed to supplement resources. It is important to vet these individuals to ensure they have the authority to make decisions and to commit their agency or their property when present in meetings. If they do not have authority to commit resources and make decisions, it can cause substantial delays in the Planning Meeting, potentially even causing last minute discussion and alternative plans to be forced upon the response. Whenever possible, the Planning Section Chief (PSC) may want to have discussions with these individuals to ensure that they have the authority that is needed.

15.4.4.4 Documentation

Documentation is an important, but often overlooked, part of incident management. By documenting meetings, a paper trail is provided which allows all working under the ICS method to have clear and concise directions. Additionally, this documentation can be used to defend the actions of the response to the media, and/or any judicial proceedings.

As has been mentioned numerous times in this book, we live in a litigious society. Documentation in every form can be used to refute tort claims, breach of contract, and equitable claims that can potentially slow or shut down a response. After an incident, there could also be the possibility that there will be accusations that money was wasted or embezzled, and the documentation can refute those rumors. Additionally, should any criminal charges be discussed or even brought against anyone, this documentation will do a lot to provide evidence of what was ordered and what was done. While these instances are extremely rare, we should always be prepared to defend ourselves.

15.4.5 Initial or Ongoing?

The beginning of the first operational period starts at the base of the Planning P, where the very top of the base of the "P" ends. As the "P" begins to transition into the circle of the letter, the entire process will be systematically completed for each operational period until the incident is completed. It should be noted that this begins a circular sequence that continually repeats itself for each operational period.

As we traverse through the continual part of the Planning P, we will identify when the meeting is different in the initial meeting versus the ongoing meeting. In those instances where there is a difference, we will look at what should be done in the initial meetings and separately what should be done in the ongoing meetings that are not the initial meeting of each task. These will be covered on a meeting by meeting basis (Figure 15.2).

The first meeting of an operational period should begin with developing and reviewing objectives. The start of the planning cycle, whether the first planning cycle or one that has been ongoing, always begins with reviewing, developing, and/or updating the incident objectives. When reviewing objectives, it is important to realize that there are two objectives that should be identified and discussed. The first is the overall or overarching objective that ends the incident. This will typically be the overarching objective for the duration of the incident. The second

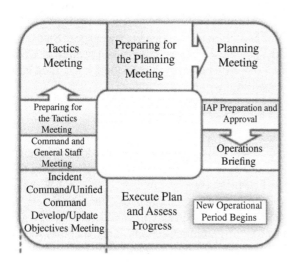

Figure 15.2 The reoccurring part of the Planning P.

objective, which may change from operational period to operational period, are the SMART objectives that are part of the planning process.

15.5 Incident Command Objective Meeting

In the initial Incident Command (or UC) Objectives meeting, the overarching objectives will be discussed, then the SMART Objectives (described in Chapter 14) will be discussed. It is important to note that the planning process utilizing the Planning P is not for the current operational period, but rather for the next operational period while monitoring and potentially adjusting the current operational period when needed.

As part of the Objective Meeting, all objectives will be reviewed or developed. As part of this process, the Situation Unit Leader (SITL) will usually be required to attend. In their capacity as the Situation Unit Leader (SITL), they will provide the Incident Commander (IC) with situational awareness by presenting a common operating picture. In the initial meeting, this will cover from the start of the incident.

The Situation Unit Leader (SITL) will essentially provide an enhanced briefing that includes items on the ICS Form 201, but they will cover the issues more in-depth. They will also provide a synopsis of any issues that were not listed in the ICS Form 201 but that were found after the transfer of command. This will assist the Incident Commander (IC) or the Unified Command (UC) in making informed decision based on the situation at that moment. Additionally, if the Situation Unit Leader (SITL) sees any issues with resources that are at risk, issues with critical infrastructure (or that may affect critical infrastructure), or any other issue that may be critical to creating objectives, then these will also be presented to the Incident Commander (IC) in this initial meeting.

The Incident Commander ([IC] or Unified Command [UC] if established) should create overarching incident objectives that will be relevant for the duration of the incident. Large or especially intricate incidents may be so multifaceted that it may take several operational periods to provide succinct and accurate overarching incident objectives. They will also develop the SMART objectives for each operational period. The SMART objectives created for an operational period should use the preferred method of creating objectives by using Management by Objectives (MBOs) as described in Chapter 14.

The overarching incident objectives will be broken down into specific tactical assignments for each operational period, and they will be designed as goals. As an example, a Goal may be to evacuate Main Street by 1400 hours. As that goal is being met, a new objective and new goals that tie in with the overarching objectives should be developed for what to do after that goal is complete.

By reviewing the objectives at the beginning of each operational period, one piece of the proverbial pie is reviewed. If we refer to the story that has been used throughout this book, you want to essentially take a small bite of that big elephant. Remembering this parable may help you remember the importance of taking small sections of the incident objectives at a time.

15.6 The Command and General Staff Meeting-The Basics

The process of planning for the next operational period will require a Command and General Staff Meeting. This meeting is extremely important; however, it becomes even more important if there is a change in the incident objectives. The meeting will usually take place at the beginning of the operational period, and the purpose of this meeting is to plan for the next operational period.

Only Command Staff and General Staff should be in attendance at this meeting. Individuals who are not part of the Command and General Staff may cause confusion, or worse yet, try to influence the Command and General Staff to change objectives in a different or wrong direction. These types of distractions can undermine a team, so the rule of thumb should be that only the core group should be present in this meeting. The meeting should not be considered a "secret" meeting, but by the same token, it should not be open to anyone other than the Command and General Staff.

The Command and General Staff meeting should be scheduled during a time when all team members can be present. This meeting should be coordinated (usually by the Planning Section Chief [PSC] or their representative) keeping in mind that all team member must be present, but prior to the planned and scheduled Tactics Meeting. Once the time for the meeting is established, it should not be publicized. Only those that should be in attendance should be notified of the time and place of this meeting.

Another reason that only core members of the team are involved in this meeting is that many consider this meeting as a time of upkeep for planning the next operational period. This meeting is the time that the Command and General Staff can identify and resolve issues that they have identified during the response and/or recovery of the incident without added outside interference.

In this meeting, the Incident Commander ([IC] or Unified Command [UC]) will usually meet with the Command and General Staff to garner ideas and input from these valued staff members. The Incident Commander (IC) will solicit input from them to address issues that need immediate review and cannot wait until the planning process is finished. While this meeting typically transpires at the beginning of an operational period, it (or one similar to it) can also take place throughout an operational period as it is needed. It is important to note that this meeting should be as brief as possible whenever it occurs outside of the daily meeting.

A primary purpose of the Command and General Staff meeting is to ensure that there is effective coordination between all team members. Ongoing cooperation, coordination, and collaboration are a vital part of incident management. It is also critical to ensure that all support elements (e.g. Communications Unit, Medical Unit) of the organization are fully functional and supporting the operations activities in an efficient manner. During the meeting, the Incident Commander (IC) will try to make sure that their staff comes away from this meeting with a clear comprehension of the direction being taken, the tasks that will need to be undertaken, and the authority that the Command and General Staff are granted in performing their duties and functions. Each is responsible for a portion of the overall planning and execution of the response, and each plays a critical role in the success.

Some may wonder why the Incident Commander ([IC] or the Unified Command [UC]) does not create these strategies on their own, without the input from others. The answer to this question is simple and has been answered previously. It is highly unlikely that one person is capable of coming up with the perfect plan, or all of the ideas on their own. Even if it were possible to do so, a complex fluid situation will have many variables, and missing just one variable could have a significant effect on the safety of personnel and the public, the stabilization of the incident, or the economic or environmental impact. Therefore, a good Incident Commander (IC) should always solicit the input of the Command and General Staff.

If Incident Commander (IC) fails to solicit input as part of the process, they could be putting themselves at risk for legal difficulties. Those legal difficulties could be in the form of a lawsuit, or if they believe the Incident Commander (IC) was negligent (and it caused harm or death), then there is the potential for criminal charges to be filed.

Another important issue to realize, there is a meeting that should occur prior to the Command and General Staff meeting that is not listed in the Planning P. This meeting is a Business Meeting, and only a select few should attend this meeting. It should be held by the appropriate Incident Management Team (IMT) members for both for the initial operational period and the ongoing incident.

15.6.1 Business Meeting Prior to Command and General Staff Meeting

The Logistics Section Chief (LSC) and the Finance/Administration Section Chief (FSC) should have a *Business Meeting* prior to the Command and General Staff Meeting. The purpose of the business meeting is to make sure that there is not a breakdown in communication concerning logistical and financial processes (and activities) as they relate to the incident. It is important for Logistics and Finance/Administration to communicate about what is working, and what needs to be revised.

It is recommended that this Business Meeting takes place on a daily basis, and the meeting should be held just prior to the Command and General Staff Meeting. This permits the Logistics Section Chief (LSC) and Finance/Administration Section Chief (FSC) to have the most current and up-to-date information for the Command and General Staff Meeting.

In order to collect the appropriate information, others within the Incident Command organization may need to attend this meeting to provide information related to their specific jobs. The individuals that may helpful in creating a more complete picture can include the Cost Unit Leader (COST), Documentation Unit Leader (DOCL), the Procurement Unit Leader (PROC), Resource Unit Leader (RESL), the Supply Unit Leader (SPUL), and the Situation Unit Leader (SITL). These individuals have a unique view of what is going right, and what is going wrong, within their specific portion of the incident. Having a unique perspective, they can help evaluate their specific area of responsibility in a way that nobody else working within the incident can.

The Business meeting should be very brief, and the topics of discussion should revolve around looking for issues and finding ways to overcome those issues. Some common issues that arise are as follows:

- Resource orders that are not equally paced with resource requests
- Current incident cost-sharing agreement(s) between parties and the efficacy of the agreements
- Funds availability (or lack thereof)
- Funding needed
- The efficiency of resource and property tracking
- Good stewardship of resources
- Wasteful or unnecessary spending
- Purchasing contracts
- Rental agreements
- A cost analysis of purchased material
- The efficacy of the bidding process

When looking at this partial list of discussion points, it is simple to see that this meeting promotes "good business" by making sure that time, energy, and resources are not wasted or misused. This Business Meeting is a chance to identify underlying problems that may exist, and to look for either the cause, or to identify a resolution to these problems.

As an example, let us say that the resource ordering is not keeping pace with the requests for resources. Essentially, supply is not keeping up with the demand of the incident. The Logistics

Section Chief (LSC) and the Finance/Administration Section Chief (FSC) can sit down in this meeting and determine where the breakdown might be. After some discussion, they might be able to identify that some (or all) personnel are not following the procedures in place for the request process. They may then decide to examine why this might be happening. After looking at the problem, they could determine that the current procedure is too cumbersome, or possibly even that it has too many desks to cross. By discussing these issues, they can work together to come up with the solutions needed to better support the incident.

The costs of the response and/or recovery should be monitored closely to ensure that care and conservancy of not only funds but also of the incident resources (including equipment) is taking place. According to the Texas Interagency Coordinating Center, "A rule-of-thumb is when costs have reached 80% of the current ceiling, the ceiling should be increased." (Texas Interagency Coordination Center & Texas A&M University, 2014a p. 24).

It is important for the Logistics Section Chief (LSC) and the Finance/Administration Section Chief (FSC) also diagnose and make predictions about funding. This includes when the allotted money will run out and reporting this to other Command and General Staff members. This Business Meeting is the ideal place to delve into this type of calculation and to determine what can be done to overcome it.

It is also important to realize that effective property tracking is an important way of gauging if resource management is adequate. Weak property tracking systems, or worse yet, no system at all, will usually lead to mysterious losses of durable equipment. This can also lead to an increased demobilization process, whereby additional time will need to be spent looking for missing resources. Having regular Business Meetings can help prevent or stop these types of issues.

Another issue that should be brought up in this meeting should be the good stewardship of the physical resources that the Incident Management Team (IMT) has been entrusted with. This involves making sure that the Logistics and Finance Section are providing the appropriate support (e.g. oil changes, fresh fuel, preventative maintenance) but undertaking this support in the most cost-effective way. In the beginning of the incident, this may be more difficult because the incident is still developing, and the Operations Section may still be trying to get the incident under control, or at the very least manageable. While it may be difficult, it does not mean that it is impossible.

Even as the incident progresses, or the emergent nature of the incident begins to slow, the individuals that attend the Business Meeting should be motivated to improve the stewardship of the resources they have been entrusted with. If the Agency Administrator appointed an Incident Business Advisor, they too should be brought into the loop of helping to improving stewardship.

Nothing will sour elected officials, and the public, more than the wasting of money on unneeded resources. If those in charge of an incident are ordering more specialized equipment while having three pieces of that same identical equipment sitting in reserve, it is a waste of time (ordering it) and money. The equipment that is sitting on the sidelines should be in use prior to ordering more. Additionally, the purchase of equipment and/or services should involve price checking.

As has been mentioned previously in this book, price checking does not always mean going with the cheapest price. Sometimes the cheaper price will be more expensive in the long run because you did not compare what equipment and/or services came with that price. It is mandatory to creating credibility that those involved with ordering and purchasing choose the best value, not the best price. This sometimes may not be possible in the emergent stage of a disaster, when resources and equipment are needed right away. As the incident begins to settle down and becomes more manageable, then the process of ensuring the best value becomes more important.

15.6.2 The Initial Command and General Staff Meeting

The initial Command and General Staff Meeting of an incident is substantially different from the ongoing meeting, and it will be important to the remainder of the overall response and/or recovery. It is considered by some as a part of the foundation of the planning process. If basic foundational issues are addressed in the initial meeting, then it clears a path for a more organized planning process throughout the entire incident. Some key issues that should be addressed include the following:

- Defining the incident objectives
- Establishing the meeting schedules
- Presenting any new or ongoing issues and concerns
- Creating operational timeframes for completion of tasks
- Reviewing functions and activated staff (including the need to expand or reduce general staff)

In the initial Command and General Staff Meeting, it is important to set the expectations that should carry on throughout the entire incident. The expectation that this meeting is intended to share information related to only incident objectives and strategies is among the first thing that should be discussed, primarily because a meeting can be easily sidetracked or derailed.

Additionally, expectations should be set for mandatory attendance of the Command and General Staff Meeting as well as set the duration and time of meetings. In some instances, the time may need to be adjusted due to a member of the team not being available at that time, but these instances should be the exception rather than the rule. The initial meeting should also work to consolidate resources into specific functional areas (Command, Planning, Logistics, and Finance/Administration) and identify any additional information that may be required on a regular basis (FEMA, 2008). It should be mentioned that the combining functional areas should not be done. Combining functions can create confusion and conflicting job responsibilities such as combining planning and logistics into one function (FEMA, 2008)

It is also important to note that the order in which individuals in this meeting make presentations may vary, and the order of those speaking should not be written in stone. It should also be noted that the initial opening and closing of the meeting should remain the same in every incident; Planning opens the meeting and lays ground rules, the Incident Commander (IC) speaks, then the Situation Unit Leader (SITL) provides an update. Also, remaining the same in the closing of the meeting, the Incident Commander (IC) will provide final thoughts, and the Planning Section Chief (PSC) closes the meeting.

The initial Command and General Staff Meeting is facilitated by the Planning Section Chief (PSC), who will also document important decisions made in the meeting. After the meeting has been called to order a role call will be done using ICS Form 207 (Incident Organization Chart). Immediately after calling the meeting to order and taking the roll call, the Planning Section Chief (PSC) will explain the ground rules. Usually, in the initial Command and General Staff Meeting, each member of the Command and General Staff will introduce themselves and identify which function they are taking responsibility for. This allows everyone to know who the key players are, and it allows them to make a mental note of who to contact in specific areas, should they need to interact with them.

Once introductions have been completed, the Incident Commander (IC) will provide opening remarks. Opening remarks should be short, brief, and to the point. These remarks should, at the very minimum, focus on the importance of safety and teamwork. As part of the initial meeting, the Incident Commander (IC), and in some instances the Planning Section Chief (PSC), will discuss the objectives that were initially developed in ICS Form-201. They will also discuss key decisions that have been made up to the time of the meeting. The Incident Commander (IC) may suggest staying with the original objectives, or they may have already determined that those objectives were not viable. If they found them ineffective or unviable, they can either discuss their new objectives or task the Planning Section Chief (PSC) with creating new objectives for the Command and General Staff Meeting.

Upon the completion of opening remarks and the key decisions made, the Situation Unit Leader (SITL) will discuss the initial situation report based mainly on ICS Form-201 (Initial Briefing Form). In most instances, a copy of the filled-out form will be provided to all Command and General Staff; however, in some instances, this information may be displayed using an overhead projector, a PowerPoint Presenter, a whiteboard, or poster boards with the same information. Whenever possible and practical, even when presented by other means, a copy of the form should be handed out to each member of the team, so they can keep it with their records and as part of their documentation. During this stage of the meeting, no questions should be asked, and the only input that should be presented should be any last-minute changes that the Situation Unit Leader (SITL) was not aware of.

The Situation Unit Leader (SITL) at this point should also discuss the incident so that all members present have a common operating picture. The Situation Unit Leader (SITL) should then discuss the population at risk (if applicable) and the economic and environmental factors that may affect the incident. The Situation Unit Leader (SITL) should also discuss other intelligence such as the current and predicted weather conditions, tides, topography, or other factors that may affect the response. If applicable, this may include identifying toxic factors, such as toxic smoke or spillage, and how the weather may affect these issues. If the resources are available, they can also provide modeling predictions of the incident. Because this is the initial Command and General Staff Meeting, they should cover the main aspects of the incident from the beginning, which is not something that they will do in ongoing meetings.

The next step in the initial meeting is to address action items. Action items are a list of actions that the Incident Commander (IC) wants to see completed. Action items are tasks that will be assigned during the meeting, and they will identify who is taking responsibility to complete that action. In assigning these action items to individuals, it allows the Incident Commander (IC) to know that the actions that they have identified are being addressed. It also allows the Planning Section Chief (PSC) to not only realize that this is an issue that is being addressed but it also provides the ability to monitor that item, but to ensure that the action is being taken.

After the action items are addressed and documented, then order in which the different Command and General staff will speak is dictated by the agenda that the Planning Section Chief (PSC) has provided. It is important to note that different agencies and different Incident Management Teams (IMTs) will often have different formats of who should present next and the order they present in.

15.6.2.1 The Command Staff Briefing Within the Command and General Staff Meeting

The Public Information Officer (PIO) will discuss the methods that they think will be most effective for information flow, as well as identify any politically sensitive issues related to the incident. In Type-1 and Type-2 incidents, there will likely be federal resources. When there are federal resources, the information flow should include creating or maintaining a Joint Information System (JIS), which will be under the supervision of the Joint Information Center ([JIC], FEMA, 2017a). They will also discuss suggested restrictions on the release of information. Should anyone involved in this meeting have concerns about public information, they will then have the opportunity to voice those questions or concerns and briefly discuss them.

The Liaison Officer (LOFR) will also provide a brief summary of their issues or concerns relating to the objectives. The Liaison Officer (LOFR) will identify the resources on hand at the time of the meeting, their abilities, and how these might affect creating effective objectives.

They will also lay out their plan for the organizational structure of liaising and where they intend to physically place the Assistant Liaison Officers (if needed) so that it is less complicated and/or cumbersome to meet up with them. They will address any VIPs that are at the incident or that plan to visit the incident within the next operational period. Upon completing their part of this briefing, those involved in the meeting will have a chance to identify any concerns and briefly discuss them.

The Safety Officer (SOFR) will provide a safety status briefing. In this briefing, they will highlight any near misses that almost caused an injury or death from the beginning of the incident, up to the time of the meeting. They will also discuss any injuries that occurred and that required medical attention beyond basic first aid. Finally, they will bring up any safety concerns they have identified, including safety issues at Incident Facilities. After presenting their briefing, there will be time for airing questions or concerns. This may include any additional safety issues that others may have noticed, or additional discussions on how to mitigate the issues that were presented.

15.6.2.2 The General Staff Briefing Within the Command and General Staff Meeting

The General Staff will also make presentations at the Command and General Staff Meeting. Their briefings will be based on their job function and any positive actions that have been taken as well as any concerns they may have. It should be remembered that this is the initial Command and General Staff Meeting. The initial meeting will be substantially different from the ongoing Command and General Staff Meetings.

The Finance/Administration Section Chief (FSC) will provide a brief overview of the current cost of the response. In doing so, they will identify any rental contracts and/or agreements that were in place when they took over the Finance and Administration position. They will also explain any potential or actual claims that are related to the incident prior to being appointed to the position. They will identify funding sources in place, potential funding sources, what the ceiling of funding is, and provide an initial projection of the burn rate (amount of funding compared to the work yet to be done). At the completion of their briefing, the participants can pose any questions that are related to the setting of objectives and how they might affect the Finance and Administration function of the response.

Similar to other functions, the Logistics Section Chief (LSC) will also provide a summary of how the objectives will affect their work. They will likely discuss the obvious shortcomings and surpluses in the supply chain (and inventory) for the response, as well as the adequacy of communications. They will also identify the current medical capabilities as well as determine what additional medical capabilities may be needed in the future. When their portion of the briefing is complete, then the participants in the meeting can discuss any issues or concerns of how the objectives may specifically affect their function and supplying the needs of the incident.

The Planning Section Chief (PSC) will give an assessment of how the objectives might affect their current situation. Often, this will entail predictions of how the incident may change using those objectives, and what objectives might be set to mitigate those changes. Upon completion of their briefing, the participants are again allowed to voice any issues or concerns that they may have.

The Operations Section Chief (OSC) will also provide a briefing as it relates to the setting of objectives and how it will affect their job function. They will likely discuss probable tactics that may be used and provide an overview of the resource requirements to meet these proposed objectives. It is important to note that these tactics may change during the Tactics Meeting. Because this briefing is the initial Command and General Staff Meeting, the basic pieces of the response are somewhat narrowed down to a broad plan by setting objectives. After the completion of their briefing, they field any questions or concerns that the participants of this meeting may have.

15.6.2.3 Optional Command or General Staff if Activated

The Investigations and Intelligence Gathering Section Chief or the Investigations and Intelligence Gathering Officer if activated will also provide a briefing (if applicable) They will provide an assessment of the current situation and how the current objectives will affect the investigation or intelligence gathering that is being undertaken. They should also provide a notice of any information that should remain confidential, as well as any other issues that may be affected by the objectives. Similar to other Command Staff and Section Chiefs, once their briefing is complete, the participants will be allowed to voice any concerns or issues that they may have with the situation. Because there may be "need to know" information, they should also ask questions about what information can or cannot be released.

15.6.3 The Closing of the initial Command and General Staff Meeting

After all the Command and General Staff has completed their briefings, the Incident Commander (IC) will provide final input, including any changes to the objectives, and deliver closing comments. It is important to note that the Incident Commander (IC) is there to increase communications among the functional staff and to give direction on how the incident should proceed. The Incident Commander (IC) is there for the overall management of their team, but they are not there to micromanage. Their job is to assign the main tasks and to allow their team members to take care of the details prior to the Planning Meeting.

At the end of the meeting, the Planning Section Chief (PSC) will set, or reiterate, the time of the next Command and General Staff Meeting and verify that everyone understands. The Planning Section Chief (PSC) will then complete the ICS Form 202 (Incident Objectives), and if necessary, ICS Form 202b (Critical Information Requirements), which is a supplemental for to ICS Form-202.

15.6.4 Preparations for the Ongoing Command and General Staff Meeting

The ongoing Command and General Staff Meeting has sometimes been referred to as the "employment" meeting" because it helps determine what the work requirements might be, and who is needed to work on the incident. The primary purpose of the meeting is for all Command and General Staff review the incident priorities, incident objectives, open actions, and other critical issues and to communicate anything that they think may be of importance to the response of the incident. This meeting is also a meeting to ensure that everyone is on the same page, or in ICS talk, it can be referred to as "unity of effort" (Texas Interagency Coordination Center & Texas A&M University, 2014b).

Because this meeting helps make sure that everyone is on the same page, Command Staff and General Staff will need to prepare information for this meeting on a daily basis. Having the necessary information available will help the meeting to progress at a quicker pace and will waste less time. Each will be described individually over the next paragraphs.

15.6.4.1 Planning Section Chief (PSC) Preparations

In preparing for the Command and General Staff Meeting, the Planning Section Chief (PSC) will create an agenda. They will also ensure that they have a current copy of the Ground Rules to help keep the meeting on track and ensure that they have a current copy of the ICS Form-207 (Incident Organization Chart). The ICS Form-207 (Incident Organization Chart) will assist them in taking role call and in answering any questions about the structure of the organization, or any changes to that may have occurred to that structure.

The general format for the agenda should be similar to the following agenda posted. Of course, as has been mentioned previously, this can be modified to make it more specific for the incident you are facing, or it can be modified to meet local protocols (if needed). The order in which participants speak is not important except for calling the meeting to order by the Planning Section Chief (PSC), followed by the Incident Commander (IC) speaking at the beginning of the meeting, and the Incident Commander (IC) providing closing remarks toward the close of the meeting followed by the Planning Section closing the meeting and handing out schedules.

Initial Command and General Staff Meeting for (Incident Name)

ISSUES, OBJECTIVES AND STRATEGIES

All Command and General Staff are required to attend this meeting.

Planning Section Chief (PSC) calls meeting to order at (insert time)

- Ground rules (Cell phones and pagers off, no sidebar conversations)

Incident Commander (IC)

- Opening remarks
- Objectives

Situation Unit Leader (SITL)

- Situation when last reported
- Current Situation

Incident Commander

- Action items

Safety Officer (SOFR)

- Resolution of safety concerns
- Ongoing or new safety concerns

Information Officer (PIO and/or IO)

- Current information situation
- Resolution of information issues
- Ongoing or new information issues

Liaison Officer (LOFR)

- Resolution of liaison issues
- Ongoing or new liaison issues

Investigations and Intelligence Gathering (I/I) Officer (if activated)

- Resolution of Investigations and Intelligence Gather issues
- Ongoing or new investigational or intelligence gathering issues

Finance and Administration Section Chief (FSC)

- Resolution of finance issues
- Ongoing or new finance issues

Logistics Section Chief (LSC)

- Resolution of previous logistical problems
- Ongoing or new logistical problems

Planning Section Chief (PSC)

- Resolution of planning issues
- Ongoing or new planning issues

Investigations and Intelligence Gathering Section Chief (if activated)

- Resolution of investigation/intelligence issues
- Ongoing or new investigation/intelligence issues

Operations Section Chief (OSC)

- Ground Operations
- Air Operations (if applicable)

Incident Commander (IC)

- Verify current planning cycle objectives are covered
- Discussion on concerns or issues related to the objectives
- Confirm that the strategic framework is established

Conclusion-Planning Schedule by Planning Section Chief (PSC)

- Reiterate time frames for the Planning Process and ensure everyone has the schedule

---

In most instances, the Planning Section Chief (PSC) will make enough copies of the ICS Form-207 and the meeting agenda for all participants of the meeting as well as have extra copies on-hand for all those present for the Command and General Staff Meeting. As a rule, it is better to have too many copies than not enough for all participants.

It should be noted that there may be other deliverables that the Planning Section, or the Planning Section Chief (PSC), may be responsible for. This can sometimes vary from incident to incident, or jurisdiction to jurisdiction. They should also work to complete and have a brief report on any open action items that they were required to address.

15.6.4.2 Incident Commander (IC)/Unified Command (UC)

The Incident Commander (IC) or Unified Command (UC) should prepare for the meeting by reviewing the current objectives and how the current tactics have led to those objectives being met. If an objective has been completed, then they will purge that objective from the objective list. If they believe that new objectives are needed to meet the incident needs, they will identify those changes, and create new objectives, or possibly even hone the current objectives.

The Incident Commander (IC) will also be responsible for completing any open action items that they assigned to themselves. If an open action item has been completed, they should prepare to give a brief statement if they decide that a statement about the closed action item is required.

15.6.4.3 Situation Unit Leader (SITL)

The Situation Unit Leader (SITL) will provide a general briefing on the incident. In doing so, they will provide a common operating picture for all Command Staff and General Staff members. In preparing this briefing, the Situation Unit Leader (SITL) will need to gather the latest and most up-to-date information about the incident. This should include what progress has been made since the last Command and General Staff Meeting, as well as any aspects that may affect the response. In looking at the various aspects that may affect the response, they will need to go to the proper individuals that are monitoring these issues (if they are not handling the issue themselves) and get the latest and most accurate reports and maps. If available, they should also meet with technical experts (prior to this meeting) to help them identify issues that could happen based on the modeling of the incident. They will need to gather this information and be ready to present it in the Command and General Staff Meeting.

Depending on the importance of the information, the Situation Unit Leader (SITL) may need to make copies of their report (or portions of their report) so that the key players have this information in written form. It should be noted that some Incident Commanders (IC) and/or Agency Administrators require copies of reports for every meeting, while others leave it at the discretion of the Situation Unit Leader (SITL).

15.6.4.4 Operations Section Chief (OSC)

The Operations Section Chief (OSC) will prepare to present all issues that the Operations Section is facing. This is done so the Command and General Staff members can help to maintain operational effectiveness. By identifying operational issues in this meeting, they can help improve the chances for operational success by allowing others to think on the challenges and potentially come up with better solutions.

The Operations Section Chief (OSC) should prepare to deliver an operational update. In preparing for the Command and General Staff Meeting, they will evaluate a range of factors that affect operations. These can include resource support issues such as insufficient (or too many) personnel, the need to replace crews, food issues, fuel issues, transportation issues, communications issues, and a whole list of other support items that are needed. Another issue that they should have on hand should be a summary of any interagency issues that have become a problem. Examples of this might be one agency does not get along with another agency, or ongoing infighting. They will also reveal any issues with support facilities such as staging areas, Helibases and Helispots (Texas Interagency Coordination Center & Texas A&M University, 2014b).

In preparation for the Command and General Staff Meeting, the Operations Section Chief (OSC) will compile all the aforementioned (or any other operational) issues and write them down. The purpose of writing them down is to ensure that nothing is forgotten. The Operations Section Chief (OSC) should also prepare a status on any action items that they were tasked with in prior meetings, and if an item was taken care of, then they should be ready to present a brief summary about that.

15.6.4.5 Safety Officer (SOFR)

In the ongoing Command and General Staff Meeting, the Safety Officer (SOFR) should prepare a condensed report that is based on the safety status of the incident. In most instances, the Safety Officer (SOFR) can create an outline that describes all the issues rather than a full report. In preparing for the meeting, they should gather information on the number of injuries of personnel and separately the general public, the number (and types) of near misses that affected the personnel or the public, the top three hazards on the incident, and what mitigation measures were or should be put in place. If required by the Incident Commander (IC) or the Unified Command (UC), they may have been tasked with creating a site health and safety plan. If tasked with this action, or any other action, they should be prepared to discuss the status of those actions items. It is also important to note that the Safety Officer (SOFR) should prepare to take copious notes in the meeting, writing down any safety issues that might be brought up in the Command and General Staff Meeting (Texas Interagency Coordination Center & Texas A&M University, 2014c).

15.6.4.6 Logistics Section Chief (LSC)

The Logistics Section Chief (LSC) will, for the most part, prepare a brief report that discusses any issues about integrating with incoming or already on-scene resources. They will document any issues they have faced with requesting resources, approval methods, and the ordering process. They also will identify any interagency issues and support facility needs. As the person who deals with the logistics of resources, they will provide a first-hand account of any issues that are affecting the Logistics Section.

This information is usually documented in a brief outline rather than a full report so that the Command and General Staff can better understand any logistics issues, and possibly provide input. It also allows the Incident Commander (IC) to make more informed decisions about such things as ordering resources, the approval process, and requesting resources, which will streamline these procedures and inform the remainder of the Command and General Staff on any new procedures that have been decided upon.

While not mandatory, it is strongly suggested that the Logistics Section Chief (LSC) and the Finance/Administration Section Chief (FSC) should daily hold a brief Business Meeting prior to the Command and General Staff Meeting. This meeting is a chance for these two important functions to identify, discuss, and come up with a game plan to address any logistics and/or finance issues that may be slowing response, or a hindrance to the overall mission. As will be described in the Business Meeting section, others may need to attend this meeting to ensure a complete understanding of any issues.

15.6.4.7 Finance/Administration Section Chief (FSC)

The Finance/Administration Section Chief (FSC) will need to prepare a report that brings up some of the same issues as the Logistics Section Chief (LSC), but from a different perspective. They will prepare for the Command and General Staff Meeting by identifying any issues from the financial and administrative perspective. This may include a one-page report about projections, how much work is left, and if there is sufficient or insufficient funding to sustain the response and/or recovery with the current funding avenues. The report will also identify what the funding ceiling is, and as necessary, advise when the incident has used up 80% of the allotted funding. They may also address the request process for resources, the approval process, and the ordering process as it relates to ordering and obtaining them in a timely manner.

In preparation for the Command and General Staff Meeting, the Finance/Administration Section Chief (FSC) should review these processes and make sure that fiscal responsibility is in place. They should also create or utilize a method to identify and document that the resources ordered were actually needed for this incident. They will need to document any financial or administrative interagency issues, such as unusual monetary requests, unusual reimbursement requests, special request from responding agencies in financial tracking, or any other similar issues.

15.6.4.8 Intelligence and Investigations ([I/I] if activated at the Command or General Staff Level)

As has been mentioned before, the addition of Intelligence and Investigations (I/I) was added as an "as-needed" function. Depending on the incident, it may be activated as a Command Staff position, a General Staff position, or it can be placed as a function of Planning. If implemented under the Planning Section of ICS, then they will not attend the Command and General Staff meeting; the Planning Section Chief (PSC) will be responsible for presenting anything that is relevant to intelligence and/or investigations in that placement.

If Intelligence and Investigation (I/I) is (or becomes) a Command Staff or General Staff Position, then the Intelligence and Investigations (I/I) Section Chief will need to be a part of the Command and General Staff Meeting. The type of intelligence or investigation that is being undertaken will usually determine what preparation is needed for the Command and General Staff Meeting. If the Intelligence and Investigation (I/I) is related to a health emergency or a pandemic, this section will typically provide reports on the spread rate, the strategies that may help contain or prevent the spread of the health emergency, or they may provide information on where vaccinations should take place and who should receive vaccinated.

If the incident involves criminal activity, or a terrorist attack, the Intelligence and Investigations (I/I) Section may provide strategies that could be enacted for preserving evidence, or the strategy or strategies being used to catch the perpetrators. They may also provide strategies for personnel to identify evidence and to mark it or document it for later investigative purposes once the scene has been made safe. If an incident involves multiple hidden explosive devices, they may prepare a strategy for keeping other personnel safe.

At the present time, there is little guidance on the preparations for the Command and General Staff Meeting as related to the Intelligence and Investigations (I/I) Section Chief or Officer. The information provided in the previous paragraphs is just basic information. It is hoped that future guidance from a government entity is presented to cover this area more in-depth.

15.6.4.9 Public Information Officer (PIO)

Prior to the Command and General Staff Meeting, the Public Information Officer (PIO) should create a brief that describes what has happened since the last Command and General Staff Meeting. This may include developing, recommending, and executing public information plans and strategies that should be presented at the meeting. They should also gather any information being presented that are misinformation or rumors, and any information that is being released by "undisclosed sources". In all instances, they should prepare not only the problems being faced but also a way to mitigate those problems.

If a Joint Information System is activated, according to FEMA Guidance (2017a, 2017b), they should provide updates on the Joint Information System (JIS) and any problems with coordinating any interagency messaging. This should include any issues in communication between the Emergency Operations Center (EOC), the Multiagency Coordination (MAC) Group, and Incident Command. This Joint Information System is most often needed in Type-1 and Type-2 incidents.

15.6.4.10 Liaison Officer (LOFR)

In getting prepared for the Command and General Staff meeting, the Liaison Officer (LOFR) will prepare a brief on the coordination of efforts. In that briefing, they will need to gather the information about the cooperating and assisting agencies, as well as any stakeholders, contractors, individuals, and/or agencies that may be involved with the response and/or recovery. The Liaison Officer (LOFR) will also need to gather the information of how liaison is organized, and the current location(s) of any Assistant Liaison Officers. This is done so that the right individuals can be easily found when there is an issue with coordinating outside resources.

The Liaison Officer (LOFR) will also create a list of any outreach efforts that have been undertaken. They will then need to create a list of any VIPs that have visited in the current and previous operational period, and they will list any VIPs that are expected to visit in the next operational period. They will also create a list that briefly describes any liaison issues that have arisen from assisting agencies or stakeholders during the last operational period.

15.6.5 The (Ongoing) Command Staff and General Staff Meeting

The ongoing Command and General Staff meeting is designed to allow the key leaders of the Incident Management Team (IMT), a chance to review the key decisions, objectives, and the action items list prior to the actual Planning Meeting. This nickname of this meeting is often *The Strategy Meeting*. This is because strategy issues are usually discussed, and honest and open discussions about the strategy are discussed among the key leaders of the incident.

In most instances, the Incident Commander ([IC] or Unified Command [UC]) will garner input or be provided immediate direction from the main players in managing the incident. It is also the opportunity to present and approve new incident objectives for the next operational period. In this meeting, valuable information will be shared in regard to managing the incident. It should be noted that in most instances, the objectives that were created in the Objectives Meeting will remain the same as developed. There are however rare instances when these objectives may need to be changed. The Objectives Meeting is the starting point where the objectives are initially determined for the next operational period, but the Command and General Staff Meeting is where the details are discussed and finely honed.

The ongoing Command Staff and General Staff meeting is often referred to a "maintenance time" for all of those involved in the decision-making process. It gives them the opportunity to regroup and discuss what is going right, as well as what is going wrong, and then come up with an overall strategy. It is also an opportunity to provide a pep talk to each other or for the Incident Commander (IC) to inspire the Command Staff and General Staff.

It is important that the agenda for this ongoing Command and General Staff Meeting is followed, however, except for the beginning and the end of the meeting, the order that the briefings occur can be substantially different. The following agenda is an example of the agenda for the ongoing Command and General Staff Meeting.

Command and General Staff Meeting Agenda for (Incident Name)

ISSUES, OBJECTIVES AND STRATEGIES

Purpose: Discuss Strategies

Participants: All Command and General Staff are required to attend this meeting.

Planning Section Chief (PSC) calls meeting to order at (insert time)

- Ground rules

Incident Commander (IC)

- Opening remarks
- Objectives

Situation Unit Leader (SITL)

- Situation when last reported
- Current Situation

Incident Commander

- Action items

Safety Officer (SOFR) briefing

Public Information Officer (PIO) and/or Information Officer (IO) briefing

Liaison Officer (LOFR) Briefing

Investigations and Intelligence Gathering (I/I) Officer briefing (If activated)

Finance and Administration Section Chief (FSC) briefing

Logistics Section Chief (LSC) briefing

Planning Section Chief (PSC) briefing

Investigations and Intelligence Gathering Section Chief (If activated)

Operations Section Chief (OSC) briefing

Incident Commander (IC)

- Verify current planning cycle objectives
- Confirm that the strategic framework is established

Conclusion-Planning Schedule by Planning Section Chief (PSC)

- Release schedule for next Command and General Staff Meeting.

Like the initial Command and General Staff Meeting, the Planning Section Chief (PSC) is the facilitator. The Planning Section Chief (PSC) will open the meeting by setting the ground rules for the meeting. They will then take a roll call to make sure that all the required functions are represented at the meeting by utilizing ICS Form-207 (Incident Organization Chart).

Upon completion of the roll call, the Incident Commander (IC) will provide opening remarks. These opening remarks should be presented in the manner of a pep talk. The Incident Commander (IC) may describe how pleased they are with progress. This could include pointing out different job functions and how their actions are showing results. The Incident Commander (IC) should briefly describe what progress has been made since the last briefing and describe any failures or issues that have come to the surface during that operational period. It is important to note that they should only lightly discuss these issues because this information will be more effectively covered by the Situation Unit Leader (SITL).

The Incident Commander (IC) should then present the priorities and incident objectives for the next operational period and clearly provide guidance on how incident operations should proceed. At this point, the Incident Commander (IC) and/or the Planning Section Chief (PSC) will announce any key decisions that have been made in the previous meeting, and then identify any incident objective changes. The review of key decisions made should only be for the previous Command and General Staff Meeting, unless a historical perspective is needed for a specific reason.

Next, the Situation Unit Leader (SITL) should give a current assessment of the situation, including all positive and negative aspects of the incident and the response to it. They will describe what the current situation is and update the Command and General Staff of any changes from the previous operational period. While the Situation Unit Leader (SITL) will share the most relevant information, they should by no means provide an extremely detailed report. While giving a situation update, they should in most instances provide relevant maps, charts, expected weather forecasts, current weather conditions, forecasts for work to be completed in the current operational period, and any other information that might be relevant.

After the Situational Unit Leader (SITL) has completed their portion of the briefing, the next step in the Command and General Staff Meeting is to address action items. As was previously explained, action items are a list of actions that the Incident Commander (IC) wants to see completed. These action items are tasks that will be assigned during previous meetings, and they identify who is assigned the responsibility to complete a specific action. It also allows the Incident Commander (IC) and the Planning Section Chief (PSC) to address any issues and to ensure that the action is being taken.

After the action items are addressed and documented, then the order in which the different Command and General staff will speak is dictated by the agenda that the Planning Section Chief (PSC) provided. Different agencies and different Incident Management Teams (IMTs) will sometimes prefer different formats of who present in what order. This order of presenters will usually be based upon the Incident Management Team (IMT) or of the Authority Having Jurisdictions (AHJs) preferences. While there is no concrete method that should be followed, the order below is the preferred order of most Incident Management Teams (IMTs).

15.6.5.1 The Command Staff Briefings

The Public Information Officer (PIO) will discuss the most effective methods for information flow. They will also identify any political sensitivities and how to approach those issues related to the incident. In Type-1 and Type 2, the Public Information Officer (PIO) may be presenting information from the Joint Information Center (JIC), or the Information Officer (IO) from the Joint Information System (JIS) may co-present with the Public Information Officer (PIO). In this presentation, they will also discuss restrictions on the release of information, what they are doing to improve internal communication flows, or anything else that is important to the response and deals with information. While information flow is important, this should be a very brief presentation, and it should not be detailed unless highly important to other functions. Should any Command or General Staff have concerns about public information, they will then have the opportunity to voice those questions or concerns and briefly discuss them.

The Liaison Officer (LOFR) will provide a brief summary of their issues or concerns relating to their job function. In doing so, they will identify the resources that are on hand, expected resources in the next operational period, and they will discuss how they will liaise with them. They will also lay out their plan for the organizational structure of liaising. If Assistant Liaison Officers have been appointed, they will identify their location and contact information so that others know where to find them if they are needed. They will include in their briefing any VIP's that are at the incident or that plan to visit the incident within the next operational period. Upon completing their part of this meeting, those involved in the meeting will have a chance to identify any concerns and briefly discuss them.

The Safety Officer (SOFR) will provide a safety status briefing. In this briefing, they will present near misses that occurred. If there were any injuries or deaths in the last operational period, they will briefly discuss the safety of how it could have been prevented. They will then bring up any safety concerns that they have identified during the meeting and discuss additional safety concerns that they have about the overall incident, including safety issues at incident facilities. Usually, they will talk about their top three safety concerns and identify what safety measures have been taken and what new measures will be implemented. After presenting their briefing, there will be time for feedback, including any questions or concerns. This may include any safety issues that others may have noticed and wanted the Safety Officer (SOFR) to be aware of or discussion on how to better mitigate the issues that were presented.

15.6.5.2 General Staff Briefings

The General Staff will also be responsible for presenting at the Command and General Staff Meeting. Their briefing, just as with all briefings in Planning P meeting, will be based on their job function. They will typically report any positive actions that have been taken, and any concerns they may have in the ongoing Command and General Staff Meeting.

The Finance/Administration Section Chief (FSC) will provide a brief overview of the current cost of the response. They will identify and briefly explain any new rental contracts and/or agreements that were signed recently and not yet discussed. Although a rare occurrence, they will also update the Command and General Staff about any claims that are related to the incident and what actions are being taken. They will speak briefly about the funding sources in place, and if applicable, potential funding sources that they are investigating or applying for. They will also share what the funding ceiling is and provide a projection of the burn rate (amount of funding compared to the work yet to be done). They will also discuss the important issues that were identified in the Business Meeting as their responsibility, and what is being done to overcome any problems identified in that meeting. At the completion of their briefing, the participants can pose any questions that are related to Finance and Administration function of the response.

The Logistics Section Chief (LSC) will typically be the next to present. They will discuss the shortcomings and surpluses in the supply chain and inventory, and they will describe what they are doing to ensure that the supply meets the demand for the response. They will usually discuss the adequacy of communications and describe any measures they are taking to overcome any communication issues. They will provide an overview of the medical capabilities and brief everyone on any additional changes that are proposed or enacted in the Incident Medical Plan. They will also discuss the important issues that were identified in the Business Meeting as their responsibility, and what is being done to overcome any problems identified in that meeting.

When their portion of the briefing is complete, then the participants in the meeting can discuss any issues or concerns with that function and supplying the needs of the incident.

The Planning Section Chief (PSC) will give an assessment of any issues that their job function has with current situation. They will usually provide predictions of how the incident may change and when applicable how they can mitigate those changes. They will often provide the relevant maps, charts, expected weather forecasts, current weather conditions, and forecasts for the next operational period. Upon completion of their briefing, the participants are again allowed to voice any issues or concerns that they may have.

The Operations Section Chief (OSC) will typically be the last of the General Staff to present. This is so they can identify and evaluate how changes being made by other Command and General Staff will affect operations. While the briefing can be on many aspects of the operational response, they will likely discuss the probable tactics that might be used. They might also provide an overview of the change in resource requirements to meet these tactics. It is important to note that these tactics may change during the Tactics Meeting because this briefing is the initial starting part for the rest of the day and many subsequent meetings. After the completion of their briefing, they field any questions or concerns that the participants of this meeting may have.

15.6.5.3 Optional Command or Staff If Activated

The Investigations and Intelligence Gathering Section Chief or the Investigations or Intelligence Gathering Officer if activated will also provide a briefing (if applicable). It is important to remember that the name used will depend on where Investigations and Intelligence gathering is placed, if this function was activated. No matter where this function was placed, if activated they will provide an assessment of the current situation and how the current objectives have affected or will affect the overall investigation or intelligence gathering that is being undertaken. They should also provide a notice of any information that should remain confidential and that should not be shared outside of the meeting. They may ask other functions to be on the lookout for specific items or anomalies that may assist their investigation or intelligence gathering efforts, and they could even ask that they document odd occurrences. In some instances, they may ask operations or other function to document information and/or take pictures. Once their briefing is complete, the participants will be allowed to voice any concerns or issues that they may have with the situation. Because there may be "need to know" information, they should also ask questions about what information can or cannot be released if they have questions about it.

15.6.6 The Closing of the Command and General Staff Meeting

After all the Command and General Staff has completed briefing everyone, the Incident Commander (IC) will provide final input and deliver their closing comments. These closing comments may provide guidance on what the Incident Commander (IC) considers priorities, issues that should be watched for, or a multitude of other final instructions or information gathering that they may want.

At the end of the meeting, the Planning Section Chief (PSC) will provide the time of the next Command and General Staff Meeting and verify that everyone understands. At the completion of the meeting, the Planning Section Chief (PSC) will then complete the ICS Form-202 (Incident Objectives), and if required, the ICS Form 202b (Critical Information Requirements), which is a supplemental for the ICS Form-202 (Incident Objectives).

15.7 The Tactics Meeting

The Tactics Meeting is facilitated by the Planning Section Chief (PSC). As the facilitator, the Planning Section Chief (PSC) should set up the room prior to the meeting and gather the required materials and visual aids that may be used in the meeting. Materials that may be needed include copies of the ICS Form-215 (Operational Worksheet) and ICS Form-215-A (Incident Action Plan Safety Analysis), maps, weather forecasts, (current) resources available, the incident objectives, and anything else that may be needed to give a complete picture of the incident.

The Planning Section Chief (PSC) may need to provide visual aids. Visual aids can include large maps, and other larger display methods with much of the information provided in handouts. These visual aids can be in the form of large poster boards, PowerPoint (or similar software) projected on a screen, or any other method of displaying information to a wide array of individuals.

The purpose of the Tactics Meeting is to review the tactics developed by the Operations Section Chief (OSC) and to finely hone them to be more effective. Tactics define the specific actions that should be performed to achieve an expected outcome. In creating these tactics, this meeting will create the who, what, when, and where of the response for that operational period. These tactics will relate directly to the implementation of strategies to achieve the incident objectives that were set earlier in the day at Objectives Meeting and discussed at the Command and General Staff Meeting.

In the Tactics Meeting, those involved will create and describe how personnel and resources are positioned and what actions they should take based on the tactical strategies developed in the meeting. It is important to note that all tactics discussed must fall in line with incident objectives. It is also important to note that work assignments for resources that are not involved in actual incident operations are not included in the tactical plan. The exception to this rule is that the Operations Section Chief (OSC) may determine that there is a convincing need to include non-field personnel. In most instances, the support teams are not included in the Tactics Meeting discussions unless they play a significant role in accomplishing a specific tactic. Additionally, work assignments for operational leaders should not characteristically be included. When describing operational leaders, we are referring to individuals such as the Operations Section Chief's, geographical and functional Branch Directors, Division Supervisors and/or Group Supervisors.

Those that should be in attendance at the Tactics Meeting include the Operations Section Chief (OSC), the Safety Officer (SOFR), the Logistics Section Chief (LSC), and the Resources Unit Leader (RESL). The meeting should be led by the Operations Section Chief (OSC). To document the Tactics Meeting, ICS Forms 215 (Operational Planning Worksheet) and ICS form 215A (Incident Safety Analysis) should be used and filled out.

Developing the Operational Planning Worksheet (ICS Form-215) is a crucial part of the incident action planning process. These worksheets can be, and usually are, central to the success of all response and recovery operations. This holds especially true on larger incidents. The Operational Planning Worksheet (ICS Form-215) helps operations personnel to identify the most viable tactics that should be successful, and then they identify resource assignments to achieve the goals and objectives of the response. Reviewing these worksheets during the Tactics Meeting will facilitate discussion, and it focuses on the review of tactics and resource assignments. Furthermore, these worksheets are widely used in the response to document the assignment of resources and supporting information. The ICS Form-215 worksheet was designed to capture work assignment information and to provide a basis for the operations review process. The ICS Form 215 will also assist in preparing ICS-Form 204 (Incident Action Plan Assignment List).

In meeting, the key players will need to review a multitude of information. This information can include current and expected weather conditions, personnel and equipment available as well as their abilities and training to effectively perform the task. They will also discuss any constraints, such as a lack of personnel or other resources, any limitations, or any other considerations that may be detrimental to the response.

The Tactics Meeting should be a brief informal meeting. While the meeting should not take a long to complete, an agenda should still be created to keep the meeting on track. By creating and adhering to the agenda, it should take even less time to complete the Tactics Meeting. There are many ways the agenda can be created, but the agenda items should be the same or similar to the sample agenda in the following.

Tactics Meeting Agenda for (Incident Name)

Purpose: Tactics for operational period on (insert date)

Participants: Operations Section Chief (OSC), Resource Unit Leader (RESL), Logistic Section Chief (LSC), Safety Officer (SOFR), Liaison Officer (LOFR)

Operations Section Chief (OSC)

- Review the established incident objectives
- Review and draft work assignments (ICS Form-215 Operational Worksheet)
- Come to an agreement on work assignments (ICS Form-215)
- Make sure that Resources identifiers are correct

Resource Unit Leader (RESL)

- Identify any resource shortcomings
- Identify excess resources

Logistics Section Chief (LSC)

- Work out and identify the logistical needs for the work assignments

Safety Officer (SOFR)

- Identify and discuss safety concerns of work assignments
- Propose safety mitigation measures for work assignments

Liaison Officer (LOFR)

- Consider and discuss the limitations and needs of outside resources support

Operations Section Chief (OSC)

- Approve the Operational Planning Worksheet(s) (ICS Form-215)

Operations Section Chief (OSC) and Logistics Section Chief (LSC)

- Update the incident map to reflect progress and new assignments

Operations Section Chief (OSC)

- Discuss other topics and/or concerns related to tactics (as needed)

While the Planning Section Chief (PSC) is responsible for facilitating the meeting, they do not lead the meeting. The Planning Section Chief (PSC) facilitates the meeting by ensuring the meeting room is set up, that the need maps are provided, the ancillary equipment that may be needed for the meeting is available, and all the needs of the meeting are in place. In some instances, the Operations Section Chief (OSC) will run the meeting. In other instance, the Planning Section Chief (PSC) will start the meeting with roll call, lay out the ground rules, and then leave, while yet in other incidents it may be required that they stay for the meeting. This is dependent on the protocols for the Incident Management Team or specific rules set forth by the Authority Having Jurisdiction (AHJ).

Informed decisions and discussion in the Tactics Meeting will lead to a more effective response. It is incumbent upon those within this meeting to openly share and discuss the latest intelligence and information. Some of the most essential information that should be shared is what resources are available for the next operational period. Planning to use resources that are not available or that do not have the skills, knowledge, or expertise to accomplish a tactic is counterproductive. Furthermore, if a tactic is assigned to a crew that does not have the required expertise and abilities, then it increases the chances that the assigned tactic will not be successful, or worse yet, someone on that crew may become injured or killed.

Additionally, the ingress and regress routes should be discussed in the Tactics Meeting. There should be a review of road closures, damaged bridges, debris in the road, flooded roadways, and any other issues that may affect tactical implementation. If a crew should have to reroute or backtrack on the way to a tactical assignment, it will in most instances slow the response, and put the tactical task assigned to this resource at risk of not being completed during that operational period. Upon identifying the kind, type, and qualification of all available resources, then any constraints or limitations should be recognized and presented so that there is a 360° picture of what the common operating picture is. Armed with a complete operating picture, a tactical plan can then be thoroughly developed.

At the planned and allotted time, the Operations Section Chief (OSC) or Planning Section Chief (PSC) will call the meeting to order. They will then read the agenda and the ground rules for this meeting. As was mentioned earlier in this chapter, ground rules are important to help keep meetings on track, thereby taking less time to take care of the business at hand. The Operations Section Chief (OSC) will then provide a Situation Report (SitRep) on the successes and failures of the previous operational period and provide a current situation report.

Upon completion of reading the agenda, the Operations Section Chief (OSC) will provide a review of the established incident objectives. The Operations Section Chief (OSC) will then explain the proposed assignments for the next operational period. After they are finished presenting the proposed assignment, there will be discussion on the proposed tactics and assignments. Participants in this meeting will identify and discuss gaps are, or that may become, an issue. They will also point out any duplication in work assignments. The reason for pointing out the duplication of resources is to prevent an additional cost time and money, as well as to prevent added confusion and chaos from conflicts or coordination issues. The Operations Section Chief (OSC) will have the Tactics Meeting participants help ensure that resources are identified correctly (resource identifier). When there are multiple resources of the same kind and type, there should be unique individual identifier assigned to each one.

The Operations Section Chief (OSC) will then ask the Resources Unit Leader (RESL) to speak. The Resource Unit Leader (RESL) will identify any shortcomings in resources with the proposed tactical plan and any additional or excess resources that may be available. Identifying shortcomings in resources will allow there to be modifications to the tactics to either overcome those shortcomings or to adjust the tactics to meet the available resources. Identifying excess resources is done so that the maximum benefit of the resources available can be achieved.

Having a resource sitting on a sideline that could be undertaking tactical work is a waste of the resources time, and a waste of the incidents available money. By outwardly discussing these issues with those in the meeting, the work assignments can often be adjusted, thereby creating an improved tactical response.

When the Resources Unit Leader (RESL) is finished, the Logistics Section Chief (LSC) will identify the logistical needs for work assignments in the next operational period. In discussing these needs, input from other meeting participants should identify any logistical support that might be needed but was not identified by the Logistics Section Chief (LSC). This is extremely important because a shortcoming could affect the response and/or the safety of the response.

After the Logistics Section Chief (LSC) is complete, the Safety Officer (SOFR) will address, identify, and discuss any safety concerns that they have with the proposed work assignments. As has been mentioned previously, the highest priority is responder safety followed by the public's safety. Any concerns that the Safety Officer (SOFR) has should be discussed, and if needed, mitigation measures should be identified to overcome these safety concerns. Additionally, any of those present in the meeting should share any safety concerns that they may have with the proposed work assignments for the next operational period. They will also fill out the ICS Form-215A (Incident Safety Analysis) for the next operational period.

Once the safety portion of the agenda is completed, the Liaison Officer (LOFR) voice any concerns that relate to the needs and any limitations of outside resources. A specific outside resource might have a limitation in how long they can work, or perhaps even what type of work they can or cannot do. They also may need a specific type of food during that operational period for religious or medical reasons. Any potential issues that might be needed by these outside resources should be considered, and plans should be made to ensure that their needs and requirements are met (within reason).

After all the concerns are voiced and worked through by all participants, the Operations Section Chief (OSC) approves the Operation Planning Worksheet(s) (ICS Form-215). While some Operations Section Chief's (OSC's) will take an official vote to approve the tactical assignment, this is not necessary and somewhat overkill. In order to approve the worksheet, they only need to ask if anyone in the meeting has any other concerns. If they do not have concerns, then the tactical plans can be approved.

Once the plans are approved, the work of the tactics meeting is not finished. The Operations Section Chief (OSC) and Logistics Section Chief (LSC) will need to update the incident map. This is done to keep the map current and to show the proposed tactical areas where personnel will be working. The updating of the map should reflect progress that has been made and any new or differing assignments.

Upon finishing the updating of their incident maps, the Operations Section Chief (OSC) should ask the meeting participants if there are any other topics and/or concerns related to tactics that they would like to discuss. No matter how large or small a topic or concern might be, this is the time in which all participants should speak up and provide honest and open feedback. If they are concerned about something, and do not say anything, then they are not properly doing their job. This portion of the meeting is not the time to hold back.

Once all concerns have been addressed, then the Operations Section Chief (OSC) (or the Planning Section Chief [PSC]) will adjourn the meeting. At the conclusion of the meeting, the Resources Unit Leader (RESL) will collect copies of the approved Operation Planning Worksheet(s), (ICS Form-215) and the Safety Analysis (ICS For-215A) and submit them to the Planning Section Documentation Unit.

15.8 Preparing for the Planning Meeting

Following the Tactics Meeting, the Planning Section is tasked with preparing all the needed items for the Planning Meeting. The Planning Meeting is a significant meeting that will need a large number of deliverables. Because there is so much to do, the Planning Section will need to provide all that is needed in a short amount of time. They will need to "Divide and conquer" the tasks to provide the deliverables by dividing up the work among several units and/or individuals. Each will be discussed based on their functional duties.

15.8.1 Planning Section Chief (PSC)

In preparation for the Planning Meeting, the Planning Section Chief (PSC) will oversee all planning functions to ensure that they are completing the needed tasks for the Planning Meeting. The Planning Section Chief (PSC) will also have their own responsibilities that they must also complete for the meeting.

After the Tactics Meeting is completed, the Planning Section Chief (PSC) will obtain a copy of the ICS Form-215 (Operational Worksheet) from the Documentation Unit Leader (DOCL) and analyze it for any issues or concerns. If any are found, they will call the Operations Section Chief (OSC) to voice those concerns. While this rarely occurs, it should not be overlooked when needed. The Operations Section Chief (OSC) will need to make a decision if the concern is legitimate, or if the plan should proceed as written. In some instances, the Operations Section Chief (OSC) may want to consult with the other participants in the Tactics Meeting and then make a decision as to whether the concern raised by the Planning Section Chief (PSC) should be addressed. The Planning Section Chief (PSC) will also obtain ICS Form-215A (Safety Analysis) and look for any safety issues not addressed. If some are found, then they will contact the Safety Officer (SOFR).

In preparation for the Planning Meeting, they will work with various subordinates in the Planning Section. In order to make informed decisions based on situational awareness, they will need to assess the current ongoing operations, and to gather the needed intelligence. They will also need to ensure that the time of the meeting is established, determine who should be participants in the Planning Meeting (with the Incident Commander [IC]), and determine where the meeting will be held and what deliverables are expected from them. They will also need to post the information in the required locations and send messages to those that should attend the meeting. In some cases, they may need to send reminders of the meeting to certain individuals.

As part of their organizational responsibilities, they will need to ensure that all planning maps, available resources, and their status are up to date for the meeting and that all of the ICS Forms that might be needed will be available at the meeting. The Planning Section Chief (PSC) will also assign someone to take notes during the Planning Meeting. Perhaps one of the most important things they do for the Planning Meeting is to ensure that there are no surprises and no delays.

15.8.2 The Situation Unit

As part of their job function, the Situation Unit collects, processes, and organizes ongoing incident information. In preparing for the Planning Meeting, they will look at the latest data and prepare a summary based on the latest intelligence. In doing so, they will help those in the Planning Meeting to have better situational awareness so that they also can make more informed decisions. The types of information that the Situation Unit may provide could include the impact the incident and/or the response is having to public health, the damage caused to natural resources, the status of evacuation efforts, progression of operational efforts (and their effects), the results from sampling and monitoring efforts undertaken, and other pertinent intelligence that might affect the tactical planning for the response.

Other deliverables that the Situation Unit should prepare for the meeting are all graphical and numerical projections that are related to the incident, and they will be required to create displays of incident information and data. These displays might include incident information relating to movement, growth, or reduction of the incident and any mitigation or intelligence activities including the successes of those efforts. They will also provide weather predictions and forecasted weather changes that may affect the response. By doing so, they help to provide (and to maintain) a common operating picture. They will also provide current and up-to-date maps.

15.8.3 The Resource Unit

The Resource Unit will be responsible for filling out the ICS Form-204 (Assignment Lists). This is normally prepared after the Tactics meeting by the Resources Unit Leader (RESL). The assignments that the Operations Section Chief (OSC), and others in the Tactics Meeting, created will be used to fill out the ICS Form 204 by transferring information gleaned from the Incident Objectives (ICS Form-202) and the Operational Planning Worksheet (ICS Form-215). Prior to being implemented, it will be possibly adjusted then approved at the Planning Meeting by the Incident Commander (IC). Prior to being approved by the Incident Commander (IC), it will most likely be reviewed by the Planning Section Chief (PSC) and/or the Operations Section Chief (OSC).

The Resources Unit will also need to create a large ICS Form 215 (Operational Worksheet) for the Planning Meeting if none is available. If a large ICS Form-215 is available, they will also fill it out. If no oversized ICS Form 215 is available, they will create this oversize poster board version and fill it out so that those in the Planning Meeting can discuss and communicate the decisions made by the Operations Section Chief (OSC) during the Tactics Meeting. This will include displaying the resource assignments and other vital needs for the next operational period.

Finally, the Resources Unit will be responsible for gathering and documenting the most current and up-to-date resource data on resources for the Planning Meeting. This is done to enhance situational awareness and ensure that the resources needed for the next operational period are available. This also helps to ensure that resources that could be used are not sitting on the sidelines doing nothing. This is also part of the fiscal responsibility that all individuals tasked with managing an incident are responsible for.

15.8.4 Technical Specialists

Technical Specialists are Subject Matter Experts (SMEs) in their field of study. As such, they can provide valuable input for the Planning Meeting in their area of practice. In some instances, they may be asked to provide an analysis and alternative strategies. They should also be present to ensure that proposed actions in the Incident Action Plan (IAP) do not cause any harm in the work field for which they are an expert. Types of deliverables that they may provide include a risk analysis (for their specific area of expertise), identify potentially useful response and/or clean-up technology that may be available (but not being utilized), epidemiological data and information, meteorological data and information, as well as a whole host of other information. It should be remembered that these Technical Specialists will sometimes be called upon to be present in the Planning Meeting to provide a better situational awareness or alternative response capabilities in their area of expertise.

In preparing for the Planning Meeting, the Technical Specialists may need to gather information and/or create diagrams to assist in providing a higher level of situational awareness that could assist with the decision-making process. Technical Specialists should make sure that any information they provide is from credible sources and that the information itself is credible. In some instances, they may be told to bring specific material to the Planning Meeting, and at other times, they may determine to bring them of their own accord. Even if they feel that they do not need to bring deliverables, and they are not told to bring any, they should be prepared to speak about their area of expertise if asked to do so.

15.8.5 Incident Action Plan Preparation and Approval

Upon writing the incident goals and objectives, the next step in the Planning P process is the preparation and approval of the Incident Action Plan (IAP). The written plan is comprised of a series of standard forms and supporting documents that convey the Incident Commanders (IC's) intent and the Operations Section Chief's (OSC's) direction for the accomplishment of the plan during the next operational period. This part of the Planning Process allows the main players an opportunity to review and validate the Operational Plan as proposed by the Operations Section Chief (OSC).

The real purpose of the Planning Meeting is to identify and correct any inconsistencies or issues that may not have been previously identified. As this is completed, those present will begin to put the final touches on ICS Form-215 (Operational Planning Worksheet). This meeting is also part of making sure everyone is on the same page, which helps to garner support for the plan created by the Command and General Staff. Subsequently, this meeting is used to gain approval from the Incident Commander (IC) for the Incident Action Plan (IAP).

Creating a final Incident Action Plan (IAP) for the next operational period is similar to how most people would imagine a battle plan being designed for the military. The Planning Meeting is figuratively the time when everyone in the battle comes together and identifies the intricacies of the plan to ensure that everyone is on the same page and that even the smallest detail is not missed. While there are often small items that may come up (or not go according to plans), the Planning Meeting helps to ensure that these instances are few and far between, rather than a normal and ongoing problem.

Because the Tactics Meeting has already taken place, the information and proposed plan should be common knowledge to most Command and General Staff prior to the Planning Meeting. Based on the premise that it will already be widely known among the participants, the Planning Meeting will usually be a brief meeting, and it will only address minor changes. It is important to keep meetings as short as possible, while still covering all of the issues that need to be discussed. By keeping the meeting short but thorough, the Command and General Staff will have more time to oversee their duties and manage their subordinates. This means they will spend less time sitting in a chair at a conference table discussing issues that (in most instances) might be irrelevant.

15.8.5.1 The Planning Meeting

As is common throughout the ICS method of incident management, the Planning Section Chief (PSC) is responsible for facilitating the Planning Meeting. They will be responsible for ensuring that the flow of information is succinct and that speakers are not long-winded. In most instances, while preparing for the Planning Meeting, most of the participants will have already completed information gathering for their specific function. By doing so, it helps the Planning Section Chief (PSC) to keep the meeting brief and on point. If the Planning Meeting is facilitated well, it should take 30 minutes or less to complete.

In the Planning Meeting, the Incident Commander (IC) will place a significant emphasis on developing the Incident Action Plan (IAP). The reason for this is that this plan will address the entire situation while detailing the multiple smaller assignments that will help to reduce the incident in size and scope. The Operations Section Chief (OSC) can also identify the types of resources they will need and how many of each type to make the next operational plan successful. The Planning Meeting allows the Resources Unit Leader (RESL) to work with the Logistics Section Chief (LSC) so they can identify and order all the resources needed for that operational period. It also allows the Logistics Section Chief (LSC) to ensure that the financing is in place for those resources with the Finance/Administration Section Chief (FSC) to provide the financial resources needed. Much like the previous analogies of the car, every system with the automobile is working together to reach their intended destination, based on the map that was created.

A good guide to keep the Planning Meeting on track while ensuring that everything is covered would be to use a Planning Meeting Agenda. In most instances, the order will be the same (or nearly the same) for the Planning Meeting, regardless of the type of incident. As with any meeting, the agenda will provide the participants with a basic outline of what should occur, and in what order. While not critical, it should be noted that the order of presenters can be used for almost any type of incident. The reason the agenda is set up in this way is to reduce unnecessary questions that would have been answered later if the agenda was in a different order. By having speakers in this order, it reduces the time that is needed to accomplish the necessary task that the meeting addresses.

(Incident Name)
PLANNING MEETING AGENDA

Purpose:

Participants:

Call meeting to order, conduct roll call, explains ground rules, and reviews agenda	Planning Section Chief (PSC)
Opening remarks and review of objectives	Incident Commander (IC)
Briefing on current situation and resources. Identifies resources at risk, weather forecast, and incident projections. Identify and specify proposed tactics for the Incident Action Plan (IAP)	Planning Section Chief (PSC) and Operations Section Chiefs (OSC)
Safety status briefing and identification of safety mitigation measures for Incident Action Plan (IAP)	Safety Officer (SOFR)
Stipulate resources that will be needed for the proposed Incident Action Plan (IAP)	Operations Section Chief (OSC), Planning Section Chief (PSC)
Identify all facilities and reporting locations	Operations Section Chief (OSC), Planning Section Chief (PSC), Logistics Section Chief (LSC)
Provide details for resource and personnel orders that meet the proposed Incident Action Plan (IAP)	Logistics Section Chief (LSC)
Update finances and provide an update on approximately how long earmarked finances for the incident will last.	Finance/Administration Section Chief (FSC)
Identify and explain any changes to the communications, medical, and traffic plans.	Logistics Section Chief (LSC), Planning Section Chief (PSC)
Finalize, approve, and implement Incident Action Plan (IAP)	Incident Commander (IC),

In looking at the agenda, it is easy to see that it starts with the foundation of the response (or recovery) by identifying what the objectives are for the incident. It then builds on that foundation by providing the proposed tactics to meet those objectives, one of the most important parts of the Planning Meeting. The players in this meeting then address the facilities that will provide support and the resource needs for the response, including the number and types of resources. Those present then cover what resources will need ordered to meet the proposed plan and how to pay for those resources. This is then followed by how they should communicate, how they will get from place to place, and what they should do if something goes wrong. Once this is done, everyone is asked if they can support the plan, and if everyone is in agreement, then it is approved and signed by the Incident Commander.

It should be noted that the Planning Meeting is held in a systematic way. It is arranged in such a way that it builds upon the previous part of the meeting. Step-by-step, you put together the Incident Action Plan (IAP) and ensure that everything you will need will be in place. Much like a pyramid, each part of the meeting builds on the next. While it can, and has been done, in a different order, utilizing this order allows each piece to serve as a foundation for the next layer of planning. In developing the Incident Action Plan (IAP), and in planning the next operational period (using this order), it provides a holistic approach that reduces the risk of potential issues that could sidetrack the response and/or recovery for an incident.

Another important factor to consider for the Planning Meeting is that everyone will need to support this plan. This support is not only physical but also mentally. This directly ties into Management by Objective as described by Drucker (1954, 2010). All parties are involved in the decision-making process, so they should have buy-in for the plan. Because they all had a part in creating the plan, they have a vested interest in making sure that it succeeds. This is the reason that just prior to closing the meeting that the Incident Commander (IC), or the Planning Section Chief (PSC) will ask each of the General Staff members, and usually the Command Staff Members present at this meeting, if they support the plan. If all agree that they can support this plan, then the Incident Commander (IC) will approve the plan and sign it. Once approved, the Planning Section Chief (PSC) will set a deadline for when all the deliverable support elements of the plan should be submitted to facilitate the next step in the process.

The Planning Section Chief (PSC) will need to provide a quick synopsis and identify who is responsible for developing each piece of content for the plan, when the content should be completed and turned in, and where it should be delivered. In doing this, they should identify the person (or persons) who should receive the completed portion of the plan. This person, and possibly the Planning Section Chief (PSC) will review all the components of the plan to ensure that the content is complete and that it has the required approvals.

The Planning Section Chief (PSC) will also discuss the Operational Period Briefing location, ensuring that it is a quiet setting, large enough to facilitate all the participants. They will ensure that it is set up in such a way that a display area can be utilized for charts and maps, and if needed that an audio system is available and can be used. When selecting the time of the Operational Period Briefing, the Planning Section Chief (PSC) will need to take multiple things into consideration. They will need to take into consideration when the operational period begins, how long it will take for those that attend the Operational Period Briefing to the Staging Area, and the time the Operational Period Briefing will take to complete.

Operational Period Briefings typically take 30–45 minutes. Once the time and location are determined (and agreed upon), then this information will be sent and/or posted so that those who should be present can be informed about the briefing immediately after the Planning Meeting. Usually, this briefing will be at the same time every day that the incident is ongoing.

Depending on how the Incident Management Team (IMT) is set up, and how that team does business, there may be one brief additional discussion. This discussion may use the entire

Incident Management Team (IMT) to determine how many copies of the Incident Action Plan (IAP) should be made. When determining this, they will need to identify all the personnel that have a supervisory role. This often includes all the way down to Unit Leaders and Crew Bosses. They will also need to take into consideration the number of facilities, both within the incident area and outside of the incident area (such as staging areas and fixed wing air bases), and the number of bulletin boards it should be posted on. They should also add to the number of outside supporting agencies and organizations that should receive copies, the number of Agency Administrators, volunteer organizations, and different disciplines that need a copy for their support roles. They may also need to discuss how many extra copies of the Incident Action Plan (IAP) should be made if the Planning Section Chief (PSC) is unsure.

After completion of these discussions, Incident Commander (IC) will provide closing remarks. After their closing remarks, the Planning Section Chief (PSC) will ask if there are any questions or comments. Once all questions or comments are completed, the Planning Section Chief (PSC) will briefly identify when the next Planning Meeting will be. They will in most instances identify who should be present at the Operational Period Briefing, and the location and time of the briefing. They will then close the meeting and begin the process of printing the Incident Action Plan (IAP).

15.9 Printing the Incident Action Plan

The Incident Action Plan (IAP) is a specific document that is issued to ensure that all personnel involved in an incident have a common understanding of what is expected of crews during the response or recovery. They will be able to look at the Incident Action Plan (IAP) that was developed, and they will be able to see the approved objectives, strategies, and tactics that should be used. The Incident Action Plan (IAP) helps to ensure the safety of personnel, the effectiveness of response operations, and provides operational direction for incident personnel.

Because incidents vary in the type, size, complexity, and unique requirements, each Incident Action Plan (IAP) will vary to meet the needs of the incident. The Incident Action Plan (IAP) provides guidance for the response using the planning process to accomplish the incident objectives. The entire planning process and the Planning P are designed to meet the needs of the planning process and to create the Incident Action Plan (IAP).

An Incident Action Plan (IAP) must be created for each operational period and distributed prior to the Operational Period Briefing. Doing so gives a roadmap to success for those working to overcome the challenges of the incident. The Incident Action Plan (IAP) must be as accurate as possible and communicate the information that was agreed upon by the Command and General Staff during the planning process, but be brief and to the point.

The Planning Cycle was established to enable the development of an effective Incident Action Plan (IAP) in an orderly and systematic manner. The Incident Action Plan (IAP) is prepared by the Planning Section Chief (PSC) with input from the appropriate Sections and Units of the Incident Management Team (IMT). While it is written at the beginning of the response, it should be frequently reviewed and updated to meet the everchanging needs.

The Planning Section Chief (PSC) supervises the development, preparation, completion, duplication, and distribution of the Incident Action Plan (IAP), and the entire plan is created by the Planning Section. The plan is created from various other Command and General Staff providing most of the pieces needed for it. When creating the Incident Action Plan (IAP) for the Operational Period Briefing, prior to printing the plan, it should be checked thoroughly by at least two individuals. Those individuals should know what should be in an Incident Action Plan (IAP) to ensure that all components are added and that plan is complete and acceptable.

The Planning Section Chief (PSC), with the written approval of the Incident Commander (IC), determines what content will be included in the Incident Action Plan (IAP) for each operational period. The content of the plan may change from operational period to operational period based on the changing needs of the incident. Once the Incident Commander (IC) approves the Incident Action Plan (IAP), then the Planning Section Chief (PSC) becomes charged with the completion and distribution of the final Incident Action Plan (IAP). They will be responsible for reviewing the entire Incident Action Plan (IAP) and for ensuring that it printed and completed before the Operational Period Briefing. In doing so, they will need to check to make sure that the Incident Action Plan (IAP) is complete, that all last-minute changes have been added, and that the information is current and up to date. They will also need to check that no pages were omitted and to check the entire document for other mistakes such as incomplete sentences, pagination, and sometimes they may even look for misspelled words. The main thing they need to check for is that it is complete, easy to read, and that all pieces of this document are clear and concise.

Incidents that generally require written Incident Action Plans (IAPs) typically involve a large number of resources, and they tend to exceed one operational period. These plans are always managed and approved by an Incident Commander (IC) who requests an Incident Action Plan (IAP) and/or agencies that mandate that one be implemented.

It is important to note that on smaller and/or less complex incidents, the Incident Commander (IC) or an assistant may perform this task. Because smaller or less complex incidents will not be expanded to include a multitude of outside resources, then the Incident Commander (IC) may choose to handle these issues on their own. While expansion of Command and General Staff is not a requirement, neither is the writing of an Incident Action Plan (IAP) on simple incidents; however, it is in the best interest of the incident, the Incident Commander (IC), and the operational workers to have written directions, or to document those directions even on larger single-agency single-resource incident. This assists in making sure that there is no breakdown in communications regarding what should be done, how it should be done, and when those tasks should be completed. Additionally, a written Incident Action Plan (IAP), even if not a formal plan like what is used on larger or more complex incidents, helps to reduce liability for all involved in the response, and more importantly it helps to reduce the potential for death or injury.

The content of the Incident Action Plan (IAP) is incident driven. This means that all the factors of the incident are taken into consideration, and the best plan to meet the needs of the incident is created. The level of detail that will be included in an Incident Action Plan (IAP) will depend on the size and complexity of the response. It is critical that the plan is accurate and that it completely transmits the information that was produced during the planning process.

The approval of the Incident Action Plan (IAP) during the Planning Meeting provides a roadmap for personnel working in and around the scene. When the information identified in the Planning Meeting is put into an Incident Action Plan (IAP), then a comprehensive and detailed roadmap of how to overcome the incident evolves. The types of actions that the plan might (or might not) specify include:

- Control objectives
- Tactics
- Resources
- Organization

- Communications Plan
- Medical Plan
- Traffic Plan
- Other appropriate information

The design of the NIMS method and the ICS companion method dictates that the Planning Section is responsible for taking the approved information and plans from the Planning Meeting and using that information to create the Incident Action Plan (IAP). This means that they will gather, arrange, and print the needed information before the beginning of the next operational period. The various pieces of information that are provided to the Planning Section by other Command and General Staff (or their subordinates) are collated, and this creates the required information for that operational period. The information will be combined into one document that should make it easy for personnel to read and understand the directions for an effective response.

When a fully staffed Planning Section is in place, and in some incidents even when every position in the Planning Section is not activated, the Planning Support Unit will be responsible for compiling the completed ICS forms for the Incident Action Plan (IAP). They will also prepare the Incident Action Plan (IAP) cover sheet. They will also undertake a quality control check on the Incident Action Plan (IAP) to ensure that it is clear, concise, and that it meets the necessary compliancy standards of NIMS, and the ICS companion method.

15.9.1 The Incident Action Plan (IAP) Cover Sheet

The incident name, date, and operational period all need to be on the cover of the Incident Action Plan (IAP). The Planning Support Unit should also check with the agency on any other policies regarding the cover or contents of the Incident Action Plan (IAP). It is important to remember that the Incident Management Team (IMT) is a guest. As such, they should respect the wishes of the Authority Having Jurisdiction's (AHJ) as their guest.

In creating the Incident Action Plan (IAP), the Planning Section creates an Incident Action Plan (IAP) cover sheet or uses one that is premade. As of the writing of this book, there was no standard ICS Form for the Incident Action Plan (IAP) cover sheet; however, there are available cover sheets from different agencies and businesses that can be used. Most of the available cover sheets that can be used are Microsoft Word or pdf fillable forms. These can in some instances be slightly different based on the type of incident they were designed for. These cover sheets can be obtained by doing an online search for "Incident Action Plan Cover Sheet". It would be in the best interest for local jurisdictions to identify the type of cover sheet that meets their standards and that will likely fit their needs in a larger incident, then download and print multiple copies before an incident occurs.

The Incident Action Plan (IAP) cover sheet that you choose should provide the mandatory (and specific) information about the plan, and the incident. While this is important, it becomes even more important when there are state and federal resources that are responding. The mandatory information that should be on this cover sheet must be consistent with, and reflect, the standards of the NIMS method, and the ICS companion method.

The first and most basic requirement that will help make the cover sheet NIMS compliant is that it must be printed on plain white paper using black ink. Logos are allowed on the cover sheet; however, they cannot be color logos and they must be in black ink as well. Additionally, this logo should not be of high resolution, or so overly graphic so that it makes the file size larger. This may pose problems with electronically sending or sharing files with those that need to be informed. To be NIMS compliant, the cover sheet must contain the following information:

- Title of the incident
- Sequential plan numbers (first plan should be 01, the second plan should be 02, etc.)
- Incident type (e.g., severe storm, terrorist attack, hurricane)
- Operational period (date and time)
- A signature block for Incident Commander (IC)

Additionally, if this is a unified response, then additional information will be needed. A unified response in this instance refers to responses that involve state and/or federal agencies, or a major assisting outside resource such as Disaster Assistance Mortuary Team [DMORT], Disaster Medical Assistance Teams (DMAT [both explained better in Chapter 16]), a preorganized Task Force, or any other specialized group or agency. If these resources are involved in the incident, then this additional information will be needed on the cover sheet for the Incident Action Plan (IAP);

- The declaration number (If a Stafford Act incident)
- Any state or other unified command group organization declaration number
- If activated, the Joint Field Office (JFO) address with US national grid coordinates
- Signature blocks for the Federal Coordinating Officer (FCO), the State Coordinating Officer, and/or a Unified Command Group

One thing that is common among almost of the cover sheet is a list of ICS Forms included in the Incident Action Plan (IAP) with checkboxes next to the ICS Forms. In using the checkbox method, it will allow the person creating the cover sheet to use the checkboxes to identify what ICS Forms and incident attachments are included in the plan, rather than typing them out each operational period. This makes it less time consuming and more efficient. Forms that are listed in the checkbox section on the cover sheet typically include the following:

- ICS Form 201-Incident Objectives
- ICS Form 202-Safety Analysis
- ICS Form 203-Organization Assignment List
- ICS Form 204-Assignment List
- ICS Form 205-Radio Communications Plan
- ICS Form 206-Medical Plan
- ICS Form 208-ICS Safety Message
- ICS Form 220-Air Operations (if Air Operations is activated)
- Traffic Plan (internal and external to the incident)
- Incident Map (top section or sketch)

It is important to note that the Medical Plan (ICS Form-206) provides important information about emergency procedures, and this should be included in every Incident Action Plan (IAP) that is released, and it should be well covered at every Operational Period Briefing. This will be discussed more in-depth in the Operational Period Briefing section in the following.

In most instances, there is a blank section with the title of Optional Information. This section is used to identify anything that is not already listed on the cover page. The types of additional information that could be included are too numerous to list. Some of the more common optional information includes Facilities Maps (especially if new facilities were added or changed location), Traffic Plan Maps, Logistics Section notes or warnings, Finance and Administration notes, and Demobilization notes.

It is important to ensure that what will be included in the Incident Action Plan (IAP) is listed on the cover page. It is also important to make sure that deliverables on the cover page are available to include in the plan, if they are found to be needed. The most common components to include are as follows:

- Incident goals.
- Operational period objectives.
- The major strategies and priorities that should be addressed, in the specified operational period to achieve the goals or control objectives.
- The priorities and the general approach to accomplish the objectives.
- The methods developed by Operations to achieve those objectives.
- The Organization list with organizational chart showing primary roles.
- Assignment list with specific tasks.
- Critical situation updates and assessments.
- Combined resource status updates.
- Health and safety plan to prevent injury or illness to personnel.
- A Communication plan.
- A Logistics support plan.
- A Medical plan.
- Incident maps.
- Additional plans which are dictated by factors of the incident.

15.9.2 ICS Forms Integration with the Incident Action Plan (IAP)

The ICS Forms that will be included in the Incident Action Plan (IAP) will provide guidance to not only those in the Operations Section but to everyone involved in the response. As has been stated before, the whole purpose of all Command Staff, General Staff, and all others that are under the umbrella of the ICS method is to support the Operations Section. These other individuals are in place so that the Operations Section can deal with the incident while everyone else works to make sure they are successful.

During the planning process, many different ICS Forms are filled out. These forms have been nothing more than pieces of paper with information on them. When added together, they become the basis for the Incident Action Plan (IAP). The Incident Action Plan (IAP) is a compilation of forms, so we will describe what purpose each form has in the Incident Action Plan (IAP).

15.9.2.1 The ICS Form 202-Incident Objectives

This is the first page of the Incident Action Plan (IAP) after the cover page. It is used to describe the basic incident strategy and control objectives. It also offers the weather information for the operational period and any safety considerations for that specific period. The Incident Commander (IC) was the responsible person who prepared incident objectives, even though the Planning Section Chief (PSC) may have been appointed to prepare the form for the Incident Commander (IC). Prior to creating the incident objectives, they had to take into account not only the total incident situation but also the political, legal, and fiscal aspects of the response. They also had to describe the strategy that should be used and how they wanted the incident controlled. In filling out the form, they gathered the information on the predicted weather and weather patterns for that operational period. They also provided information on the basic safety considerations. This information was based on the operational period for the day that it would be used as the plan.

15.9.2.2 The ICS Form 203-Organization Assignment List

This list provides personnel with information on the units that are activated during the operational period. It also provides the names of the personnel that are staffing each position or unit. The information that is in this form is used to assist in completing the ICS Form

207-Incident Organizational Chart, which will be posted at the Incident Command Post (ICP). This list is prepared and maintained by the Resources Unit Leader (RESL), who maintains all resource information under the direction of the Planning Section Chief (PSC). The ICS Form 203-Organization Assignment List is duplicated and attached to the ICS Form 202-Incident Objectives Form and then is provided to all recipients as part of the Incident Action Plan (IAP).

15.9.2.3 The ICS Form 204-Assignment List

The ICS Form 204 (Assignment List) is used to provide the Operation Section (and their corresponding personnel) of the incident assignments for each resource. At the completion of the Planning Meeting, the assignments are agreed upon by the Incident Commander (IC) and the General Staff. This is added to the Incident Action Plan (IAP) so that assignment information can be given to the appropriate Units, Divisions, and Groups who will then apply the assignment to each resource. The Incident Action Plan (IAP) is formatted in such a way that each Division or Group will have their own page so that there is no confusion about who should do what.

The ICS Form 204-Assignment List will identify the supervisors and contacts of the Division or Group for that operational period. The Assignment list will identify the type of resource, the kind of resource, and the number of resources assigned to each Division or Group for that operational period. This Assignment List will also identify the method(s) of transportation, the work location, and at what times they will be required to work. It will also specify what will be done by whom, and it will list any special instructions or information for the given Division or Group for an operational period.

More often than not, the hazard mitigation measures that are acknowledged in the ICS Form 215A (Incident Action Plan Safety Analysis) are provided on the specific Division or Group Assignment List as it relates or affects them. The ICS 204-Assignment List is typically attached to the Incident Objectives as part of the Incident Action Plan (IAP). This Assignment List is duplicated and attached to the Incident Objectives and then provided to all recipients of the Incident Action Plan (IAP).

15.9.2.4 ICS Form 205-Communications Plan

The ICS Form 205-Communications Plan is used to provide the spectrum of radio frequency assignments (in one document) for an operational period. This plan is written from the information that was filled out in the ICS Form 216 (Radio Requirements Worksheet) and the ICS Form 217 (Radio Frequency Assignments Worksheet) by the Communications Unit Leader (COML). After completing these worksheets, the Communications Unit Leader (COML) will provide the frequency assignments on the Communications Plan. This is done so that it is widely known what frequencies are for work, emergencies, command, and various other important jobs. These frequencies (and who should use each) are almost always inserted into the ICS Form 204-Assignment List. When added to the Incident Action Plan (IAP), it is disseminated to all individuals who should receive a copy of the plan.

15.9.2.5 ICS Form 206-Incident Medical Plan

The ICS Form 206-Incident Medical Plan is created to provide information and locations of incident medical aid stations, transportation services, hospitals, and the medical emergency procedures for the incident. Prior to the Planning Meeting, the ICS Form 204-Medical Plan is prepared by the Medical Unit Leader (MEDL) and reviewed by the Safety Officer (SOFR) to ensure completeness. It is approved at the Planning Meeting and is added to the Incident Action Plan (IAP).

The Medical Plan should be disseminated to all individuals who receive an Incident Action Plan (IAP). When inserting it into the Incident Action Plan (IAP), it is usually an attachment to the Incident Objectives. Specific information from the medical plan regarding incident medical aid stations and medical emergency procedures might also be taken from the plan and placed on Assignment Lists so that it is readily available in an emergency.

15.9.2.6 ICS Form 208-Safety Message

The ICS Form 208-Safety Message is an optional form for the Incident Action Plan (IAP). This is used for information that needs to be better explained from the ICS Form 215-A (Incident Action Plan Safety Analysis) on that form. The ICS Form 208-Safety Message can be included to provide further guidance on safety measures or in cautions and potential safety threats. It will typically be added if it is decided at the Planning Meeting that there needs to be more clarification or more detailed information regarding safety (beyond the ICS Form 215A-Incident Action Plan Safety Analysis). If additional pages are needed (beyond the first page), then the same form should be used for those additional pages, and it should be put in proper sequence, and the pages should be numbered identifying the page and the total pages (e.g. 1 of 3 pages).

When used, this form will be submitted by the Safety Officer (SOFR), and this form is used specifically for the Incident Action Plan (IAP). The ICS Form-208 (Safety Message) will be printed and inserted into the plan. It will be given to all personnel that receive an Incident Action Plan (IAP) and should be attached to the individual Unit or Group assignment list.

15.9.2.7 ICS Form 220-Air Operations Summary

When air operations are in use during an incident, the ICS Form 220 (Air Operations Summary) is essentially a quick reference list that provides the Air Operations Branch with the number, type, location, and specific assignments of air resources. This includes rotary, fixed wing, and in some instances UAV (drone) aircraft. The Air Operations Branch Summary (ICS 220) is filled out by the Operations Section Chief (OSC) or the Air Operations Branch Director (AOBD) prior to, or during, the Planning Meeting. When air resources are during an incident, the assignment information will be gleaned from the Operational Planning Worksheet (ICS Form 215), which is completed during the Planning Meeting. Specific call signs and designators for the assigned air resources will be provided by the Air and Fixed-Wing Support Groups.

After the ICS 220 is completed, the form will be given to the Air Support Group Supervisor (ASGS) and the Fixed-Wing Coordinator personnel to assign aircraft to a designated aircraft. These personnel complete the form by indicating the designators of the helicopters and fixed-wing aircraft who are assigned missions during that operational period. This information is provided to Air Operations personnel who, in turn, give the information to the Resources Unit. Usually, only those in the air support group will receive this in their copy of the Incident Action Plan (IAP).

It is important to note that the ICS Form 220 should not include aircraft that are utilized for rescue or medical purposes. If aviation resources will be assigned for rescue or medical transport, they should be referenced in the Medical Plan (ICS 206) and be coordinated with the Medical Unit Leader (MEDL). As such, these types of aircraft will be identified on the ICS Form 206-Incident Medical Plan.

15.9.2.8 Incident Action Plan Map

The Incident Action Plan (IAP) map is a specialized map that is usually prepared by the Situation Unit Leader (SITL), a Display Processor (DPRO), or a Geographic Information System Technician (GIST). The reason that is considered specialized is that it provides detailed information about the incident. Types of information that should be identified on the map include the following:

- Locations/areas of impact
- Locations/areas under control
- Locations/areas in the recovery process
- Where the clean-up process is ongoing
- Locations sampled for contaminants
- Locations that are being monitored and potentially the status of those locations
- Debris drop-off locations
- Debris removal status
- Restoration status
- Search and rescue status
- Reoccupation status of affected areas
- Plume dispersion
- Other important information

The Incident Action Plan (IAP) map should use approved NIMS symbols to identify various locations and their status. This map will be important to supply situational awareness for all who receive and work within the incident.

15.9.2.9 The Traffic Plan

The Traffic Plan is important to ensure the safety of personnel and the public. The Traffic Plan establishes traffic control strategies to facilitate various work and to reduce the risk during evacuation or in working along roadways. These Traffic Plans will also be created to reduce traffic congestion and to provide quick access to operational support when needed. It also identifies strategies to move large numbers of personnel into and out of the incident.

Whenever there is an evacuation of the population, roads can become congested, and it can reduce the ability of personnel to access the incident. With a Traffic Plan, control of traffic can allow specific routes to be identified that allow an easier access for personnel while keeping most of the evacuees to take a specific route. While not all evacuees will remain on the evacuation route, this strategy can greatly reduce the influx of traffic that is evacuating, thereby allowing fewer problems with the transportation of personnel into the incident.

In incidents where large numbers of personnel are needed, but their vehicles are not needed, the Traffic Plan can identify loading areas for busses and other transportation forms to be embarked, as well as unloading areas near the scene where those transportation vehicles should park and be disembarked while still allowing traffic flow. By doing so, these plans reduce risk for personnel, help traffic not to be blocked, and allow emergency traffic to maneuver the incident. The plan will also identify control points to manage the flow of traffic and identify what information has been communicated to the public about traffic control and the signage as it relates to evacuation.

The Traffic Plan will identify optimal routes for evacuation and operational re-entry with special consideration given to locations of pick-up points, reception centers, and shelters. These plans identify the strategies for staging and deploying personnel and the traffic considerations for the influx of additional resources. The plan will identify what routes can be taken if the evacuation corridors need to be shut due to hazards, and it identifies places of refuge as last resort. The plan will usually be developed and prepared by the Ground Support Unit Leader (GSUL) who will describe the routes of travel. The Situation Unit will be involved with transferring the Traffic Plan information to a finished map product. In larger incidents, technical experts can be used to provide more traffic flow during an emergent situation. The distribution of the traffic plan should be to all Command and General Staff as well as in the Incident Action Plan (IAP) that is given to all Units, Divisions, and Groups.

15.9.2.10 Weather Forecast

The weather forecast is important to almost every incident. For this reason, a weather forecast should be included in every Incident Action Plan (IAP). The weather forecast should either be prepared by a meteorologist or it should be obtained from the National Weather Service (NWS). The weather forecast should provide current weather conditions and the predicted weather conditions for the operational period. If needed, an expanded safety message should be prepared by the Safety Officer (SOFR) that covers specific concerns during weather event and what to do during those weather conditions (based on what is expected).

15.9.2.11 Additional Information

It is important to note that each incident is unique and has distinct circumstances that are affiliated with the incident. Because of the distinctiveness of each incident, the previously mentioned forms and documents that are typically included in an Incident Action Plan (IAP) may or may not be useful. Furthermore, additional information, plans, or even additional ICS Forms may need to be added to the Incident Action Plan (IAP). Those that are involved in the planning process will need to determine what is needed to save lives, reduce risk, prevent injuries, and mitigate the effects of the incident. When adding additional information to the Incident Action Plan (IAP), it is important to keep the information direct and to the point rather than wordy, and it should be easy to understand.

15.9.3 Early Distribution of Incident Action Plan

Whenever possible, the Incident Action Plan (IAP) should be released as soon as it is completed to individuals who will need the information earlier to allow preparation time for specific units and individuals. For this reason, as soon as the plan is produced, it should be given to the following:

- The Ground Support Unit Leader
- The Food Unit Leader
- The Supply Unit Leader
- Air Operations
- Agency Dispatchers
- The Medical Unit
- The Communications Unit

The listed Units can use the additional time of having the Incident Action Plan (IAP) to locate and/or order supplies, to identify strategies, and to plan for variances in the number of personnel that affect their work. By providing the Incident Action Plan (IAP) to these Units and individuals early, it allows them to reduce waiting time for the necessities of the response and/or recovery.

15.9.4 Regular Distribution of the Incident Action Plan

When the Incident Action Plan (IAP) is printed, an adequate number of copies should be produced to provide at least one copy to all supervisory personnel at the Section, Branch, and Leader levels, as well as each member of the Command and General Staff. Due to the distinctiveness of each incident, the distribution of the Incident Action Plan (IAP) varies with each situation. In some cases, the number of personnel and the total number of copies will play a factor in determining how the plan should be distributed. As a rule of thumb, it is acceptable to add an additional 10% more of the plans than the number of recipients to cover newcomers and for misplaced plans. It is also important to remember to keep an original copy and a corrected copy for the Documentation Unit.

During the vast majority of incidents, the same plan will be used by all personnel. It is important to remember that a limited number of Incident Action Plans (IAP) may require controlled distribution. This holds especially true if Investigations and Intelligence Gathering (I/I) is part of the incident at the operational level. In these instances, there may be a special Incident Action Plan (IAP) for Investigations and Intelligence Gathering (I/I) and a separate plan for the remainder of the response. Also, in some incidents, it may be determined that Air Operations should only have the plan for that function as well. An option to consider for distribution is the bundling of Incident Action Plans (IAPs) for specific groups. Whether a special Incident Action Plan (IAP) is needed, or if the plan is the same across the board, the plan can be bundled by the number that require the plan, or a different plan in those groups.

In some incidents, the Incident Action Plan (IAP) can be mostly distributed at the Operational Period Briefing. The Operational Period Briefing is immediately before the operational period begins. In most instances, the vast majority of individuals who should receive an Incident Action Plan (IAP) will be present at this briefing, so a large portion could be handed out. They could also be separated and bundled and handed out to the correct individuals based on their function at the Operational Period Briefing.

Once the Incident Action Plans (IAPs) have been distributed, the Planning Section should evaluate the effectiveness of the procedures to create the plan and the distribution of the plan. By doing so, improvements and/or adjustments might be made to make the process quicker, the plans more complete, and the distribution more efficient. After the Operational Period Briefing, the entire Planning Section should prepare to do it all again.

15.9.5 The Operational Period Briefing

Just prior to an operational period, the day will start for supervisors with an Operational Period Briefing. This briefing can also be referred to as the Operations Briefing or the Shift Briefing. Often, both supervisors and tactical personnel will receive the Incident Action Plan (IAP), and then they are briefed on that plan. In this briefing, the participants will be informed about any critical operational or safety issues and be provided logistical information. The participants also have the chance to ask questions of the people who created the plan because these individuals will almost always be present for that briefing.

The Planning Section Chief (PSC) is responsible for facilitating the briefing and, most of the time, the Command and General Staff who created the Incident Action Plan (IAP) will provide a short presentation. These presentations are provided to clarify the procedures that should be followed and in what way. This also allows questions to be asked that were not answered or that were unclear in the plan.

The Planning Section Chief (PSC) and the Operations Section Chief (OSC) share the responsibility for the Operational Briefing. The Operations Section Chief (OSC) has the responsibility to ensure that all operations personnel are adequately briefed. Depending on the Incident Management Team (IMT) and the incident, there may (or may not) be an agenda for this meeting. If an agenda is used, the most widely used order would be the following:

- The Planning Section Chief (PSC) sets ground rules and discusses the agenda.
- The Incident Commander (IC) or Planning Section Chief (PSC) will present the incident objectives or confirm the existing objectives.
- The Situation Unit Leader (SITL) provides the current incident situation.
- The Operations Section Chief (OSC) provides an assessment of the current conditions and recent accomplishments.

- The Operations Section Chief (OSC) covers the work assignments and staffing of Divisions and Groups for the upcoming operational period.
- When needed, the Safety Officer (SOFR) may provide a safety briefing.
- The Logistics Section Chief (LSC) will provide a presentation on transportation, supplies, and possibly communications (if the Communications Unit Leader [COML]) is not present.
- The Finance/Administration Section Chief (FSC) will present any fiscal issues that need to be covered.
- The Public Information Officer (PIO) presents information on public information issues, including information that should not be released.
- The Liaison Officer (LOFR) provides a presentation on any interagency issues.
- When needed, the Communication Unit Leader (COML) will discuss important issues or changes to the Communication Plan.
- Usually the Medical Unit Leader will provide a brief presentation about changes to the Medical Plan and locations of medical assistance.
- When demobilization is imminent, the Demobilization Unit Leader (DMOB) will present a briefing.
- The Planning Section Chief (PSC) identifies any corrective actions to the Incident Action Plan (IAP).

If the incident is operational 24-hours per day, then the incoming Operations Section Chief (OSC) will provide the work assignments and staffing of Divisions and Groups for the upcoming operational period. Similarly, the incoming Logistics Section Chief (LSC), Safety Officer (SOFR), Finance/Administration Section Chief (FSC), Public Information Officer (PIO), Communications Unit Leader (COML), Medical Unit Leader (MEDL), and Liaison Officer (LOFR) will provide the briefing for the operational period that will be starting.

When the Operational Period Briefing is completed, those in attendance will be released. They will then travel to the Staging Area (or other predesignated area) where they will meet their personnel. The supervisors will then brief their assigned personnel on their respective assignments.

15.9.6 Beginning of the Operational Period

It should be noted that some Incident Management Teams (IMTs) consider the briefing to personnel as the beginning of the operational period, while other teams consider the end of the briefing as the start of the operational period. The beginning of the operational period will actually be based on a specific time; however, the personnel briefing will be held before the operational period, while others start the briefing at the designated time for a new operational period. Once the personnel briefing is over, they will execute the plan and assess the progress throughout the day. Assessing the plan should be a continual process for all personnel working on the incident, and any concerns should be passed up the chain of command.

15.9.7 Special Planning Meetings

Not all meetings that will be held as part of planning are listed in the Planning P. The reason these are not covered in the planning process is because these are occasional meetings, most of which revolve around special circumstances. These special circumstances may be a transitioning from one Incident Management Team (IMT) to another team (or returning command to the local jurisdiction), debriefing/close out meetings, demobilization meetings, and other various meetings. While not every type of meeting will be covered, a brief synopsis of most meetings will be given in the following.

15.9.7.1 Transition Meetings

Transition Meetings are conducted when the incident is transitioning from one Incident Management Team (IMT) to a different Incident Management Team (IMT). A Transition meeting is also held when the incident is being turned over to local jurisdiction who was the Authority Having Jurisdiction (AHJ). In most instances, the Incident Commander (IC) will assign the Planning Section Chief (PSC) from the existing team to facilitate a Transition Meeting. The Transition Meeting follows the same principles and transfer of command that was described early in this chapter for a Transfer of Command

15.9.7.2 Debriefing/Close-Out Meeting

Debriefing/Close-out Meetings are usually held with the Agency Administrator and/or the Authority Having Jurisdiction. In some instances, this will be held in combination of or within a Transition Meeting. The Planning Section Chief (PSC) is usually assigned by the Incident Commander (IC) to facilitate this meeting.

Debriefing/Close-out Meetings are the time when the incident has come to an end. The Incident Management Team (IMT) reports their accomplishments during their tenure. In most instances, the Incident Management Team (IMT) will provide an Executive Summary or Close-out Narrative which will describe various aspects of the work they did during the incident. No matter what the name that is used, the document should include the successes that the Incident Management Team (IMT) had, the challenges they faced, and any recommendations for the future. The recommendations could be mitigation measures to prevent future incidents, incident management improvements, ways to better facilitate outside resources, or a whole host of other issues that may be helpful in the future. It is important that the document is put together not as a way of pointing fingers, but rather in helping the agency prevent or manage future events.

15.9.7.3 Public Meetings/Press Conferences

Depending on the situation, there may be times when public meetings or press conferences are needed. If there is certain information that is classified or that should not be known by the general public, then there may need to be a meeting to plan the public meeting or press conference. In preparing for this meeting, those that will be involved should have a brief meeting with the Public Information Officer (PIO) and/or representatives from the Joint Information Center (JIC) if it is activated, to lay out an agenda and to determine what information should, and should not, be released. It is important to note that all incident management personnel should be on the same page so as to not cause public concern or mistrust.

15.9.7.4 Demobilization Planning

Demobilization planning can involve members of the Command and General Staff meeting with the Demobilization Unit Leader (DMOB), and usually does. It is important to note that demobilization will usually occur throughout an incident. Early in the process of incident management, a meeting to determine the procedures and priorities for demobilization should be scheduled. While the Demobilization Meeting is not a requirement, the best plans are developed when the Demobilization Unit Leader (DMOB) provides the basics for the plan, and Command and General Staff have input on the plan and the intricacies that may be involved. When the Command and General Staff are involved in the planning for demobilization, it allows them to better understand the procedures and priorities. There may need to be additional Planning Meetings for demobilization when large numbers of resources are released at one time, such as the end of the incident.

Real-life Incident

During the Deepwater Horizon incident in April of 2010, it was easy to see that the effects would cover a large geographical area. This was a monumental task for the personnel and the Command Staff that responded to the incident. It was reported that the Emergency Support Function eight (ESF-8; Public Health and Medical Services) was not represented at all of the planning meetings. According to the Florida Department of Public Health (2011), this led to a lack of communication which slowed response for mental health, medical conditions made worse by the spill, and more.

The After-Action Review (AAR) suggested that all individuals who play a key role in planning or their representative should be present during the planning process. This will allow them to have greater input on the Incident Action Plan (IAP), enhance communications, and allow a representative from every ESF function to have a hand in planning how their function can integrate into the overall plan (Florida Department of Public Health, 2011)

Chapter 15 Quiz

1 The _____ refers to a diagram of the Planning Process that is shaped in the form of the alphabetical letter.

2 A _____ is a tool to remove emotions and fears from the decision-making process and it allows the Incident Commander (IC) to make decisions based on the facts.

3 A _____ ___ _____ is the process of moving the responsibility for incident command from one Incident Commander (IC) to a different Incident Commander (IC).

4 If the structure of the incident needs to expand, the initial Incident Commander (IC) will usually fill out the ICS Form _____.

5 When creating the _____ ___ _____ _____, it is important to realize that the incoming Incident Commander (IC) or the Unified Command (UC) needs to have enough authority to do their job.

6 **True or False:** There may be times when the Agency Administrator may even need to approve something that is outside of the Delegation of Authority (DOA). These approvals should be done verbally to save time.

7 **True or False:** While identifying the particulars of a Delegation of Authority (DOA) is important, it is equally important that the authority given to the Incident Management Team (IMT) is returned to the Agency Administrator when the incident is over.

8 Having an Incident Management Team (IMT) assist is the procedure of moving the responsibility for incident command from one Incident Commander (IC) to another is referred to as:

9 **True or False:** After assuming command, a broadcast should be sent out over every radio frequency used on the incident, announcing a formal transfer of command.

10 As the facilitator, _____ _____ _____ should make it clear to all participants prior to the meeting that they should come prepared for the meeting.

11 **True or False:** To prevent meetings from getting sidetracked, they should always start on time and ground rules should always be given at every meeting

12 _____ _____ allow the facilitator to advise those involved in the meeting what is expected of them, and what is not acceptable.

13 As part of the _____ _____, all objectives will be reviewed or developed. As part of this process, the Situation Unit Leader (SITL) will usually be required to attend.

14 SMART objectives that are created for an operational period should use the preferred method of creating objectives by using _____ __ _____.

15 In this meeting, the Incident Commander (IC) or the Unified Command (UC) will usually meet with the Command and General Staff to garner ideas and input from these valued staff members.

16 The purpose of the _____ _____ is to review the tactics developed by the Operations Section Chief (OSC) and to finely hone them to be more effective.

17 In this meeting, you put together the Incident Action Plan (IAP) and ensure that everything you will need will be in place.

18 The _____ _____ _____ is a specific document that is issued to ensure that all personnel involved in an incident have a common understanding of what is expected of crews during the response or recovery.

19 **True or False:** The Incident Action Plan (IAP) should always be released to everyone at the same time.

20 In this briefing, the participants will be informed about any critical operational or safety issues and be provided logistical information. The participants also have the chance to ask questions of the people who created the plan because these individuals will almost always be present for that briefing.

Self-Study

FEMA Incident Action Planning Guide (2012). Retrieved from https://www.fema.gov/media-library-data/20130726-1822-25045-1815/incident_action_planning_guide_1_26_2012.pdf.

San Joaquin Operational Area Emergency Preparedness Committee (2015). The Incident Planning Process For Extended Operations: An All-Hazards Approach. Retrieved from https://www.sjgov.org/ems/pdf/ippparticipanthandbook10.26.15.pdf.

16

Integrating Incident Management into Hospitals

It is not how much we do; it is how much love we put into the doing

Mother Teresa

With the United States Fire Service seeing remarkable success in managing large incidents using the Incident Command System (ICS) method, hospital planners and hospital emergency managers began to think of how something similar to ICS could apply to hospitals. In the mid-to-late 1980s, they began to recognize that hospitals might reap rewards by integrating an Incident Management System (IMS) that was consistent with ICS principles. They began to investigate what options they had, and they began to look at how they could make ICS work for them.

It was not long before they came up with something that they called the Hospital Emergency Incident Command System (HEICS). This would later transform into the Hospital Incident Command System (HICS), which would also integrate with National Incident Management System (NIMS) and the ICS method. Much like ICS is a companion for the NIMS method of incident management, HICS has also become a companion method to NIMS. Because HICS integrates with the NIMS method, it provides all the benefits of state and federal resources, as well as other benefits, in the event they are needed. While incident management has come a long way from where it began, it is important that we understand where it came from, and certain historical perspectives. When we look to the past, we can see where we have been, which should help us plot a course for where we are going.

16.1 Hospital Emergency Incident Command System (HEICS)

The initial version of emergency incident management in hospitals was named the Hospital Emergency Incident Command System (HEICS). This system helped create the foundation of an IMS method for hospitals in the United States, and eventually in most of the world. While it would eventually become the Hospital Incident Command System (HICS), HEICS played a significant role as the foundational structure (California Emergency Medical Services Authority, 2014).

The designers of HEICS were involved in hospital emergency management, and they were keen to find an IMS method that would help them to prepare for, and respond to, a wide array of incidents. The initial problem with creating such a method was that it would need to be widely accepted. By being widely accepted, hospitals could integrate with each other as well as the varying disciplines might also interact with them in a disaster.

Emergency Incident Management Systems: Fundamentals and Applications, Second Edition.
Mark S. Warnick and Louis N. Molino Sr.
© 2020 John Wiley & Sons, Inc. Published 2020 by John Wiley & Sons, Inc.
Companion Website: www.wiley.com/go/Warnick/EIMS_2e

This goal was problematic because most hospitals fall into two categories; independent (for profit) hospitals and nonprofit organizations. Essentially, each hospital had its own management structure that met the needs of their hospital. In most instances, there was a staunch difference from hospital to hospital (San Mateo, 1998). There was no consistency in management or command structure from one to the other.

Even more problematic was the person who would usually be responsible for the overall running of the hospital was often an administrative person, not the first responder type. These individuals usually did not have any knowledge, experience, or skills in emergency response and integrating resources with other agencies. Interacting with other medical facilities was usually counterintuitive to business as usual. Based on these factors, the original designers of HEICS had many discussions related to whether an administrator with no emergency or response management knowledge would be capable of managing an emergency or disaster type of incident. Moreover, they questioned if an administrator was really the best person to manage an emergency incident.

Similar discussions revolved around what common management tasks were most equivalent to the management of day-to-day operations. Essentially, those designing this system wanted to identify similarities between the hospital job descriptions for daily operations and those functions that would be required in the command structure used in an emergency incident. It was quickly realized that not every command position in a hospital needed to be filled on every incident. Discussion continued among hospital emergency managers and others for many years. In all the discussion, there was a definite recognition of the value and importance for a hospital-based IMS system.

In 1991, the Orange County Emergency Medical Services (EMS) Agency received a grant from the California Emergency Medical Services Authority (n.d.). While HEICS had already been designed, there needed to be a Guinea Pig to test the system and to ensure that it worked. This iteration of HEICS was loosely based on the ICS structure being utilized in the FIRESCOPE program.

The pilot program was tested at Orange County EMS. Administrative staff who would function as managers in an emergency incident were introduced and trained on the HEICS method over a period of three months. At the end of the three months period, they would take on a full-scale exercise to identify the efficacy of the system in relationship to the hospital and managing the incident. Some were concerned about the relatively short timeframe that was used to prepare these individuals (90 days). Despite the concerns about the short timeframe and lack of practicing HEICS, positive results were witnessed in the full-scale exercise. While there were minor issues that needed to be worked out, the overall method tested in such a way that those involved knew they were on the right track.

At the end of the full-scale exercise, the system, and how it performed, was reviewed. It was also discussed how HEICS could be implemented in other medical facilities using the 90-day training model. Some concerns were raised about three months being an insufficient timeframe to train and implement HEICS into an institution, so a discussion and debate ensued. It was ultimately determined that the length of the implementation timeframe for each healthcare facility would vary, and it would likely be dependent upon a variety of factors. Those factors included the size of the facility, the amount of people devoted to the venture, the funding avenues available to sponsor the implementation, and the level of support (buy-in) for the system by upper and mid-level management. Consideration for these factors was then built into the program design for implementation at multiple and varying types of facilities.

Amazingly, the HEICS method quickly became the commonly accepted and utilized IMS method used in healthcare settings. The system was widely accepted because this ICS method provided a commonality of mission and utilized a mutual language. This is one of the reasons that the system was found to be usable across a multitude of healthcare facilities. Another positive aspect was that it allowed workers and senior managers from different healthcare facilities to seamlessly integrate into each other's facility and to be effective when doing so. The designers of HEICS also incorporated an organizational chart with generic position titles that closely mirrored most management circles. This helped to make this method in both private and public entities.

One of the key concepts that made HEICS method effective and acceptable to administrators was the responsibility matching from daily jobs to key positions in the command structure. This was achieved by creating an organizational structure that addresses many reoccurring characteristics of an emergency, then identifying how daily duties coincided with those positions. These responsibilities helped to provide a manageable scope of supervision in all positions and functions. It also allowed for accountability of resources and a manageable span of control. This is the number of individuals that one person can manage. By using this system properly, the healthcare facilities could also meet the compliancy standards for disasters, and later the NIMS compliancy that was ordered by Homeland Security Presidential Directive (HSPD) 5 and HSPD-8.

The HEICS method also prioritized responsibilities using Job Action Sheets. These Job Action Sheets provided job descriptions that identified a prioritized list of emergency response tasks that would usually need to be undertaken. These Job Action Sheets also helped to promote reporting by putting an emphasis on incident documentation. Standardized forms were created to thoroughly encompass the documentation of actions taken by both personnel and the facility as a whole. The Job Action Sheets helped pave a pathway that would lead to initiating detailed records. The benefits of these detailed records were that it provided accountability of, and for the staff, reduced the liability for the healthcare facility, and reduced financial costs related to the incident. This comprehensive documentation also created a way to account for financial expenditures that were above and beyond normal daily operations. This was especially important when the incident had recoverable costs from state or federal agencies that would partner with healthcare facilities that suffered a disaster.

HEICS was designed in such a way that many in the daily management structure of a hospital had a parallel task in the command structure. This was done through matching the job qualifications and duties with the specific role within the HEICS method. The previously mentioned Job Action Sheets provided guidance to ensure that if, and when, the person was assigned a specific position (in the Command Structure), then the specific tasks that were critical and required in managing an incident were not overlooked. Additionally, the organizational chart for incident management would be quite similar to the organizational chart for day-to-day operations.

Much like ICS method counterpart, the HEICS method was also appropriate across a diverse range of incidents. It did not matter if the incident was planned or emergent, it was relevant and effective in all instances. It was flexible enough that it could allow the command structure to expand or collapse based on the magnitude of the incident. The command functions were structured enough to provide effective management, yet flexible enough to meet the specific needs of any incident. It was also simple for someone trained in the HEICS method to transform from an administrative position to a valued member of the Incident Management Team (IMT) seamlessly and immediately.

Another positive aspect and selling point for the use of HEICS was that this method allowed for the rapid transfer of resources. Mutual aid resources within a healthcare system could easily transfer personnel and/or materials from one facility to another using less effort and with less red tape. Prior to the HEICS method being implemented, many healthcare facilities would need to cut through a multitude of red tape to share or utilize another facility's resources. With common Mutual Aid Agreements (MAAs), the sharing of materials and personnel was facilitated with less effort, and it created a way that transferring resources did not need to be a long, drawn-out, affair. Moreover, the HEICS method provided a common and consistent system of managing an incident, and it standardized the most commonly used terms. Standardizing the terms of the items used helped to ensure that the healthcare facility requesting resources received exactly what they wanted and what they needed, thereby reducing confusion.

While HEICS was structured, it also provided the freedom to create additional sections or branches in the organizational chart as might be required by incident needs. If a section or branch was needed in one specific area of the response, it could be added without activating a multitude of other sections or branches. HEICS was resilient enough that it allowed the healthcare facility to be customized to meet the needs of an individual incident.

In most IMS methods used around the world today, a Section is the part of the organizational level that has the responsibility for managing the incident at the upper levels of the organizational chart (Operations, Planning, Logistics, Finance/Administration). When assigned to a Section level, the Section Chief reports directly to the Incident Commander (IC). In most IMS methods, a Branch is the organizational level having functional, geographical, or jurisdictional responsibility for major parts of operations within an incident. The Branch level was organizationally just below the Section and reports directly to the Section Chief. When looking at an organizational chart of the ICS method, HEICS used the same Command and General Staff positions to lead an incident in a healthcare facility. In fact, almost all the structure is the same, utilizing the same names for positions (Section, Branch, Group, Unit, etc.). The structure is the same, and the structure of the command and general staff are usually an exact match; however, when it gets to the Branch or Group level, the naming and tasks become specific to healthcare facilities and the functions that coincide with managing a healthcare facility.

HEICS was an Emergency Incident Management System that was primarily based on ICS; however, it was adapted to be specific to healthcare facilities. Much like ICS, the HEICS method was designed to:

- Create a clear chain of command that is effective in all types incidents (both emergent and prearranged), regardless of size or type, by establishing a clear chain of command.
- Integrate personnel from different agencies or departments into a common structure to create a structure that can effectively address issues, and delegate tasks.
- Create logistical and administrative support that permits operational personnel to be more effective.
- Guarantee that key functions are not missed.
- Eliminates duplication of efforts.

As with any effective IMS method, the planning process plays a critical role in ensuring that the response is effective and orderly. As we saw in Chapter 15, a lack of planning creates more chaos, uncertainty, and complexity. In the HEICS (and later the HICS method), the incident planning process is and was undertaken for any event, and that process is not dependent on the size of the incident or complexity that an incident might present. The planning process in the HEICS method contained six essential steps:

- Understanding the hospital's policy and direction.
- Assessing the situation.
- Establishing incident objectives.
- Determining appropriate strategies to achieve the objectives.
- Giving tactical direction and ensuring that it is followed (e.g. correct resources assigned to complete a task, performance monitored.)
- Providing necessary support for the incident as needed (e.g. assigning more or fewer resources, changing tactics, assigning individuals with specialized knowledge.)

HEICS was a specific incident management method that identified some specific roles that were needed in the organizational chart. Each of these roles had a specific mission during an emergent situation, and there was a list of individual job descriptions which guided who should be appointed for that position.

HEICS provided more coordination between hospitals and other institutions involved in emergency incidents by utilizing logical management structures and duty descriptions. HEICS also created a clear path for reporting as well as evolving the system into a common and simple terminology system. This helped to reduce confusion. It appeared almost from the beginning of its use that the HEICS system was a success.

16.2 HICS

While HEICS was the initial iteration of HICS methods, this method was continually evaluated to ensure that the system was meeting the needs of incident response. HICS, which was formerly HEICS, has undergone multiple revisions since its introduction, the most recent being updated in 2018.

Its predecessor, HICS has proven to be effective and resilient, and it has been used for almost 30 years in hospitals and healthcare facilities throughout the United States. In fact, it has become so relied upon that it is used in many places around the world. As with most IMS methods based in the United States, HICS has a revision process that continually evaluates this method, and then incorporates changes based on the lessons learned. These lessons learned stem from the system being used in actual incidents, and when it is tested and practiced in exercises.

Changes to the HICS method have not been undertaken by an individual, and changes have not been made by individuals who have not used the program. HICS updates have been a representation of the feedback and evaluation of those who have actually used the system, the practitioners. In its latest iteration, HICS has become an incident management method that can be used by any healthcare facility. It can be used to manage a multitude of threats, to undertake large (or small) planned events, or it can be used to manage any type of emergency incident. HICS provides an organizational structure for incident management, and it provides guidance for the planning process. HICS also provides guidance for building both physical structures (e.g. tent emergency rooms, temporary decontamination) and organizational structures (all levels of supervision and tasks). Moreover, it provides guidance for adapting or modifying these structures.

Much like driving a car, the more it is used, the more proficient the user becomes. For those that work within the healthcare system, no matter what position you hold, it is strongly recommended that you take the course or courses, become certified, and that you become proficient in using HICS. Training courses are available online from the Emergency Management Institute (FEMA, n.d. Training), and it is offered at no charge. The more you understand the HICS method, and the more you use it, the more proficient you will be at using it when it really matters.

HICS has become such a mainstay that it is strongly suggested when healthcare facilities have planned events, that they use the HICS method to completely manage that event. Using HICS for planned events can help keep the skills fresh in the user's mind, and it can help enhance those skills. It can also help the user to identify any issues that they may have with using this method in their specific facility, and how to overcome those issues. By doing so, when a large-scale incident transpires, the user will be better prepared to utilize it in an effective and efficient way. This may be especially important when lives are on the line. The time to practice the skills you have learned in training with the HICS method is not during a life-threatening incident.

The latest version of HICS, much like the previous versions, is based on the Incident Command System (ICS) component of the National Incident Management System (NIMS). The primary difference is that it has been adapted to meet the needs of healthcare. Unlike the original version of HEICS, HICS applies to all hazards and all tasks affiliated with NIMS principles including prevention, protection, mitigation, response, and recovery.

HICS provides a system for managing all types of incidents. Because it is grounded in crucial (time proven) components, the HICS method is useful, relevant, and helpful to both public and private sector healthcare organizations. HICS creates a standard design which is not only successful but also immediately recognizable to those trained in ICS. This is important when an incident requires outside resources such as police and firefighters to be integrated into an ongoing incident. While HICS is uniquely designed to afford incident management to healthcare facilities, it is similar enough to the ICS component of NIMS, that other agencies and disciplines who are not familiar with healthcare (e.g. fire, law enforcement, emergency management, non-profits) can be easily integrated into the response. This can be done providing they understand the principles of ICS. This is primarily because the HICS system mirrors ICS in so many ways.

Similar to its predecessors, and other current IMS methods (including ICS, NIMS, and HEICS), it is a way of providing incident management that is only specific to a healthcare setting. It is designed in such a way that any hospital or healthcare facility can utilize this method. With HICS, it does not matter where the healthcare facility is located, the volume of patients that are being attended to, or the patient's sensitivities or perception. The HICS method is easily adaptable to meet the needs of emergencies and special events. It also does not matter what type of incident or severity of the threat being faced because the HICS method is designed to expand in order to meet the demand. Also, similar to its predecessor, HICS can be expanded to meet the needs of the incident, or the user can contract or intentionally collapse the organization (and the organizational chart) if the incident is overstaffed. The same can be said for ramping up during an event and staffing when it is needed or shrinking the amount of staff and the structure as the incident comes to a close.

By utilizing the HICS method, healthcare facilities implement a nationally recognized standard and an organized method that promotes successful incident management. This IMS method can operate within a healthcare facility, and it is set up in such a way that it supports the integration with community response partners. By the same token, if community response partners need healthcare in the field, those trained in the HICS method will be able to integrate with the ICS method. Healthcare providers from across the nation can understand and integrate with their counterparts, even if they are from 1500 miles away. Utilizing HICS, they will be able to respond and understand the hierarchy and the structure.

Because HICS is recognized and used by many healthcare facilities around the world, in most instances, international healthcare workers can also integrate into the system seamlessly. While HICS was created and primarily used in the United States, it is an IMS method and strategy that is used in various countries around the world, including modified versions in Iran and Australia. This means that healthcare personnel could travel to other countries that use HICS, and they can easily become part of the solution when it comes to managing a healthcare system disaster.

What makes HICS so effective is that it is an all-inclusive, all-hazards, incident management method that guides and assists healthcare organizations to manage multiple resources. Moreover, the HICS method is designed in such a way that it can be used in emergency incidents and day-to-day operations. As was previously mentioned, it can also be used for managing special events and basic hospital management.

The same basic principles that are used for managing a large contingency of individuals working on differing parts of a response can also be used for transporting patients daily. The same system that is used for a major disaster can also be used to manage patients in an Emergency Department. While HICS can be integrated into some daily settings within a healthcare facility, it is not meant to replace the basic management structure because it may not be conducive to the way business is typically accomplished in healthcare facilities. Still, healthcare facilities should consider utilizing HICS when possible. This sets up a more organized and understood response when a critical or catastrophic event occurs. This is primarily because the principles will be fresh in the memory of personnel, and it will be the routine they have become used to.

16.2.1 Triage Briefly Described

While the vast majority of individuals reading this book will know about triaging patients, it is also realized that there will be new students to the field who may need a brief explanation of triage. Triage is the process of sorting and classifying the injured. Triage is used external to the hospital in Mass Casualty Incidents (MCI), and it is used internally at healthcare facilities in emergency rooms and worse yet, when the unthinkable happens in a healthcare facility. It is important to note that when triaging patients inside a healthcare facility, the triage officer may have to triage staff members from the facility. In order to save more lives, an honest triage evaluation should be undertaken in the same manner for staff members.

In most instances, whether in the prehospital setting or at the facility, the task of triaging patients is relegated to an individual with an advanced level of emergency medical training such as a triage nurse, a paramedic, or an emergency (medicine) medical doctor. The proper triaging of patients should take the emotion out of making decisions, and evaluating patients should be based on their ability to survive and the severity of their condition. Someone without specialized knowledge in emergency medicine may have a more challenging time of sorting patients in this manner.

Over the years, there have been numerous methods developed for triaging patients in the prehospital setting. As of the writing of this book, the most common prehospital method in use is the START system. START stands for Simple Triage and Rapid Treatment, and when there is a multitude of casualties, it groups patients into four (sometimes five) categories. These four categories have a color-coded system that allows first responders, and upon arrival at a facility the medical staff, to identify what level of care is needed with just a glance. There is a large contingency of triage tools on the market, including color-coded triage tarps, triage flags, triage kits, triage wristbands, and more. The most typical prehospital triage method is to attach a large tag, in a visible location on the person's body. The tag will be written on so that anyone reading the tag can gain information quickly and easily. The color-coding system is standard across the United States.

Similar to the START method of triaging patients, there is also the JumpSTART method of triaging pediatric patients. The JumpSTART method is for patient between infancy and eight years old. It is currently being used around the United States for Mass Casualty Incidents (MCI)

When looking at triage tags, if the patient has a black tag, they are either deceased or they are beyond help. When we talk about being triaged as beyond help, we need to discuss what that means, and how that decision is made. In some instances, the triage officer will need to evaluate whether the extensive care that a single patient needs would take resources away from multiple other patients. When this is the case, and the available resources are limited, then they will black tag a patient.

As an example, let us say that one critically injured patient could possibly be saved with the limited resources available, but their injuries are so extensive, and it would likely take 10–12 trained individuals to do so. At the same time, those same resources (10–12 individuals) could potentially save four patients that were in critical condition but savable. Using this method of evaluation, the triage officer will need to make the decision based on how the most people can be saved with the resources on hand. In this instance, they may have to black tag the most critical patient to save others.

If a patient is given a red tag, then the patient needs immediate care. With immediate treatment, the patient will have a moderate to good chance of survival. If they do not receive immediate care, they likely will become deceased. Using the START triage system, these patients would become a high priority for the limited prehospital resources.

The yellow tags are used for those who require observation and do not need immediate care. Because their condition may deteriorate, they should be re-triaged at 5- to 10-minute intervals until transported. At the time of triage, their condition was considered stable, so they were not in immediate danger of death. While these patients will require medical care, they are stable enough to wait for that care at the time of triage.

Patients with green tags are considered the "walking wounded." These are the patients that can wait longer periods of time for medical care. At some point, they will still need medical care, but their injuries are minor compared to the injuries of others. Their condition is stable, and they can wait until those with more critical injuries have been treated. Even though they are stable, they too should be monitored every 10–15 minutes to ensure that the condition of green tag patients does not deteriorate before transport.

In some instances, there is a fifth tag that is used, especially when there is a community-wide disaster. When used, a white tag is given to patients with very minor injuries. These injuries might be cuts or scrapes, and they will usually not require a doctor's care. In many instances, the wounds can be washed and covered with a dressing, and in some instances, these may be taken care of in the field. While there is not an urgency based on life-threatening issues, in some instances, it may be prudent to assign one or two people at a receiving facility to care for these individuals upon arrival at the facility (in the initial stages) so that more room can be made to care for others that are in more serious condition. Rather than adding to the stress and confusion at the healthcare facility, it should be considered opening a temporary aid station either on the grounds of the facility (but not near where more serious injuries are arriving) or in another location such as a parking lot or another area of the facility.

While the prehospital services will bring in patients based on their triage, the receiving facility will further triage them upon arrival. The most common method for triaging patients in a facility is the Emergency Severity Index (ESI). The Emergency Severity Index (ESI) rates patients on a one to five scale, with one being the most serious patients and five being the least serious. This triage is based on the type of acute health care problems the patient has, and the number of resources the care is anticipated to require.

In a disaster or surge situation, patients would first be START triaged in the field to identify the priority of transport. Upon arrival at the facility, patients would be triaged using the Emergency Severity Index (ESI) method to further evaluate the priority for each patient. Triage tags are left in place, and at some facilities, the Emergency Severity Index (ESI) number will be added to the triage tag. In especially large surge incidents, the healthcare facility number will be added to the triage tag. In especially large surge incidents, the healthcare facility will also set up specific areas for specific triage numbers.

As an example, a minor wound care could be set up in close proximity to the hospital triage officer, potentially even in a cordoned off area for patients as identified as a five on the Emergency Severity Index (ESI), thereby freeing up space for more critical patients inside in the triage area. Rather than standing and potentially walking around in the triage area, they could be sent elsewhere, out of the way of more critical patients and staff that are working to save lives. They could be treated and released without adding to the chaos, confusion, and uncertainty. Similarly, those that were rated as a one on the Emergency Severity Index (ESI) might be placed close to areas where a staff member will be at all times so they could be more closely watched.

A key concept to remember is that triage can be initiated in the field (external to a healthcare facility), or in the facility (internal to the healthcare facility) should an incident occur in that facility. Triage tags can also be used when a surge of patients from personally owned vehicles inundates the healthcare facility. It is imperative that all medical staff in healthcare facilities understand triage and can be proficient with its use. By utilizing a triage system in connection with an emergency event, the HICS method is supported by better organizing priorities.

16.3 HICS Does Work for Incident Management

There is no doubt that the HICS method can (and does) save lives. HICS helps to manage patients and all the other problems that occur in an emergency event. HICS is a method that can help internal and external healthcare providers to treat patients, often within the first hour. In recent years, there have been multiple instances where hospital staff have used HICS and triage to better manage Mass Casualty Incidents (MCI).

When surge events occur, these strategies have proven to reduce the patient's wait time during these incidents. Healthcare providers need to realize that if a disaster or emergency strikes a healthcare facility, those involved and potentially injured will need to be triaged and taken care of in the most efficient manner. If the patients suffered a trauma external to the facility and were assessed in the field, they will need to be re-assessed upon arrival to a healthcare facility.

Numerous incidents have occurred where these methods have been used to save more lives in a hospital, including the Joplin MO tornado and the Las Vegas mass shooting incident. Because experience is often the best teacher, a brief review of a major incident in a healthcare facility is in order. This historical review will look at a major incident, and some of the actions the hospitals (and other healthcare facilities) took.

16.3.1 Joplin MO Tornado

On 22 May 2011, at approximately 5:40 pm, an EF-5 (200+ mph winds) multivortex tornado struck Joplin MO. This mile-wide tornado killed 158 people, injured approximately 1150, and it destroyed Mercy Hospital with a direct hit (Houston et al. 2015). St. Johns Mercy Hospital was an 890 000 square feet hospital with 220 rooms. Four people died in the hospital during the tornado, with others dying in the subsequent days and weeks (Silvis, 2015). An additional 83 physicians' offices that were affiliated with Mercy Hospital were also destroyed.

The damage to the hospital was catastrophic. Windows and Walls were blown out, and a sizeable percentage of the roof was ripped off. The tornado also did damage to emergency systems related to the hospital. Perhaps the most devastating and dangerous part of the incident was that the tornado ripped the large air conditioning units off the roof, and one of them landed on the back-up generator, totally destroying it (Houston et al. 2015)

The tornado was so fierce, the steel was twisted off its foundation more than four inches in one section of the hospital. Water sprinklers, water pipes, gas, and sewer pipes were damaged in the storm, and after the tornado passed patients, staff, and rescuers alike had to wade through water and sewage from broken pipes. Broken gas pipes leaked natural gas into the facility. Liquid oxygen tanks that provided the facility oxygen for patients were also damaged, and they were leaking into the atmosphere. Making matters worse, all methods of communication to the outside world were also destroyed. Phones, the PBX system, radios, antennas, and even cell phones that were left behind were nothing more than paperweights (Reynolds, n.d.).

The tornado was so violent that it blew most of the windows and doors out (Adler & Bauer, 2011). The only windows that were not blown out were those on the mental health ward because they were covered with a shatterproof film. From small piles to intermittent large piles, debris covered the hallways, staircases, and the entire premises. At the time of the tornado struck, there were 117 employees present, and 183 patients in the hospital. Twenty-four of the patients were located in the emergency department, one patient was in surgery, four were in the postanesthesia care unit, and 28 were in a critical care unit. Due to advanced warning systems, most of the patients were evacuated to the hallway prior to being struck by the F5 tornado. The one exception was the critical care unit. Most of the patients in the unit were so critical that they could not be moved. Even with the majority of patients and visitors sheltering in the hallways, when the storm blew out virtually every window, the hallways only provided minimal protection. The high-powered winds ripped IVs out of numerous patient's arms, and the IV poles became flying debris. There were also reports that some patients and visitors were thrown about like rag dolls, and many individuals were injured by the flying debris. These flying debris injuries happened both inside and outside of the hospital because the tornado also wiped out a large part of the City of Joplin (Adler & Bauer, 2011). The storm was so violent that CNN reported that gurneys were found more than five blocks from the hospital, and X-rays were eventually found more than 70 miles away in a totally different county (CNN Wire Staff, 2011).

Just moments after the tornado passed, it was obvious to the survivors (left inside the facility) that the hospital would need to be evacuated. In many cases, they initiated the evacuation with no central direction. Staff and visitors began to evacuate patients immediately. Emergency Department medical doctors and nurses began to evaluate the patients that were in their area and then they began to evacuate their patients outside. Once outside, ambulance radios were the only form of communications that the hospital had until a communications unit was delivered.

While the hospital itself had suffered a direct hit, so had the City of Joplin. Within minutes, immediate life-threatening injuries started arriving. The injured people of the community knew to go to the hospital in hopes of finding someone that could save their, or their loved ones, life. The majority of the injuries were life-threatening and included arterial bleeds, impaled objects, open head wounds and more. One person has a stop sign that impaled their body. To treat the arriving patients with life-saving medicines, staff members had to break into drug-dispensing machine. The storm had rendered them inoperable. Within the first five minutes, someone arrived with a deceased patient and asked where to put the body (Reynolds, n.d.). Inside the structure, it was obvious that the uninjured staff at the hospital could not quickly and effectively evacuate the hospital by themselves, and they could also not turn away severely injured patients. Without action, these people would die, but there was no way to communicate.

Off-duty hospital staff self-deployed to the hospital within minutes to see what they could do to help. While nobody knows when or how public safety was notified, or when the first contact was made, EMS, firefighters, police, and volunteers responded to the hospital. Even with all the help that showed up, it still took 90 minutes to evacuate the nine-story hospital.

At the same time evacuation was ongoing, other staff members were treating immediate life-threatening injuries. Individuals that were evacuated were initially sent to one of three locations on hospital grounds. The triage center on the west side of building, which was just outside the Emergency Department, the triage center on the east side of the building, and a staging area for the walking wounded at the Conference Center.

Inside the hospital, operations were slowed by the debris left in the aftermath. The tornado was so violent that many of the doors were ripped from their hinges, steel water lines broke, ceilings fells, and what was clean and orderly just minutes before was nothing more than a debris field. Patients and rescuers alike had to work through a maze of debris to evacuate, and there were dangers and hazards everywhere (Reynolds, n.d.). While there is no documentation to the effect, looking at the damage that was done to the hospital, at least some patients and staff had to wonder if the building would collapse.

The outside grounds of the hospital were also covered with debris. Cars were tossed around like matchbox toys, and the hospital's helicopter was on its side and severely damaged. From the first glance at that medevac unit, it was obvious that it could not be used for transporting critical patients. Both living and deceased victims were trapped in their cars, some crying for help. In total, over 18 000 cars were destroyed by the tornado throughout the entire city (Reynolds, n.d.).

Because the devastation was citywide, it quickly became obvious that the standard equipment that might be used in an emergency would be in wide demand. There were so many injured that all the equipment resources were used up in no time. There was a limited supply of backboards, stoke baskets, and various other stabilizing devices typically used to move injured patients were unavailable, almost from the onset. Nontraditional and available traditional types of equipment were used to evacuate patients, injured staff, and visitors. Doors, wheelchairs, plywood, mattresses, and other equivalent items were used to help carry patients out (Reynolds, n.d.). Off-duty nurses and volunteers with trucks began evacuating patients, helping the already overwhelmed ambulance service (Adler & Bauer, 2011).

Community members continued to arrive at the hospital with more arterial bleeds, collapsed lungs, more impaled objects, skin and muscles ripped open to the bone, and much more. Lives continued to be saved while the hospital was being evacuated. Outside resources were integrated into the evacuation as well as the treatment of new patients (Patterson, 2014).

Mercy Hospital had gone above and beyond in preparing for a disaster, but the hospital disaster trailer was destroyed by the tornado. It would later be found two blocks away, ripped to pieces. Triage tags, emergency paperwork, and many of the supplies that would be helpful in evacuating patients and treating incoming patients was all destroyed or carried away. The predetermined Incident Command Center (ICC) and the alternate Incident Command Center (ICC) were also destroyed and littered with debris.

Personnel adapted to the situation they faced and came up with a new game plan. Within 10 minutes of the tornado passing, a temporary Incident Command Post (ICP) was set up at the onsite Rehab Building to facilitate the evacuation of patients. Additional Incident Command Posts (ICPs) were set up at the local Emergency Operations Center (EOC), a clinic in Neosho, MO, Freeman Hospital in Joplin, Mercy Hospital in Springfield MO, and various other locations. With the tornado destroyed or carried of the supplies needed and the documentation papers, the tasks at hand become substantially more difficult. The task of identifying which patients were evacuated, their condition, and where they were transported became a major task (Reynolds, n.d.).

An ambulance staging area was set up. Along with volunteers in trucks, patients were transported to one of many locations based on the patient's condition. Only one of those locations was a local hospital. With transportation of patients being a major priority, ambulances were requested from nearby communities. The first of these ambulances began arriving in the first half-hour. Over the first two weeks after the disaster, over 100 ambulances from seven different states would arrive to assist in treating and transporting patients (Patterson, 2014). This was facilitated by utilizing the HICS method, and it proved highly effective.

Patients were triaged, and those that were walking wounded or stable but placed in wheelchairs were sent to the Catholic High School or to Memorial Hall, a local community building. Patients that had critical injuries were sent to the Freeman Health System Emergency Room, just two miles away. Even while the evacuation was ongoing, even more patients were coming to the hospital with life-threatening injuries. These individuals with tornado-related injuries were being treated on a street just outside of Mercy Hospital, then transported to Freeman Hospital or other locations.

The tornado strike to the hospital created a multitude of other issues. Beyond the preplanned Command Posts being rendered inoperable due to the storm, the hospital's Emergency Operations Center (EOC) was filled with debris, making it inoperable. Parts of the facility had unsecured radioactive materials, and hazardous materials were scattered around the facility. Within hours of the incident, looters tried to break into the hospitals pharmacy to steal drugs.

HICS was used to set up and to direct resources to cover all of the issues that arose. During and after the evacuation, it was used to integrate outside and inside resources with the overall response. Resources that the hospital utilized included Missouri's Disaster Medical Assistance Team (DMAT) who was on scene within two hours, the Missouri National Guard, the Missouri State Department of Health, the Department of Mental Health, local and outside law enforcement officers, firefighters, emergency managers, and multitude of other agencies.

In the aftermath of the tornado, command, control, communication, coordination, and collaboration were so effective and so well managed, that in only six days after the incident, a 60-bed Mobile Medical Unit was fully operational. This allowed the hospital to offer almost all the services that were available prior to the tornado. This would not have happened so quickly if not for the HICS method integrating with outside resources. In four and a half months, a temporary heated and cooled, hard-walled facility was built by the Missouri National Guard. This was built and installed on the premises, so the hospital could continue to serve patients through the upcoming winter (Mercy, 2012). A little less than four years after the tornado struck, a brand-new 220 bed Mercy Hospital was built with disaster in mind (Silvis, 2015). None of this would have been possible in such a short amount of time if not for the ability of HICS to help manage the incident in response and especially the recovery.

The nearby Freeman Hospital also suffered damage, although not as much damage as Mercy Hospital. The facility lost power, including the back-up generator, and a handful of windows were blown out. A small portion of the roof was ripped off, and six rooms became flooded from the storm, and the patients in those rooms had to be evacuated. While still operational, there was also damage to the oxygen system in the hospital. The use of HICS also facilitated (as much organization as possible for the response and recovery for Freeman Hospital). This allowed them to treat 467 patients in just a few short hours while keeping everyone safe and the facility open (The Joplin Globe, 2011).

Because of the widespread devastation, Freeman Hospital initially did not know that magnitude of the disaster. Most of Freeman's communication systems were also damaged or destroyed. In the initial aftermath, they had no idea that Mercy Hospital took a direct hit and that it was essentially destroyed. Within 5–10 minutes of the tornado hitting Joplin, the 40-bed Emergency Department saw 130 patients come through their doors (South Dakota Department of Health, 2011).

In an effort to provide the best care possible, administration and staff personnel began taking actions utilizing the HICS method that would allow them to care for more patients. The staff lounge was almost immediately turned into a wound care center, so that those patients that suffered non-life-threatening wounds could be seen by just a few staff members, treated, and quickly released. They also cleared all the waiting areas, so they could accommodate more patients. As soon as it was realized that this was a major incident, staff and administration also began to discharge patients that were well enough to care for themselves.

While discharging would seem like a logical and viable solution, they soon realized that most of these patients either had no transportation available or they had nowhere to go because their homes were destroyed (South Dakota Department of Health, 2011). Taxi and bus services were also either disabled or overwhelmed.

While Mercy Hospital suffered severe damage to the hospital, and Freeman hospital suffered moderate damage, other healthcare facilities also suffered damage. Three nursing homes were struck by the same tornado that hit Mercy Hospital. Greenbriar Nursing Home was completely destroyed, and 10 residents and one staff member lost their life. In an interview, staff members recounted how the roof was taken off by the tornado and that they looked up to see cars and trucks swirling above the roof. Not long thereafter, several patients were picked up and carried off by the tornado.

Two other nursing homes also had significant damage, but no lives were lost. Thanks to an effective Incident Management System (IMS), the HICS method working in combination with the ICS method, patients from all three of these facilities were evacuated from the damaged or destroyed nursing homes. Residents were systematically placed in a myriad of temporary and permanent facilities in an (mostly) orderly fashion (McKnight's News, 2011).

It should be stated that after researching this disaster extensively, and in talking with multiple people who were present during the response and recovery in the hospitals, or that responded to the hospitals, it was found that only minor issues arose during this response and recovery. The use of HICS facilitated ongoing response for days, as well as being used for months and sometimes years for the recovery process. It was said by multiple people in personal interviews that the triage system greatly assisted in saving lives. Many lives were saved using the HICS method to manage total chaos, confusion, and uncertainty that the Joplin Tornado delivered.

16.4 The Fundamental Elements of HICS

Just like its predecessor ICS, the principles that guide the HICS method call for assigning specific positions based on the scope and the magnitude of the incident that the healthcare facility is responding to. No matter what the incident, the fundamental elements allow the healthcare facility to be engaged and effective in their response. The incident may be internal (e.g. a fire, earthquake, hurricane, direct hit from a tornado) or external (pile-up on the interstate, HazMat spill, plant explosion, mass shooting), but HICS provides the structure needed to respond effectively and efficiently, no matter how large or small the healthcare facility is. HICS is a consistent IMS method because of the fundamental elements that make up this system.

One of the fundamental elements that makes HICS effective is a predictable chain of command. Not only is the chain of command predictable but it also identifies how many people should be managed by one person, also known as span of control. When an incident occurs, all personnel know how the hierarchy is arranged, who reports to whom, and the maximum and minimum number of people that will be on their team. They also know the individual who is higher or lower than them in the chain of command and that the organizational structure will remain the same for the most part (California Emergency Medical Services Authority, 2014).

Another fundamental element is an accountability of position as well as accountability of teams. More specifically, this role-based accountability makes the individual take ownership of their specific positions, and not only the work they do but also the work that their subordinate team members accomplish. This accountability is supported by Job Action Sheets that identify the specific tasks that should be completed. By having this accountability in place, if there is a failure in an area of a healthcare incident, it can be identified and corrected quickly. When individuals recognize their roles, and they assume the accountability to fulfill that role, it adds efficiency to the task they are assigned (California Emergency Medical Services Authority, 2014).

Another important principle is the use of common language. The use of plain English is extremely important when integrating outside resources. While each healthcare facility may have a similar vernacular, they will undoubtedly have different names or acronyms for parts of the facility, or perhaps special names for certain departments. When thinking about outside resources being brought in to assist during a crisis, it makes perfect sense that interagency communication should be plain English (California Emergency Medical Services Authority, 2014).

Imagine if you will, an earthquake has severely damaged a hospital structure and volunteers, law enforcement, firefighters, and EMS have responded to help with the evacuation. If the individual organizing the evacuation with outside resources tells an outside resource that they need take their crews to R-3 and search for any patients or staff, unless they are extremely familiar with the hospital, they would not know where that was. If the supervisor tells them that R-3 was on the third floor in an add-on wing that provides chemotherapy treatment and that the "R" stood for radiation, then they would be more likely to understand. If they provided directions that told the outside resource to, "Go up to the third floor from the south stairwell. When you get to the third floor, go north four or five doors and take the hallway to the left. When you get to that hallway, search that area all the way to the west stairwell", it would likely be less confusing. By telling them this information, it would be more likely that the correct area would be searched. By the same token, if they needed a specific piece of medical equipment from another nearby healthcare facility, and they did not use the proper plain English name, they could get the wrong piece of equipment that was of no use to them.

The fundamental principle of being a flexible and scalable method that identifies planning and response needs is extremely important as well because the HICS system is applicable to any size of healthcare facility. Due to HICS flexible and scalable design, the planning and the response needs can be customized to fit the need of the healthcare facility (California Emergency Medical Services Authority, 2014). It is universally applicable to a 12-bed hospital or a 1200-bed facility.

HICS uses the fundamental principle of modular design, which makes it adaptable to nearly any type of incident (California Emergency Medical Services Authority, 2014). Modular design allows planning and management of not only emergency situations but also nonemergent events including a public relations crisis. Modular design was borrowed from the business world and refers to operation management whereby operations are standardized. Much like the business world, component parts of HICS are subdivided into modules. Modules can easily be replaced by other modules, or new modules can be added as the incident requires. By using a modular design, it makes it easier to diagnose issues and to identify the remedy for any problems (California Emergency Medical Services Authority, 2014).

As an example, if the IC realizes that cost may be a factor due to an expanding incident, they can add the module of Finance and Administration to manage costs. If logistical issues are a problem, or they may be a problem, they can activate the Logistics Section. If basic Logistics need to be expanded, they can add modules under the supervision of Logistics that can be added with a full complement of staff members.

Unity of response efforts is another guiding principle of HICS. Because the system integrates with NIMS, and it also integrates with the ICS component of NIMS, it meets HSPD-5 and HSPD-8. This provides a structure that integrates with all response partners, including local government, state governments, the federal government, and trained volunteer organization. HICS is also a requirement for accreditation agencies in healthcare settings. In almost every instance, the accreditation agencies cite a unity of effort with community response partners as a primary reason for implementing HICS.

HICS also incorporates the fundamental principle of effective planning of response and recovery efforts. The planning method used in the planning process is called Management by Objectives (MBOs). As was discussed in Chapter 14, Management by Objective (MBO) was first used in business world. This process takes the larger problem and breaks it down into smaller, more manageable tasks based on smaller objectives. As these smaller tasks are completed, it leads to the overall management of the incident or event. When an issue is encountered, it is evaluated, and a plan to overcome this issue is identified. By utilizing SMART goals, the larger issue is managed more easily. SMART goals will be discussed in the following Planning Process.

The California Medical Service Authority created and maintains guidebooks for HICS. For those that wish to take the training at no charge, they can also visit https://www.calhospitalprepare.org/hics-courses. This training program that was developed, also uses tabletop exercises to ensure students being trained learn how HICS works, and they learn practical experience when it is not critical or in an emergent situation.

16.5 Chain of Command

It should be noted that this entire Chain of Command Section was gleaned from the California Emergency Medical Authority website (www.emsa.ca.gov) Job Action Sheets. The California Emergency Medical Authority is a comprehensive resource for those that wish to become more proficient and more knowledgeable about the HICS method of incident management. Without the California Emergency Medical Authority's willingness to make this information readily available, there is no doubt that substantially fewer individuals would be proficient with the use and utility of the HICS method.

The chain of command in the HICS system closely resembles that of the ICS component of the NIMS method and the ICS companion method. While there are a few distinct differences between HICS and the ICS components, it would be redundant to go in-depth into every position because it was discussed throughout this book. Instead, this section will only briefly describe the Chain of Command except for those areas that are distinctly different from ICS.

The purpose of Command is multifunctional. Command is used to oversee the general management of an incident. In doing so, they will work with their staff to create incident objectives, priorities, and goals. This is done through a multitude of meetings whereby information and intelligence about the incident is shared, then strategies are created that meet the needs of the incident. Once these strategies are approved, then staff members ensure that the strategy is carried out. This process continues until the incident is complete.

16.6 Command and General Staff

At the top level of the organizational chart, there are three distinct functional areas. The Incident Commander (IC), the Command Staff, and the General Staff. The Incident Commander (IC) is the individual who is responsible and answerable for all aspects of the incident or planned event. Their primary responsibilities include developing incident objectives in a fast and efficient manner, the efficient use of resources, ensuring the safety of personnel, and in overseeing and managing all incident operations. Just like with the ICS method, this is not to say that the Incident Commander (IC) is in the middle of an incident watching over every move made. While they may watch every move in a smaller incident, in larger incidents it is typically quite the opposite.

In larger incidents, the Incident Commander (IC) delegates authority to others, then make decisions on information that is collected from the field. The Incident Commander (IC) is supported by the Command Staff and the General Staff, and they are ultimately responsible for making and/or approving final decisions. Command Staff in the HICS system is very similar to the ICS component. As with the ICS component, these functions can be managed completely by the Incident Commander (IC) in smaller incidents. If the Incident Commander (IC) determines that they cannot effectively manage these tasks, they can expand the Command Staff, giving specific responsibilities to other individuals. This allows the Incident Commander (IC) to focus on other issues, which typically are more pressing.

Much like ICS, there is a Public Information Officer (PIO) who manages public information (including the media), a Safety Officer (SOFR) who oversees all safety issues related to an incident, and a Liaison Officer (LOFR) who is the point of contact for outside agencies (including private organizations) and other governmental entities. Where HICS differs from the ICS component of NIMS is that there is an additional Command Staff position that may be added called the Medical/Technical Specialist.

The Medical/Technical Specialist can be assigned as a Command Staff Advisor or they may be assigned to the Operations Section Chief (OSC). Their primary job is to counsel the Incident Commander (IC), or if placed in operations, they will report to the Operations Section Chief (OSC) and will advise on issues that are related to the medical staff. The person relegated to this position is typically a Physician, and they usually provide physician related advice. They may also communicate staffing needs, staffing capabilities, identify issues that may arise with illness within responders (including support staff), and provide an overview of credentialed medical staff working within the incident. Essentially, they are a liaison specifically dedicated to integrating healthcare facility staff into the incident.

The General Staff positions are a specific group of personnel that are organized according to function they serve. They work with each other but report directly to the Incident Commander (IC). The General Staff, if fully activated, typically consists of the Operations Section Chief (OSC) who oversees all tactical operations, the Planning Section Chief (PSC) who manages all planning considerations for the incident, the Logistics Section Chief (LSC) who manages logistical needs (e.g. facilities, services, people, and materials), and the Finance/Administration Section Chief (FSC) who manages all financial, administrative, and cost analysis aspects of an incident.

16.6.1 HICS Operations Section

The Operations Section will usually be substantially different from the ICS method Operations Section. Subordinates that will report to the Operations Section Chief (OSC) might include the following:

- The Staging Manager
- The Medical Care Branch Director
- The Infrastructure Branch Director
- The Security Branch Director
- The HazMat Branch Director

- The Business Continuity Branch Director
- The Patient Family Assistance Branch Director

While all these positions can be activated at the same time, the occasion to do so is very rare. Even if they were all activated at the same time, the span of control for the Operations Section Chief (OSC) would still be manageable, primarily because there are seven subordinates. The maximum people that one person can supervise in the HICS method is seven.

16.7 Staging Manager

The Staging Manager is a position in the HICS method that has a similar duty in the ICS method. The Staging Manager has the responsibility of organizing and managing the delivery and placement of additional resources, including personnel, vehicles, equipment, supplies, and medications. Much like the Staging Manager in the ICS method, they organize resources in preparation of deploying those resources where they are needed (or assigned). In a major incident, or in an incident where a multitude of additional resources are needed, they could have up to four subordinates, each with a very specific job.

- Personnel Staging Team Leader
- Vehicle Staging Team Leader
- Equipment/Supply Staging Team Leader
- Medication Staging Team Leader

As can be seen, these are self-explanatory jobs, so no further explanation will be given. Because there are often so many working parts in a health care facility, especially hospitals, the Staging Manager is the go-to person when personnel, vehicles, equipment and supplies, or medications are needed. While they will not usually deal directly with these resources, they will maintain a list of what is available and make priority decisions of where they should be deployed when multiple requests are made.

16.7.1 Medical Care Branch Director

The Medical Care Branch Director organizes and manages the delivery of emergency, inpatient, outpatient, and casualty care and clinical support services. In doing so, they will initially evaluate various areas of medical care for the ability to effectively provide Inpatient Care, Outpatient Care, Casualty Care, Mental Health, Clinical Support Services (e.g. lab, radiology, pharmacy) and Patient Registration.

To properly do their job, they will need to evaluate whether new patients are being rapidly assessed and moved to their proper locations, in an acceptable timeframe. The proper locations mentioned can include admissions to a facility, surgery, discharge from the facility, or transferred to another facility. They will also make sure that patients who were already admitted prior to the surge or disaster incident are receiving the proper care and comfort. They are essentially responsible for assessing problems and needs in these areas. As resources are needed, they will coordinate with others to provide and further manage those resources.

They also have supervisory responsibilities such as ensuring that their subordinate personnel are complying with safety procedures. They will also instruct and/or teach all Unit Leaders how to assess equipment that is in the facility, supplies, medication inventories, and staffing needs in collaboration with the Logistics Section Branches. When the HICS method is expanded, they will have Unit Leaders that report to them, which will be explained in the following.

16.7.1.1 In-Patient Unit Leader

The In-Patient Unit Leader is responsible for ensuring the proper treatment of inpatients. This Leader is also responsible for managing the inpatient care areas and ensuring that these areas are safe, secure, as well as maintained. They are also responsible for overseeing that discharges are controlled and warranted.

16.7.1.2 Outpatient Unit Leader

The Outpatient Unit Leader is the individual charged with preparing outpatient service areas to meet the needs of in-house and newly admitted patients. During surge events, this could be a challenging position. We only need to look at the Joplin Tornado to see that the influx of patients was so large that Freeman Hospital opted to turn the staff lounge into an outpatient treatment center. The Outpatient Unit Leader will oversee the treatment of outpatients and identify ways to properly care for outpatients and discharge them in a short amount of time.

16.7.1.3 Casualty Care Unit Leader

The Casualty Care Unit Leader is usually positioned in the Emergency Department. When this position is activated, they will typically identify the patient receiving area and implement the use of triage by designating specific areas for each level of triage. Their job function is to guarantee the delivery of emergency care to patients that arrive at the facility during a surge event.

In some instances, this Unit Leader will need to assist with the creation of treatment areas in additional or new locations when necessary. They will also be responsible for identifying problems in the receiving area, and any identified issues with treatment needs in each area. When issues are found, they will coordinate additional staffing and supplies to meet needs of the incident.

16.7.1.4 Mental Health Unit Leader

Disasters and stressful events can aggravate already diagnosed mental health issues and reveal previously not known mental health issues to come to the forefront. There will often be patients that were injured in the incident who may be suffering mental health issues for a multitude of reasons. When these stressful events occur, there will usually be a surge in mental health issues that require treatment. Additionally, surge events also can take their toll on staff members as well.

The Mental Health Unit Leader is the person who will oversee the mental health emergency response. They will be responsible for managing the mental health care area, and in overseeing, supervising, and psychologically triaging of patients to identify the level of mental health help that will be needed. When needed, they will coordinate mental health response activities with other units to ensure that their staff provide the proper care of mental health issues.

16.7.1.5 Clinical Support Services Unit Leader

Clinical support refers to individuals who work closely with medical staff. This may include a long list of individuals such as lab technicians and others that assist with the diagnosis. The Clinical Support Unit Leader helps to organize and manage clinical support services. They will work with clinical support staff to help providing the best delivery of these services and functions. They will also monitor the use and maintenance of both personnel and diagnostic/equipment resources.

16.7.1.6 Patient Registration Unit Leader

Patient registration is an important part of managing patients. If patient registration is not undertaken, then there is no paper trail, and no way to truly identify who was seen. The Patient Registration Unit Leader Coordinates the inpatient and outpatient registration with their staff, and the ensure that the process is completed as quickly as possible while still obtaining all the information needed.

16.7.2 Infrastructure Branch Director

The Infrastructure Branch Director plays a critical role in making sure that the infrastructure to support the healthcare facility is operational and safe. In this capacity, they will organize and manage the required services to maintain and repair the hospital's infrastructure operations including contacting outside services if needed. The infrastructure they may be responsible for could include the following:

- Facility heating and air conditioning
- Power
- Telecommunications
- Potable and nonpotable water
- Medical gas delivery
- Sanitation
- Road clearance
- Damage assessment and repair
- Facility cleanliness
- Vertical transport
- Facility access

In most instances, they will utilize specialists at the Unit level to ensure that all the facility infrastructure is at optimal efficiency. Unit Leaders that they could utilize include the Power and Lighting Unit Leader, the Water and Sewer Unit Leader, the Heating Ventilation and Air Conditioning (HVAC) Unit Leader, the Building and Grounds Unit Leader, and the Medical Gases Unit Lead. As can be seen by the names of these Units, they are self-explanatory what each is responsible for.

16.7.3 Security Branch Director

The Security Branch Director is responsible for the safety and security of the facility, the staff, and patients. Depending on the circumstances, they may need assistance from local law enforcement or a contract security firm. This holds especially true when a disaster such as the Joplin Tornado strikes and there are insufficient resources to restrict entry and control points. Depending on the type of hospital and how it is configured, the security staff that the Security Branch Director oversees may have powers of arrest, or they may have no powers of arrest, and they are considered unarmed security.

In a disaster, or surge event, they may need to address a multitude of issues. If a pandemic, they may need to either restrict visitation or potentially even be required to lock down a facility. In any type of incident, they may need to ensure that their staff maintain traffic control, and potentially personal belongings management. If the incident was a criminal or terrorist act, as part of the personal belongings management, they may be required to bags items from patients and ensure that there is a chain of custody for evidence if law enforcement is not present. In some instances, they may need to address crowd control, searching for missing individuals, and a whole host of other safety and security issues. When activated, this Branch can have multiple Units, and the names of those units are pretty much self-explanatory as to their duties.

16.7.3.1 Access Control Unit Leader

The Access Control Unit Leader, if activated, will be tasked with ensuring only authorized personnel can access the building or buildings. If necessary, they may also be tasked with ensuring access control to certain parking lots, or even the entire facility. In fulfilling this task, they may utilize hospital security, contracted security agencies, or they may ask law enforcement to integrate with their personnel to obtain additional assistance.

16.7.3.2 Crowd Control Unit Leader

There can be a multitude of reasons why crowd control may be needed at a healthcare facility. Sometimes, it does not involve a surge event or disaster. The reason that there needs to be a team controlling these crowds is that it can go from peaceful to lethal in just a few minutes if the crowd is not monitored and intervention is not taken appropriately. Additionally, during a surge event, crowds of individuals could interfere with patient care, access-restricted areas, or even physically interfere with staff that are trying to save lives. When activated, the Crowd Control Unit Leader will manage the teams from Security Branch resources to effectuate crowd control in a safe, efficient, and professional manner.

16.7.3.3 Traffic Control Unit Leader

During a surge event or a disaster, healthcare facilities can be inundated with traffic. Ambulances coming in and leaving, Personally Owned Vehicles (POVs) rushing patients in to be seen, and concerned friends and family members rushing to the facility. As this traffic flows in and out of the facility, these individuals are only thinking about what their priority is, and some will have no concern for the difficulties that could be caused by blocking lanes, where they park, and a whole multitude of other concerns that could decrease traffic flow. There may also be damaged areas of the facility that could prove dangerous, in which it is decided that traffic should not be in that location. Concerns should also be addressed when it comes to the flow of traffic and the safety of pedestrians.

The Traffic Control Unit Leader will implement a traffic control strategy to maintain the highest level of efficiency for the pickup and delivery of patients, the safety of visitors, and the safety of staff. They may also set up traffic control points to essentially sort vehicles with emergency traffic going one direction and visitors in another direction. This could be done with orange signs, personnel, or a combination of both. Depending on the circumstance, the Traffic Control Unit Leader may also need to assign team members to guide visitors in where to park, and possibly even open temporary parking in places other than parking lots or grassy areas. They will do whatever needs done to ensure that traffic can flow because this is their primary concern. In undertaking this task, this Unit Leader may need to utilize hospital security, contracted security agencies, or they may ask law enforcement to integrate with their personnel to obtain additional assistance. In small rural communities, resources may be limited, so the Security Branch should not overlook volunteers and nonessential staff that could be put in a vest and used as traffic control personnel in a significant event.

16.7.3.4 Search Unit Leader

There are a number of reasons that a Search Unit Leader may be activated under the HICS method. It could be a suspected bomb or Improvised Explosive Device (IED) that has been reported, a missing or kidnapped person or child, or the facility could be damaged and there may need to be a search for survivors. These are just the more obvious reasons as to why a Search

Unit would be activated. In nearly all these instances, the Search Unit Leader will need to work with and most likely integrate with outside resources (e.g. bomb squad, law enforcement) and other Unit Leaders (e.g. Access Control Unit Leader, Crowd Control Unit Leader) to ensure that resources become part of the solution rather than part of the problem. The Search Unit Leader under the direction of the Security Branch and the Incident Commander (IC) will typically only be activated to undertake a systematic and efficient search.

16.7.3.5 Law Enforcement Interface Unit Leader

The Law Enforcement Unit Leader is an important task when law enforcement is brought in to assist with a situation. The Law Enforcement Interface Unit Leader will be the point of contact with outside law enforcement agencies. They will become a liaison with the Liaison Officer (LOFR) from the law enforcement agency(s), and these two individuals will work together in a collaborative and cooperative manner.

This task usually entails securing a workspace that allows easy access for outside agencies (not in the middle of the facility). This workspace should meet the needs of law enforcement agencies, including sufficient space for command personnel and if needed a negotiation team. In undertaking this task, the Unit Leader may also need to work with the Infrastructure Branch, gather requested facility information when it is requested by these outside resources they are working with. This Unit Leader also will work out any radio frequency compatibility issues with these resources and ensure that best possible communication between these resources and the facility is enabled.

The Law Enforcement Interface Unit Leader will coordinate all information that is needed by law enforcement agencies and confirm that the information is correct. This may include the number of patients and the patient care status. It is important to note, however, that information should be first approved by the healthcare facility Incident Commander (IC) and the healthcare Security Branch Director.

16.7.4 HazMat Branch Director

For those that are just learning these terms, HazMat stands for hazardous materials. The HazMat Branch Director cannot be supervised and handled by just anyone. It is an area whereby there must be at least a cursory understanding of hazardous material identification, response, and recovery, as well as understanding the necessary actions that may need to be taken to protect staff, patients, and visitors. This individual will need to be someone that can effectively and safely organize and direct the incident response. Depending on the circumstances, this could include a multitude of tasks all related to HazMat response. As needed, they may need to activate specific Units that are specialized in this type of response, or work with outside HazMat response resources. As part of the modular design in the HICS method, the subordinate Units that the Branch Director might activate include the following:

- Detection and Monitoring Unit Leader
- Spill Response Unit Leader
- Victim Decontamination Unit Leader
- Facility/Equipment Decontamination Unit Leader

Because the names of these Units are self-explanatory, they will not be further discussed. One important thing to note when managing the decontamination of equipment, especially if that equipment is the reason for the HazMat incident, is that in managing the incident, the Facility/Equipment Decontamination Unit may need to remove while trying not to contaminate other parts of the facility.

16.7.5 Business Continuity Branch Director

The Business Continuity Branch is important, especially when disasters strike. This Branches primary function is to make sure that the business functions of the facility are maintained, restored, or augmented. By doing so, they will increase resiliency of the facility and minimize the negative financial impacts that can be caused by business interruptions. As part of their job, they will need to task individuals and/or Units with identifying issues with business systems. If it is found that these systems are not operating, or operating in a limited capacity, then they will need to find mitigation measures that will overcome or fix problems with Business Continuity failures in an expedient manner. They may also need to identify alternate locations where business personnel can relocate and undertake business matters, or potentially identify methods that can assist in overcoming the issues faced.

In recent years, this Branch has become even more important. With almost everything in a facility that relates to the facility doing business becoming computerized, it has sped up the process, maintained more complete records, and ensure that all business is documented, but it has also introduced risks for the facility. We only need to look at ransomware that has been released and how it has affected some healthcare facilities to see that risk. We can also see the effect that the Joplin Tornado had on the facilities business. These risks are possible and real. That is why certain measures should be taken prior to an incident ever occurring such as Continuity of Operations Planning (COOP). The Units and their responsibilities and/or tasks will be discussed in the following.

16.7.5.1 IT Systems and Applications Unit Leader

The IT Systems and Applications Unit Leader will manage personnel who will assist in identifying IT issues. Doing so, they will need to work collaboratively and cooperatively with a multitude of Sections, Branches, and Units. This collaboration and cooperation efforts will speed recovery of business systems.

They will work with the Infrastructure Branch, as needed, to evaluate damage to data centers. Whenever significant damage has occurred, they may be required to manage personnel or outside resources to identify salvageable equipment and initiate the needed repairs. They might also be required to collaborate with the Logistics Section and the Information Technology/Information Services (IT/IS) Equipment Unit to secure the needed equipment and services. This Unit Leader may also be required to convey what specialized personnel and resource needs that the Unit may have to facilitate a speedy diagnoses and recovery through notifying the Business Continuity Branch Director. They may be required to work with, or have their personnel work with, nearly anyone in the command structure to restore services to specific areas.

While the IT Systems and Applications Unit Leader will start with an initial strategy for restoring services in relation to Business Continuity, that strategy can quickly change based on the needs of the incident and personnel working within it. For that reason, they must regularly update Unit personnel with strategy changes as needed. They must also ensure that data and application recovery are prioritized in the Business Recovery Plan. Recovery should include patient records, payroll records, contracts, payroll, and similar data or as directed by the Business Continuity Branch Director, including Computer recovery, system recovery supporting major platforms, and various different applications, network recovery (of the intranet), internet connectivity, and any issues with digitally stored media.

They may also need to confirm that data access and security procedures are in place and functional. While this is only a small amount of the tasks that they may need to perform or manage, it should be realized that it would be nearly impossible to list every part that this Unit Leader might need to undertake and/or supervise in an incident.

16.7.5.2 Services Continuity Unit Leader

The Services Continuity Unit Leader is the individual who makes sure that any business, clinical and support services are operational. In doing this task, they will maintain, restore, or augment the support services that are needed to meet the objectives and goals at hand. Their primary goal is to minimize interruptions to permanency of essential business operations. The vast majority of their work will be identifying issues within the services of the facility and ensuring that those essential services are operational and stay as such.

16.7.5.3 Records Management Unit Leader

During a disaster, especially large disasters that cover a wide geographical area, the transfer of records may become increasingly important. The Records Management Unit Leader is responsible for ensuring that vital medical (and business) records are preserved and kept current and that there is little or no interruption in information requests. While this may seem like a simple task, it is not.

When this position activated, the Records Management Unit Leader will undertake that Hospital Record Preservation Plan as ordered and/or as needed. In doing so, they will prioritize the recovery and/or preservation of documents under the direction of the Business Continuity Director. In some instances, this may be required when an evacuation and relocation is needed, while in other incidents, it may be needed when there is damage to the facilities infrastructure. Prioritizing these documents, in collaboration and direction of the Business Continuity Branch Director, they will typically make a prioritization list that may include paper-based medical and laboratory records, electronic medical records, business contracts and financial records, billing records, personnel records, and library materials.

The Records Management Unit Leader may also be required to evaluate the need for relocation of critical records and make that recommendation to the Business Continuity Director. If it is determined that a relocation is needed, they will need to coordinate with the Logistics Section, identifying space and staff needs. They will also need to coordinate medical records to travel with any evacuated or transferred patients.

If damage has already been done, they will likely be responsible for documenting that damage with the Compensation/Claims Unit (under the control of the Finance and Administration Section). This could include still pictures, videos, security video, written documentation, and more. Additionally, if damage has been done, they will provide an assessment of whether the salvage can be done in-house with facility staff, or if a consultant or disaster recovery service should be contracted.

16.7.6 Patient Family Assistance Branch Director

While the primary concern of hospital staff will be the patients, we also need to realize the needs of their families. Having a loved one being admitted to a healthcare facility under emergency circumstances is stressful enough, add to it the fact that it may be part of a disaster and that will only add to the stress. In some instances, families may be in a state of shock, and they are having problems meeting their own needs, or they have not even thought about them because of the situation. The Patient Family Assistance Branch Director is a position within the HICS method who oversees the organization and management of assistance to meet patients family care needs. This could include providing for their communications needs, lodging needs, food needs, health care needs, and the spiritual and emotional needs of the family. In undertaking this task, they have the capability of activating two modular Units to support them; The Family Reunification Unit and the Social Services Unit.

16.7.6.1 The Family Reunification Unit Leader

In the early stages of a disaster, family members sometimes cannot find other family members. They do not know if they injured, deceased, or hospitalized or perhaps injured and trapped beneath rubble. When facilities face a surge event, a large number of families may not be initially informed that their loved one was transported to that facility prior to arrival. The Family Reunification Unit is responsible for ensuring that the patient and their family are reunited. The Family Reunification Unit Leader is responsible for managing and overseeing staff that facilitate the reunification.

Often, the Family Reunification Unit Leader will activate predetermined protocols and measures for reunification of patients with their loved ones. This may include detailed including identification and tracking of the patient, documentation about the patient, and locating and communicating with family and staff about that patient. They will work to provide reunification resources to children, families, and those with Access and Functional Needs (AFN). As part of communicating with families, they will ensure that personnel know the protocols for communicating with families, especially protocols about sharing information on the patient's status and location. They will also make sure that the cultural and spiritual needs of the patients and their families are considered and if needed, provide interpreter services so that communication with the patient, medical staff, and/or family is not hindered. In disaster situations, they may be required to coordinate through the Liaison Officer (LOFR) with outside resources, such as the local Emergency Operations Center (EOC) and the American Red Cross (ARC), to identify the point-of-contact that can integrate facility reunification effort with the overall community tracking and reunification efforts.

Additionally, they may need to identify any transportation needs. When looking at these needs, their staff and the Family Reunification Unit Leader will need to be cognizant of special circumstances to ensure that all meets are met. This might include special consideration for those with special needs and overall disability access.

When transportation is needed, they will likely need to coordinate with the Transportation Unit which is under the leadership of Logistics Section. Before releasing the patient, they will need to make sure that proper procedures are followed for safe release of patients are considered, including the special needs of minors, non-English speaking patients, and those in custody. If they are unsure if a situation is safe, they may need to speak with the Medical-Technical Specialists to discuss the best option to meet their needs. While these are not all of their duties, there should be enough of an overview to identify the type of assistance that this Unit, and the Unit Leader, provide.

16.7.6.2 Social Services Unit Leader

The Social Services Unit Leader is responsible for organizing and managing support resources that will meet the patients social service requirements during a disaster. In doing so, they and their staff will coordinate with community and government resources to identify specific resources that are available. In providing social services, they will evaluate the patient(s) for specific social service needs, then coordinate the use of hospital resources (including associate resources like medical supply companies), and community resources to ensure the patients' needs are met. They will also make sure that social services resources are made available to children, families, and those with special needs when they are needed.

In order to accomplish this, they will usually meet with the Patient Family Assistance Branch Director to plan, project, and coordinate the specific social service needs of the patient. While discussing the patient with the Patient Family Assistance Branch Director, they will typically provide direction and suggestions to meet the patient and family's needs. They may also need initiate communications with the patients family in some instances.

16.8 HICS Planning Section

The Planning Section of the HICS method is nearly identical to the Planning Section of ICS. There are some differences however in two of the Divisions. In the Resources Unit, there are two other Divisions that can be added if needed. These are the self-explanatory divisions of Personnel Tracking Division led by the Personnel Tracking Manager, and Material (Material) Tracking Division led by the Material Tracking Manager. The second change from the normal ICS method is the addition of Divisions under the Situation Unit. Added to these were the Patient Tracking Division, which is led the Patient Tracking Division Leader and the Bed Tracking Division led by the Bed Tracking Manager. Once again, these jobs are self-explanatory, but extremely important in managing a healthcare facility, especially when utilizing the HICS method.

16.9 HICS Logistics Section

Much like the ICS method, the HICS method separates logistics between a Service Branch and a Support Branch. They are divided into these Branches in such a way that all the services provided by the Logistics Section comes from the Services Branch and the Support Branch Supports those working on the incident.

16.9.1 Services Branch

While both the ICS method and the HICS method share using the Communications Unit and the Food Unit, the HICS method also adds the IT/IS Equipment Unit. This Unit Leader is responsible for providing computer hardware, applications, and acquiring and installing the IT needs for the facility.

The IT/IS Equipment Unit is responsible for identifying and developing expected computer needs, network equipment needs, and applications needs. In doing so, they will also assist in budgeting for these anticipated needs and purchase process. They will also be responsible for placing emergency orders for equipment and/or applications through utilizing protocols, or if needed following the special procedures that have been designated by the Procurement Unit, which is under the Finance/Administration Section. When placing these orders, they are required to notify the Service Branch Director. They are also responsible for coordinating the delivery, installation, and set up, of tele-triage or tele-medicine equipment in the areas where they are needed and attain and install more computers and peripherals in the Hospital Command Center (HCC) as they are needed.

16.9.2 Support Branch

The Support Branch of the HICS method shares the Transportation Unit and the Supply Unit with the ICS method; however, there are three distinct and very important Units that are unique to the HICS method. These distinct areas are focused on employee health and wellness, credentialing and the labor pool, and the care of family employees. Two of the three are related to caring for taking care of the employees and their family, with the third being focused on patient care.

16.9.2.1 Employee Health and Well-Being Unit Leader

In the short version of what the Health and Well-Being Unit Leader does, they are activated to make sure that employees and their families/dependents have logistical support, psychological support, and medical support during an incident. In other words, they take care of the caretakers and those they care about, but this Unit does much more than that.

The Employee Health and Well-Being Unit Leader and their staff will make sure that wounded staff (and volunteers) receive care when it is needed. This, at times, can be difficult because many healthcare workers are so focused on caring for their patients, that they forget to care for themselves. In certain incidents, this Leader and their staff will need to estimate the potential injury and illness impacts of an incident and the share that information with the Medical Care Branch Director who is under the direction of the Operations Section.

Depending on what the incident entails, the Employee Health and Well-Being Unit Leader may be required to develop a medical care plan for staff. This would entail assigning staff members with levels of care (needed), and they will identify any personnel and resources that might be needed to accomplish the plan. Once the plan is complete, they will document that on the standardized HICS Form 206-Staff Medical Plan and then submit to the Support Branch Director. This is done so that approval of the plan can be garnered, and it can be incorporated into the Incident Action Plan (IAP). By incorporating it into the Incident Action Plan (IAP), caring for employees becomes part of the mission rather than an afterthought.

The Employee Health and Well-Being Unit Leader will also coordinate with the Compensation and Claims Unit, who is under the direction of Finance and Administration Section, regarding any claims that have been submitted. They will also track staff illness and absenteeism and identify any trends in coordination with the Medical Care Branch Director. In some instances, especially if trends are seen, they may be required to implement intervention plans to address these and other identified issues.

As part of the monitoring of staff, the Employee Health and Well-Being Unit Leader may be required to institute monitoring programs. This holds especially true for staff exposed to biological, chemical, or radioactive agents, but it does not stop there. They may be required to initiate behavioral health services for employees and volunteers when they are needed, especially in extremely stressful situations. They should monitor during stressful issues such as long work hours, separation from family, experiences of injuries and illness, and recurrent negative patient outcomes and various other stressful times and then create a plan to address the psychological effect that these may have. As part of this plan, but not exclusively for the plan, they will make sure that a process exists to refer personnel to resources that may help. These could be programs such as Employee Assistance Programs, faith-based services, and counseling. As another part of dealing with stressful issues, they will work with the Behavioral Health Unit (under the direction of the Operations Section) to assign therapists to various locations where there is easy access for staff, and in making sure that the Hospital Incident Management Team (HIMT) has access as well.

The Employee Health and Well-Being Unit Leader may also need to create and implement Staff Prophylaxis Plan if needed, which would include planning for staff augmentation of Unit staffing to provide a Continuity of Care and prepare a Point of Dispensing (POD) location. In that plan, they may need to determine medication, dosage, and quantity with the Operations Section-Medical Care Branch Director and identify the methods needed to obtain and distribute the medicines. It may be required that staff members loved ones may also need prophylactic medicines as well. These are just a few of the Employee Health and Well-Being Unit Leader (and their staff) duties, and it is strongly recommended that proper training and practice is undertaken on this position as well as the entire HICS method before trying to use or implement it in a healthcare facility.

16.9.2.2 Labor Pool and Credentialing Unit Leader

The Labor Pool and Credentialing Unit Leader is responsible for the Labor Pool and ensuring their readiness. In this position, they coordinate a process to initiate staff call backs and assist the various department managers to implement a staff recall. They may be required to work with the Security Branch to undertake background checks and to provide temporary identification for those that are not employees of the facility; however, they have medical training and are willing to help. They will also implement an emergency credentialing process, to ensure that volunteer medical staff or community members are vetted, and they will document it on the HICS Form 253-Volunteer Registration.

Once a volunteer has been vetted, the Labor Pool and Credentialing Unit Leader will coordinate the needed orientation for individuals who are working at their facility for the first time. This orientation should include an overview of safety and security issues, infection control issues, rest and nutrition services, and who may supervise them. Once completed, they will be made available (if needed) to the Staging Manager or provided the location of their assignment.

An important part of the Labor Pool and Credentialing Unit Leader is monitoring the performance of volunteers and to make changes when needed. They should also closely monitor the efficacy of the emergency credentialing process, and if there are problems identified, they should make changes to improve it. As with the rest of the positions previously identified, this was only a brief synopsis of this position.

16.9.2.3 Employee Family Care Unit Leader

The Employee Family Care Unit helps to ensure that employees have their mind on the work at hand rather than also worrying about their family. Their primary task is to make sure medical, logistical, behavioral health, and day care are available for the families of staff members. They also may be required to coordinate prophylaxis, vaccination, or immunization of family members if it is needed.

This critical Unit will make sure that short-term childcare and elder care are provided, including all of their needs such as safety and security, recreation, rest and hygiene, food, and even overnight family accommodations (to name a few). In doing so, they may need to contract outside services such as hotels, shelters, childcare centers, elder day care, pet shelters, etc.; however, this would be done in coordination with Support Branch Director.

They will also need to consider that not all things that matter to staff are human. Depending on the length of the incident, the Employee Family Care Unit may also need to identify which employees have pets and/or livestock that need cared for. In most instances, they may need to contract this work out. Even so, the well-being of these animals should not be forgotten because they are important to the employee, and this will allow these staff members to focus on patients rather than being distracted.

16.10 Finance and Administration Section

The Finance and Administration hierarchical organizational chart is set up essentially the same way in the HICS method as it is with the ICS method of incident management. While there may be some adjustments from facility to facility, the concept is the same.

It should be noted that this was only a brief overview of the HICS Command Staff. More in-depth information which will develop more understanding about Command and General Staff responsibilities can be found in previous chapters. You can also take classes in the HICS method, which will provide more in-depth understanding.

It should also be noted that this entire Chain of Command Section was gleaned from the information provided on the California Emergency Medical Authority website (www.emsa.ca.gov) Job Action Sheets. The California Emergency Medical Authority is a comprehensive resource for those that wish to become more proficient and more knowledgeable about the HICS method of incident management.

16.11 The Planning P/The HICS Planning Process

The planning process in the HICS system mirrors the ICS component in almost every way, including the use of the Planning P described in Chapter 15. The planning process is an essential part of effective IMS system. Effective planning is key to providing a successful response and recovery from any incident. Planning for each phase of a response also provides a systematic method of identifying the needs and matching the best resources to complete the task.

By developing and applying an Incident Action Plan (IAP), the Command and General Staff essentially provide a blueprint for how an incident should be handled. When planning is applied, it provides a detailed account of the goals, strategies, and tactics, thereby providing strategic direction. This strategic direction is known by everyone in the organizational chart, from top to bottom, but it applies mostly to the Operations Section.

An important component of planning is the process of using Management by Objective (MBO) which was discussed in Chapter 14. It is important to note, even though it has already been discussed previously, that the Management by Objective (MBO) framework was first introduced in by Drucker (1954, 2010). Drucker's (1954) work identified common attributes that allows the Management by Objective (MBO) method to be effective and efficient. This Management by Objective (MBO) framework has been adapted to a multitude of businesses and organizations, and it has also been used in managing a multitude of emergency incidents. The widespread adoption of Management by Objective (MBO) has been due to how flexible this planning method is. The process of dividing larger tasks into multiple smaller jobs that lead up to completing the larger task. These smaller jobs are a part of a basic process where success builds on success. By taking a larger process, and breaking it into smaller, more manageable and more achievable goals, each part leads to the completion of the overall larger project. If you have not already read Chapter 14, you should take the time to do so before trying to understand the planning process.

The HICS planning process is an effective way of managing smaller, short-term incidents as well as the more complicated long-term incidents. Much like the ICS method, in the HICS method, the Incident Action Plan (IAP) is scalable to meet the needs of the incident. HICS does have one slight difference from ICS when it comes to the planning process. HICS has developed a *Quick Start* worksheet for managing and planning an incident. This Quick Start worksheet is a short but effective form that combines HICS Forms.

- Form 201-Incident Briefing
- Form 202-Incident Objectives
- Form 203-Organization Assignment List
- Form 204-Assignment List
- Form 215A-Incident Action Plan Safety Analysis

The Quick Start planning worksheet can be used in place of using the extended version of the forms. It is primarily used to document the initial actions that have been taken, or in shorter incidents, it can be used to document the particulars of the incident. It is also important to note that HICS forms are nearly identical to ICS Forms. The reason for the similarities pertains to the fact that many of these forms are standard forms taken from ICS that were modified to be more applicable to the healthcare setting.

As was previously mentioned, Incident Action Planning (IAP) is an important part of the process of managing an incident. This planning process helps direct the incident and the organization in a strategic direction, and it helps to maximize all the available resources. The planning process also helps to prevent a duplication of efforts while making sure that nothing is overlooked. It helps the Command and General Staff to gather and distribute information, and it improves and increases the effectiveness of communication. The planning process also plays a major role in reducing the associated costs of an incident, and it provides a written historical record of problems and actions that occurred during the incident, not to mention reducing the risk of successful litigation.

Whenever an incident occurs, the Incident Action Planning (IAP) process should be initiated. In most instances, this will begin by activating the hospital's Emergency Operations Plan (EOP) and initiating the HICS method of incident management. Whether the planning process (and Command staffing) will be carried out by the Incident Commander (IC) alone, or if planning and command requires partial or full staffing, will be dependent on the information that has been initially gathered. Information and intelligence gathering are critical to the process of both staffing and planning. Without current and credible intelligence about a situation, the Incident Commander (IC) will be forced to make decisions while lacking complete knowledge of the incidents complexity and depth. An in-depth situational awareness is extremely important, just as it is in the ICS method of incident management.

The swift gathering, verifying, and validating of information about the incident is also critical to initial decisions. The Incident Commander (IC) will need to know the status of hospital, the hospital systems, and the current abilities and capabilities of hospital staff. This will assist them in determining what outside resources may be needed to perform operations. Planning based on credible information gathering will assist in establishing incident objectives, providing direction for the response, and establishing incident priorities.

Information gathering will assist in activating and assigning the Hospital Incident Management Team (HIMT). With credible information, the individual team members can be chosen based on the most qualified person for the task at hand, and their experience in a given position. By doing this, the most qualified available person for a task will manage that task and provide better support for the Incident Commander (IC) and the incident.

As an example, if there were an explosion that injured multiple people and damaged the structure, it would make sense to put someone with knowledge about structural safety (engineering), and someone from EMS or the Emergency Department on the Hospital Incident Management Team (HIMT), even if they were listed as a technical advisor. These individuals would also be extremely helpful in the planning process.

A key factor in the planning process is to create Objectives, based on SMART goals. In the ICS method, SMART is an acronym for Specific, Measurable, Achievable (enough resources to accomplish the goal), Relevant (to the overall mission), and in a specific Timeframe. In the HICS method, the latest version of SMART goals is listed differently than the ICS method. In the HICS method, SMART goals are described as Specific, Measurable, Action Oriented, Realistic, and Time sensitive. No matter which way you choose, it still leads to the same ending. SMART goals relating to the ICS method were also discussed in greater detail in Chapter 14.

No matter which version of SMART you utilize, these goals and objectives will assist the Incident Commander (IC) and the Command and General Staff (if activated) in developing effective strategies and tactics needed to address the incident. The planning process will also be instrumental in creating the specific plans and actions that will need to be taken in order to complete each objective. The planning process will provide guidance in projecting resource needs and assigning resource requirements. The response and recovery will be more organized because predictions of the needs and assignments will be based on continual situational awareness and ongoing evaluations of the incident, planning for each task.

The planning process will also help establish communications and aid in ensuring a unified response with external agencies. Moreover, the planning process helps to identify capabilities of external responders, which may have not been previously known. This allows the Incident Commander (IC) and the Command and General Staff to plan tactical work based on NIMS resource typing. By utilizing the NIMS resource typing tools, they will be able to order for their exact needs and utilize outside resources based on their skills, knowledge, and their abilities, then assign those resources to fulfill the tasks that need to be accomplished.

As an example, if a nursing home suffered a failure in their air conditioning on a 100-degree day, the fire department will likely be equipped with specialized ventilation fans. These high-powered fans could help to move substantial amounts of air throughout the facility as the facility is being evacuated, thereby keeping residents cooler. Additionally, the fire department or law enforcement may have a professional Public Information Officer (PIO) that already has contacts within the media and can get the word out to let the family of patients know or potentially ask for assistance from the public.

Utilizing the planning process can help to identify who may need to receive alerts about the incident, and the best method is to send notifications. There is also an outside chance that an external agency has a maintenance man knowledgeable about heating and air conditioning or maybe one of the responding agencies has someone on staff that has a part-time job as heating and air conditioning repair person. With following the planning process, even a shortened version for smaller incidents is a way of making the complicated process simple.

16.12 Emergency Operations Plan

Hospitals and some other healthcare facilities are required to have an Emergency Operations Plan (EOP). An Emergency Operations Plan (EOP) is a guide that is created to designate how a facility will respond to, and recover from, all hazards. It essentially is part of incident management because it already plans for the incident before it ever happens. It also provides guidance when a healthcare facility has had its world torn apart.

This Emergency Operations Plan (EOP) is a guide for the hospital based on six specific elements that the Joint Commission's Emergency Management Standards identify as critical when responding to a healthcare emergency. Those specific areas are communications, resources and assets, safety and security, staff responsibilities, utilities, and clinical support activities.

While the Joint Commission's Emergency Management Standards apply to hospitals and only certain other healthcare facilities, the Center for Medicaid and Medicare Services (CMS) also requires standards for healthcare facilities if they provide services for Medicare and Medicaid patients. In 2017, the Center for Medicaid and Medicare Services (CMS) mandated a new, and more comprehensive, level of preparedness mandate. The new rules require more specific plans, and more training and testing of those plans (Stone, 2017).

Training on basic responses is given in HICS training. To ensure that the students grasp the concepts of HICS in a classroom setting and that they learn practical experience in a nonemergent situation, HICS training includes tabletop exercises that (somewhat) mimics areas that should be covered in the Emergency Operations Plan (EOP). The tabletop exercises provided in the training, cover a multitude of incidents within a healthcare setting, including the following:

- Chemical Tabletop Exercise
- Evacuation Tabletop Exercise
- Hostage or Barricaded Individual Tabletop Exercise
- Improvised Explosive Device (IED) Tabletop Exercise
- Loss of Power Tabletop Exercise
- Loss of Water Tabletop Exercise
- Earthquake Tabletop Exercise

The use of these specific tabletop exercises at one point led to concerns that healthcare facilities might consider using the instruction they learned (in the training) as their facilities Emergency Operations Plan (EOP). Healthcare facilities should be aware, and cautioned, that the HICS Guidebook and any exercises associated with the training should not be used in place of an Emergency Operations Plan (EOP). Using this as an Emergency Operations Plan (EOP) would likely set the healthcare facility up for failure if a real-life incident occurred.

Resources regarding how to create an effective Emergency Operations Plan (EOP) can be found in a multitude of places on the internet. Perhaps one of the best overall resources is the Comprehensive Preparedness Guide (CPG) 101, which gives a complete overview of the planning process and creating an Emergency Operations Plan (EOP). It should be noted that this resource is not specifically for healthcare facilities. Even so, if the emergency planner(s) utilize the concepts in CPG 101, they will create a much better plan and have substantially more understanding about what is needed in the Emergency Operations Plan (EOP) and why it is needed.

Arguably, one of the most useful ways to create an Emergency Operations Plan (EOP) for a healthcare facility is by visiting a specific website. The United States Department of Health and Human Services (HHS) Office of the Assistant Secretary for Preparedness and Response (ASPR) provides many resources for creating an effective Emergency Operations Plan (EOP). The Office of the Assistant Secretary for Preparedness and Response (ASPR) sponsors the Technical Resources, Assistance Center, and Information Exchange (TRACIE) commonly referred to as ASPR-TRACIE. The information is divided into three main categories.

The ASPR TRACIE Technical Resources include two main mechanisms for information. The first is the Resource Library and the second is Topic Collections. These resources provide a collection of articles, books, and research from numerous contributors, including the United States National Library of Medicine (NLM). It also includes access to databases from the National Library of Medicine (NLM) and other government agencies. The Technical Resources section encompasses up-to-date disaster medical and healthcare system preparedness, as well as public health preparedness materials.

The ASPR Assistance Center provides Technical Assistance Specialists to answer any questions you may have about preparedness. If you have a question or concern, you can submit a request or question to subject matter expert through telephone, email, or electronically between Monday to Friday from 9 am–5 pm (Eastern Time). You can obtain support on topic-specific resources such as how to manage volunteers, credentialing volunteers, hazard vulnerability assessments, etc., and they will need to connect you (or your facility representative) with a subject matter expert.

The ASPR TRACIE Information Exchange allows involved with preparedness for healthcare facilities to connect and discuss issues with others who have, or are preparing, their healthcare facility for disaster. This system allows you to connect and participate in conversations that share plans, discuss practices, and to identify and share pending and actual health threats. In order to gain access to the Information Exchange, you must be involved in healthcare system preparedness and be validated prior to approval.

Another resource that may be useful is the International Association of Emergency Managers (IAEM) Healthcare Caucus. While a membership with the International Association of Emergency Managers (IAEM) is required, the information and experience that is available to healthcare facility (emergency) planners is astounding. The sheer number of individuals with specific knowledge and the number of individuals that are willing to share information make this resource well worth becoming involved with the organization. The healthcare caucus (as well as other caucuses) hold regular conference call meetings and webinars to discuss disaster preparedness in healthcare facilities. In these meetings, they regularly share information, share plans, share new and evolving methods, and they identify and discuss new issues and threats.

There are also resources available through the Center for Disease Control (CDC). In fact, the Center for Disease Control (CDC) provides resources that can assist during a public health emergency based on the type of healthcare facility. These include physicians' offices, pediatric offices, pediatric hospitals, outpatient clinics, and urgent care centers. It also includes long-term, acute, and chronic care facilities as well as hospitals, healthcare systems, first responders, and the community. It should be realized that the Center for Disease Control (CDC) does not provide guidance on Emergency Operations Plans (EOPs) but rather on public health emergencies. This means that this resource could help planners identify best practices for health emergencies when writing the Emergency Operations Plan (EOP).

16.12.1 An All-Hazards Plan

The Joint Commission Emergency Management Standards are very specific regarding the requirements to have a hospital Emergency Operations Plan (EOP). While this will not translate to all healthcare facilities, it would serve other types of facilities well to do their best to meet these standards. Some of the Joint Commission requirements actually go beyond an Emergency Operations Plan (EOP) and enter into mitigation and preparedness tenets. Even so, preparedness and mitigation measures are a good thing when preparing for the response and recovery from a disaster.

It is important to note that each healthcare facility has unique challenges and staffing requirements. For this reason, each facility should create their own *All-Hazard Emergency Operations Plan*, based on their specific weaknesses, strengths, and their exclusive needs. The all-hazards EOP should be created in such a way that it designs facility-specific functional annexes to address special situations that may occur and to project specific threat scenarios which may affect the facility. While a template may be helpful, using only a template rather than making the Emergency Operations Plan (EOP) specific to the facility is a recipe for disaster.

An Emergency Operations Plan (EOP) is a living document that describes how the healthcare facility will respond to, and recover from, an incident. It serves as a healthcare facility guide for disaster. It helps the individual facility to build resilience by identifying specific threats and hazards. Once these threats and hazards are identified, then contingency plans are designed how the facility staff should respond and operate in a safe, efficient, and effective manner. While these plans may provide guidance, they also need to be flexible, not rigid. The reason for the flexibility in the plan is based on the theory that not every type of incident is exactly the same.

16.12.2 Who Should Create the Emergency Operations Plan (EOP)?

To create Emergency Operations Plans (EOPs) for a healthcare facility, it is critical to realize that a single individual should not be responsible for developing these plans. A diverse committee of employees and usually subcommittees made up of employees should be utilized. This allows the Planning Committee to have a complete view from the different perspectives of diverse group of staff members. When designing Emergency Operations Plan (EOP), every job undertaken in the healthcare facility should have at least one representative involved in the planning process. While this may seem like a daunting task, this method helps to ensure that there are no issues overlooked. It also helps to ensure that all staff have responsibilities in an emergency, even if that responsibility is to self-evacuate.

Further explaining the Emergency Operations Plan (EOP) process, while a doctor or a nurse may have a view about what is important in an emergency, perhaps security, engineering, or even housekeeping may have a different view that neither the doctor nor nurse has thought of. Perhaps the administrator will be worried about a critical issue that none of the previously mentioned personnel had even considered. Therefore, there should be committees and subcommittees utilized in the Emergency Operations Plan (EOP) design process.

Additionally, the committees and subcommittees should not overlook outside resources, which may be able to guide the committee in the designing of an Emergency Operations Plan (EOP). Local Emergency Managers would be an excellent outside resource. They usually can provide substantial guidance. They know what resources the local government can provide and estimated time frames in how long it would take to obtain those resources. This might include equipment, personnel, and specialized advisors. Healthcare facilities should also consider talking with local ambulance companies and/or medical transit companies who might be needed in an evacuation. Other companies and agencies that might have specialized input include snowplow companies, heavy equipment companies, crisis communications consultants, and a multitude of other companies and agencies may be able to help provide guidance. The end goal is to think through every scenario, then preplan what resources are available and useful, and then create the best possible plans to protect patients and staff. In order to create the best possible plans, there are a multitude of areas that need addressed.

16.12.2.1 Acknowledgement of the Incident

One area that should be addressed in the Emergency Operations Plan (EOP) is the acknowledgement that an incident has occurred or is occurring. If there is no recognition that a mass shooting might send an additional 20 patients to the facility, how can the facility initiate their Emergency Operations Plan (EOP)?

Another crucial factor is how to activate alerts and notifications to staff. If extra staff are needed from other parts of the facility, and no alert or notification went out, then there will be a failure. Healthcare facilities also need to identify what prompts an activation of the Emergency Operations Plan (EOP), who has the authority to start an activation, and how will the Emergency Operations Plan (EOP) will be implemented. The Emergency Operations Plan (EOP) should also lay out any associated policies and procedures that apply to an activation as well as guidance on demobilization and how to recover from an activation.

16.12.2.2 How an Incident Should be Managed

The Emergency Operations Plan (EOP) should provide guidance for a multitude of tasks that might be needed during an incident. The primary task of the Emergency Operations Plan (EOP) should include how the incident should be managed, based on the type of incident it is. As an example, if a tornado warning is issued, the Emergency Operations Plan (EOP) might stipulate that all patients are to be moved to the hallway and that all room doors should be closed. It might also discuss all surgeries not underway should be delayed until the warning is lifted, and it might stipulate special precautions for patients in the Intensive Care Unit (ICU). There may be considerations for evacuating the Emergency Departments waiting room, and where those patient and visitors should be placed to ensure maximum protection.

16.12.2.3 Preplanning Actions and Resources Needed

The Emergency Operations Plan (EOP) should undertake considerations for preplanning of actions and resources that will likely be needed. Much like planning for how to keep patients and visitors safe, planning should include what actions might be needed if the healthcare facility is damaged or destroyed. Contingency plans need to address how to evacuate all patients and visitors, how to track patients (both with and without technology), where patients might be evacuated to, and how will they be transported to that location. This type of planning might also need to secure Memorandum of Understanding (MOU) or a written Mutual Aid Agreement (MAA) to ensure that there are no miscommunications. The more in-depth the planning is under nonemergent situations, it improves the chances that all obstacles will be overcome when a real incident occurs.

Other resources that are available in a disaster should also be considered in creating the Emergency Operations Plan (EOP). The Health and Human Services Office of the Assistant Secretary for Preparedness and Response (ASPR) has taken the initiative to organize various teams that would be common for healthcare facilities to request and need in an emergency or disaster. One such resource would be a Disaster Medical Assistance Team (DMAT). A Disaster Medical Assistance Team (DMAT) is a group of professional and para-professional medical personnel designed to provide medical care during disasters, public health emergencies, and/or national security events. Disaster Medical Assistance Teams (DMATs) are staffed with medical professionals who can help the area health systems respond by providing expert patient care. Disaster Medical Assistance Team (DMAT) members can supplement a healthcare facility with advanced clinicians, medical officers, registered nurses, respiratory therapists, paramedics, pharmacists, safety specialists, logistical specialists, information technologists, communication specialists, and administrative specialists. Disaster Medical Assistance Teams (DMATs) also have access to temporary structures which can be used to treat patients evacuated and/or to help handle overflow patients ("NDMS Teams," n.d.).

Another outside resource that the Department of Health and Human Services can provide is a Trauma and Critical Care Teams (TCCTs). Individuals that make up these teams are capable to provide critical care, operative care, and emergency medical care during and after a disaster or emergency. These teams are also used to supplement federal, state, local, tribal, and territorial agencies when they are requested by local agencies ("NDMS Teams," n.d.).

Disaster Mortuary Operational Response Teams (DMORTs) may also be a resource to consider in a disaster or large emergency. It is another resource that can be requested of the Department of Health and Human Services. Disaster Mortuary Operational Response Teams (DMORTs) contain a multitude of professionals, which may include funeral directors, medical examiners, pathologists, forensic anthropologists, fingerprint specialists, forensic odontologists, dental assistants, administrative specialists, and security specialists. They are often used to enhance federal, state, local, tribal, and territorial resources, and they are deployed at the request of local authorities.

One consideration that is often overlooked by healthcare facilities is the Victim Information Center (VIC) Teams. These Department of Health and Human Services Teams work in the aftermath of a mass casualty event. They collect data that can help identify those killed. That data can include pictures, fingerprints, dental X-rays, and other medical records that might help identify the deceased. These Victim Information Center (VIC) Teams also acts as a liaison to the families of those with missing or deceased victims ("NDMS Teams," n.d.).

The healthcare facility has access to a multitude of agencies that are specifically designed to respond to emergencies and disasters, should they be needed. The United States Public Health Service can provide certain specialized medical teams within 12 hours. Healthcare facilities can also request the resource of the Medical Reserve Corps (MRC) and/or the National Disaster Medical Service (NDMS) volunteers. The military, the national guard, and the Veterans Administration may also be able to provide staffing resources. The list of resources available to a healthcare facility is innumerable. Healthcare facilities should research these resources prior to an incident so that the facility can request the exact resources that are needed for their unique facility and the unique circumstances.

16.12.2.4 Situational Awareness

The Emergency Operations Plan (EOP) should also address situational awareness in multiple ways. The first and foremost type of situational awareness should include identifying safety issues that may affect patients and/or staff. Safety should always have the highest priority when managing any type of incident. Situational awareness should also include landmarks that identify when an incident is beginning, when more assistance will be needed, when an incident is winding down, and when to terminate the response. Situational awareness should also relate to gathering intelligence and information that will assist in the planning process.

Intelligence and information gathering are crucial to making informed decisions. Those in charge require having a full situational awareness of what is damaged, what is operational, the capabilities of the facility, and all of the needs. In taking command, the Incident Commander (IC) must have the most recent intelligence or information so that they can make decisions based on current information. The latest information will allow the management team to form alternative strategies based on changes caused by the incident. During a major incident, the Incident Commander (IC) may want to appoint someone to gather intelligence on the evolving situation. The person appointed for this task would typically be assigned to collecting, evaluating, disseminating, and using incident data to make more informed decisions.

16.12.2.5 Roles and Responsibilities

The Emergency Operations Plan (EOP) should also identify the roles and responsibilities for varying incidents. These roles and responsibilities should be based on departmental roles (and responsibilities), the entire organization's roles (and responsibilities), as well as the roles and responsibilities of staff members. These roles and responsibilities should encompass what happens before, during, and after emergency. It is important to note that many of these roles and responsibilities may be based on the type of incident.

As an example, if the incident is a radiation leak or some other hazardous material incident, only those with this type of specific training and experience should have roles and responsibilities or support roles (e.g. security) in this type of incident. Similarly, if there is a hostage situation or a barricaded individual, only those with specific training should be considered to be at the forefront of this type of incident.

When planning for roles and responsibilities, it is also important to remember that an individual with a specific role or responsibility may be absent at the time of a disaster or emergency. For that reason, an order of succession, sometimes called delegation of authority, should be mapped out. This order of succession/delegation of authority will help to provide what is more commonly known as a Continuity of Operations Plan (COOP).

Continuity of Operations Plan (COOP) is a method that ensures the facility is capable of continuing essential functions under austere conditions. This is effective in a wide range of circumstances, simply by having multiple successors for each role. This means the facility should identify multiple individuals who can take responsibility of critical jobs when the primary person is not present or capable of doing so.

Perhaps a better way of thinking about Continuity of Operations (COOP) is to think of various healthcare personnel taking a scheduled vacation. While these individuals are on vacation, facility operations do not stop. This is usually because someone has been assigned to temporarily fill the job of the person who is on vacation. But what happens if the individual filling in for the person on vacation becomes sick? Is there another person that can take over those responsibilities? Not only have they been trained in how to do the job but are they competent at that job? Do they have the institutional authority to make decisions for this job position? Then consider a disaster which incapacitates some of the upper management or administration personnel, who will fill the positions of those that are incapacitated, or worse yet, killed.

By creating a Continuity of Operations Plan (COOP), not only do you train successors in how to do the job but also you need to make sure they are proficient. Additionally, you need to make sure that they have the organizational authority to actually do the job. This flexibility allows the healthcare facility to continue effective operations and to ensure that critical jobs are staffed and operational.

16.12.2.6 Communications

Another important part of an Emergency Operations Plan (EOP) is communications. When we look at communications, we need to be cognizant of not only internal communications but also external communications. The ability to communicate what should be done, who should do it, and what priorities are important can be extremely critical in an emergency. This, of course, would be internal communications. In the best of circumstances, this could be done by telephone, but what happens when phone lines are damaged? The facility needs to consider what to do when phone lines are down. There should also be considerations for when face-to-face communications may not be possible, such as the structure is damaged, and the facility is littered with debris, such as in the Joplin Tornado incident.

External communications may be critical as well. How can the facility communicate the need for outside assistance? How can they request off-duty staff to come in? How do they request additional staff from another healthcare facility? How do they communicate the need for local or state government assistance? Whenever possible, the healthcare facility should rely on their own method of communication rather than bringing in outside help from law enforcement (or potentially fire personnel) similar to what was done in the Oklahoma City Bombing. When planning for external communication, it is important to remember that in most instances, the disaster will not only disrupt the facility but it will also likely affect the entire community.

16.12.2.7 Staffing

Staffing is another important consideration in the Emergency Operations Plan (EOP). Many of those in a healthcare facility, or brought to a healthcare facility, are vulnerable. Unfortunately, vulnerable populations are more likely to die in a disaster (Hoffman, 2009). Having a sufficient amount of staff in place will provide more personnel to tend to the needs of vulnerable individuals, even if they are temporarily vulnerable from the effects of a disaster.

When considering staffing, the Emergency Operations Plan (EOP) should identify where staff could, or should, come from. The Emergency Operations Plan (EOP) planning committee will need to consider how many staff members will be available for recall if a disaster occurs. They also need to assess what other healthcare facilities might be able to supplement staffing in a disaster or emergency. If they determine that they could benefit from it, they could also create and sign a Mutual Aid Agreement (MAA) with other facilities so that they support each other in disaster.

16.12.2.8 Credentialing

In the event of a disaster or major emergency, it is highly likely that individuals who are not employed by the healthcare facility, but have specific medical expertise, might offer to volunteer their services. Employed staff already have privileges, and they are considered credentialed by the healthcare facility. Those who volunteer will probably not already have privileges at the facility, so they will need to undergo an emergency or disaster credentialing process, or perhaps go through a credentialing system.

This process (or system) to temporarily credential medical personnel should be written into the Emergency Operations Plan (EOP), and then it should be streamlined as much as possible. The purpose of having a system already in place is primarily to reduce confusion and to lessen the time it will take to utilize volunteers. This will also reduce delays in focusing on the top priority, providing patient care.

The primary reason healthcare facilities should have and do have a credentialing process is to ensure patient safety. In the credentialing process, information is gathered to ensure that the person presenting themselves as a volunteer is (in fact) the person they say they are and to verify that they are a trained and accredited medical professional. In the credentialing process, it is important to realize that some basic information and verifications are needed so that the facility can verify the status of the medical professional and complete a background check. Whenever possible, the credentials being provided by the volunteer should be copied and/or scanned so that there is a paper trail in the event that the individual or the healthcare facility is called into question.

The first verification would be to ensure the identity of the person. This would include a government-issued identification. These government-issued documents could include a driver's license, a VISA, a military identification card, or a similar piece of government identification. In addition, they should be required to present at least one or more of the following pieces of documentation:

- Documentation of a current active license, certification, or registration.
- Current hospital picture identification card with their professional designation (e.g. Internal Medicine Physician, X-ray Technician, Registered Nurse).
- Verification of licensure, certification, or registration (if required to practice).
- Disaster Medical Assistance Team (DMAT), Medical Reserve Corps (MRC), or other recognized state or federal response organization identification.

In exceptional circumstances, especially incidents that cover a large geographical area and that has an enormous number of patients, special authority can be granted for providing patient care, treatment, and/or services under these circumstances. The special authority will typically be granted by a federal, state, or municipal government. Even if authority is provided by another agency, the legitimacy of the special authority should be determined on a case by case basis, usually by state healthcare licensing bodies. If a person presents themselves under a special authority, they too should have identification and/or documentation that the individual has been granted special authority to provide patient care.

While the initial intake of outside medical personnel may require a few pieces of documentation, the healthcare facility should also do their due diligence and undertake further verifications and background checks within 72 hours. When initially checking in, personnel should be provided with an application form. These applications are widely available on the internet, and it would serve the healthcare facility well to have hard copies of these applications in a disaster kit prior to an incident. They are often called *Emergency Volunteer Privileges Application*.

Except in extremely rare circumstances, hospital administrators and their staff should verify the information that the volunteer provided. This verification process should confirm the volunteer's license, certification, or registration, and verify their current competency. As with any medical personnel that would be hired under the nondisaster process, methods of verification may include Drug Enforcement Agency (DEA) registration, National Practitioner Data Bank (NPDB), and potential verification of Board Certification through the American Board of Medical Specialties and/or American Osteopathic Association Specialty Boards. While we want to think the best about volunteers, there is also the possibility that a very miniscule amount of them may have ulterior motives or that they have a nefarious background. For this reason, a criminal background check and potentially the Office of the Inspector General (OIG) List of Excluded Individuals/Entities (LEIE) check should be done.

When writing the Emergency Operations Plan (EOP) for credentialing, it is important to remember that any medical volunteers must be credentialed as well as nonmedical staff. This includes volunteers from a multitude of agencies, plus state and federal resources. The types of medical professionals that may need credentialed include the following:

- Physicians
- Psychiatrists
- Epidemiologists
- Medical residents
- Physician assistants
- Psychologists
- Nurse practitioners
- Behavioral health practitioners
- Nurses
- Pharmacists
- Medical Technologists
- Radiological technologists
- Respiratory therapists
- Laboratory staff
- Phlebotomists
- Nurse assistants
- Paramedic Emergency Medical Technicians (EMT-P)
- Advanced Emergency Medical Technician (EMT-A)*
- Intermediate Emergency Medical Technicians (EMT-I)*
- Emergency Medical Responder (EMR)*
- Basic Emergency Medical Technicians (EMT-B)
- DOT Medical First Responders
- Infection control practitioners

It should be noted that this is not a comprehensive list and that those with asterisks are either being phased in or phased out nationwide, as of the writing of this book. With a multitude of specialized medical staff, this list could be expanded substantially. This list of medical professionals should provide (at least) a basis for the types of medical credentialing that may be required. Even beyond medical personnel, the facility should also include nonmedical staff. This should include engineering and maintenance personnel, material management personnel, environmental services personnel, cafeteria/food services personnel, and anyone else who will work in the facility. Because they will be working in close proximity to vulnerable individuals, primarily injured, disabled, elderly, and incapacitated patients, it is critical that every step is taken to ensure that they are protected.

Disaster privileges for medical staff are generally granted when the facility activates the emergency operations plan (EOP). The person who will usually have the authority to activate the plan is the Chief Executive Officer (CEO), the Chief of Staff (COS), the Medical Director, or their designee(s). Decisions about granting privileges to any volunteers during a disaster should be made on a case-by-case basis by the facility. In the initial 72 hours, the medical staff oversees the performance of each volunteer licensed independent practitioner. Based on its oversight of each volunteer licensed independent practitioner, the critical access hospital determines within 72 hours of the practitioner's arrival if granted disaster privileges should continue.

16.13   Volunteer Management

When a large-scale disasters or extended health emergencies occur, healthcare facilities may need to supplement their teams with volunteers. This is often done to ensure that patient care is a priority. If not managed properly, spontaneous volunteers can become the disaster after a disaster. That is why it is so important to develop protocols and procedures for the use of volunteers prior to a disaster or healthcare emergency.

Healthcare facilities should describe when volunteers should be used and in what way volunteers will be used to supplement staff. The protocols should also prearrange how volunteers will be identified, how they will be credentialed, and how they will be assigned. The protocols for the Emergency Operations Plan (EOP) should spell out how the volunteers workload will be managed to ensure that the healthcare facility's standards are met and how they will be evaluated for performance.

Another consideration for the management of volunteers in the Emergency Operations Plan (EOP) is to create (or annually update), a handbook that plainly spells out the Policies and Procedures for Emergency Volunteers. Direction for this handbook should be based on facility's Emergency Operations Plan (EOP) and the emergency volunteer management policies and procedures. To maintain the same level of care and proficiency, it should also incorporate the hospital employee policies and procedures. The Emergency Operations Plan (EOP) should also spell out and develop an internal training program that allows current hospital staff to effectively learn how to utilize volunteers in a disaster. Hospital staff should also learn how to train volunteers for job-specific tasks that might be required.

If, and when, volunteers are used, they should receive a brief orientation. For this reason, the EOP for the healthcare facility should create or develop an orientation plan for volunteers. Of course, due to the fact that the volunteer has probably never been to the facility or worked there before, the facility should provide basic information that the volunteers need to know if they are working in your healthcare facility, especially in an emergent situation. Information they should know based on an emergency or disaster situation should include knowing the incident objectives, the chain of command, security procedures, safety information, and emergency codes (if plain language is not used). Regardless if it is an emergent or nonemergent situation, consideration for how to get around the facility should also be addressed. This can usually be handled by providing the volunteer with a map of the campus and/or a facility map. The orientation should also include how volunteers can obtain logistical support, medical support, mental health services, and other supportive services that may be available.

As part of managing volunteers, the Emergency Operations Plan (EOP) should also address performance of the volunteer. During the first 72 hours, volunteers should be partnered with hospital staff. This allows them to practice on their own but still have a supervisor that oversees the work that they do. At the end of 72 hours, the hospital staff responsible for a volunteer should voice any concerns they have through the appropriate channels. If it is deemed that there are problems, especially if the need for volunteers may be for an extended time, then corrective action can be taken. Corrective action could be anything from discussing issues with the volunteer to dismissing them.

16.14 Health and Medical Operations

Health and medical operation can encompass a multitude of different activities. Depending on the type of healthcare system, these activities could be both internal and external to the facility. This holds especially true when that healthcare facility is a hospital with an emergency department. If the hospital is an EMS provider within the community, health and medical operations could be extremely active outside of the hospital doors while still being managed internally by the hospital. For this reason, health and medical operations should be thoroughly covered both internally, and if applicable, externally to the facility.

Addressing health and medical operations in the Emergency Operations Plan (EOP), the healthcare facility establishes a framework for a variety of differing activities which need to integrate with each other. If done properly, the facility will accomplish tasks as if it were as a single unit rather than a mix of smaller specialized areas. In creating plans for health and medical operations, the Emergency Operations Plan (EOP) will need to consider what services will need to be maintained in a disaster or emergency.

Depending on the circumstances, the Emergency Operations Plan (EOP) may need to address coordinating and integrating EMS, hospital, public health, environmental health, mental health, and mortuary services when planning for a disaster. Depending on the type of facility, there may be other special tasks that may need addressed. When planned properly, protocols and policies will guide the response actions of each facility and assist in making health and medical operation run substantially smoother.

In almost all instances, a Medical Command Post (MCP) will need to be established, and a Medical Director will need to be appointed. In appointing a Medical Director to manage health and medical operations, the facility should choose someone with advanced knowledge, skills and experience, especially if the facility is a hospital. They should also be familiar with the way things are done, or the methods used, in the facility. This usually leads to choosing a medical doctor that has been working at the facility for several years.

In a hospital setting, there could be a multitude of tasks that may need overseen, and the Emergency Operations Plan (EOP) should provide guidance on how these tasks should be approached. After the Mercy Hospital was destroyed in the 2011 tornado, Freeman Hospital (in Joplin) saw a surge of patients. Over 1000 patients presented at the 133-bed hospital, and 22 surgeries were performed in 12 hours. In the After-Action Review (South Dakota Department of Health, 2011), it was stated that a standardized (the same type used by all hospitals) triage tag system was needed to reduce confusion. It should also be mentioned that triage tags should be left on the victim until they are released or admitted.

As part of the health and medical operations, there also needs to be plans for medical care and transportation of injured individuals. Even if the healthcare facility is not involved in EMS or transport, there is the possibility that patients and/or staff may be injured during the disaster or emergency. In the case of a tornado, hurricane, or earthquake strikes, a large quantity of injured people may need to be moved in a short amount of time. By making agreements and creating methods (and possibly alternative methods) of medical transportation in the Emergency Operations Plan (EOP) and signing Mutual Aid Agreements (MAAs) with those companies, it will create a pathway for expedited services.

Additionally, the Emergency Operations Plan (EOP) should identify potential holding and treatment areas for surge events. When a major disaster occurs, those that are injured just know to go to the hospital. Past events such as the Las Vegas shooting and the Joplin Tornado have confirmed that a large influx of patients in a short amount of time can seriously overwhelm a healthcare facility. Spelling out what should be done in the Emergency Operations Plan (EOP) will once again prevent those in charge from being overwhelmed with no plan in place.

16.14.1 Fatality Management

Another facet of Emergency Operations Plan (EOP) is what should be done with those that are deceased. During a major disaster or a pandemic, there may be a large number of deceased individuals. This may hold especially true in a pandemic situation, and it might be overwhelming with how many individuals pass away.

To be prepared for such an incident, the Emergency Operations Plan (EOP) should spell out how the deceased should be identified and transported during such an event. It should be considered whether these bodies may need to stay onsite, or if they will need to be transported to another facility or business. This type of planning prevents the healthcare facility from being overwhelmed with trying to figure out what to do with cadavers while still dealing with those that are alive.

If possible, the healthcare facility should try to preidentify where the deceased should be held, potentially even making prior arrangements for refrigerated trucks to hold a large quantity of corpses. The healthcare facility should also identify the maximum number of deceased they can handle in-house, the maximum number that could be housed in temporary morgues, and security procedures needed to prevent unauthorized personnel from accessing these corpses.

It is important to remember that while the facility may be able to summons outside help from a Disaster Mortuary Operational Response Team (DMORT) or other entity, the requested assistance may not arrive for several days. This means that everything will need to be handled at the local level until resources arrive.

16.14.2 Decontamination

Another factor that should be addressed in the Emergency Operations Plan (EOP) would be matters that relate to Chemical, Biological, Radiological, Nuclear, and Explosive (CBRNE) devices. It should be clearly written in the Emergency Operations Plan (EOP) how these patients should be isolated, decontaminated, and treated. When victims of hazardous chemical or infectious disease arrive at the Emergency Department, those who will be in charge of this type of incident should not have to rewrite the book on what should be done. All the components and most of the decisions of how to accomplish the task at hand should already be in place. Additionally, protocols should be in place for identifying the type(s) of hazardous chemicals or infectious diseases, controlling their spread, and reporting what is known to appropriate State and Federal health or environmental authorities.

16.14.3 Health and Medical Advisories

The Emergency Operations Plan (EOP) should also provide standard protocols for issuing health and medical advisories to the public. This may include contamination of water supplies, the location of emergency water supplies, proper waste disposal methods during a disaster or emergency, and where mass feeding services for patients, visitors, and staff could be located. Other medical advisories may include the trajectory of where the incident may spread to, immunizations to prevent illness (and where to get them), proper ways to disinfect to reduce the spread of an incident, and other similar advisories. It should be preplanned on how to effectively get the information out to the general public in a relatively short amount of time.

Another factor in health and medical advisories should be when to enact the advisories. Whenever possible, the Emergency Operations Plan (EOP) should spell out landmarks of an incident that would initiate that advisories be given. If an advisory is given too early, it may cause unnecessary fear and/or panic, and if it is given too late, potentially already overwhelmed healthcare facilities may see a larger influx of patients.

16.14.4 Interjurisdictional Relationships

Mutual Aid Agreements (MAAs) should also be preplanned whenever possible. This includes how to request health and medical assistance to, or from, neighboring jurisdictions. These resources might be found inside the state (in which the facility resides), or possibly jurisdictions from outside of the state, should they be required. Whenever creating an Emergency Operations Plan (EOP), it is extremely important that it is consistent with local, tribal, regional, state, and national plans, including the National Incident Management System (NIMS) and the Department of Homeland Security (2013c). If the committee is new to creating an Emergency Operations Plan (EOP), or they have new members who do not fully grasp the concept, the Comprehensive Preparedness Guide (CPG) 101 (2010) is a helpful resource.

16.14.4.1 Patient Management

Patient management begins at the time a patient arrives at the healthcare facility. Patients who present themselves at a healthcare facility will need to be quickly triaged and most likely sent to a treatment location for medical care. If this is a mass casualty event, triage tags or bands should be used to identify the priority level of the patient. The method to be used for tagging patients should be spelled out in the Emergency Operations Plan (EOP), and the triage tags should be discussed with other local hospitals to ensure that confusion is reduced.

Additionally, it is important to create a paper trail for the patient, so that all information pertaining to the patient is documented. A swift but consistent registration process should be created, including what the activation requirements are to activate expedited registration and the criteria for returning to the normal registration. Returning to the normal registration process should be contingent upon incident stabilization and acceptable staffing levels. Preplanning this process and writing it into the Emergency Operations Plan (EOP) will help avoid delays in medical care, while still allowing the tracking of patients.

Another specific area of patient care that should be covered in the Emergency Operations Plan (EOP) would be Hazardous Materials (HazMat) incidents. Patients who have been involved in a HazMat incident can arrive at the facility through being transported by ambulance or from a Personally Owned Vehicle (POV). If they arrive by ambulance, in most instances, they will have been decontaminated prior to being loaded in an ambulance. Additionally, healthcare staff typically have advanced warning from the EMS crews prior to the patient's arrival.

When arriving in a Personally Owned Vehicle (POV), things can get complicated very quickly. Patients arriving via Personally Owned Vehicles (POVs) typically arrive with no warnings, and in many instances, it is not known that they were contaminated by (or even involved with) a HazMat until the patient crossed the threshold to the medical care facility. These actions by the patient typically end up with contamination of the receiving facility.

When a victim arrives at a facility, contaminated with a HazMat substance, proper actions must be taken to protect other patients from the HazMat and to decontaminate anything that has been contaminated. The Emergency Operations Plan (EOP) should spell out who should be considered properly trained personnel, as well as what level of protection should be incorporated into the decontamination process. Additionally, using a standardized and well-practiced decontamination procedure before these individuals are allowed into the main hospital is an important issue that should be addressed.

In most situations that involve HazMat situations being introduced into healthcare facilities, only life-saving interventions should be even considered prior to, or during, decontamination. Anything more than this puts staff and patients at risk. The preferred method that should be identified and defined is care that is provided in the healthcare facility after decontamination is completed. By creating protocols in advance of an incident, the Emergency Operations Plan (EOP) and the protocols should help prevent the needless exposure of staff and patients to any hazardous material that may come into the facility.

In HICS, The Medical Care Branch coordinates both inpatient services and outpatient services. This usually includes behavioral and/or mental health services, clinical support services, and patient registration services. To better serve the patients that the facility will serve during a disaster or emergency, the Emergency Operations Plan (EOP) should clearly spell out the chain of command for patient care and management. This should also spell out that the Medical Care Branch Director should be required to work with the Logistics Branch to ensure that personnel, equipment, medication, and supplies are delivered to the areas of the facility when and where they are needed. By spelling out how, and who, makes the wise decisions on how to manage resources when they are limited are extremely important. While overall direction should come from Command (who should also consult the Emergency Operations Plan [EOP] for guidance), it can be more than challenging, and basic direction of how to handle this type of conundrum will likely be helpful.

The Emergency Operations Plan (EOP) should also identify local resources that may be helpful to enhance or maintain patient care. This is especially true if outside resources have committed with a Memorandum of Understanding (MOU), a Mutual Aid Agreement (MAA), or any other agreement that might help supplement patient care (including supplying logistical needs). All resources and the type of resources that can be provided through these agreements should also be documented.

16.14.4.2 Logistics

The Logistics Section is similar to gasoline and oil that is needed to keep a car running smoothly. Because the Logistics Section provides for all the support needs of the incident, it is in the best interest of a facility to preplan resources for the logistical needs during an activation of HICS and the Emergency Operations Plans (EOPs). The Emergency Operations Plan (EOP) should be written in such a way that it clears a path for attaining resources from both internal and external sources. Preplanned acquisition procedures for supplying operation needs during an emergency are extremely important.

Whenever possible, the Emergency Operations Plan (EOP) should spell out when and how to request resources from other healthcare facilities, corporate partners, and the local Emergency Operations Centers (EOC). Depending on the way that the facility is managed and where it is located (in the United States or potentially another country), the Regional Hospital Coordination Center (RHCC) (or their equivalent) can also be incorporated.

The Emergency Operations Plan (EOP) should also identify how every resource request made within the facility should flow or at the very least, how it will be reported to the Logistics Section. By identifying and utilizing predetermined ordering procedures when Hospital Incident Command (HICS) or the Emergency Operations Plan (EOP) is activated will allow better oversight in a disaster or emergency.

Another critical item to write into the Emergency Operations Plan (EOP) is how to request items from outside sources. The first thing that should be addressed is the method of contact for ordering. The facility should prearrange the method of how requests are to be made to various agencies, businesses, and resources. They need to identify whether requests should be made via email, fax, phone, or some other method, and it should be documented in the Emergency Operations Plan (EOP). Additionally, there should be a contingency plan in place in case the original method of contact has been compromised or becomes inoperable.

When ordering from outside sources, it is important that the Emergency Operations Plan (EOP) spells out that the Logistics Section be specific about what the facility needs rather than making blanket statements of need. While some may be tempted to make a blanket statement such as "We have a shortage of everything, send us what you can spare," this type of request can do more to undermine a response than it does to help it. The Emergency Operations Plan (EOP) should require those in logistics to be specific about what is needed to reduce instances of unwanted or unusable items and/or equipment.

Remember, unwanted and/or unusable items will still need sorted, cataloged, and stored. This means that personnel that will probably already be in short demand will be pulled away to deal with unwanted and unusable items while in the middle of a disaster or emergency. It also means that it will take up space, which may be better utilized storing what is truly needed or treating patients.

It is also extremely important to use common terminology and exact language of what is needed. Without utilizing common terminology, or in ordering the exact item, there could be even more confusion because a common vernacular was not used. In almost all instances, the logistical needs of the facility will not be decided by the Logistics Section, but this section will be responsible for ensuring that the exact item ordered is obtained in a timely manner.

16.14.4.3 Finance and Emergency Spending Authorizations

When your normal world is turned upside down, one of the last things on most people's mind during a surge or disaster is financial accountability, but it is an important factor. Financial expenditures and emergency spending authorizations should have protocols that provide basic guidance, as well as identify who has the authority to make financial decisions. Lack of a protocol for finance and emergency spending can lead to abuses and/or unnecessary expenses.

While the method for overall expenditures may be prearranged, it is quite possible that the daily financial reporting requirements will be revised. A method of approving expenditures, created in advance of an incident in the Emergency Operations Plan (EOP), will cut down on the time needed to gain approval during an incident. Additionally, if there may be a reimbursement from state and/or federal officials, then it is likely that the facility will need to be accounted for in a different manner from the normal operational accounting already undertaken by the facility. Part of the preplanning efforts should preidentify the state and federal financial aid documents that will need to be completed in order to receive reimbursement; however, when making finance and emergency expenditure decisions, the potential of reimbursement should not play a part in the decision to spend. This could lead to a serious deficit for the facility if costs are denied.

As a part of accountability and the need for funding, the Emergency Operations Plan (EOP) should require that the Finance and Administration Section monitors all the costs that are related to the incident. This would include accounting for procurement of equipment and supplies, as well as time recording. It should also be required that a daily cost analysis of how much more funding will be needed. It is important that the costs that are specifically associated with the disaster or emergency response must be accounted for from the beginning of the incident.

16.14.4.4 Resource Management

Resource is a broad term used in incident management to describe facilities, equipment, personnel, communications, and procedures. Resource management is integrated into almost every facet of incident management. In fact, it is so important, that it is discussed at length in NIMS documents. The Emergency Operations Plan (EOP) should provide guidance on how to manage resources that are in compliance with state and federal standards by using a method widely accepted by healthcare facilities in their country. When an incident occurs, the healthcare facility must be prepared to manage all resources in a very short amount of time. Failure to do so will result in more confusion, chaos, and uncertainty.

In order to accomplish resource management, the healthcare facility should adhere to common methods for describing resources, keeping an up-to-date inventory (of *all* resources), and ensuring that all resources are ready to be used. Additionally, there should be protocols in place in regard to sending resources to where they are needed, tracking those resources, and recovering usable resources back when the incident has finished. This resource management should be in place maintaining resources brought into your facility and when resources are sent to other facilities.

16.14.4.5 Donations Management (Solicited and Unsolicited)

Donations management, if not managed properly, can be another area where it is considered a disaster after a disaster. Managing a huge influx of donations can be a daunting and labor-intensive endeavor. After donations arrive, healthcare facility staff and volunteers are faced with the monumental logistical task of sorting, cataloguing, storing, and distributing these donations. This can and will divert energy away from addressing the more urgent needs after a disaster or emergency.

When writing the protocols in the Emergency Operations Plan (EOP) for managing donations, it is important to make sure that you write how to provide clear expectations to media sources and the general public about the types of donations that are needed. If the facility does not address the types of donations that are needed (including stating that no donations are desired), then a good-hearted public will send whatever they feel may be important. If the public is intent on sending a specific type of donation, it can be written in the Emergency Operations Plan (EOP) that suggestions can be made to potential donors, or those donations should be sent to a local charity.

To help you better understand how this might work, we will look at what happened after Superstorm Sandy. As the Episcopal Relief and Development Team began to assist those affected by Superstorm Sandy, they began to ask for donations. While food, housing, and mucking/clean-up were a priority, they received a call from a congregation who felt that a donation of teddy bears to comfort storm survivors was important. Knowing that feeding and housing needs were critical, the team suggested that each teddy bear that was donated could include gift cards that could be used at local restaurants. Several days later, they received over one hundred bears all hugging a gift card% By making this suggestion, the teddy bears provided emotional support for children, and the gift cards helped address the physical requirement of feeding survivors (Episcopal Relief & Development, n.d.).

While it is unlikely that a healthcare facility might not need this type of donation, we must also realize that there are times when donations will come in, even if they are not solicited. By creating written protocols in the Emergency Operations Plan (EOP), managing donations can clear the path for fewer personnel and less headache. If donations will be accepted, the preplans should include how they will be sorted, catalogued, and stored so that those managing donations do not have to reinvent the wheel during or after a disaster or emergency.

16.14.4.6 Infrastructure Management (Building, Grounds, Utilities, Damage Assessment)

Suffice it to say that a large portion of tending to and treating patients is dependent on the infrastructure that a healthcare facility provides. In a disaster, it is critical that maintenance crews evaluate the facility to identify any issues that can affect patient and staff safety. This is extremely important because the maintenance staff supports the healthcare facility's operations. The facility, and those in the maintenance section, supports activities to meet the medical care needs of the patients and protect staff.

When identifying who has the responsibility for the of the facility structure in HICS, it is important to realize that Infrastructure Branch (located in the Operations Section) should oversee these important tasks. This branch is responsible for maintaining and (if need be) expanding operating capacity. This Branch is also tasked with identifying and repairing any failures in utilities or the structure itself. Although a rare occurrence, the Infrastructure Branch can condemn parts or all of the facility for patient care, providing they feel it will put patient care and/or safety (as well as staff safety) at the facility at risk.

The Emergency Operations Plan (EOP) should predetermine what areas of the facility should be evaluated as well as create a priority list of what should be evaluated first. This is done so that the most critical to patient care and the safety of staff and patients is completed first. Areas that will likely need to be evaluated include the following:

- Power and lighting (HICS Power/Lighting Unit)
- Water and sewer (HICS Water/Sewer Unit)
- Heating, ventilation, and air-conditioning (HICS HVAC Unit)
- Medical gases (HICS Medical Gases Unit)
- Building and grounds (HICS Building/Grounds Damage Unit).

The individuals who should be tasked with evaluating the facility are the staff personnel that normally handle these issues on a daily basis (e.g. maintenance crews). If outside contractors are used to maintain a facility, then the Emergency Operations Plan (EOP) should spell out which contractors are responsible for each facet of maintenance, evaluation, and/or repairs. If possible, a Memorandum of Understanding (MOU) should be signed with these companies, and it should be required that they take HICS classes and potentially ICS classes. The mandate for HICS classes is important. If they are not familiar with HICS and an incident occurs, these contractors will need to be informed about how HICS works, thereby wasting valuable times and resources in the midst of a disaster.

16.14.4.7 Evacuation

In a disaster or emergency, evacuation of the facility may be necessary. In creating the Emergency Operations Plan (EOP) for evacuation, it should be realized that the plans should match other protocols in the plan, and there should be considerations for sheltering in place, partial evacuations, and a full evacuation of the facility. The Emergency Operations Plan (EOP) should provide guidance on what should trigger each type of evacuation, where evacuees may be transported, and methods of transportation for evacuation.

As one might imagine, evacuating a healthcare facility is not as simple as evacuating most other buildings. This holds especially true if there are patients who cannot care for themselves, who have mobility issues, are on life support, are postoperative patients, and various other injuries and ailments that may have brought these individuals to the facility. Due to the complexity of an evacuation at a healthcare facility, the Emergency Operations Plan (EOP) should assign who should decide to undertake the evacuation and who should be responsible for each part of the evacuation.

As with any part of an Emergency Operations Plan (EOP), there must be definitive criteria for what activates an evacuation and who has the authority to activate the evacuation or sheltering in place. It should also spell out how to communicate that an evacuation has been ordered, including the integration of the evacuation with HICS. It should be realized that when an evacuation is ordered, it will likely take the help from outside resources, especially if the facility is a moderate to high patient count facility.

When the evacuation procedures are worked out, the facility will need to train staff members on the plan, specific roles, and responsibilities, how to operate and use evacuation equipment, proper techniques for lifting and carrying patients when rolling beds are not available, and they will need to provide training on all evacuation routes.

Once training is complete, then the plan should be tested. It is important to remember that this testing should not be tested using actual patients, but rather actors. Additionally, after a training or real-life incident, an After-Action-Review (AAR) should be undertaken to identify how improvements could be made.

While this book could provide a myriad of tips and protocols that could or should be undertaken, many requirements are already mandated, and the healthcare facility should follow the guidance provided by the Center for Medicare and Medicaid Services (CMS), state requirements, and local requirements. Additionally, Assistant Secretary for Preparedness and Response (ASPR) sponsors the Technical Resources, Assistance Center, and Information Exchange (ASPR-TRACIE), which provides a myriad of guidance on how to prepare for evacuation.

In most instances, the guidance available for the Emergency Operations Plan (EOP) evacuations covers mainly patient care. It is important to realize that the facility will not only have to address and plan for the various forms of evacuation but they should also create a plan for an influx of patients that will typically come after a disaster. The plan will also have to identify how to transport patients from the facility, how to secure and/or transport medicines, how to secure radioactive materials, where patients should be taken, and a whole host of other issues that come with evacuation. A disaster or an emergency that requires evacuation will create multiple additional problems that should be addressed in the Emergency Operations Plan (EOP).

16.14.4.8 Safety and Security

When we discuss safety and security in a healthcare facility, immediately our minds go to security guards or law enforcement. In the sense of the Emergency Operations Plan (EOP) and a healthcare facility immediately, we must think beyond these individuals and look also the staff that might be able to help fill these roles. This is not to say that staff should stop patient care and take on a different role; because in most instances, they should not. However, it should be realized that everyone should make a priority of identifying and reporting safety issues.

The immediate priority after a disaster or emergency will be safety. In most instances, the Safety and Security Team will look for areas of concern and ensure that those managing the infrastructure are aware of any safety issues that are seen. Members of the Safety and Security Team can, and probably should, include a mix of non-healthcare employees as well. This will allow more issues to be identified more quickly, and those in charge of infrastructure maintenance can spend more time evaluating potential issues and less time looking for these issues. There is also the case where medical staff may see something unsafe than the maintenance people may not. Of course, this will be dependent on (and directly affected by) the type of disaster. Depending on the emergency or disaster, the infrastructure could take a hard hit, or it may be totally unaffected.

The Emergency Operations Plan (EOP) should identify what tasks should be completed by Safety Team members and who should be responsible for specific tasks. As an example, you might have custodial staff as part of the Safety Team, and they could identify broken pipes, fallen ceilings, or signs of potential structural damage. While custodial crews may be well suited to look for these issues, having them evaluate the safety of radiological equipment or identifying issues with telemetry and durable medical equipment would likely be unsuccessful.

For this reason, the Emergency Operations Plan (EOP) should spell out which individuals should be chosen for the Safety Team. This does not mean they should be named as a person, but rather they should be identified by the job they do on a daily basis. An example, perhaps the charge nurse for each floor would be responsible for ensuring that durable medical equipment is still operational and that medicines are not compromised. This is not to say they must do it themselves because they can delegate authority and have each nurse on the floor, or even perhaps medical technician check the equipment and then report back.

It is important to realize that there is much more than the structure that will need evaluated for safety. The accessibility to the facility will need to be evaluated as well as exterior areas. Water and sanitation will need to be evaluated, even beyond the healthcare facility. Did a sewer pipe break somewhere outside of the facility and is it now backing up? Were water lines broken somewhere away from the facility and has it contaminated the water? What about the Emergency Department, is it still operational? What about the Neonatal Intensive Care Unit (NICU) and the Intensive Care Unit (ICU)? Are they still operational? Was there damage, and if so, how critical is it to their work? What about the Operating Room (OR), the morgue, the different wards, and radiology? Are they safe and operational? Is dialysis, the blood bank, the pharmacy, and the lab all safe and operational? What about the kitchen and laundry? Are there gas leaks, broken pipes, exposed electrical or some other issue? There are many aspects of a healthcare facility that will need to be inspected for safety. Because each healthcare facility is unique, each facility will need to identify what the safety issues may be in their facility. After a disaster or emergency strikes a healthcare facility, it is critical to realize that the daily protocols for safety and security will be challenged.

While safety is the top priority, security of the healthcare facility should also be addressed. Some of the information in this section may seem over the top (or unbelievable), we need to realize that during desperate times, desperate people can do desperate things. While most facilities will have no major security issues after a disaster, this does not mean that the Emergency Operations Plan (EOP) should plan for minimal security issues. In fact, it is best to hope for the best and to plan for the worst.

Many healthcare facilities have items that nefarious individuals might want. These items might include drugs, electronics, appliances, and even furniture may become a target for looters. If the facility has a radiological department, there is also the possibility, albeit a slim possibility that someone may want to steal radioactive material. If the facility cannot be secured, there will likely be individuals who will take advantage of the healthcare facilities misfortune. There could be looting, theft (of almost anything), and most likely the theft or attempted the stealing of drugs.

There may also be patients and patient's families that are upset or downright angry. In these instances, there will likely need to be interventions. Realizing the reason for their anger might be helpful in diffusing such situations, but security should be used immediately if someone feels threatened. It could be that they feel that the facility did not do enough to protect them (or their loved one), or it could be that they feel that they need urgent help, and they are not getting it. It might be that their loved one was deemed an individual that was beyond help, and they are angry because nothing is being done to help their loved one to survive. It could be that they are just plain upset because of the circumstances. These are just a few of the many reasons that the Emergency Operations Plan (EOP) should be proactive and preplan for security in disasters and emergencies.

It is also important to note that safety and security might cover even more issues, depending how the healthcare facility is administratively set up. Security might include receiving and communicating any threats that could affect the hospital and the continuity of care. Security might also involve looking for a missing patient or child, or in a major disaster, multiple missing patients or children. Another task that security could be tasked with could include hazardous materials spills and any associated decontamination that has occurred on the facility's property. Security might also be responsible for overseeing fire suppression and fire prevention duties. This is just a few of the many areas that may be covered under safety and security section of the Emergency Operations Plan (EOP).

16.14.4.9 Coordination with External Agencies

Coordination with external agencies is a critical part of mitigating a disaster or emergency. It is likely that at some point, a disaster or emergency will happen within the healthcare facility, and outside resources will be needed. Having the proper mechanisms in place will help the integration of outside resources to become almost seamless.

Unlike some aspects of the Emergency Operations Plan (EOP), coordination with external agencies should not only be on paper. To create an effective coordination, those in charge of emergency management within the facility must be actively engaged in meeting with their emergency management colleagues from various professions that will likely be needed. This should include local EMS agencies, law enforcement agencies, fire departments, local public health agencies, local behavioral health agencies, local elected officials, nearby area hospitals, local nonprofits, and local businesses and other nongovernmental agencies. It is important to realize that this coordination with external agencies should not be relegated only to the city or county that the facility is in. This holds especially true in healthcare facilities that serve rural areas. Coordination with the outside resources should include nearby areas that the healthcare facility typically serves.

As an example, Macon County Illinois has two hospitals located in the city of Decatur. Because the majority of other hospitals are more than 45 road miles away, it is fair to assume that most people will likely go to the nearest hospital in an emergency. This would mean that individuals from 20 to 25 miles away would likely go to one of the hospitals in Decatur, IL. If the medical problem was related to a new outbreak, a pandemic, or some various other issues, the hospital may be required to notify the county in which the patient lives.

For this reason, the hospitals in Decatur should try to forge agency relationships with not only Macon County but also Logan County, Shelby County, Sangamon County, Moultrie County, Piatt County, DeWitt County, and Christian County. By building relationships in these seven counties that they serve anyway, they could potentially have five to seven times the resources in a localized disaster. If this were a regional disaster, it is likely that little to nothing would be gained.

The same would hold true for the nearest hospitals, in the event of a disaster that is local to a city such as Decatur, IL, the county seat. Building relationships with nearby hospitals could prove to be extremely beneficial. If a local manufacturing plant had an incident that injured multiple victims, the local hospital could request resources from other hospitals within the region. In doing so, they could potentially have additional supplies and/or personnel within an hour or two. Of course, any agreements that would be made or relationships that would be forged would have to be reciprocal. This also does not consider building relationships with urgent care facilities that are not affiliated with the hospital, and it does not consider other healthcare facilities that may be able to provide resources in a major incident.

Similarly, any healthcare facility could benefit from building these types of relationships. A nursing home that suffered a mass illness among staff, who has good relationships with other nursing homes in the vicinity, might be able to borrow staff from other nursing homes. Of course, this might be dependent on the type of relationship that was forged. Perhaps a nursing home suffered a fire or flood. With the proper relationships being forged, and proper coordination in place, the nursing home might be able to find open beds at other nearby nursing homes, or at the very least be able to borrow staff to help evacuate the nursing home. The same might hold true for a doctor's office or a whole host of other medical facilities. By building relationships and coordinating with these external agencies, the healthcare facility will essentially create more options that may be able to help in times of need.

While it is important to create relationships with external agencies, it is equally important to plan, train, and respond with these outside agencies, including non-healthcare agencies. If healthcare facilities choose to not do so, then they will probably not be effective when an incident occurs. When looking at different disciplines outside of healthcare, there needs to be a fundamental understanding of the Emergency Operations Plan (EOP) and what role outside agencies may be needed to fill. The Emergency Operations Plan (EOP) should provide clear direction on who to call and when to call them. The response partners should have clear guidance and understanding, as well as what resources may be required to assist the healthcare facility after a disaster.

It is important to note that each facility is unique, as is every region. Some healthcare communities use the selection process of appointing a Duty Officers, while other agencies utilize a RHCC or some other type of cooperative coalition. Both methods help to facilitate information sharing and response coordination among hospitals. These efforts are usually undertaken in cooperation and collaboration with the local health department or emergency medical services agency as well as other agencies.

In most instances, healthcare response is often focused on hospitals. Because a community's healthcare is almost always provided by a system of hospitals, clinics, urgent care centers, surgical centers, physician offices, prehospital care, long-term care facilities, nursing homes, home health agencies, hospices (and other similar agencies), it is important to recognize that all healthcare stakeholders should be included. They should be invited and involved in mitigation and preparedness efforts, including cooperative planning, training, and all methods of exercises. By doing so, it can help to provide a clear understanding of each entity's roles and responsibilities, the resources that each can provide, and it can help identify potential shortages in personnel and supplies. This cooperation and collaboration can also strengthen response skills and help to better facilitate a more streamlined patient transfer, should it be needed.

Active coordination that takes place prior to a disaster will help ensure that important mechanisms are part of the answer rather than part of the problem. This is especially important when the local hospital or other health care facility is pushed beyond their capacity and they need help fast. For this reason, the stakeholders for healthcare in each community should regularly meet to discuss and undertake planning and training issues. These planning and training meetings should focus on the unique needs of the community and should lead to drills and exercises that include all stakeholders.

It is important to note that these meetings should not be used to catch up on the latest news, but rather they should work toward a more integrated response in the event of a disaster or emergency. One area these meetings should focus on is parts that often lead to confusion or misinterpretation. Terminology is perhaps one of the more common areas where there is confusion among outside agencies. Additionally, the standardization of plans and procedures among all stakeholders in a community can facilitate how limited resources can be shared.

In many instances, successful planning will lead to emergency assistance agreements among hospitals and other healthcare providers within the community as well as agreements with neighboring communities. These may take shape as an automatic mutual aid agreement, a (requested) Mutual Aid Agreement (MAA), or Memorandum of Understanding (MOU). These agreements should be set in ink and established among participants long before disaster strikes.

Any agreement that is forged should address the procedure for requesting assistance, how resources will be shared, acceptable methods of credentialing staff members and volunteers, and how to facilitate patient transfers. Additionally, these agencies should discuss and agree on how stakeholders will be reimbursed for responding, as well as potentially creating standardized reimbursement forms.

Some stakeholders in various communities have even taken this a step further, to help better facilitate communications and coordination. In these communities, all stakeholders have joined together to purchase equipment and supplies. In doing so, they have saved considerable money by mass purchasing and at the same time created a familiarity of operating equipment with all stakeholders. When the exact same equipment is purchased by all facilities in the area, there is no learning curve due to the distinctiveness of operation between one brand of equipment and another. This increases operational effectiveness through increased capability to share common resources.

Another important stakeholder in external coordination for healthcare facilities would be prehospital care. This could include fire department, Emergency Medical Services (EMS) such as ambulance providers, air ambulance services, and the governing authority for prehospital care. These agencies play significant roles in day-to-day emergency care and would likely play an even more substantial role during an incident or emergency. In the last 20 years, many fire departments have expanded their responsibilities to include basic life support (BLS), Advanced Life Support (ALS), or some combination thereof. Depending on the fire department, the entire fire crew may be paramedic trained and have all the necessary equipment to provide Advanced Life Support (ALS) to their patients, while other fire departments may have one paramedic and several lower level EMTs, and yet others may just be manned by basic medical responders. While the exception rather than the rule, some fire departments will not provide prehospital care. In some instances, fire departments have even incorporated ambulance transport into their responsibilities rather than using a private ambulance service.

Even if a fire department does not provide prehospital care, they can be an important part of a health care facility's Emergency Operations Plan (EOP). While healthcare facilities need to have the capability to decontaminate patients, it should also be noted that fire departments are often known for their capability of dealing with hazardous materials (HazMat). In the event that a HazMat occurs, and the healthcare facility is receiving patients that are contaminated, fire departments are often the first line of defense to prevent further contamination. Depending on the number of individuals that are contaminated, decontamination can be as complex as a self-contained decontamination unit or a system of large pop-up tents. The healthcare facility should realize that the fire departments will be the experts in decontamination. It should be written into the Emergency Operations Plan (EOP) that either the fire department should be the lead in decontamination during an incident or that a unified command should be created to look at all aspects.

Fire departments are also typically experts in search and rescue, including wildland search and rescue. In the event that a facility is hit by a tornado or earthquake, the fire department (or in some areas, rescue squads) will be relied upon to search for survivors and missing individuals. This will take an exorbitant amount of coordination, primarily because the fire department will not know where the majority of people will be during a given time of day. Hospital staff will be vastly more familiar with where most individuals will be based on the time of day.

Another area that may need advanced coordination with a healthcare facility is private ambulance companies and municipal ambulance service. In prehospital care, these services may provide basic life support (BLS), Advanced Life Support (ALS) services, basic transport services, interfacility transportation, and standby services for scheduled events. They will also usually provide services to fire and/or law enforcement for HazMat events, search and rescue, and disaster response, and in some parts of the United States, ambulances will be on standby for structure fires and disaster response.

Emergency Medical Services (EMS) is typically governed by local or, in some instances, state entities. These governing bodies provide medical licensing to prehospital emergency medical technicians and paramedics. They also provide licensing oversight for ambulances, which will likely include annual inspections, and approval and/or agreement for where they may operate. While local or state agencies are used for the licensing oversight, a local medical doctor who specializes in emergency medicine is typically appointed to be the local Medical Director of prehospital care. This EMS oversight usually extends to the interface between prehospital and hospital emergency services, and they provide protocols for EMS providers. In some cases, this Medical Director will fill the role of Medical Unit Leader during a disaster and be tasked with creating a medical plan. Because of the flexibility and similarities between ICS and HICS, as well as the capability to seamlessly interface with NIMS, these varying but similar IMS methods can work cohesively together.

Another area where outside coordination is critical (in the planning process) is air ambulance. It does not matter whether the air ambulance is rotary (e.g. helicopter) or fixed wing (e.g. plane), the healthcare facility should create relationships with air ambulances that may serve their area. Predisaster coordination will help the healthcare facility to understand the needs of each specific type of air ambulance. By knowing the needs to land the aircraft, the healthcare facility can better utilize these resources during a disaster or emergency.

As an example, if a Lear Jet 35 is needed to transport a single patient long distance, then it will require a runway that is 2200 feet long. If a Beechcraft King Air is needed to transport two patients, it will need 2074-feet-long runway. Many municipal airports can meet the needs of these aircraft; however, you also need to know their needs in case runways are damaged. By the same token, rotary aircraft typically require a 100 × 100-feet Landing Zone (LZ). While a 100 × 100-feet area does not seem like a large area, there are other mitigating factors that may restrict the landing of a rotary air ambulance. Uneven ground cannot be used, and in most instances, it is preferred to use concrete, asphalt, or grass to land on. Power lines should be avoided, and if they are near the Landing Zone (LZ), the pilot should be made aware of this. There also should be no debris that can be stirred up and sucked into the intake the helicopter. This includes rocks, leaves, trash, plastic, and other types of debris that may be stirred up by the rotors. Depending on the type of disaster, it may be nearly impossible to find such an area, especially after a tornado.

Because hospital staff will probably not have a hands-on approach when setting up Landing Zones (LZs), part of the coordination that will be needed to meet the needs of the Emergency Operations Plan (EOP) is training for those that will be the likely ones to identify a Landing Zone (LZ), set it up, and to assist the pilot in landing the aircraft. These individuals may be law enforcement personnel, fire department personnel, EMS personnel, emergency management personnel, Community Emergency Response Team (CERT) members, and more. Any person that may be required to set up the Landing Zone (LZ) and/or coordinating with the pilot should be provided with the specialized training they need. This training and the needed coordination to facilitate this training should be written into the Emergency Operations Plan (EOP).

During a disaster, it is probable that EMS will transport a substantial number of patients to hospitals and potentially other healthcare facilities of the impacted population to the hospital for medical care. Coordination should be written into the plan for communicating and information sharing procedures. These plans should be provided and be well known by all stakeholders.

These plans will need to be dependable, redundant, and interoperable, meaning that it should be frequencies that can be used by all or most stakeholders. This will allow EMS and fire crews to provide early notification of the number of patients that will be transported and an overview of the extent of their injuries. In a disaster, there may be instances when competing hospitals may need to coordinate with each other to identify who will take each patient, and early notification and patient updates will provide guidance for that decision. In most parts of the United States, there are Regional Medical Communications Centers (RMCCs) that will do this; however, healthcare facilities should also try to be ready to take on this responsibility if necessary.

As an example, if a disaster involved multiple gunshot wounds to patients, one hospital may be overwhelmed if they were to receive all the patients. If the two hospitals worked in cooperation to determine where these patients would be transported, then there would not be an overload of patients at one facility. By the same token, patients may have to wait longer times before being taken to surgery (due to limited surgeons or operating rooms) if all or most of the patients were taken to one facility. Through coordinating with prehospital stakeholders as well as other healthcare facilities, there is less of a chance of overwhelming one facility. For this reason, protocols should be written into the Emergency Operations Plan (EOP) that identifies when to activate coordination with other facilities and how to divide the patients among these facilities.

Plans for responding to disasters and emergencies (especially where the hospital may be the actual scene of the incident) should be developed and periodically exercised in coordination with all stakeholders. By co-creating these plans, the healthcare facility ensures that everyone is on the same page. By practicing and exercising these plans, gaps in planning and errors in judgment can be identified before a disaster or emergency ever happens.

The coordination needed can be an overwhelming task for someone new to disaster preparedness. Those already listed are some of the more common entities that the healthcare facility will need to coordinate with. In all honesty, to create the relationships and the coordination that is needed for a healthcare facility, especially a hospital, it will probably take many years of someone with just the right personality to complete this task. Even then, they will look for more relationships to build and search for ways to increase coordination.

16.15 Conclusion

It is imperative that hospitals integrate with other patient care stakeholders. By applying HICS and in adapting it to an individual healthcare facility needs is critical to managing a disaster. Equally important in managing a disaster in a healthcare facility is creating an Emergency Operations Plan (EOP). Both are critical factor in incident management. It is important to realize that these types of incident management methods are not a "one and done" proposition.

To implement HICS, the healthcare facility will need to provide regular education and training. This education and training (with knowledgeable instructors) will produce proficiency and competency among the staff. Once it is learned and comprehended, it provides an easy-to use framework that integrates with outside resources, which can be used to manage relatively any incident. Whether the disaster or emergency is an internal or an external incident impacts, the healthcare facility will be able to effectively respond. This is critical for the safety of patients, visitors, staff, and outside resources. It also provides continuing service or reopening of healthcare services as quickly as possible after a disaster or emergency.

The capability of hospitals to effectively function during a disaster, while working austere conditions, is reliant on a number of factors. Some of those factors include effective coordination, the structural stability of the hospital, the amount of staff that can work during and after a disaster, and more. The creation of an Emergency Operations Plan (EOP) can, and does, provide guidance on how to more effectively manage a disaster or emergency, thereby becoming an important part of incident management. This is why it was included as part of the IMS system in relationship to HICS.

Disasters are not an everyday occurrence for healthcare facilities. A disaster will require an increase in patient care, and it will likely affect the internal and external operation processes used daily. HICS will create order out of the chaos, confusion, and uncertainty that is created by a disaster or emergency. Through the implementation and use of HICS, healthcare facilities will be more capable of resourcefully responding to an incident, and successfully coordinating with external stakeholders. The healthcare facility that undertakes these steps will become part of the solution in a community-wide disaster. Their role in the response to a disaster will be reinforced, and the patients they serve and their staff will be beneficiaries of their thoughtful planning.

Real Life Incident

On 1 October 2017, a lone gunman at the Mandalay Bay hotel in Las Vegas opened fire on a crowd at the Route 91 Harvest Music Festival. In the aftermath, 58 individuals were pronounced dead, and over 500 were injured due to the gunfire (Las Vegas Review Journal, 2017).

Almost immediately, this was identified as a Mass Casualty Incident (MCI), and hospitals were notified. The emergency room physician, Dr. Kevin Menes, at Sunrise Hospital was stunned by the news and even turned to a police officer that was in the emergency department and asked if it was real. Menes regularly prepared for this moment. He was well known for quizzing nurses and staff on what they would do in a Mass Casualty Incident (MCI).

Once it was realized that this was not a drill, Menes jumped into action. He first told three other emergency physicians, including Dr. Dave MacIntyre, that they needed to get everything ready. While Menes told secretaries to call in all staff and surgeons they could find, MacIntyre began preparing trauma carts, unlocking medicines, and preparing the supplies that would likely be needed. Menes had housekeeping bring all the clean linens to the emergency department, while nurses threw open curtains and cleared the emergency room of as many patients as possible. Menes called for every gurney and wheelchair that could be found to be brought to the receiving door of the emergency department. Black permanent markers and triage tags were also brought to the receiving area, so that there was effective triage for incoming patients, no matter what mode of transportation brought them. All physicians were also given permanent markers, so they could write information on the foreheads of their patients. A pharmacist in the hospital unlocked all the drugs used to intubate patients. Nurses began to stuff the pockets of their scrubs with this medicine, along with all the O-negative blood available and IV bags. In less than 15 minutes, they had initially prepared themselves for the onslaught of patients to come, but they were waiting on more help.

Resident surgeons were put on alert and readied to work on patients with little to no supervision. From the beginning, it was known that more surgeons would be needed, so available secretaries were lighting up the phone lines of every surgeon that was affiliated with the hospital. This included plastic surgeons, pediatric specialists, orthopedic surgeons, and ear, nose, and throat surgeons. Fortunately, six neurosurgeons were available, and they would be at the hospital in just 30 minutes. These neurosurgeons performed delicate operations that saved lives.

As Dr. Menes walked out of the receiving doors to inspect preparations and to wait on the first incoming patients, the thought crossed his mind that this might be a terrorist attack. He turned to an officer and voiced his concern. Within minutes, the hospital had a string of police cars that created a perimeter around the hospital. Menes began waiting for the first patient to arrive, which did not take long.

Menes was stationed outside as the Triage Officer. He described how the patients seemed to come in waves. First, patrol cars with two patients at a time arrived. Then came pickup trucks and cars, carrying as many people as they could. Soon ambulances began to arrive, often with five patients at a time, even though this was far more patients than was acceptable by their protocols. Some of the time-consuming EMS protocols were thrown out the window in an effort to save as many lives as possible. Mixed in with these waves of vehicles were the walking wounded who came through the sliding doors rather than the receiving area.

Dr. Menes triaged patients as they arrived. Standard triage tags that were color-coded were used to identify the severity of the injuries. Red tags were taken directly to one of four trauma rooms. It did not take long before these rooms were filled beyond capacity. Yellow tags were put in curtained cubicle rooms and monitored by a large crew of nurses. Green tags were asked to sit in plastic chairs or on the floor while more critical patients were treated. Because there was not enough clerical staff to keep up with the pace of these incoming patients, foreheads and arms became the place to write information. If their name was not known, and the patient could not verbally say their name, they began a system of alphabetically giving them a first name, with their last name being "trauma." First was Adam Trauma, then Becky Trauma, and so on.

Each trauma room had specific injuries. One trauma room was filled with head wounds, another with chest wounds, another with belly wounds, and the fourth room was filled with critical wounds that did not fit into any of the previous categories. As surgeons arrived, Dr. MacIntyre could continue working on patients while verbally telling the incoming surgeon what trauma room they should focus their attention on, based on their surgical specialization.

The team worked together for the good of the patients. Preplanning had definitely played a role in the response to this tragedy. Everyone did what was needed. Environmental services were continually mopping up blood and changing bloody sheets. Transporters freed up doctors and nurses by moving patients from triage to their designated area (based on their triage tag) and in transporting patients to surgery, x-ray, or wherever the patient was sent.

It was six hours from the time that the shooting began, until the last patient was treated. In those six hours, a total of 199 patients were triaged at Sunrise Hospital. Ten people arrived with no pulse during those six hours. In total, 16 people were pronounced deceased; however, it is not known if any that came in without a pulse were saved. Planning had played a significant positive role in the outcome of this incident, which was evident by how many people were saved (Woods, 2017).

Chapter 16 Quiz

1   **True or False**: The HICS method for hospital incident command was inspired by the ICS method used by firefighters.

2   **True or False**: HICS does not have a revision process.

3   **True or False**: HICS has become such a mainstay that it is strongly suggested when healthcare facilities have planned events, that they use the HICS method to completely manage that event.

4   HICS can be _____ to meet the needs of the incident, or the user can contract or intentionally collapse the organization (and the organizational chart) if the incident is overstaffed.

5 _____ is the process of sorting and classifying the injured.

6 What color-coded item is usually attached to a patient to quickly and easily identify their condition in the prehospital setting, and often during a surge event internal to the hospital?

7 The most common method for triaging patients in a facility is the _____ _____ _____.

8 The _____ _____ can be assigned as a Command Staff Advisor or they may be assigned to the Operations Section Chief (OSC). Their primary job is to counsel the Incident Commander (IC), or if placed in operations, they will report to the Operations Section Chief (OSC) and will advise on issues that are related to the medical staff.

9 The _____ _____ has the responsibility of organizing and managing the delivery and placement of additional resources, including personnel, vehicles, equipment, supplies, and medications.

10 The _____ _____ _____ Director organizes and manages the delivery of emergency, inpatient, outpatient, and casualty care, and clinical support services.

11 The _____ _____ _____ Leader is usually positioned in the Emergency Department. When this position is activated, they will typically identify the patient receiving area and implement the use of triage by designating specific areas for each level of triage.

12 The _____ _____ _____ Leader is the person who will oversee the mental health emergency response. They will be responsible for managing the mental health care area, and in overseeing, supervising, and psychologically triaging of patients to identify the level of mental health help that will be needed.

13 The _____ _____ _____ Leader helps to organize and manage clinical support services. They will work with clinical support staff to help providing the best delivery of these services and functions.

14 The _____ _____ _____ Leader Coordinates the inpatient and outpatient registration with their staff and ensure that the process is completed as quickly as possible while still obtaining all the information needed.

15 The _____ _____ Director plays a critical role in making sure that the infrastructure to support the healthcare facility is operational and safe.

16 The _____ _____ Director is responsible for the safety and security of the facility, the staff, and patients.

17 This Branches primary function is to make sure that the business functions of the facility are either maintained, restored, or augmented.

18. The _____ _____ _____ _____ Director is a position within the HICS method who oversees the organization and management of assistance to meet patients family care needs. This could include providing for their communications needs, lodging needs, food needs, health care needs, and the spiritual and emotional needs of the family.

19. **True or False**: The Finance and Administration hierarchical organizational chart is essentially set up the same way in the HICS method as it is with the ICS method of incident management.

20. **True or False**: The Quick Start planning worksheet can be used in place of using the extended version of the forms. It is primarily used to document the initial actions that have been taken, or in shorter incidents, it can be used to document the particulars of the incident.

21. The _____ _____ _____ is a guide for the hospital based on six specific elements that the Joint Commission's Emergency Management Standards identify as critical when responding to a healthcare emergency. Those specific areas are communications, resources and assets, safety and security, staff responsibilities, utilities, and clinical support activities.

Self-Study

California Emergency Medical Services Authority (2014). *Hospital incident command system guidebook*, (5 ed.). Retrieved from https://emsa.ca.gov/wp-content/uploads/sites/71/2017/09/HICS_Guidebook_2014_11.pdf.

Las Vegas Review Journal (2017). Causes of death released for 58 killed in Las Vegas shooting. Retrieved from https://www.reviewjournal.com/crime/homicides/causes-of-death-released-for-58-killed-in-las-vegas-shooting.

McKnight's News (2011, May 25). Nursing home residents dead after Missouri tornado. *McKnight's Long-Term Healthcare News* (25 May). Retrieved from https://www.mcknights.com/news/nursing-home-residents-dead-after-missouri-tornado.

Mercy (2012). Mercy opens new hospital in Joplin. Retrieved from https://www.mercy.net/newsroom/2012-04-09/mercy-opens-new-hospital-in-joplin.

Reynolds, (n.d.). The Joplin tornado: the hospital story and lessons learned (presentation). Retrieved from https://c.ymcdn.com/sites/www.leadingagemissouri.org/resource/resmgr/annual_conference/wednesday_joplin_tornado_les.pdf.

References

Abramson, A. (2017). *These 9/11 families still don't have their relatives' remains 16 years later.* Time Magazine. (11 September). Retrieved from http://time.com/4932331/911-anniversary-remains-families/

Adler, E., & Bauer, L. (2011). *Condition gray: Inside the hospital as the Joplin Tornado hit.* The Kansas City Star. Retrieved from https://www.kansascity.com/news/special-reports/article298810.html

Alyn, K. (2011). Love what you do: When you have passion, you inspire people. *Firehouse*, 36, 34. Retrieved from http://www.firehouse.com/article/10463504/love-what-you-do

An, L. H. (2014). *Country report of Cambodia Disaster Management. Assistant to Secretary General.* National Committee for Disaster Management (NCDM), Royal Government of Cambodia & Visiting Researcher of Asian Disaster Reduction. Retrieved from http://www.adrc.asia/countryreport/KHM/2013/KHM_CR2013B.pdf

Arbuthnot, K. (2015). Incident Command in the UK's FRS [Blog]. Retrieved from https://www.linkedin.com/pulse/incident-command-uks-frs-kevin-arbuthnot/.

ASEAN Disaster Workshop in Thailand. (n.d.). ASEAN disaster medicine workshop in Thailand: Brunei Darussalam presentation [PowerPoint slides]. Retrieved from http://www.niems.go.th/th/upload/file/255705211114517884_8qnws51mcrejhxoi.pdf.

Asian Disaster Preparedness Center. (2009). Regional training course on Incident Command System for disaster management. Retrieved from http://www.adpc.net/v2007/Downloads/2010/Mar/Brochure_ICS2.pdf.

Asian Disaster Preparedness Center. (2014a). Training on hospital incident command system for hospital leadership held in Dhaka. Retrieved from http://www.adpc.net/igo/contents/media/media-news.asp?pid=591&topic=#sthash.ZGdRFl8n.dpbs.

Asian Disaster Preparedness Center. (2014b). Hospitals well equipped for emergencies in Manila. Retrieved from https://www.adpc.net/igo/category/ID684/doc/2014-n63Umy-ADPC-Hospitals_are_well_equipped_for_emergencies_in_Manila_(ADPC_Version).pdf.

Backer, H. (2016). Hospital Incident Command System [Memorandum]. Cordova, CA. Retrieved from https://emsa.ca.gov/wp-content/uploads/sites/71/2017/07/3-16-16_CommisionAgenda.pdf.

Backer, H. (n.d.). *Update on EMSA initiatives.* California Hospital Association. Retrieved from https://www.calhospital.org/sites/main/files/file-attachments/cha_2015_backer_dl.pdf

Barry, D. (2001). *A nation challenged: The site; at the scene of Random devastation, a most orderly mission.* The New York Times. (31 July 2010)

Benson, W. F. (2007). *CDC planning goals: Protect vulnerable older adults.* Atlanta, GA: Center for Disease Control-Healthy Aging Program: U.S. Center for Disease Control. Retrieved from http://www.cdc.gov/aging/pdf/disaster_planning_goal.pdf

Emergency Incident Management Systems: Fundamentals and Applications, Second Edition.
Mark S. Warnick and Louis N. Molino Sr.
© 2020 John Wiley & Sons, Inc. Published 2020 by John Wiley & Sons, Inc.
Companion Website: www.wiley.com/go/Warnick/EIMS_2e

Bermuda Department of Communications. (2013). Government 'prepared' after mock oil spill drill. Retrieved from http://www.bermudasun.bm/MobileContent/NEWS/News/Article/Government-prepared-after-mock-oil-spill-drill/24/270/64763.

"Best Practice". (n.d.). Retrieved from https://www.hsdl.org/?view&did=765527.

Bipartisan Committee. (2006). A failure of initiative. Retrieved from https://www.nrc.gov/docs/ML1209/ML12093A081.pdf.

Brady, C., & Woodward, O. (2008). *Launching a leadership revolution: Mastering the five levels of influence*. New York: Obstacles Press.

Brodie, M., Weltzien, E., Altman, D., Blendon, R. J., & Benson, J. M. (2006). Experiences of hurricane Katrina evacuees in Houston shelters: Implications for future planning. *American Journal of Public Health*, 96(8), 1402–1408. doi: 10.2105/AJPH.2005.084475

Brooks, R. B. (2017). British & American Strategies in the Revolutionary War [Blog post]. Retrieved from http://historyofmassachusetts.org/revolutionary-war-strategies/.

Broward County Sheriff's Office. (2017). Fort Lauderdale-Hollywood International Airport (FLL) active shooter/mass evacuation critical incident: critical incident report. Retrieved from http://www.trbas.com/media/media/acrobat/2017-10/94837910-10093411.pdf.

Cabag, J. S., Jr., (2012). *PHL adopts Incident Command System for disaster response*. The Daily Guardian. Retrieved from https://thedailyguardian.net/nation/phl-adopts-incident-command-system-for-disaster-response/

California Emergency Medical Services Authority (2014). *Hospital Incident Command System Guidebook* (5th ed.). Retrieved from https://emsa.ca.gov/wp-content/uploads/sites/71/2017/09/HICS_Guidebook_2014_11.pdf

California Emergency Medical Services Authority. (n.d.). Hospital Incident Command System – Welcome!. Retrieved from https://emsa.ca.gov/disaster-medical-services-division-hospital-incident-command-system-resources/.

California Fire Siege. (n.d.). Retrieved from http://www.fire.ca.gov/downloads/2003FireStoryInternet.pdf.

Carter, H. R. (2008). *Leadership: A view from the trenches*. Victoria, BC, Canada: Trafford Publishing.

Center for Strategic and International Studies. (2006). Iraqi force development: A current status report. Figure 49. Retrieved from http://www.comw.org/warreport/fulltext/060214cordesman.pdf.

Chew, V. (2009). *Severe Acute Respiratory Syndrome (SARS) Outbreak, 2003*. Singapore: National Library Board. Retrieved from http://eresources.nlb.gov.sg/infopedia/articles/SIP_1529_2009-06-03.html

CNN Wire Staff. (2011). Damaged Joplin hospital almost the only building left standing in area. Retrieved from http://www.cnn.com/2011/US/05/23/missouri.tornado.hospital/index.html.

Cohen, B. (2010). *Remembering Buffalo Soldiers' bravery during fires of 1910*. The Missoulian. Retrieved from https://missoulian.com/news/local/remembering-buffalo-soldiers-bravery-during-fires-of/article_fddc9ee8-ae72-11df-b008-001cc4c03286.html

Community Emergency Response Team. (2011). NYC CERT helps build emergency response teams in Haiti. *Community Emergency Response Newsletter*, 3, 3. Retrieved from https://www.fema.gov/media-library-data/20130726-1844-25045-0097/cert_newsletter_january_2010.pdf

Compton, D. (2010). *Progressive leadership principles, concepts, and tools: The fire officers guide to excellence in leadership*. Stillwater, OK: Fire Protection Publications.

Coordinated Incident Management System (2014). *The New Zealand coordinated incident management system* (2nd ed.). Committee for Domestic and External Security Coordination

Department of the Prime Minister and Cabinet. Retrieved from https://www.civildefence.govt.nz/assets/Uploads/publications/CIMS-2nd-edition.pdf

Covart, E. M. (2014). Planning the final action: George Washington and Rochambeau, May 1781. *Journal of the American Revolution*. Retrieved from https://allthingsliberty.com/2014/05/planning-the-final-action-george-washington-and-rochambeau-may-1781/

Crawford, B. (2010). The pain of leadership: increased authority brings increased stress. *Fire Rescue*, 28, 90.

Creamer, T. (2005). Multiagency response to WMD. *Fire Engineering*. Retrieved from https://www.fireengineering.com/articles/print/volume-158/issue-11/wmd-supplement/multiagency-response-to-wmd.html

Curd, H. (2013). Incident Command System (ICS) [Presentation]. Retrieved from http://www.kystverket.no/contentassets/2a7f03b30c864207bcbd3641287ff857/1015_1040-incident-command-system_v4.pdf.

Davis, E., & Mincin, J. (2005). *Nobody left behind: Incorporating special needs populations into emergency planning and exercises*. Lawrence, KS: Research and Training Center on Independent Living University of Kansas. Retrieved from http://www.nobodyleftbehind2.org/findings/pdfs/JMFinal072105.pdf

Dienstvorschrift. (2007). Leadership and command in emergency operations: Command and control system [translated]. Retrieved from https://www.bbk.bund.de/SharedDocs/Downloads/BBK/DE/FIS/DownloadsRechtundVorschriften/Volltext_Fw_Dv/FwDV-100%20englisch.pdf?__blob=publicationFile.

Department of Homeland Security. (2013a). NIMS: Intelligence/investigations function guidance and field operations guide. Retrieved from https://www.fema.gov/media-library/assets/documents/84807 (Retrieved 2 May 2017).

Department of Homeland Security. (2013b). Homeland Security Exercise and Evaluation Program (HSEEP). Retrieved from https://www.fema.gov/media-library-data/20130726-1914-25045-8890/hseep_apr13_.pdf.

Department of Homeland Security (2013c). *National response framework* (2nd ed.). Washington DC: Government Printing Office.

Department of Homeland Security. (2016). *TechNote: GIS and event modeling for disaster planning*. Retrieved from https://www.dhs.gov/sites/default/files/publications/GIS-and-Event-Modeling-TN_0816-508.pdf.

Djalali, A., Hosseinijenab, V., Peyravi, M., Nekoei-Moghadam, M., Hosseini, B., Schoenthal, L., & Koenig, K. L. (2015). *The hospital incident command system: Modified model for hospitals in Iran*. Public Library of Science. Retrieved from https://doi.org/10.1371/currents.dis.45d66b5258f79c1678c6728dd920451a

Doctors Without Borders (2018). *Doctors without borders organises regional emergency response workshop in Thailand*. Press Release. Retrieved from https://www.msfindia.in/doctors-without-borders-organises-regional-emergency-response-workshop-thailand

Donaldson, R. (2012). After action review for January 16, 2012 Port-Au-Prince mass casualty incident. Retrieved from https://trekmedics.org/wp-content/uploads/2014/02/Delmas33_AAR_Intl-Medical-Corps-Jan12.pdf.

Drucker, P. F. (1954). *The practice of management*. New York: Harper & Row.

Drucker, P. F. (2010). *The practice of management*. New York: Harper & Collins.

Edison Electric Institute. (n.d.). Mutual assistance. Retrieved from http://www.eei.org/issuesandpolicy/electricreliability/mutualassistance/Pages/default.aspx.

Emergency Management Assistance Compact. (n.d.). Retrieved from https://www.emacweb.org/.

Emergency Management Assistance Compact Act. (1996). Emergency Management Assistance Compact Act of 1996 § 104, 321 U.S.C. §110 Stat. 3877.

Emergency Management Institute. (2012a). ICS resource center: Incident management team position task books. Retrieved from https://training.fema.gov/emiweb/is/icsresource/positionchecklists.htm.

Emergency Management Institute. (2012b). *NIMS ICS all-hazards position specific training program for incident management teams*. Retrieved from https://ahimta.org/Resources/Documents/NIMS%20ICS%20All-Hazards%20Position%20Specific%20Training%20Program%20for%20Incident%20Management%20Teams.pdf.

Emergency Management Magazine. (2010). *Effective disaster management strategies in the 21st century*. Retrieved from http://www.govtech.com/em/disaster/Effective-Disaster-Management-Strategies.html.

Evacuteer (2011). *Five years later: Emergency preparedness improvements in New Orleans, Louisiana since Hurricane Katrina*. New Orleans: A collaborative effort of Evacuteer, the University of New Orleans, and Louisiana State University. Retrieved from http://www.evaccenter.lsu.edu/pub/10-02.pdf

Episcopal Relief & Development. (n.d.). Tips & lessons: working with donations after a disaster. Retrieved from http://www.episcopalrelief.org/uploads/EducationFileModel/167/file/Post%20Disaster%20Donations%20Management.pdf.

Federal Bureau of Investigation. (n.d.). National Crime Information Center. Retrieved from https://www.fbi.gov/about-us/cjis/ncic.

Federal Emergency Management Agency. (2004a) *Dear Governor*. Retrieved from https://www.fema.gov/txt/nims/letter_to_governors_09082004.txt.

Federal Emergency Management Agency. (2004b). *NIMS and the Incident Command System*. Retrieved from https://www.fema.gov/txt/nims/nims_ics_position_paper.txt.

Federal Emergency Management Agency. (2007). NIMS implementation activities for hospital and healthcare systems implementation FAQ's. Retrieved from https://www.fema.gov/pdf/emergency/nims/hospital_faq.pdf.

Federal Emergency Management Agency. (2008). Command and General Staff functions for local incident management teams: student manual. Retrieved from http://fire.nv.gov/uploadedFiles/firenvgov/content/bureaus/FST/CGSFLIMT-StudentManual.pdf.

Federal Emergency Management Agency. (2012). FEMA incident action planning guide. Retrieved from https://www.fema.gov/media-library-data/20130726-1822-25045-1815/incident_action_planning_guide_1_26_2012.pdf.

Federal Emergency Management Agency. (2013a). IS 200b: ICS for single resources and initial action incidents (ICS 200): Unit 4 [visual]. Retrieved from https://training.fema.gov/emiweb/is/is200b/visuals/pdf/04_visualsics200b_october2013.pdf.

Federal Emergency Management Agency. (2013b). ICS 402: overview for executives/senior officials G0402 [Presentation]. Retrieved from https://training.fema.gov/gstate/xcr3wnlp/g0402%20-%20ics-402%20-%20incident%20command%20system%20(ics)%20overview%20for%20executives%20and%20senior%20officials/ics-402_ig.pdf.

Federal Emergency Management Agency. (2013c). *National incident management system: Intelligence/investigations function guidance and field operations guide*. Retrieved from https://www.fema.gov/media-library-data/1382093786350-411d33add2602da9c867a4fbcc7ff20e/NIMS_Intel_Invest_Function_Guidance_FINAL.pdf.

Federal Emergency Management Agency. (2014). *Federal emergency management agency incident management handbook: FEMA B-761*. Washington, DC: Government Printing Office.

Federal Emergency Management Agency. (2015a). *Fact sheet: Mount weather emergency operations center*. FEMA Press Release. Retrieved from https://www.fema.gov/media-library-data/1433170060726-5272c667a842f8a56ef9bae118c2ba0a/MWEOCFactsheet.pdf

Federal Emergency Management Agency. (2015b). 9526.1 hazard mitigation funding under section 406 (Stafford Act). Retrieved from https://www.fema.gov/95261-hazard-mitigation-funding-under-section-406-stafford-act.

Federal Emergency Management Agency. (2015c). *FEMA aid reaches $16.9 Billion for New York's Hurricane Sandy recovery*. Press release No. NR-013. Retrieved from https://www.fema.gov/news-release/2015/10/21/fema-aid-reaches-169-billion-new-yorks-hurricane-sandy-recovery.

Federal Emergency Management Agency. (2015d). *FY 2015 emergency management performance grant program*. Retrieved from https://www.fema.gov/media-library-data/1427284768817-b62b93d48b12617f423c0e8fbfde562b/FY2015EMPG_NOFO.pdf.

Federal Emergency Management Agency. (2017a). *National incident management system* (3rd ed.). Retrieved from https://www.fema.gov/media-library-data/1508151197225-ced8c60378c3936adb92c1a3ee6f6564/FINAL_NIMS_2017.pdf

Federal Emergency Management Agency. (2017b). Incident Management Assistance Teams. Retrieved from https://www.fema.gov/incident-management-assistance-teams.

Federal Emergency Management Agency. (2017c). Incident Management Assistance Teams. Retrieved from https://www.fema.gov/media-library-data/1440617086827-f6489d2de59dddeba8bebc9b4d419009/IMAT_July_2015.pdf.

Federal Emergency Management Agency. (2017d). Fact sheet: What is FEMA's individual assistance program? Retrieved from https://www.fema.gov/disaster/4294-4297/updates/fact-sheet-what-femas-individual-assistance-program.

Federal Emergency Management Agency. (2018a). Residential building fire trends: 1977–2016. Retrieved from https://www.usfa.fema.gov/downloads/pdf/statistics/res_bldg_fire_estimates.pdf.

Federal Emergency Management Agency. (2018b). *NIMS implementation objectives for local, state, tribal and territorial jurisdictions*. Retrieved from https://www.fema.gov/media-library-data/1527847820319-0c604c12c628b5a8119fb8d08c4ed07c/NIMS_Implementation_Objectives_FINAL_20180530.pdf.

Federal Emergency Management Agency. (2019). *G 0402 NIMS overview for senior officials (executives, elected, & appointed)*. Retrieved from https://training.fema.gov/gstate/xcr3wnlp/g0402%20-%20ics-402%20-%20incident%20command%20system%20(ics)%20overview%20for%20executives%20and%20senior%20officials/03%20sm/g402_complete_sm.pdf.

Federal Emergency Management Agency. (n.d.a). Incident Command System 100-b. Unit five lesson summary. Retrieved from https://emilms.fema.gov/IS100b/ICS0105summary.htm.

Federal Emergency Management Agency. (n.d.b). ICS 219: Resource Status Card (T-Card). Retrieved from https://www.fema.gov/media-library-data/20130726-1922-25045-2119/ics_forms_219s.

Federal Emergency Management Agency. (n.d.c). ICS 201: Forms used for the development of the Incident Action Plan. Retrieved from https://emilms.fema.gov/is201/ics01summary.htm.

Federal Emergency Management Agency. (n.d.d). FEMA Director battle book. Retrieved from http://tacsafe.net/resources/Emergency/FEMABattleBook.pdf.

Federal Emergency Management Agency. (n.d.e) *Demobilization Form (ICS 221)*. Retrieved from https://emilms.fema.gov/IS201/assets/ICS%20Forms%20221.pdf.

Federal Emergency Management Agency. (2010). *Developing and maintaining emergency operations plans: Comprehensive Preparedness Guide (CPG) 101 version 2.0*. Washington, DC: Government Printing Office. Retrieved from http://www.fema.gov/media-library-data/20130726-1828-25045-0014/cpg_101_comprehensive_preparedness_guide_developing_and_maintaining_emergency_operations_plans_2010.pdf

Final Report. (1996). *Final report: Alfred P. Murrah Federal Building Bombing April 19, 1995*. Stillwater, OK: International Fire Service Training Association.

Fire Engineering. (1995). Report from fire chief. *Fire Engineering*, 148, 10. Retrieved from https://www.fireengineering.com/articles/print/volume-148/issue-10/features/report-from-fire-chief.html#

FIRESCOPE. (2014). Incident Command System position manual: Fireline emergency medical technician ICS-223-10. Retrieved from http://www.firescope.org/ics-pos-manuals/ICS%20223-10.pdf.

Florida Department of Public Health. (2011). 2010 Deepwater Horizon oil spill response: ESF 8 after-action report and improvement plan. Retrieved from http://www.floridahealth.gov/programs-and-services/emergency-preparedness-and-response/training-exercise/_documents/deepwater-aar.pdf.

Fordyce, E., Sadiq, A., & Chikoto, G. L. (2012). Haiti's emergency management: A case of regional support, challenges, opportunities, and recommendations for the future. In D. A. McEntire (Ed.), *Comparative emergency management: Understanding disaster policies, organizations, and initiatives from around the world*. FEMA, U.S. Department of Homeland Security. Chapter 29, Retrieved from http://training.fema.gov/EMIWeb/edu/CompEmMgmtBookProject.asp

Forrester, C., Dixon, R., Judy, C., et al. (2008). *Managing mass fatalities: A toolkit for planning*. Santa Clara County Public Health Department.

Fuady, A., Pakasi, T., & Mansyur, M. (2011). Primary health centre disaster preparedness after the earthquake in Padang Pariaman, West Sumatra, Indonesia. *BMC Research Notes*, 4, 81. doi: 10.1186/1756-0500-4-81

Future Learn. (n.d.). The Gold-Silver-Bronze command structure. Retrieved from https://www.futurelearn.com/courses/systems-thinking-complexity/0/steps/20392.

Galvin, J. (2007). *The big burn: Idaho and Montana, August 1910*. Popular Mechanics. (29 July). Retrieved from https://www.popularmechanics.com/science/environment/a1961/4219853/

Garfield, L. (2017). *$20 for a gallon of gas, $99 for a case of water — reports of Hurricane Harvey price-gouging are emerging*. Business Insider. (2 September). Retrieved from https://www.businessinsider.com/price-gouging-in-texas-gas-prices-hurricane-2017-9

Global Security. (2001). Mount Weather high point special facility: Mount Weather emergency assistance center. Retrieved from http://www.globalsecurity.org/wmd/facility/mt_weather.htm.

Government of India. (2002). Guidelines for hospital emergency preparedness planning. Retrieved from http://asdma.gov.in/pdf/publication/undp/guidelines_hospital_emergency.pdf.

Government of the People's Republic of Bangladesh. (2010). Standing orders on disaster. Retrieved from https://www.preventionweb.net/files/18240_sodapprovedbyndmb.pdf.

Grimm, D. (n.d.). *The human factor: Lessons learned from the haiti response*. Disaster Resource Guide. Retrieved from http://disaster-resource.com/index.php?option=com_content&view=article&id=1697&Itemid=141

Guthrie, V. H., Finucane, M. J., Keith, P. E., & Stinnett, D. B. (2017). After action review of the November 28, 2016, firestorm. Retrieved from http://gatlinburgtn.gov/pdf/planning/Wildfire/AAR%20of%20the%20Nov%2028%202016%20Firestorm.pdf.

Hausweld, M., Richards, M. E., Kerr, N. L., & Schmidt, T. A. (2010). The Haitian earthquake and academic emergency medicine. *Society for Academic Emergency Medicine*. doi: 10.1111/j.1553-2712.2010.00803.x

Health Emergency Operations Plan. (2018). Republic of Maldives Ministry of Health. Retrieved from http://www.health.gov.mv/Uploads/Downloads//Informations/Informations(124).pdf.

Herbosa, T. (2007). *Hospital Preparedness for Emergencies (HOPE) course instructors guide*. U.S. A.I.D. Retrieved from https://www.researchgate.net/publication/200533788_Hospital_Preparedness_for_Emergencies_HOPE_Course_Instructors_Guide

Hewitt, C. (2003). *Understanding Terrorism in America: From the Klan to al Qaeda* (pp. 106). Routledge. ISBN: 978-0-415-27765-5

Hirschkorn, P. (2005). *Identification of 9/11 remains comes to end*. CNN.

Hoffman, S. (2009). *Preparing for disaster: Protecting the most vulnerable in emergencies* (Vol. 42, pp. 1491). Davis: University of California. Retrieved from https://lawreview.law.ucdavis.edu/issues/42/5/articles/42-5_Hoffman.pdf

Hospital Incident Management System. (n.d.). *Hospital incident management team training*. Metro Fire. Retrieved from http://www.metrofire.com.au/hospital_incident_team_training.php

Houston, J. B., Spialek, M. L., Stevens, J., First, J., Mieseler, J., & Pfefferbaum, B. (2015). 2011 Joplin, Missouri tornado experience, mental health reactions, and service utilization: cross-sectional assessments at approximately 6 months and 2.5 years post-event. *Public Library of Science*, 7. doi: 10.1371/currents.dis.18ca227647291525ce3415bec1406aa5

Howitt, A. M., Hayashi, H., Akiyama, H., Giles, D. W., & Leonard, D. (2013). An incident management system for Japan? *Crisis Response Journal*, 1(9), 17–19.

ICS Canada. (n.d.). What is ICS Canada. Retrieved from http://www.icscanada.ca/en/about+ics+canada.html.

"ILEAS". (n.d.) Illinois law enforcement alarm system. Retrieved from www.ileas.org.

Illinois Terrorism Task Force. (2011). *Incident Command System 300: Intermediate ICS*. Washington DC: Government Printing Office.

Incident Information System. (2018). The Mendocino Complex Fire. Retrieved on October 19, 2018 from https://inciweb.nwcg.gov/incident/6073/.

Joint Emergency Services Interoperability Principles. (n.d.). JESIP the programme. Retrieved from https://www.jesip.org.uk/jesip-the-programme.

Juillet, E. (1993). Evacuating people with disabilities, the world trade center bombing: report and analysis. *Fire Engineering*, 146(12), 100–103.

Kapucu, N., & Garayev, V. (2011). Collaborative decision-making in emergency and disaster management. *International Journal of Public Administration*, 34, 366–375.

Kolva, J. R. (2002). National water summary on wetland resources. United States Geological Survey Water Supply Paper 2425. Retrieved from https://water.usgs.gov/nwsum/WSP2425/flood.html.

Kouzes, J. M., & Posner, B. Z. (2012). *The leadership challenge* (5th ed.). San Francisco, CA: Wiley Press.

Las Vegas Review Journal. (2017). Causes of death released for 58 killed in Las Vegas shooting. Retrieved from https://www.reviewjournal.com/crime/homicides/causes-of-death-released-for-58-killed-in-las-vegas-shooting/.

Leonard, H. B., Howitt, A. M., Cole, C., & Pfeifer, J. W. (2016). *Command under attack: what we've learned since 9/11 about managing crises*. The Conversation. Retrieved from http://theconversation.com/command-under-attack-what-weve-learned-since-9-11-about-managing-crises-64517

Lindsay, B. R. (2011). *Social media and disasters: Current uses, future options, and policy considerations*. Washington DC: Congressional Research Service. Retrieved from https://www.nisconsortium.org/portal/resources/bin/Social_Media_and_Dis_1423591240.pdf

Liqiang, H. (2018). *New authority focuses on emergency response*. China Daily. (30 March). Retrieved from http://www.chinadaily.com.cn/a/201803/30/WS5abd8f19a3105cdcf6515441.html

Litman, T. (2006). Lessons from Katrina and Rita: What major disasters can teach transportation planners. *Journal of Transportation Engineering*, 132(1), 11–18.

MABAS. (n.d.). Mutual aid box alarm system. Retrieved from http://www.mabas-il.org/Pages/default.aspx.

Manzi, C., Powers, M. J., & Zetterlund, K. (2002). *Critical information flows in the Alfred P. Murrah building bombing: A case study*. Chemical and Biological Arms Control Institute. Retrieved from https://www.ncjrs.gov/pdffiles1/Digitization/194411NCJRS.pdf

McEntire, D. (2007). *Disaster response and recovery* (1st ed.). Hoboken, NJ: Wiley.

McKnight's News. (2011). *Nursing home residents dead after Missouri tornado*. McKnight's Long-Term Healthcare News. (25 May). Retrieved from https://www.mcknights.com/news/nursing-home-residents-dead-after-missouri-tornado/

Mercy. (2012). Mercy opens new hospital in Joplin. Retrieved from https://www.mercy.net/newsroom/2012-04-09/mercy-opens-new-hospital-in-joplin/

Mexico–United States Joint Contingency Plan. (2017). Preparedness for and response to emergencies and contingencies associated with chemical hazardous substances in the inland border area. Retrieved from https://www.epa.gov/sites/production/files/2016-01/documents/us_mexico_joint_contingency_plan.pdf

Molino, L. N. (2006). *Emergency incident management systems*. Hoboken, NJ: Wiley.

Moore, J. (2015). *Security deployed at super bowl is largest in Minnesota history*. Star Tribune. Retrieved from http://www.startribune.com/security-deployed-at-super-bowl-largest-in-minnesota-history/471452694

Multnomah County. (2015). Reynolds High School active shooter response: An analysis of the response to the Reynolds High School Shooting on June 10, 2014. Retrieved from https://multco.us/file/57742/download.

Murakami, H. (2000). *Underground: The Tokyo gas attack and the Japanese Psyche*. New York, NY: Penguin Random House.

National Disaster Management Authority. (n.d.). *Incident response system*. Government of India. Retrieved from https://ndma.gov.in/en/irs-training/introduction.html

National Disaster Response Plan. (2016). Retrieved from https://www.preventionweb.net/files/62898_nationaldisasterresponseplanforeart.pdf

National Health Service England. (2015). NHS England emergency preparedness, resilience and response framework. Retrieved from https://www.england.nhs.uk/wp-content/uploads/2015/11/eprr-framework.pdf

National Health Service Scotland. (2015). Business continuity: a framework for NHS Scotland. Retrieved from https://www.sehd.scot.nhs.uk/EmergencyPlanning/Documents/BusinessContinuity.pdf.

National Health Service Tameside Hospital. (2016). Major incident plan. Retrieved from https://www.tamesidehospital.nhs.uk/documents/MajorIncidentPlan.pdf?fbclid=IwAR0d7Q4rFn4IQDMH0_FG10J8BebCyebMhI3f7BpnWzWnYh3sHLG_ideHlmA.

National Institute for Occupational Safety and Health. (2018). *Volunteer assistant chief killed and one fire fighter injured by roof collapse in a commercial storage building-Indiana: Report # F2014-18*. Center for Disease Control, NIOSH Fire Fighter Fatality Investigations and Prevention Program. Retrieved from https://www.cdc.gov/niosh/fire/pdfs/face201418.pdf

National Organization on Disability. (2007). Report on Special Needs Assessment of Katrina Evacuees (SNAKE). Washington, DC. Retrieved from http://www.nod.org/Resources/PDFs/katrina_snake_report.pdf.

National Volunteer Fire Council. (2012). *A proud tradition: 275 years of the American volunteer fire service*. Tampa, FL: Faircount Media Group. Retrieved from https://www.nvfc.org/wp-content/uploads/2015/10/Anniversary_Publication.pdf

National Wildfire Coordinating Group. (2001). NWCG task book for the position of incident communications technician. Retrieved from https://www.hsdl.org/?view&did=774419.

National Wildfire Coordinating Group. (2009). NWCG task book for the position of Documentation Unit Leader (DOCL). Retrieved on February 24, 2017 from https://www.nwcg.gov/sites/default/files/products/training-products/pms-311-25.pdf.

National Wildfire Coordinating Group. (2008). *Documentation unit leader: J-342*. Retrieved from https://www.nwcg.gov/sites/default/files/products/training-products/J-342.pdf.

National Wildfire Coordinating Group. (2014a). *Wildland fire incident management field guide: PMS-210*. Washington DC: Government Printing Office.

National Wildfire Coordinating Group. (2014b). *GIS standard operating procedures on incidents: PMS 936*. Retrieved from https://www.nwcg.gov/sites/default/files/publications/pms936.pdf.

National Wildfire Coordinating Group. (2016). Interagency helicopter operations guide. Retrieved from http://ordvac.com/soro/library/Aviation/Operations/2016%20IHOG.pdf.

National Wildfire Coordinating Group. (2017). Resource advisor guide: PMS 313. Retrieved from https://www.nwcg.gov/sites/default/files/publications/pms313.pdf.

National Wildfire Coordinating Group. (2018a). Medical unit leader. Retrieved from https://www.nwcg.gov/positions/medl/position-qualification-requirements.

National Wildfire Coordinating Group. (2018b). Field observer. Retrieved from https://www.nwcg.gov/positions/fobs.

National Wildfire Coordinating Group. (2018c). Infrared interpreter. Retrieved from https://www.nwcg.gov/positions/irin.

National Wildfire Coordinating Group. (2018d). Fire behavior analyst. Retrieved from https://www.nwcg.gov/positions/fban.

National Wildfire Coordinating Group. (2018e). Long-term fire analyst. Retrieved from https://www.nwcg.gov/positions/ltan.

National Wildfire Coordinating Group. (2018f). Strategic operational planner. Retrieved from https://www.nwcg.gov/positions/sopl.

National Wildfire Coordinating Group. (2018g). NWCG standards for interagency incident business management: PMS 902. Retrieved from https://www.nwcg.gov/sites/default/files/publications/pms902.pdf.

Nazarov, E. (2011). *Emergency response management in Japan: Final research report*. Asian Disaster Reduction Center. Retrieved from http://www.adrc.asia/aboutus/vrdata/finalreport/2011A_AZE_Emin_FRR.pdf

"NDMS Teams". (n.d.). Retrieved from https://www.phe.gov/Preparedness/responders/ndms/ndms-teams/Pages/default.aspx.

Nordberg, M. (2010). *EMS revisited: The big one*. EMS World. Retrieved from https://www.emsworld.com/article/10319737/ems-revisited-big-one

Northouse, P. G. (2009). *Introduction to leadership: 2009: Concepts and practice*. Thousand Oaks, CA: SAGE Publications.

Nuñez, E. (2009). NIMS and emergency management lecture [Web log post]. Retrieved from http://www.capella.edu.

Okada, A., & Ogura, K. (2014). Japanese disaster management system: recent developments in information flow and chains of command. *Journal of Contingencies and Crisis Management*, 22. doi: 10.1111/1468-5973.12041

"Oklahoma City National Memorial and Museum". (n.d.). Community response: crisis management. Retrieved from https://oklahomacitynationalmemorial.org/wp-content/uploads/2015/03/okcnm-community-response-crisis-management.pdf.

"Oklahoma City National Memorial Museum". (n.d.). Retrieved from https://oklahomacitynationalmemorial.org.

Oklahoma Department of Civil Emergency Management. (n.d.). After action report: Alfred P. Murrah Federal Building bombing 19 April 1995 in Oklahoma City, Oklahoma. Retrieved from https://www.ok.gov/OEM/documents/Bombing%20After%20Action%20Report.pdf.

Ontario Hospital Association. (2006). OHA emergency management toolkit: Developing a sustainable emergency management program for hospitals. Retrieved from https://www.oha.com/Documents/Emergency%20Management%20Toolkit.pdf.

Pangi, R. (2002). Consequence management in the 1995 sarin attacks on the Japanese subway system. *Studies in Conflict & Terrorism*, 25, 421–448.

Parr, A. (1987). Disasters and disabled persons: An examination of the safety needs of a neglected minority. *Disasters*, 11(2), 148–159.

Patterson, R. (2014). The story of Mercy and Joplin: Hospital recovery-sustainability. Retrieved from https://www.chicagohan.org/documents/14171/88419/Peterson_Joplin.pdf/11f267c5-b6d3-4bf3-88ed-61805be676e0.

Pender, K. (2005). *The true cost of Katrina*. San Francisco Gate. (27 September). Retrieved from http://www.sfgate.com/business/networth/article/The-true-cost-of-Katrina-2567118.php

Penuel, K. B., Statler, M., & Hagen, R. (2013). *Encyclopedia of crisis management*. Thousand Oaks, CA: Sage Publications.

Petersen, J. (1994–1995). *The West is burning up*. Evergreen Magazine. Idaho Forest Products Commission archive. Retrieved on 10/11/2018 from https://web.archive.org/web/20001031083216/http://www.idahoforests.org/fires.htm.

Pollak, A. (Producer), & Ives, S.(2015). *The big burn*. The American Experience. [Documentary]. PBS

Preserve America. (2008). Preparing to preserve: an action plan to integrate historic preservation into tribal, state, and local emergency management plans. Retrieved from https://www.doi.gov/sites/doi.gov/files/migrated/pmb/oepc/rppr/upload/12-18-08-Preparing-To-Preserve.pdf.

Public Intelligence. (2010). Mount weather. Retrieved from https://publicintelligence.net/mount-weather/.

PUBLIC LAW 104–321—OCT. 19, 1996 110 STAT. 3877. (1996). Retrieved from https://www.congress.gov/104/plaws/publ321/PLAW-104publ321.pdf.

Purchasing & Procurement Center. (n.d.). 12 vendor selection criteria. Retrieved from http://www.purchasing-procurement-center.com/selecting-a-vendor.html.

Ramos, B. T. (2012). *Submission of accomplishment for the FHA Biennial progress report: 2011–2013 [Memorandum]*. National Disaster Risk Reduction and Management Council. Retrieved from http://www.ndrrmc.gov.ph/attachments/article/401/MEMO_No_13_s_of_2012_re_Submission_of_Accomplishments_for_HFA_Biennial_Progress_Report_2011.pdf

Research Brief (2008). *Recent changes to emergency preparedness mandates and funding*. National Association of Public Hospitals and Health Systems. Retrieved from https://essentialhospitals.org/wp-content/uploads/2014/10/EP-Brief-Update-Final-hires.pdf

Reynolds, M. (n.d.). The Joplin tornado: The hospital story and lessons learned (presentation). Retrieved from https://c.ymcdn.com/sites/www.leadingagemissouri.org/resource/resmgr/annual_conference/wednesday_joplin_tornado_les.pdf.

Rhode Island Department of Environmental Management. (2016). Procurement unit leader job aid, p. 1. Retrieved on March 18, 2017 from http://www.dem.ri.gov/topics/erp/8_6_3.pdf.

Risoe, P., Schlegelmilch, J., & Paturas, J. (2013). Evacuation and sheltering of people with medical dependencies: Knowledge gaps and barriers to national preparedness. *Homeland Security Affairs*, 9(2), 1–10.

Roffey, R. (2016). Russia's EMERCOM: Managing emergencies and political credibility. Retrieved from file:///C:/Users/SamsClub6334/Downloads/http___webbrapp.ptn.foi.se_pdf_1a28d90c-651f-4673-90a6-ba1fd5b8a5f6%20(4).pdf.

Samsuddin, N. M., Takim, R., Nawawi, A. H., & Alwee, S. N. (2018). Disaster preparedness attributes and hospital's resilience in Malaysia. *Procedia Engineering*, 220, 27–29.

San Mateo County Health Services Agency Emergency Medical Services. (1998). Hospital incident command system. Retrieved from http://medipe2.psu.ac.th/~disaster/disasterlast/HEICS98a.pdf.

Sargent, C. (2006). *From buddy to boss: Effective fire service leadership*. Tulsa, OK: Penwell Publishing.

Sauer, L., Catlett, C., Tosatto, R., & Kirsch, T. D. (2014). The utility of and risks associated with the use of spontaneous volunteers in disaster response: A survey. *Disaster Medicine and Public Health Preparedness*, 8(1), 65–69. doi: 10.1017/dmp.2014.12

Shariat, S., Mallonee, S., & Stidham, S. S. (1998). *Summary of reportable injuries in Oklahoma: Oklahoma City bombing injuries*. Oklahoma State Department of Health. Retrieved from https://www.ok.gov/health2/documents/OKC_Bombing.pdf

Silvis, J. (2015). *New Mercy Hospital Joplin is built "just in time"*. Healthcare Design. (12 May). Retrieved from https://www.healthcaredesignmagazine.com/projects/acute-care/new-mercy-hospital-joplin-built-just-time/

Skudder, H., Druckman, A., Cole, J., McInnes, A., Brunton-Smith, I., & Ansaloni, G. P. (2016). Addressing the carbon-crime blind spot: A carbon footprint approach. *Journal of Industrial Ecology*. doi: 10.1111/jiec.12457

Smeby, L. C. (2013). *Fire and emergency services administration: Management and leadership practices* (2nd ed.). Burlington, MA: Jones & Bartlett Learning.

Smith, M. R. (2010). *Guardmembers remember Oklahoma City bombing*. U.S. Army News. (25 October). Retrieved from https://www.army.mil/article/37587/guardmembers_remember_oklahoma_city_bombing.

Soe, K. N., Othman, S. D., Salazar, V. A., & bin Haji Shahari, M. K. A. (2015). Toward a disaster-resilient ASEAN: Building the capacity of Brunei Darussalam on disaster management. Retrieved from http://www.rcrc-resilience-southeastasia.org/wp-content/uploads/2017/12/2015_soe_et_al_building_the_capacity_of_brunei_darussalam_on_disaster_management.pdf.

South Dakota Department of Health. (2011). Medical response to Joplin tornado May 22, 2011. Retrieved from https://doh.sd.gov/documents/Providers/Prepare/Joplin.pdf.

State Committee for Civil Defence. (2009). The Russian Corp of Rescuers was formed: Day of the rescuer in Russia. Retrieved from https://www.prlib.ru/history/619850.

Stockwell, M. (n.d.). *Grand strategy*. National Library for the Study of George Washington. Retrieved from https://www.mountvernon.org/library/digitalhistory/digital-encyclopedia/article/grand-strategy/

Stone, A. (2017). *Big changes coming for hospital emergency managers*. Emergency Management Magazine. (10 February). Retrieved from https://www.govtech.com/em/health/Big-Changes-Coming-for-Hospital-Emergency-Managers.html

Texas Interagency Coordination Center & Texas A&M University. (2014a). Incident command system job aid: Finance/admin section chief. Retrieved from https://ticc.tamu.edu/Documents/IncidentResponse/AHIMT/JobAid/FSC_JA-07012014.pdf.

Texas Interagency Coordination Center & Texas A&M University. (2014b). Incident command system job aid: operations section chief. Retrieved from https://ticc.tamu.edu/Documents/IncidentResponse/AHIMT/JobAid/OSC_JA-07012014.pdf.

Texas Interagency Coordination Center & Texas A&M University. (2014c). Incident command system job aid: Safety officer. Retrieved from https://ticc.tamu.edu/Documents/IncidentResponse/AHIMT/JobAid/SOFR_JA-07012014.pdf.

Tham, K. Y. (2003). *Incident Command System (ICS) for Severe Acute Respiratory Syndrome (SARS) [PowerPoint Presentation]*. Singapore: Tan Tock Seng Hospital. Retrieved from http://www.disaster.org.tw/chinese/ACTIVE/SARS/Tham%20KY%20ICS%20SARS%20Sep%2003.pdf

"The Great Fire of 1910". (n.d.). The fire. Retrieved from https://www.fs.usda.gov/Internet/FSE_DOCUMENTS/stelprdb5444731.pdf.

The Joplin Globe. (2011). More than 1,000 people suffer injuries. Retrieved from https://www.joplinglobe.com/news/local_news/more-than-people-suffer-injuries/article_dc6bd926-2646-5985-a3d9-d205885d2293.html.

Thomas, R. S., & Bode, T. J. (2011). *DNA identification of the missing after the WTC attacks: A cooperative public/private effort*. Forensic Magazine. Retrieved from http://www.forensicmag.com/article/2011/08/dna-identification-missing-after-wtc-attacks-cooperative-publicprivate-effort

Tierney, L. (2018). *The grim scope of 2017's California wildfire season is now clear: The danger's not over*. Washington Post. (4 January). Retrieved from https://www.washingtonpost.com/graphics/2017/national/california-wildfires-comparison/?noredirect=on&utm_term=.633902f621ae

Tippett, J. B. (2013). *Communication & personnel accountability for IC's*. Fire Rescue Magazine. (26 October). Retrieved from https://www.firerescuemagazine.com/articles/print/volume-8/issue-11/incident-command-0/communication-personnel-accountability-for-ics.html

Treaster, J. (2005). *At stadium, a haven quickly becomes an ordeal*. New York Times. September 1

Turner, L. (n.d.). The use of the Incident Control System (Australasian Interagency Incident Management System) in emergency management [video]. Retrieved from https://slideplayer.com/slide/4742888/.

Tzu, S. (1772). The art of war.

United Nations Development Programme. (2017). Palestinian disaster risk management system. Retrieved from https://info.undp.org/docs/pdc/Documents/PAL/00088464-DRM%20Final%20report%20Nov.%202017.pdf.

United Nations Economic and Social Commission for Asia and the Pacific. (2010). Results report: TTF-02. Retrieved from https://www.unescap.org/sites/default/files/TTF02.pdf.

United Nations General Assembly. (2002). International strategy for disaster reduction: A/RES/56/195. Retrieved from https://www.unisdr.org/files/resolutions/N0149261.pdf.

United Nations Office on Drugs and Crime. (2013). The specialized module 3 on: Security and incident management for trainers. Retrieved from https://www.unodc.org/documents/middleeastandnorthafrica/Publications/SECURITY_AND_INCIDENT_MANAGEMENT_MANUAL_-ENGLISH.pdf.

United Nations Strategy for Disaster Reduction. (2005). Hyogo framework for Action 2005–2015: Building the resilience of nations and communities to disasters. Retrieved from https://www.unisdr.org/files/1037_hyogoframeworkforactionenglish.pdf.

United States Agency for International Development. (2016). *Success story: Strengthening disaster management systems in Burma*. ReliefWeb. Retrieved from https://reliefweb.int/report/myanmar/success-story-strengthening-disaster-management-systems-burma

United States Agency for International Development. (2017). Incident command system performance evaluation: Indonesia country report. Retrieved from https://pdf.usaid.gov/pdf_docs/PA00MZHZ.pdf.

United States Coast Guard. (2014). Incident command system: Communications unit leader. Retrieved from https://homeport.uscg.mil/Lists/Content/Attachments/2916/COML%20Job%20Aid%20Oct14.pdf.

United States Coast Guard. (2017). United States Coast Guard (USCG) job book: Air operations branch director. Retrieved from https://homeport.uscg.mil/Lists/Content/Attachments/2916/AOBD%20Job%20Aid%20July2017.pdf.

United States Department of Labor. (2008). *Hazardous waste operations and emergency response*. United States Department of Labor: Occupational Safety and Health Administration. Retrieved from https://www.osha.gov/Publications/OSHA3114/OSHA-3114-hazwoper.pdf

United States Embassy in Vietnam. (2015). United States and Vietnam expand partnership to improve disaster response and coordination. Retrieved from https://vn.usembassy.gov/united-states-and-vietnam-expand-partnership-to-improve-disaster-response-and-coordination/.

United States Fire Administration. (2008). Emergency incident rehabilitation. Retrieved from https://www.usfa.fema.gov/downloads/pdf/publications/fa_314.pdf.

United States Forest Service. (n.d.). Overview of the Incident Command System. Retrieved from https://nctr.pmel.noaa.gov/education/ITTI/ics/ICS_Overview_In_an_International_Context.pdf.

United States Geological Survey. (2011). *Haiti dominates earthquake fatalities in 2010*. ScienceDaily. (17 January). Retrieved November 9, 2018 from www.sciencedaily.com/releases/2011/01/110117142732.htm

Urby, H. (2014). Emergency management in Mexico: A good beginning, but additional progress needed. In D. A. McEntire (Ed.), *Comparative emergency management: Understanding disaster policies, organizations and initiatives from around the world (Chapter 27)*. Retrieved from https://training.fema.gov/hiedu/aemrc/booksdownload/compemmgmtbookproject/

USAID Newsletter. (2015). Haiti launches ICS training program. Retrieved from https://www.usaid.gov/sites/default/files/documents/1866/lac_newsletter_january2015.pdf.

Wang, J. (2017). *More than 100 Texas gas stations slapped with price-gouging violations*. Dallas Morning News. (October). Retrieved from https://www.dallasnews.com/news/harvey/2017/10/30/100-texas-gas-stations-slapped-price-gougingviolations

Warnick, M. S. (2015). *Quantitative analysis: education and experience effects upon emergency preparedness for special needs*. ProQuest Publications database. Doctoral dissertation. Publication No. 3724949

White, J., & Whoriskey, P. (2005). *Planning, response, are faulted* (pp. A.02). Washington Post. ISSN 0190-8286.

Wildland Fire Assessment System. (n.d.). Fire effects monitor. Retrieved from https://www.wfas.net/index.php/component/content/article/30-femo/59-fire-effects-monitor.

"William B. Greeley". (n.d.). Retrieved from https://www.worldforestry.org/wp-content/uploads/2016/03/GREELEY-WILLIAM-B...pdf.

Williams, G. W. (2005). *The USDA forest service: The first century*. The United States Department of Agriculture, Forest Service. Retrieved from https://www.fs.fed.us/sites/default/files/media/2015/06/The_USDA_Forest_Service_TheFirstCentury.pdf

Wilma, D. (2003). *Forest fires in Idaho and Montana burn three million acres of timber and kill 85 people beginning on August 20, 1910*. History Link. (essay 5488). Retrieved from http://www.historylink.org/File/5488

Woods, A. (2017). *'Is this real?': Seven hours of chaos, bravery at Las Vegas hospital after mass shooting*. Arizona Republic. (30 October). Retrieved from https://www.azcentral.com/story/news/nation/2017/10/30/seven-hours-chaos-bravery-las-vegas-sunrise-hospital-after-mass-shooting/796410001/

Worksafe BDA. (n.d.). Safety training. Retrieved from http://www.worksafebda.com/safety-training/#ics100.

"World Bank". (2017). Vietnam: Disaster risk management project. Retrieved from http://www.worldbank.org/en/results/2013/04/09/vietnam-disaster-risk-management-project.

World Health Organization. (2011). Assessment report: Safe hospitals in emergencies and hospitals' resilience to climate change. Retrieved from http://www.wpro.who.int/vietnam/publications/safe_hospitals_hospitals_resilience_to_CC_assessment_report_en.pdf.

Yung, K. S. (2012). *Emergency response in Singapore: Perspectives from the SCDF incident management system [PowerPoint Presentation]*. Singapore Civil Defence Force. Retrieved from https://www.scdf.gov.sg/docs/default-source/scdf-library/fssd-downloads/fsm-briefing.pdf

Index

a

Action Cards   90, 91
Adaptable   34, 129, 130, 133, 135–136, 142, 175–176, 460, 468
Advocacy Groups   111–112, 115, 117
After-Action-Review (AAR)   19, 48, 50, 60, 76, 77, 161, 166, 261, 290, 291, 359, 452, 501
Agency Administrator   28, 132, 162, 205, 296, 349–367, 372, 400–405, 423, 440, 451
Agency Administrator Representatives   351–356
Agency Representative   122–123, 157, 166, 190, 194–195, 217, 232, 286, 296, 301, 307, 308, 320, 349–367, 399, 406
Alberta Emergency Management Agency   72
America Burning   17, 20
American Red Cross   2, 46–47, 50–53, 113, 117, 121, 162, 194, 219, 256, 336, 478
Arkansas   53
Army Corps of Engineers   52, 253
ASPR   485–488, 501
Assistant Cooks   238
Association of Southeast Asian Nations (ASEAN)   67, 68–69, 70–71, 77, 82, 87, 92
Australia   67–68, 92, 135, 460
Australian Inter-Service Incident Management System (AIIMS)   67–68, 92
Auxiliary Emergency Communicators   248–249

b

Bangladesh   67, 69–70
Baptist Child and Family Services   75
Bermuda   68, 69
Branches   9–10, 196, 199, 200, 208, 213–217, 218, 220, 226, 229, 230, 231, 257, 267, 317, 323, 327, 458, 472, 476, 479
Broadwater Farm riot   88
Brunacini   24, 87
Buffalo Soldiers   7, 8
Bureau of Alcohol, Tobacco, and Firearms (BATF)   107, 317
Burma   67, 68–69, 92
Bush, George W.,   25, 26, 27, 29
Business   30, 32, 39, 49, 58, 59, 68, 75, 79–81, 84, 106, 110–111, 117, 147, 151, 153, 161–162, 177, 206, 219, 221, 224, 233, 248, 251, 258, 283, 285, 305, 310, 331, 353–355, 373, 377–378, 379, 380, 403, 409, 411, 415–417, 424, 429, 433, 439, 442, 456, 461, 468–469, 482, 495, 498, 503
Business Continuity Branch Director   471, 476–477

c

Cambodia   67, 71
Canada   67, 72, 219
Canadian Interagency Forest Fire Centre (CIFFC)   72
Caribbean Disaster Emergency Management Agency (CDEMA)   74–76
Center for Disease Control (CDC)   52, 87, 107, 112, 115, 317, 486
Center for Medicare and Medicaid Services (CMS)   484, 501
Chain of Command   10, 16, 41, 56, 57, 171, 173, 196, 198, 233, 278, 288, 337, 380, 407, 450, 458, 467, 469, 482, 493, 497
Chaos   2, 11, 13, 30, 38, 39, 40, 42, 44, 62, 70, 74, 75, 79, 88, 92, 95, 98, 100, 103, 104, 116, 125, 133, 138, 139, 157, 164–165, 170, 172, 173, 174, 179, 183, 193, 198, 207, 220, 232–233, 350, 357, 369–370, 376, 380, 381, 387, 392, 433, 458, 463, 467, 499, 508
Chemical, Biological, Radiological, Nuclear, or Explosives (CBRNE)   3, 57, 62, 124, 495
China   72, 87
Civil Air Patrol   53, 349

Emergency Incident Management Systems: Fundamentals and Applications, Second Edition.
Mark S. Warnick and Louis N. Molino Sr.
© 2020 John Wiley & Sons, Inc. Published 2020 by John Wiley & Sons, Inc.
Companion Website: www.wiley.com/go/Warnick/EIMS_2e

Civil Contingencies Act 89
Claims Specialist (CLMS) 301, 302–303
Collaboration 7, 9, 16, 19, 23, 26, 33, 42–44, 50, 57, 58–60, 65, 67, 71, 77, 81, 82, 90, 130, 177, 189, 223, 251, 253, 258, 269, 273, 281, 313, 315, 326, 333, 341, 353, 414, 466, 472, 476, 477, 504
Command 1, 41, 68, 95, 129, 169, 187, 213, 230, 267, 295, 313, 349, 369, 391, 455
Command and General Staff 180, 181, 192, 195, 197, 199, 200, 202, 207, 208, 209, 304, 339, 353, 358, 359, 361–365, 371, 387, 393, 394, 395, 400–403, 407, 413–430, 437, 440–442, 447–449, 451, 458, 469–471, 482–484
Command Staff 48, 89, 105, 154, 165, 171, 173, 174, 187–210, 232, 251, 255, 267, 269, 274, 282, 287, 315, 318–321, 349, 358–361, 365, 369–370, 380, 397, 399, 406–408, 414, 418–420, 423, 426–430, 439, 444, 469, 470, 482, 483
Command Staff and General Staff Meeting 426
Common Terminology 18, 83, 161, 175, 178–180, 183, 498
Communication(s) 7, 43, 68, 95, 129, 170, 191, 221, 229, 269, 326, 359, 379, 400, 464
Community Emergency Response Team (CERT) 76, 112, 113, 349, 506
COMP 298–303
Company G 7
Company I 7
Compensation Claims Unit 205, 206, 298, 301, 302, 353, 477
Compensation for Injury Specialist 301–302
Complexity 17, 22–23, 26, 33, 34, 38, 40, 98, 129, 132, 133, 136, 138, 139, 158, 160, 164–165, 174–177, 207, 315, 323, 327, 333, 335, 338, 340, 344, 357, 370, 380, 392–395, 407, 440–441, 458, 483, 500
Comprehensive 1, 4–5, 7, 20, 26, 34, 69, 84, 89, 119–120, 129, 133–136, 137, 140, 177–178, 182–183, 205, 243, 251, 270, 305, 307, 314, 345, 358–359, 362, 395, 406, 441, 457, 469, 482, 484–485, 492, 496
Comprehensive resource management 178, 182
Continental Congress 4
Continuity of Operations 476, 489
Control 1, 41, 67, 95, 129, 170, 188, 213, 230, 267, 315, 351, 370, 400, 457
Cooks 238
COOP 476, 489–490

Cooperation 7, 11, 16, 26, 34, 41–45, 49–50, 54–55, 57, 58, 62, 65, 71, 75, 77, 86, 95, 101, 114–125, 129, 139, 141, 143, 150, 156–157, 163, 177, 189–191, 202, 224, 258, 315, 341, 353, 360–361, 414, 476, 504, 507
Coordinated Incident Management System (CIMS) 83–84
Coordination with external agencies 503–507
Core Planning Meeting Principles 408
Cost Unit 205–206, 298, 303–304, 306, 415
CPG 101, 137, 485, 496
Cycle of Preparedness 41, 45, 137–138, 142, 164

d

Decontamination 52, 62, 90, 115, 116, 141, 148, 155, 459, 475, 495, 497, 503, 505
Deepwater Horizon 7
Delegation of Authority 296, 350–351, 394, 395, 400–413, 489
Delegation of Authority Briefing 401–403, 405
Demobilization Unit 204, 266, 277, 286–290, 300, 327, 450, 451
Department of Defense 29, 107, 221
Department of Energy 7, 107, 124
Department of Interior 5
Dienstvorschrift 100, 73–74
Disaster Medical Assistance Teams (DMAT) 443, 466, 488, 491
Disaster Mortuary Operational Response Team (DMORT) 52, 337, 443, 488, 495
Display Processor 272, 446
Divisions 5, 39, 52, 62, 116, 150, 196, 199–200, 202, 203, 208, 214–217, 220–225, 254, 268, 269, 278, 280, 282, 323, 335, 349, 360, 366, 394, 431, 445, 447, 450, 479
Doctors without Borders 71
Documentation Unit 194, 202, 204, 240, 247–248, 259, 266, 281–286, 307, 309, 415, 434, 435, 449
Donations Management 52, 499
Drills 27, 30, 33, 77, 109, 138–139, 141–144, 387, 504, 508
Drop Points 231–232
Drug Enforcement Agency (DEA) 49, 50, 492
DV 100, 73–74, 84

e

EMERCOM 86
Emergency Control Ministry 86

Emergency Management Institute (EMI) 32, 68, 159, 197–206, 316, 459
Emergency Medical Services Authority (EMSA) 47, 81, 455, 456, 467–468
Emergency Operations Center (EOC) 46, 53, 59, 89, 106, 110, 132, 134, 135, 141, 149, 152, 156–164, 182, 218–219, 247, 253, 350, 426, 465, 466, 478, 497
Emergency Operations Plan (EOP) 137, 158–160, 350, 483–508
Emergency Spending Authorizations 498
Emergency Support Function (ESF) 28, 52
Equipment Time Recorder (EQTR) 309–310
Evacuation 8, 17, 23, 103, 108, 120, 147, 148, 151, 158, 188, 218, 219, 246, 247, 271, 280, 282, 361, 384, 398, 436, 447, 464–466, 468, 477, 485, 487, 500–501
Exercises 27, 30, 33, 45, 46, 55–56, 68, 71, 77–78, 103, 138–160, 183, 193, 456, 459, 469, 485, 504, 507
Expanding 37, 207–208, 213–226, 229–261, 265–291, 295–310, 320, 322, 323, 362, 394–395, 405, 468, 500

f

Federal Bureau of Investigations (FBI) 49–51, 53, 54, 59, 107, 115, 119, 124, 183, 317, 329, 334, 393
Federal Coordinating Officer (FCO) 51, 52, 174, 443
Federal Emergency Management Agency (FEMA) 7, 25, 46, 86, 102, 132, 174, 223, 231, 253, 285, 295, 352
Feed the Children 51
Field Observer 268–270, 272–274
Finance and Administration 180, 205, 210, 288–289, 295–310, 355, 419, 429, 443, 468, 477, 480, 481–482, 498
Finance/Administration Section Chief (FSC) 100, 201, 205–206, 295–305, 307, 308, 327, 353, 356, 360, 399, 415–416, 419, 422, 424–425, 427, 429, 438, 450, 470
Fire Behavior Analyst 273–274
Fire Effect Monitor 273
Fireline Emergency Medical Technician (FEMT) 240–243
FIRESCOPE 3, 20, 24, 34, 70, 87, 240–243, 456
Five-C's 124–125, 132, 148, 210

Flexibility 31, 73, 80, 130, 134–135, 181, 188, 202, 208, 210, 215–217, 226, 306, 314, 319–320, 342, 379, 486, 490, 506
Flexible 21, 34, 73, 80, 83, 129–130, 133–135, 175–176, 195, 208–209, 213, 215, 216, 222, 243, 256, 314, 317, 319, 327, 338, 344, 362, 365, 457, 468, 482, 486
Food Unit 201, 236–238, 255, 448, 479
Freeman Hospital 465–467, 472, 494
Full-Scale Exercise (FSE) 68, 138, 141–144, 456
Functional Exercise (FE) 138, 141–142

g

Games 97, 137–141, 183, 193, 226, 424, 465
General Assembly 65–66, 313
General Services Administration (GSA) 52, 253
General Staff 31, 48, 53, 83–84, 106, 123, 154, 171, 173–174, 180, 181, 187–210, 214, 215, 226, 229, 245, 251, 255, 274, 287, 297, 304, 315, 318–319, 321, 329, 339, 351, 353, 358, 359, 361–365, 369–371, 380, 393–395, 397, 399–403, 406–409, 413–431, 437, 439–442, 444–445, 447–449, 451, 458, 469–470, 482–484
Geographic Information System Specialist (GISS) 270–272, 291
Germany 73, 74
Goals 4, 5, 7, 10, 12, 13, 27, 43, 53–54, 57, 61, 66, 73, 86, 96, 114–115, 117, 119, 124, 131, 135, 151, 158, 162, 175, 181, 187, 189, 208, 210, 252, 266, 291, 296, 304, 325, 342, 369–388, 408–409, 413, 431, 437, 444, 456, 469, 477, 482–484, 487
Gold–Silver–Bronze 88–89
Government of Haiti Department of Civil Protection 75–76
Government Stakeholders 107–108, 111
Greelcy, William 6–8
Ground Support Unit 200, 204, 235, 269, 308, 447, 448
Group(s) 3, 41, 79, 96, 132, 173, 188, 214, 235, 278, 308, 318, 354, 375, 394, 458

h

Haiti 74–76
Hazard Mitigation 256, 445
Hazmat Branch Director 470, 475
Health Emergency Operations Plan (HEOP) 82
Helispot 182, 220, 231–232, 235, 269, 288, 423
Help Desk Operator 258
Hereford 87

HHS 485
HICS Logistics Section 479–481
HICS Operations Section 470–471
HICS Planning Process 482–484
Homeland Defense 12
Homeland Security Presidential Directive (HSPD) 26, 27, 33, 129, 170, 172, 315, 457, 468
Hospital Control Team (HCT) 90–91
Hospital Emergency Incident Command System (HEICS) 69, 70, 77, 85, 87, 91, 455–460
Hospital Incident Command System (HICS) 2, 68–72, 74–83, 85–87, 92, 133–134, 154, 455, 458–471, 472, 474, 475, 477, 479–485, 497, 500–501, 506–508
Hospital Incident Management System (HIMS) 68, 77
Hospital Incident Management Team (HIMT) 480, 483
Houses of Worship 112–113, 117, 177
Hurricane Katrina 7, 11, 29, 31, 111–112, 125, 145–146, 164, 173, 174, 192, 196, 218, 223, 297, 313, 315, 316, 333, 335, 378–379
Hyogo Framework 66–67, 69, 71, 82, 86, 87

i

Illinois Law Enforcement Alarm System (ILEAS) 117, 150
Immediately Dangerous to Life and Health (IDLH) 55
Incident Action Plan (IAP) 22, 132, 172, 189, 231, 265, 297, 352, 371, 393, 480
Incident Action Plan Preparation 437–440
Incident Base 54, 90, 159, 231–233, 255, 277, 288, 354, 393, 395
Incident Briefing 197, 198, 202, 206, 267, 268, 278, 358–359, 361, 391, 396–400, 406, 482
Incident Camp 231, 233–234, 288
Incident Command Post (ICP) 48–49, 157, 165, 174, 182, 183, 189, 203, 231, 232–233, 245–246, 260, 265, 268, 269, 281, 288, 366, 445, 465
Incident Command/Unified Command Meeting 407–408
Incident Commander (IC) 1, 38, 67, 97, 132, 170, 187, 213, 229, 265, 295, 315, 350, 369, 391, 458
Incident Communications Center 244–247
Incident Communications Plan 201, 204, 359
Incident Communications Technician 249–250, 252, 253
Incident Control System 67–68
Incident Controller 67
Incident Helibase 231, 288
Incident Message Center 247–248
Incident Objective Priorities 373–379
Incident Response System (IRS) 76–77
India 76–77, 366
Indonesia 67, 77–78
25th Infantry Regiment 7
Infrared Interpreter 272–273
Infrastructure Branch Director 470, 473
Infrastructure Management 500
INJR 301–302
In-Kind Donations Coordination Team 52
Integrated communications 104, 178, 180–181
Internal Communications Manager 255–256, 257–259
International Association of Emergency Managers (IAEM) 486
International Maritime Organization 92
International Strategy for Disaster Reduction (ISDR) 65–66
Investigations and Intelligence (I/I) 194, 220, 265, 313, 360, 420
Investigations and Intelligence Gathering 190, 194–195, 199–200, 202, 206–207, 209, 220, 313–345, 360–361, 420, 430, 449
Investigations and Intelligence Gathering Officer (IIO) 194–195, 320–321, 420
Investigations/Intelligence Section Chief (ISC) 206–207
Iran 67, 78–80, 460
Iraq 10, 80
Ireland 92

j

Japan 40–43, 59, 62, 67, 80–81, 92, 385
Job Action Sheets 457, 467, 469, 482
Joint Emergency Services Control Centre (JESCC) 88
Joint Emergency Services Interoperability Principles (JESIP) 89
Joint Information Center (JIC) 4, 51, 92, 108, 119–121, 135, 163–164, 182, 192, 287, 418, 428, 451, 481
Joint Information Systems (JIS) 134–135, 156, 164, 418, 426, 428
Joplin Tornado 173, 467, 472, 473, 476, 490, 495

k

Kansas 53

l

Landing Zone (LZ)   234, 240, 506
Leadership and Command in Emergency Operations   73
Liaison Officer (LOFR)   101, 121–122, 187–188, 190, 193–194, 200, 232, 361, 399, 418–419, 426, 429, 434, 450, 470, 475, 478
List of Excluded Individuals/Entities (LEIE)   574
Logistic Section Chief (LSC)   99, 154, 200–201, 204, 221, 226, 229–231, 235, 240–241, 243, 245, 254, 359, 399, 415–416, 419, 424, 429, 431, 434, 438, 450, 470
Logistics Branch   230, 497
Long-Term Fire Analyst   274

m

Major Incident Declared   90
Major Incident Resource Pack   90
Major Incident Standby   90
Malaysia   67, 82
Maldives   67, 82
Management by Objectives (MBO)   319, 365, 369–388, 413, 439, 469, 482
MARD   91
Mass Casualty Incident (MCI)   2, 40, 42, 47, 76, 88, 98, 336, 375, 461, 463, 508
Matsumoto   43–44
McVeigh, Timothy   45
Media Stakeholders   108–109
Medical Care Branch Director   470, 473, 480, 497
Medical Command Post (MCP)   494
Medical Director   493, 494, 506
Medical Plan   197, 201, 204, 206, 239, 241, 243, 245, 299, 359, 361, 400, 429, 442–446, 450, 480, 506
Medical Reserve Corps (MRC)   489, 491
Medical Unit   201, 236, 238–240, 243, 302, 414, 445, 446, 448, 450, 466, 506
Memorandum of Understanding (MOU)   91, 115, 117, 145, 147–148, 150, 151, 219, 306, 488, 497, 500, 505
Mental Health   50, 51, 219, 277, 289, 335, 336, 452, 464, 466, 471, 472, 493, 494, 497
Mercy Hospital   463, 465–467
Message Center   244, 245, 247–248, 254, 256, 258–259
Mexico   6, 67, 82–83
Mexico-United States Joint Contingency Plan   82
Military   1, 3–5, 7, 9–12, 17, 29, 42, 48, 51–53, 57, 59, 69, 75, 78, 86, 95, 135, 144, 148, 235, 317, 339, 355, 437, 489, 491

Ministry of the Russian Federation for Affairs for Civil Defence, Emergencies and Elimination of Consequences of Natural Disasters   86
Minnesota National Guard   2
Mitigation   20, 21, 23, 28–31, 33–34, 40, 44, 60, 79, 83, 95–97, 99–101, 109, 111–114, 118–119, 130, 133, 137–139, 145, 164–165, 173, 181, 231, 256, 271, 302, 310, 318, 321, 343, 349, 351–353, 357, 361–363, 374, 398, 424, 434, 436, 445, 451, 460, 476, 486, 504
Mobile Emergency Response Support   260–261
Mobile Medical Team   90
Modular organization   178, 180, 208–210, 215, 222, 226
Mount Weather   253–254
Multi-Agency Coordination Center (MACC)   51
Multi-Agency Incident Management System   89, 102, 189
Mutual Aid   6, 10–11, 13, 14, 16–18, 46–47, 52, 76, 100, 116–117, 130–131, 145, 171, 173, 226, 261, 308, 349–350, 366, 380, 457
Mutual Aid Agreement (MAA)   6, 110, 114, 116, 145, 147–151, 154, 219, 306, 350, 354, 392, 457, 488, 490, 494, 496–497, 504–505
Mutual Aid Box Alarm System (MABAS)   116–117, 150
Myanmar   67–69, 92
Myriad Convention Center   50–51

n

National Agency for Disaster Management   78
National Archives and Records Administration (NARA)   281, 285
National Curriculum Advisory Committee   25
National Disaster Coordinating Council (NDCC)   85
National Disaster Management Authority (NDMA)   76–77
National Disaster Medical System (NDMS)   75, 488, 489
National Disaster Risk Reduction & Management Council (NDRRMC)   85
National Fire Academy (NFA)   21, 24, 25, 67
National Health Service (NHS)   89–91
National Incident Management System (NIMS)   1, 65, 113, 129, 170, 194, 221, 231, 266, 313, 350, 379, 442, 455
National Interagency Incident Management System (NIIMS)   24–26, 67
National Library of Medicine (NLM)   485

National Police Agency   42–43
National Response Framework (NRF)   28, 31, 239
National Response Plan (NRP)   28
National Standard Operation Procedures (NaSOP)   71
National Weather Service   51, 53
Network Manager   251–253
Network Specialist   252–253
New Zealand   67, 83–84
NHS Tameside   90–91
Nichols, Terry   45
nongovernmental organizations (NGO)   27–29, 31, 32, 114, 130, 132

o

Objectives   3, 38, 86, 96, 131, 171, 189, 215, 244, 266, 298, 318, 350, 369, 394, 458

Office of Civil Defense   85
Office of the Inspector General (OIG)   492
Office of U.S. Foreign Disaster Assistance   76
Oklahoma Baptists   52
Oklahoma City   40–54, 56–62, 313, 315, 337, 345, 490
Oklahoma City Communications Center   46
Oklahoma City Fire Department   46–49, 53, 54, 61
Oklahoma City Memorial Museum   45
Oklahoma City Police   48, 51, 53
Oklahoma City Public Works   49
Oklahoma City Street Department   50
Oklahoma City Traffic Management Department   50
Oklahoma County Sheriff's Department   53
Oklahoma Department of Civil Emergency Management (ODCEM)   45–50, 58–59, 61
Oklahoma Department of Education   53
Oklahoma Department of Health   53
Oklahoma Department of Human Services   52, 53
Oklahoma Department of Public Safety   49, 50–51, 53
Oklahoma Funeral Directors Association (OFDA)   50, 76
Oklahoma Medical Examiner   50
Oklahoma Military Department   51, 53
Oklahoma National Guard   48–50
Oklahoma Restaurant Association   50
Operations Section   50, 79, 195, 213, 229, 267, 315, 365, 416, 470

Operations Section Chief (OSC)   99, 171, 189, 213, 229, 268, 331, 352, 387, 399, 470

p

Palestine   84–85
Patient Family Assistance   471, 477, 478
Patient management   496–497
Personnel Time Recorder (PTRC)   309
Philippine Islands   85–86
Pinchot   5–7
Planning and Intelligence   201, 220, 240, 265–291
Planning Meeting   4–5, 51, 89, 122, 123, 194, 203, 207, 267, 273, 274, 282, 296, 297, 357–360, 365, 370, 384, 407, 411, 426, 435–440, 441, 445, 446, 450–452
Planning P   73, 362–364, 366, 385, 386, 391–452, 482–484
Planning Process   84, 145, 154–155, 158, 164, 171, 181, 189, 208, 319, 349–367, 369, 372, 384–388, 391, 408–409, 411, 413–414, 417, 431, 437, 440–441, 444, 448, 450, 452, 458, 459, 469, 482–485, 487, 489, 506
Planning Section Chief (PSC)   100, 199, 201–204, 226, 267, 269, 274, 276, 283–285, 287, 315, 352, 353, 359–360, 363, 365–366, 384, 387, 399, 408–411, 414, 417–422, 425, 427, 428, 430, 431, 433–441, 444–445, 449–451, 470
Preparedness   26–31, 34, 41, 45–47, 60, 66–67, 69–71, 78, 82, 89–90, 95–97, 99–101, 109, 111–114, 118–119, 133, 134, 136–152, 156, 164–165, 173–174, 177, 183, 210, 253, 349, 357, 484–486, 488, 501, 504, 507
Presidentially Declared Disaster (PDD)   7, 107, 252–256, 258–260, 352, 356
Printing   256, 282, 284, 440–451
Private Sector   31, 85, 114, 117, 118, 130, 147, 148, 161, 250–251, 330, 460
PROC   283, 304–305, 307
Procurement Unit   205–206, 283, 298, 304–307, 353, 415, 479
Public Assistance (PA)   256
Public Information Officer (PIO)   4, 100, 108, 119–121, 135, 148, 163–164, 187–188, 190, 192–194, 200, 226, 232, 282, 287, 331, 335, 361, 382–383, 399, 425, 428, 450, 451, 470, 484
Putin, Vladimir   86

r

Radio Networks   105–106, 251
Radio Operator   75, 114, 235, 245–248

Rapid Intervention Team (RIT)   55, 171
Recovery   1, 7, 23, 27–31, 33–34, 44, 45, 48, 51–52, 54–55, 58, 62, 72, 75, 83, 85, 89, 92, 95–97, 99–102, 109–113, 115–116, 130, 133–135, 137–139, 143–145, 154–157, 161–165, 170, 172, 191, 196, 205, 213, 217, 222, 223, 229, 250–251, 256, 260–261, 271, 285, 287, 289, 303–306, 313, 315, 319, 331, 337–338, 353, 357, 363, 365, 374, 375, 387–388, 391, 409, 414, 416, 417, 425–426, 431, 439–440, 447–448, 460, 466–467, 469, 475–477, 482, 484, 486
Recovery Service Center   52
Regional Hospital Coordination Center (RHCC)   497, 504
Resource Management   11, 26, 131, 134, 136, 144–152, 154, 178, 182–183, 276, 349, 399, 416, 499
Resources Unit   200, 204, 238, 266, 275–278, 280–281, 366, 431, 434, 436, 438, 445, 446, 479
Responder Rehabilitation Unit   240, 243–244
Restock resource   154, 156
Rio de Janeiro   92
Risk Assessment   52, 87, 97, 137–138, 271, 366
Robert T. Stafford Act   256, 258
Rochambeau   4
RSChS   86

s

Safety and Security   501–503
Safety Officer (SOFR)   101, 143, 171, 180, 187–188, 190–192, 194, 196, 199, 204, 207, 214, 270, 301–302, 342, 352–353, 358, 360–361, 366–367, 399, 419, 424, 429, 431, 434, 435, 445–446, 448, 450, 470
Salvation Army   2, 47, 50, 113, 117, 121, 162, 194, 336
Saudi Arabia   92
School of Medicine   43, 81
Scotland   89–90
Secretary of Homeland Security   27
Security Branch Director   470, 473, 475
Self Defense Force   42–43, 58
Seminar   71, 78, 138, 139
Sendai Framework   66–67
Service Branch   230, 235–240, 244, 261, 479
Severe Acute Respiratory Syndrome (SARS)   87
Shell Oil Company   92
Singapore   67, 87
Singapore Civil Defence Force (SCDF)   87

Singapore Disaster Response Force (SDRF)   87
Single Resources   196, 200, 214–215, 220, 223–224, 239, 277, 357–358, 392, 393, 441
SitRep   232, 241, 362–363, 433
Situation Report   92, 232, 241, 330, 360, 362–363, 418, 433
Situation Unit   204, 266–270, 272–275, 291, 344, 413, 415, 417–418, 423, 428, 435–436, 447, 449, 479
SMART Objectives   189, 359, 370, 381–388, 413
Social media   96, 100, 103, 109–110, 193, 272, 326
Span of control   83, 166, 173, 178, 181–182, 188, 196, 198, 207–209, 214–217, 220–223, 225, 230, 246, 267, 276, 322–323, 327, 333, 335, 338, 344, 351–352, 457, 467, 471
Stafford Act   256–258, 443
Staging Area   47–48, 142, 182–183, 197, 199, 226, 231, 233–234, 247, 277, 280, 288, 339–341, 353, 423, 439–440, 450, 464–465
Staging Manager   234, 340, 470–479, 481
Stakeholder Communications   106–114
Standardization   130–132, 135
Standing Orders on Disaster   69
State Coordinating Officer   51
State Medical Examiner   50
Status Recorder   276–281
Stove-piping   41
Strategic Operational Planner   290–291
Strike Team   62, 75, 123–124, 196, 199, 214–215, 222–225, 277, 282, 289, 334, 342, 344, 381, 394
Substance Abuse Services   51
Supply Unit   200, 230, 236, 249, 283, 288, 297, 305–306, 309, 415, 448, 479
Support Branch   230–236, 323, 479–481
Switchboard Operator/Receptionist   90, 257–260

t

Tabletop exercise   46, 71, 138–140, 469, 485
Tactics Meeting   199, 360, 365, 414, 419, 430–437
Task Force   20, 59, 62, 104, 123–125, 196, 199, 200, 214, 215, 222–226, 231, 236, 257, 276, 277, 278, 282, 289, 319–320, 334, 342, 344, 394, 443
T-Card   278, 280
Technical Resources, Assistance Center, and Information Exchange   485, 501
Technical Specialist   122, 245, 266–268, 274–275, 285, 352, 403, 436–437, 470, 478
Telecom Manager   253–254

Texas   53, 260, 416, 420, 423, 424
TIME   308–309
Time Unit   205–206, 298, 308–310
Tokyo   40–58, 60–62, 81
TRACIE   485–486, 501
Training Specialist   291
Train-the-Trainer   67, 78, 82
Trauma and Critical Care Teams   488
Triage   46–47, 101, 117, 182, 220, 375, 461–465, 467, 472, 479, 494, 496, 508–509
Typhoon Ketsana   85
Typing Resources   151–152

u

Uncertainty   38, 40, 42, 44, 45, 57, 62, 74–75, 79, 88, 92, 95, 98, 125, 133, 138, 157, 164, 165, 173–175, 207, 370, 376, 380, 387, 392, 458, 463, 467, 499, 508
Unified Command (UC)   5, 27, 43, 46, 62, 96, 119–122, 132–133, 142, 163–164, 174, 177, 180, 183, 187–190, 192, 215, 232, 315, 316, 318, 343, 345, 351, 358, 361, 369–370, 385, 387, 391, 392, 396–397, 399–400, 402, 406–408, 413–414, 423–424, 426, 443, 505
Unified Emergency Prevention and Response State System in Russia   86
United Nations (UN)   65–67, 69, 75, 77, 85, 92, 313
United Nations Development Programme (UNDP)   77, 85
United States Agency for International Development (USAID)   67, 69, 76–78, 91, 92
United States Department of Health and Human Services   485
United States Fire Administration   243
Unity of Effort   130–132, 158, 163, 174, 250, 420, 469
University of East Ramon Magsaysay   85
Untrained Spontaneous Volunteers   45, 172, 173
Urban Search and Rescue   25, 47, 48, 51–53, 116–117, 223, 231
Utility Companies   32, 110, 121, 194, 220

v

Veterans Administration   52, 53, 489
Victim Information Center (VIC)   488–489
Vietnam   67, 91–92, 170
Volunteer Organizations   49, 113–114, 137, 147, 162, 172, 194, 440, 468

w

Washington   4–6, 8, 253, 260
World Health Organization (WHO)   91
World Trade Center   66, 218, 315, 333, 334, 337

y

Yeltsin, Boris   86